Progress in Botany 66

66 PROGRESS IN BOTANY

Genetics
Physiology
Systematics
Ecology

Edited by

K. Esser, Bochum
U. Lüttge, Darmstadt
W. Beyschlag, Bielefeld
J. Murata, Tokyo

Springer

With 49 Figures

ISSN 0340-4773
ISBN 978-3-642-42942-2 Springer-Verlag Berlin Heidelberg New York

Springer is a part of Springer Sciences+Business Media
springeronline.com

Softcover re-print of the Hardcover 1st edition 2005

Cover design: Design & Production, Heidelberg
Typesetting: M. Masson-Scheurer, Neckargemünd
31/3150 - 5 4 3 2 1 0 - Printed on acid-free paper

Contents

Review

Genetics

Recombination:

Mutation:

Physiology

Curriculum vitae

1930 Hans Mohr was born on a farm in the Black Forst/Germany

1949 Graduation from a German Gymnasium under French supervision

1950 Undergraduate student at the University of Tübingen/Germany: Philosophy or science?

1953 Graduate student in biology with Professor Erwin Bünning

1955 Submission of a Ph. D. thesis to the Faculty of Sciences at Tübingen University

1956 Postdoctoral research fellow at the Plant Industry Station at Beltsville, Md., U.S.A.

1957 Married to Dr. Iba Kraut, brilliant biochemist and wonderful partner

1959 Habilitation at Tübingen University

1960 Professor of Biology, University of Freiburg; Visiting Professor, University of Massachusetts

1966 Member of the German Academy of Sciences (Leopoldina)

1982 Member of the Heidelberg Academy of Sciences and Humanities

1982 Honorary Editor of the International Journal of Plant Biology, Planta

1992 Director at the Governmental Insitute of Technolgoy Assessment in Stuttgart

1992 Member of the presidency of Leopoldina

1992 Member of the 'Innovationsbeirat' (State Government, Baden-Württemberg)

1995 Member of the 'Technologiebeirat' (Federal Government, Bonn)

1997 Retirement

2000 Honorary Member of the German Botanical Society

2002 Member of the University Council, University of Konstanz

Many honours, including honorary doctorates, the Cothenius medal in gold, the Max Born medal and the Bundesverdienstkreuz, 1st class

Author/Editor of 18 books and approximately 400 articles

A Life History Between Science and Philosophy

Hans Mohr

1 Philosophy or Science?

When I graduated from an old-fashioned German Gymnasium under French regime my interests were divided between science (physics) and philosophy, with an intuitive preference for the latter. However, by the end of my first semester at Tübingen University I realized that I needed a more solid education in science to study that branch of philosophy I was particularly interested in, namely epistemology. By that time I considered philosophy as primarily an epistemological subject.

Epistemology investigates the origin, nature, methods and limits of human knowledge. Since in the modern world the sciences had become the major source of positive knowledge, a deepened introduction to the basic sciences was obviously a prerequisite for any career in epistemology.

Moreover, I was dissatisfied with the prevailing German academic tradition of teaching philosophy as a history of philosophical thoughts. My deep interest in modern fields, such as modern logic and analytical philosophy, was hardly met. As far as the leading philosophical fashions of the time were concerned, neither Heidegger's existentialism or fundamental ontology nor the late Husserl's transcendental phenomenology attracted me.

Fortunately, some of the famous science professors in Tübingen had strong philosophical minds: Walter Kossel in physics, Max Hartmann in natural philosophy and Erwin Bünning in botany. Bünning, who by that time had established the concept of the physiological clock, had just published a superb treatise on *Theoretische Grundfragen der Physiologie* (Bünning 1949) which I studied with keen interest. Moreover, I enjoyed reading a well-thumbed copy of A.J. Ayer's brilliant book of 1936 *Language, Truth and Logic*. This ideally accessible, lucidly written work stabilized my plans to become a real scientist before considering any career in philosophy. Ayer, the genuine philosopher, had argued convincingly: "If you want a philosopher to be constructive..., then I think you've got to marry him to a scientist." At the end of *Language, Truth and Logic*, Ayer saw the future of

Progress in Botany, Vol. 66

philosophy only in its being the logic of science. In retrospect, I gratefully acknowledge that Bünning and Ayer put me on the right track.

Equipped with a stipend of the Studienstiftung des Deutschen Volkes, I studied physics and biology. At the graduate level, I preferred biology even though quantum physics and thermodynamics remained favorite topics. Fortunately Erwin Bünning accepted me as candidate for a doctorate in 1953. He was a great mentor and a fine person.

2 Early Steps in My Scientific Work

2.1 Towards Phytochrome

The action spectrum of a photobiological response represents, with certain assumptions, the absorbance spectrum of the effective absorbing substance (photoreceptor). Bünning wanted me to identify by means of action spectroscopy the photoreceptor involved in the germination of fern spores. After some preliminary studies I chose the spores of the common male fern, *Dryopteris filix-mas*, for the following reasons:

- These spores never germinate in complete darkness.
- The light requirement can be satisfied by a short light treatment of the fully imbibed spores. That is, germination can be 'induced' by light. If the spores are placed in darkness after the light treatment, complete germination takes place.
- In the Botanical Garden in Tübingen I detected a population of cloned *Dryopteris* sporophytes. They had been derived in the 1930s from a single rhizome by the late Prof. Lehmann, who was interested in the appearance of somatic mutations. From this clone I collected the most homogenous spore population you can imagine.
- The spore material could easily be germinated and inspected on a thin agar medium. I have only very rarely observed contaminations within the time span required for germination.

As far as my experimental equipment was concerned I was equally lucky: interference filters had just become available, and I, together with my friend G. Schoser, could construct an interference filter monochromator unit for photobiological purposes. This type of irradiation device has had decisive advantages compared with prism- or grating-equipped monochromators. Measurement of the photon flux of the monochromatic light beams posed a problem since photocells turned out to be a poor choice. However, Bünning provided the money to buy an expensive bolometer, and Kossel offered me an extremely sensitive thermopile which I could use over night

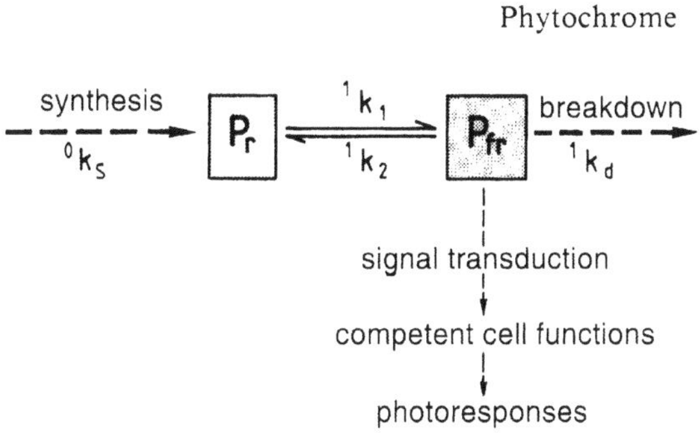

Fig. 1. Scheme of the phytochrome system. Explanation: phytochrome appears in two forms, P_r and P_{fr}. Without light only P_r, the physiologically inactive form, is made. Under the influence of light P_r changes into P_{fr}, the physiologically active form. The photoconversion P_r–P_{fr} is photoreversible; it follows first-order kinetics $^1k_1$, $^1k_2$ in both directions. The signal induced by light is passed on from P_{fr} (signal transduction) and received by cell functions competent for this signal, e.g. by promoter regions of competent genes. Specificity of photoresponses is determined by the spatial and temporal competence pattern for P_{fr}. Light (P_{fr}) has no influence on development of this competence. Phytochrome is a predominantly hydrophilic chromoprotein which is easily isolated from the cell. The absorption maximum of P_r is 665 nm in vitro, i.e. in the red (R); that of P_{fr} is 730 nm in the far-red (FR). (Mohr and Schopfer 1995)

(since it was needed in the Physics Department during the day). In fact, with the help of my girlfriend Iba (who fortunately agreed to marry me later), I could produce and measure with high precision all types of monochromatic beams which I needed for my work. The result was worth the effort: I was able to elaborate a very precise action spectrum, and we found the 'reversible red/far-red photoreactive system', which was later named 'phytochrome' (Mohr 1956).

In the meantime, however, the reversible red/far-red photoreactive system had been discovered and described by the Beltsville research group in studies on light-induced seed germination. A modern version of the phytochrome system is reproduced in Fig. 1.

Even though I was only second, I received the Research Prize of the University – the first time in my life I owned 2,000 Marks – and, more important in the long run, I was invited by H.A. Borthwick to join the Beltsville group as a postdoctoral research fellow.

2.2 The Beltsville Group

The discovery of the reversible red/far-red control of plant growth and development and the subsequent in vivo identification and isolation of the photoreceptor pigment phytochrome constitutes one of the great achievements in modern biology (Sage 1992). It was primarily a group of investigators at the Plant Industry Station, Beltsville, Maryland, USA, headed by the botanist H.A. Borthwick and the physical chemist S.B. Hendricks, who made the basic discoveries and developed a theoretical framework on which the progress in the field of (molecular) plant development has been largely based.

I joined the Beltsville group in mid-1956. The mode of cooperation at the Plant Industry Station opened my eyes to the benefits of teamwork, and the wisdom, humility and helpfulness of the two senior scientists was an unforgettable experience which has been a constant inspiration throughout my research career.

2.3 Photomorphogenesis

Photomorphogenesis (Mohr 1972) has remained the major theme of my scientific efforts after my return to Germany, following Bünning's advice to try for habilitation, by that time a prerequisite for an academic career in Germany. By 'photomorphogenesis', we designate the fact that light controls growth and differentiation (and therewith development) of a plant independently of photosynthesis. In order to grasp the full importance of this phenomenon we must recall that the specific development of any living system depends on its particular genetic information and on its environment. In the case of higher plants, the most important environmental factor is light. Of course, light does not carry any specific information with regard to plant development. Rather, light – operating via photoreceptor molecules – must be regarded as an elective factor which influences the manner in which those genes that are contained in the particular organism are being used. In this sense, the study of photomorphogenesis (Fig. 2) became central to a worldwide program to investigate the influence of the environment on the development of higher organisms, including man. I contributed to this latter aspect with a book chapter in *Freiburger Vorlesungen zur Biologie des Menschen* (Mohr 1979a).

3 My Academic Career – A Short Story

I stayed at Tübingen University only for a little while. In February 1960, some months before my 30th birthday, I was offered the traditional chair for botany at the University of Freiburg to succeed Professor Friedrich Oehlkers, an eminent cytogeneticist.

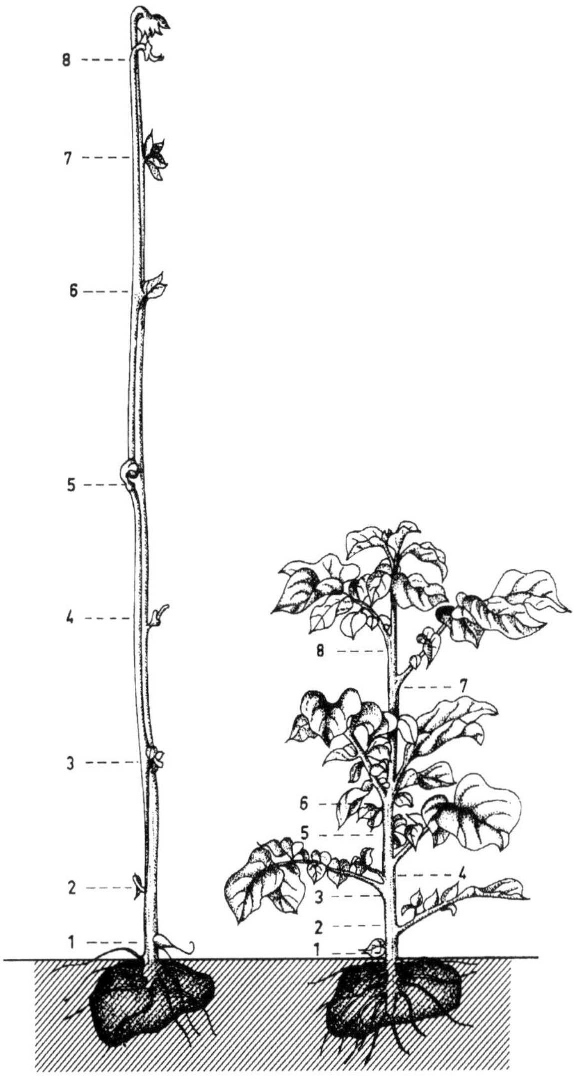

Fig. 2. Both potato plants (*Solanum tuberosum*) are genetically identical. *Left* An etiolated plant grown in darkness; *right* the normal light-grown plant. Scotomorphogenesis (etiolation) is characteristic of development of plants under light-deficient conditions; the alternative development in light is called photomorphogenesis. It is noteworthy that light affects the expression of patterns, the specification of which is light independent. In the present case, patterns of leaf arrangement (called phyllotactic patterns) are precisely the same in etiolated plants as in light-grown plants. However, development from leaf primordia to leaves takes place only in the light. Etiolation is an adaptive response of plants, because as long as the plant is grown in the dark, the limited supply of storage substances is predominantly invested in production (growth) of the shoot axis. This is the most likely way of ensuring that the plumule reaches the light before reserves are depleted. This scotomorphogenesis may be interpreted as a strategy for survival. Photomorphogenesis, on the other hand, is the suitable strategy for development in light (affluence strategy). (Mohr 1972)

To make the right choice was not easy, since I had planned firmly to return to the United States after habilitation, at least for a couple of years. Eventually, my wife and I decided to accept the position (Professor and Department Head) in Freiburg. An important factor at that time was that the administration agreed that I could spend at reasonable intervals a couple of months at American institutions to maintain close personal contacts with my colleagues in research and teaching.

Freiburg University turned out to be an excellent place. Once we had established a new faculty and moved into new buildings, the plant biology department was ready for top research and new kinds of teaching. In 1968, we were chosen by the Deutsche Forschungsgemeinschaft to become a center of excellence (SFB). This implied that we could count on sufficient support provided that we could meet the strict requirements in 3-year intervals.

During the political turmoils in the period 1968–1972, my wife and I reconsidered emigration to the USA. However, since 1972, I have had no doubts that Freiburg was the place where I wanted to work and to live. Only in 1991 did I accept an offer by the State Government of Baden-Württemberg to become a director at the newly established Institute for Technology Assessment in Stuttgart. This meant, in early 1992, the final departure from the laboratory and from regular academic teaching.

4 Some of my Research Topics in Freiburg

4.1 Photosensors in Photomorphogenesis

In order to react optimally to the light conditions in their environment, higher plants require various sensor pigments. Based on molecular physics, you can predict that phytochrome alone is not sufficient to measure all the relevant solar spectrum (290–800 nm) with the required accuracy. Today, it is known that three types of photosensors are involved in the process of photomorphogenesis in higher plants: phytochromes (> 520 nm; red/far-red), cryptochromes (340–520 nm; blue/UV-A) and the UV-B photosensor (290–350 nm).

We have tried over the years to understand at the physiological level the mode of coaction of the photosensors in bringing about photomorphogenesis, including control of gene expression. It can be seen from an early diagram (Fig. 3; Mohr 1987) how we felt the three photosensors worked together. It appeared in all cases that phytochrome (P_{fr}) is the effector proper whereas the blue/UV-A and the UV-B photoreceptors (together with phytochrome) determine the plant's responsiveness to P_{fr}. Even though in

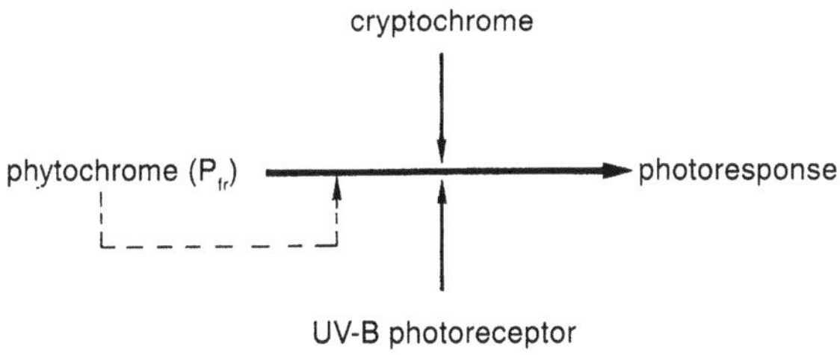

Fig. 3. Schematic of cooperation between phytochrome and the blue/UV-absorbing photosensors. Light absorption in cryptochrome and the UV-B photoreceptor determines the sensitivity of a photoresponse towards P_{fr}. *Dashed line* indicates that light absorbed by phytochrome can also increase the efficiency of P_{fr} or, expressed differently, it can increase the sensitivity of plants towards P_{fr}. (Mohr 1987)

the meantime the experimental approaches have become 'molecular' rather than 'physiological', the basic message contained in Fig. 3 has remained: in addition to the transduction chain triggered by P_{fr} and influencing responsive promoters by direct contact with transcription factors, "extensive cross-talk between signaling cascades downstream of multiple photoreceptors has become apparent..." (somewhat confusing, I admit, but this is the tribute of our field to becoming 'molecular') (Frankhauser and Staiger 2002).

4.2 Multiple Effects of Phytochrome

Obviously, P_{fr} has multiple effects at the organ level (Fig. 2). Multiple effects of P_{fr} can also be demonstrated at the cell and tissue level (Fig. 4). Subepidermal cells of mustard hypocotyls synthesize large amounts of anthocyanin under the influence of P_{fr}. Other cell layers of the axis do not form anthocyanin, even though they all react to P_{fr} with respect to their longitudinal growth. Some epidermal cells (trichoblasts) grow under the influence of P_{fr} into long hairs, but do not form anthocyanin, etc. The obvious multiple effect of P_{fr} at the level of tissues and cells can only be explained by the assumption that cells are differently competent for P_{fr} and that this pattern of competence exists before active P_{fr} is first formed. Epidermal trichoblasts of mustard hypocotyls (see Fig. 4) react differently to P_{fr}: on the one hand, they grow into long hairs, whilst on the other hand, their longitudinal growth is inhibited. Subepidermal cells show a corresponding pattern, i.e. one may say that they react positively by the synthesis of anthocyanin to P_{fr}, whilst they react negatively to P_{fr} with respect to their growth. We must inevitably assume specifically competent cell functions.

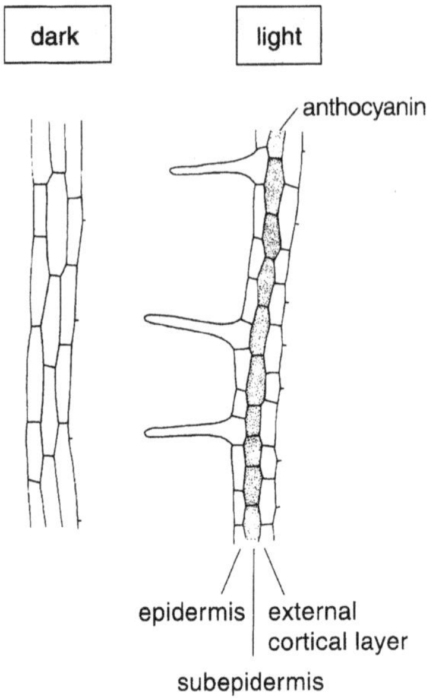

Fig. 4. Drawings representing the three outer cell layers of a mustard seedling hypocotyl in longitudinal section. *Left* Dark-grown seedling; *right* seedling kept for 24 h in light. (Mohr 1972)

4.3 Control of Gene Expression by Phytochrome

In 1966, we reported the induction of enzyme synthesis by phytochrome (Durst and Mohr 1966). Unexpectedly, the scientific community responded quite reluctantly. When I gave a seminar on the subject at Beltsville (October 1966) I could not convince colleagues that they should join us in investigating photomorphogenesis in terms of control of gene expression. Though readily accepted in Germany, the gene regulation hypothesis faced an icy reception in the USA. I received ironic (insulting?) comments throughout my lecture tour, even by some prominent colleagues who only a few years later enthusiastically joined the crowd once molecular physiology had become popular, and nobody any longer opposed the concept that phytochrome operates on development via control of gene expression. In principle, I was not very concerned because I knew that I was right, and I made precise plans while travelling in the USA to substantiate the gene regulation hypothesis after my return to the Freiburg laboratory. It worked!

A first step was to demonstrate that within the same tissue – we had chosen mustard seedling cotyledons where cell number and DNA content remained constant during the experimental period – phytochrome (P_{fr}) could simultaneously induce (Dittes et al. 1971) and repress (Oelze-Karow et al. 1970) enzyme synthesis while syntheses of some marker enzymes of the basic metabolism were not affected at all (Karow and Mohr 1967).

Unfortunately we were not able in the late 1960s to elucidate the nature of the signal transduction cascade from the cytosol to – what we assumed – competent promoters.

Our studies of nuclear-encoded chloroplast enzymes have finally demonstrated that phytochrome in fact regulates gene transcription (Schuster and Mohr 1990). However, we did not suggest that gene expression is controlled only at the transcriptional level. Rather I pointed out: "Full gene expression means the appearance of a final direct gene product – a protein – active at its physiological site of action... In principle, there are many steps between the initiation of transcription and the accumulation of the gene product at its functional location where gene expression could be regulated." As an example, in the case of nitrite reductase phytochrome produces the mRNA, whereas in order to make the enzyme out of the mRNA you need nitrate (Schuster and Mohr 1990). So you have a beautiful two-step control, transcriptional as well as post-transcriptional, which we could take apart.

4.4 Nitrate Assimilation and the 'Plastid Factor'

In 1982, I was elected member of the venerable (and well-endowed) Heidelberger Akademie der Wissenschaften. From 1986 onwards, the Academy financed a research unit in Freiburg to study the formation of the apparatus of nitrate/ammonium assimilation during the development of chloroplasts. The final goal of the research was to breed plants with an improved potential to assimilate nitrate (Mohr and Neininger 1994).

Since the research unit attracted a couple of excellent graduate students we could establish within a few years a consistent model for the formation of the apparatus of nitrate assimilation during the development of chloroplasts (Mohr 1990a). A fascinating result of this research may be mentioned briefly: the plastid factor (Oelmüller and Mohr 1986). Research from different angles (defect mutants, photo-oxidative damage of plastids) led to the conclusion that there is a plastid signal which acts as a transcription factor on nuclear genes. This signal (plastid factor) informs genes in the nucleus, which code for plastid proteins, that plastids are receptive to their protein products (Fig. 5). If the signal is missing, for example as a consequence of photo-oxidative damage to the plastids, transcription of nuclear genes coding for plastid proteins is blocked. In this case phytochrome is ineffective as an inducer of transcription. The plastid factor, the molecular nature of which is still unknown, is thus at a higher level of the regulation of transcription than is phytochrome (and nitrate) (Rajasekhar and Mohr 1986). The gene expression of typical cytosolic enzymes is not affected by

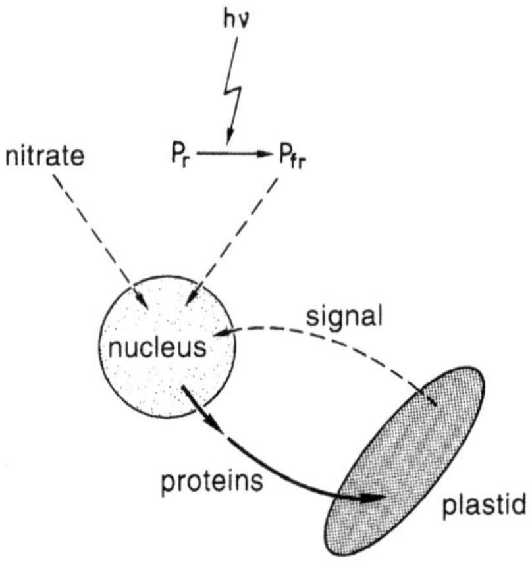

Fig. 5. Schematic showing the significance of plastid factor (*signal*) for expression of nuclear genes coding for plastid proteins. The plastid factor can be interpreted as an unspecific transcription factor without which neither light (via phytochrome) nor nitrate can become effective at the level of transcription. (Mohr and Schopfer 1995)

the lack of plastid factor. Under experimental conditions where, for example, SSU-mRNA and LHCII-apoprotein-mRNA disappear completely, synthesis of representative cytosolic, mitochondrial and glyoxysomal enzymes proceeds normally (Oelmüller and Mohr 1986).

5 A Textbook on Plant Physiology

In the 1960s a modern textbook of plant physiology was urgently needed to support teaching and to improve learning. Erwin Bünning persuaded me in the mid-1960s to publish my lectures on plant physiology which I had delivered in Freiburg and in part abroad. The original version of the textbook was written in German (Mohr 1969). Later, once the text was well established and Peter Schopfer had agreed to join me as an author, new editions as well as English (Mohr and Schopfer 1995) and Japanese translations followed.

To write a textbook is very different from preparing lecture notes! Expressed differently: to transform lecture notes into a coherent, consistent and balanced text requires an enormous concentration, in particular if the only time during the week left over for this task was from Friday night to Sunday afternoon. To achieve this goal you need stable health and a tolerant family. Fortunately, I had both.

The text was readily accepted by the students – and even by most of my colleagues. This was by no means a matter of course, since I left no doubt in my 'concepts of physiology' that physiology is not something like immature biochemistry but a science of its own, namely the science of organismic regulatory and control processes (Mohr 1988). Moreover, in my view, physiology is a quantitative (or exact) science. By analogy with physics the aim of physiology is to elaborate general statements (laws).

This was not my personal hybris! Physiologists have always wished to postulate at least some statements with the same authority and validity as our colleagues in physics. The recent successes in biochemistry and molecular biology have significantly reduced the self-confidence of physiologists and even, at times, produced a kind of neurosis, leading to the statement "that physiology has moved to the periphery of the problems." I am still convinced that my view of physiology as sketched above is right, but I am no more convinced that the present generation of physiologists can prevent what I called some years ago "the molecular collapse of quantitative physiology." Of course, I appreciate and enjoy the amazing discoveries made in molecular plant biology during the last 15 years, but the limits to reductionism may not be ignored (Mohr 1989). A very important challenge in the next decades will be constructing an interface between genomics and whole plant and animal physiology (Melvin 2003).

6 Steps in my Philosophical Thinking

I do not claim to be a professional philosopher of science. Rather, I consider myself a natural scientist with a profound interest in the nature of scientific thought and in the significance - including the cultural significance - of science. Since nobody can afford time to follow up every interesting idea, I had to select philosophical problems with a high probability of a pay-off for a practicing natural scientist. As a consequence, my 'philosophy', including my 'political philosophy', has remained closely connected with the progress of the sciences.

6.1 Structure and Significance of Science

The discoveries of science had a profound effect on man's philosophy, ethics and spiritual beliefs. I had planned for many years to organize my ideas on this subject matter. The opportunity of writing a treatise was made possible by a Visiting Professorship granted to me by the University of Massachusetts during the autumn term of 1975. The published text is based on a series of 15 lectures that I delivered at the University. I am still grateful to the students, and to my colleagues and friends at the University, for the cordial reception, continuous interest and constructive criticism. It was the positive response of my class and the fascinating intellectual climate at

Amherst, Maas., that encouraged me to revise the lectures for print (Mohr 1977a).

I dedicated the book to Erwin Bünning and to Walter Kossel. As I mentioned above, Bünning's book *Theoretische Grundfragen der Physiologie* was a major determinant in my decision to become a biologist (Bünning 1949). The late Walter Kossel introduced me to physics. He was not only a great physicist, but also a fascinating philosopher and an admirable personality. I had some bad feelings when I submitted the final text for print. I could only hope that the professional philosophers would forgive me if my treatise did not always respect the conventional division of labor between science and philosophy. I fully agreed with David Hull who had just criticized some noted scientists who tried their hands at 'philosophizing': "Just as scientists are entitled to established standards of competence for their undertakings, philosophers have a right to expect at least minimal competence in theirs" (Hull 1975). On the other hand, I felt that it was legitimate to base a reflection about 'structure and significance of science' primarily on the self-understanding of the practicing scientist. My deep-rooted respect for philosophy in toto and for epistemology in particular would hopefully prevent me from becoming chauvinistic in favor of the scientific world view.

The influential Anglo-Saxon schools in the philosophy of science have generally equated philosophy with epistemology, treating ethics as not properly part of academic philosophy. Since I was not obliged to any philosophical school, but looked at the problems from the point of view of a practicing scientist, I did not follow the tendency of excluding anything from consideration that might raise moral problems. Rather, I intended to emphasize this aspect. Moreover, I took the liberty of looking at some traditionally epistemological problems, such as empiricism and rationalism, from the point of view of scientific knowledge.

Another point was that by this time (1975) most philosophers of science, in particular within the dominant positivist school, took the Comtean view of physics as the paradigmatic science and of biology as a relatively immature and secondary study. Even as an enthusiastic biologist who was proud of his trade, I could not ignore this tendency since there is some truth in it.

While there is no principal difference between physics and biology, the general approach in both fields and the nature of physical and biological theories and laws obviously differ to a considerable extent. I often referred to physics rather than to biology not only for the sake of simplicity, clarity and brevity, but also for the reason that physics has a far wider scope than biology. Physics deals with the properties of all matter whereas biology is only concerned with living systems or with ecosystems in which living systems play the major part. All living systems are physical objects, but only a very small number of physical objects are considered to be living systems.

In the treatise I have often used the term 'responsibility'. This term implies, and I did emphasize this at the very beginning, that we are responsible for our acts. Indeed, I presume that moral responsibility is part of human nature, irrespective of the century-long discussion on determination, free will and moral responsibility. Determination to a scientist conveys the general proposition that every event has a cause. Whether this general proposition is true is a difficult question to decide, but it is certainly assumed to be true by most scientists. Otherwise science, in particular prediction, explanation and purposive action, would not be possible. On the other hand, we presume that we are responsible for our acts. It is implied as a matter of course that moral responsibility is an integral part of human nature. Indeed, we all believe that moral responsibility is real.

Since moral responsibility implies free will and self-determination (in the sense that we can create de novo determinants for our conduct and thus break causal continuity), the very serious and difficult question arises of whether moral responsibility (which implies free will) is compatible with our scientific knowledge, which plainly says that the concept of a breach in causal continuity is not acceptable. From the point of view of science, the reality of free will cannot be conceded. On the other hand, as human beings, we depend on the belief that at least some of our actions (called 'willed actions') are preceded by deliberation and choice and that our choice can be influenced by consideration of consequences.

Of course, I could not solve the paradoxes of free will (Stent 2002), but a thorough description of the paradoxes turned out to be a great advantage when I analysed – later in the book – the principle of causality and the structure of teleological action.

6.2 Epistemology and Evolution

Einstein once stated that for him the most unintelligible thing about the world is that it is intelligible. Why can we use the axioms and theorems of Euclidian geometry to reason about the physical world (Mohr 1977b)? Why is it legitimate to apply to a wheel (a physical object) the mathematical formula derived for a circle:

$$c = 2\pi r$$

As you all know, the use of diagrams is not essential to geometry. Geometrical reasoning per se is purely abstract. If diagrams are introduced it is only as an aid to our reason. In any case, a circle and a wheel are totally different things, but nevertheless the wheel obeys the formula obtained for the circle.

I have written here another formula, derived by Gauss, for a purely mathematical relationship between two variables:

$$y = \frac{1}{\sigma\sqrt{2\pi}} \cdot e^{-(x-\bar{x})^2/2\sigma^2}$$

Why is it possible to describe and treat the frequency distribution in biological populations with the help of this relationship, e.g. the frequency distribution of intelligence test scores in a human population? In brief: why is mathematics, a purely deductive system of axioms and theorems, applicable to nature? Galileo stated in 1623 that "nature is written in mathematical language" (and this phrase has been repeated and followed by scientists ever since), but he could not give any naturalistic explanation why this is so.

We may extend our question to the whole of logic (I consider mathematics to be part of logic). Why can we rely on syllogistic reasoning? It is probably that the principle of the syllogism was formulated not before but long after the usefulness and validity of syllogistic reasoning was discovered by man. Logic is the theory of deductive argument, not its source. Why does the real world obey logic? Wittgenstein (in his early phase) was very concerned about this question. As he puts it, our justification for holding that the world could not conceivably disobey the laws of logic is simply that we could not say of an illogical world how it would look. This is obviously not a good argument. Rather, it is a sign of perplexity and ignorance.

Or remember Ayer in *Language, Truth and Logic* (Ayer 1936): His second class of propositions were the formal propositions of mathematics and logic, and they were held to be tautologies. Ayer thought of them (as did Wittgenstein) "as being merely rearrangements of symbols which did not make any statement about the world" (Magee 1971). As Ayer pointed out:

The empirist does encounter difficulty...in connection with the truths of formal logic and mathematics. For whereas a scientific generalization is readily admitted to be fallible, the truths of mathematics and logic appear to be necessary and certain to everyone. But if empiricism is correct, no proposition which has a factual content can be necessary or certain. Accordingly, the empirist must deal with the truths of logic and mathematics in one of the two following ways: he must say either that they are not necessary truths, in which case he must account for the universal conviction that they are; or he must say that they have no factual content, and then he must explain how a proposition which is empty of all factual content can be true, useful, and surprising. If neither of these courses proves satisfactory, we shall be obliged to give way to rationalism. We shall be obliged to admit that there is some truth about the world which we can know independently of experience; that there are some properties which we can ascribe to all objects, even though we cannot conceivably observe that all objects have them. And we shall have to accept it as a mysterious inexplicable fact that our thought has this power to reveal to us authoritatively the nature of objects which we have never observed. (Magee 1971)

We know today that the fact Ayer is referring to is neither mysterious nor inexplicable. Ayer (and nearly all philosophers so far) did not take into account that experience has been accumulated and preserved as genetic information during biological evolution. From the point of view of the individual, this inherited foreknowledge about the structure of the world has the character of synthetic judgements a priori; from the point of view of evolution, however, the same statements must be regarded as synthetic judgements a posteriori, based on experience.

From the standpoint of the individual, a synthetic judgement about the world that we can know independently of experience is a synthetic judgement a priori in a strict sense. However, from the point of view of evolution, the same judgement is a synthetic judgement a posteriori; it is based on experience, namely on the experience of our phylogeny, which is preserved and stored in the genetic information, in the peculiar nucleotide sequence of the genetic DNA we have inherited from our parents. Kant's dictum that, although there can be no doubt that all our knowledge begins with experience, it does not follow that it all arises out of experience, can no more be maintained. The fact is that we combine in our individual life two kinds of experiences: the genetically inherited experience of our ancestors and the experience we have made in our personal life, including the experience transmitted to us by cultural tradition and social imitation. This is, in brief, the message of what has been called evolutionary epistemology (Vollmer 1975).

Having introduced this new branch of epistemology, I want to return to Ayer (and to Kant). Ayer (1936) writes in *Language, Truth and Logic*:

...the admission that there were some facts about the world which could be known independently of experience would be incompatible with our fundamental contention that a sentence says nothing unless it is empirically verifiable...the fundamental tenet of rationalism is that thought is an independent source of knowledge, and is moreover a more trustworthy source of knowledge than experience; indeed some rationalists have gone so far as to say that thought is the only source of knowledge. And the ground for this view is simply that the only necessary truths about the world which are known to us are known through thought and not through experience....

From the point of view of evolutionary epistemology, the venerable confrontation of empiricism and rationalism is a fictive problem. In reality, it does not exist, since synthetic judgements a priori are also based on experience. The rationalism vs. empiricism debate is a striking example of a philosophical discussion that scientific advance – in this case, the theory of biological evolution – has rendered pointless.

Let me close my apologetics of this new branch of philosophy with a personal remark: in my opinion, further progress in science and in epistemology will depend on the strong and steady, i.e., not only occasional, interaction of both fields. Epistemology must respect and consider seriously the actual, genuine knowledge elaborated by scientific disciplines. Science has proved that it is capable of producing genuine knowledge, although the theory of how knowledge can actually be obtained, epistemology, may not yet be satisfactory. On the other hand, those scientific disciplines, in particular those fields that have advanced far beyond the realm of common sense, such as classical quantum physics, elementary particle physics or molecular genetics and neurobiology, must consider epistemology as part of their endeavor. Fortunately, a considerable number of scientists have realized that epistemology is part of their trade. Professional philosophers, in particular in Germany and France, are more reluctant, following Wittgenstein at least in this regard, who firmly believed that the dominance of scientific thought since the Renaissance has been a disaster. Following the late Ayer, I do not see how one is to effect changes of this attitude except by changing the conceptual outlook among modern philosophers, and I do not think one can do this except in conjunction with scientific theories (Magee 1971).

Philosophy and science diverged in the nineteenth century, partly in consequence of the romantic movement and partly because science got too difficult, and now we must utilize the outlook that they are coming together again.

6.3 The Normative Code of Science (Mohr 1979b)

Scientists consider science as a systematic attempt of the human mind to obtain genuine knowledge. However, scientific knowledge in a true sense is 'public knowledge', i.e. knowledge shared by a scientific community. Repeatedly, I have investigated the normative prerequisites, the intrinsic values of science, which enable the scientific communities to obtain genuine objective knowledge. As a rule the intrinsic values of science are being shared by all members of a scientific community. This moral commitment is the reason why science became the prime cultural force of our age. Upon questioning, most scientists will tell you that they consider 'freedom to inquire into the nature of things' a great privilege. To what extent, however, is the individual scientist 'free'? In fact, he or she is not free at all. Rather, the activity of every scientist is subject to a strong social control. He/she is faced with a normative code which itself is based on the intrinsic values of science.

Why does the scientist accept this normative code as compulsory? Because scientists need and desire recognition by other scientists they cannot but conform to the goals and norms of the scientific community. The striving for professional recognition is in fact the major motivation for scientists to conform to the normative code.

The normative code is a heterogeneous complex. It consists of at least two parts: basic assumptions that are rigorously shared by the members of the scientific community and actual commandments. Since usually the assumptions as well as the actual commandments are not written up explicitly, the following lists are possibly not complete in the eyes of some scientist.

Among the basic assumptions, we may discriminate two sets:

1. There is a real world (negative version: the notion of solipsism is not acceptable); the real world is intelligible at least in part; formal logic (including mathematics) is valid, without any restriction, in the description of the real world; there is no break in causal continuity.
2. Freedom of thought and freedom of inquiry must indeed be guaranteed (this does not necessarily imply freedom in the choice of any particular goal; it implies, however, that the result of a scientific inquiry may not be influenced by factors extrinsic to science); genuine, objective knowledge is good, i.e. it is superior to ignorance under all circumstances (this implies that there is no code of forbidden knowledge). Complete freedom to publish cannot be assumed in reality. The industrial scientist as well as the scientist working on classified governmental projects must be aware of the possibility that he may not be permitted by his employer to publish the results of his research without delay.

A tentative list of the actual commandments includes:

- Be honest.
- Never manipulate data.
- Be precise.
- Be fair (e.g. with regard to priority of data and ideas).
- Be without bias (e.g. with regard to data and ideas of your rival).
- Do not make compromises, but try to solve a problem.

This list of fundamental commandments could be extended by more explicit formulations:

- Use words and symbols with explicitly defined meaning.
- Try to improve singular and general propositions with regard to inner perfection and degree of credibility.
- Make accurate predictions and indicate the range of error.
- Consider observational (experimental) data as the ultimate court of appeal.
- Be ready, any time, to modify or replace a theory in view of inner inconsistencies or experimental refutation.
- Always keep in mind that the members of the scientific community must depend on each other for reliability of material and intellectual methods, data, conclusions and theories.
- Regard simplicity as of high value. Do not create new constructs if unless absolutely necessary.

It is expected by the scientific community that the normative code is obeyed as a matter of course whenever a person works as a scientist. Misconduct and deliberate fraud are extremely rare in scientific practice, since every scientist is aware of the fact that penalties for misconduct are more severe in science than in any other profession.

6.4 Contributions to Economics

My amateur contributions to economics concerned the theory of evolutionary economics as compared to the biological theory of evolution (Mohr 1990b), and the concept of sustainable development. I published a carefully investigated book on the latter subject under the title *Qualitative Growth as a Survival Strategy* (English translation) (Mohr 1995). Even though I still consider it as my best book, it was not a success. My ideas and suggestions were indeed appreciated and even hailed by many readers, but the book had no detectable influence on political decisions or economic practice.

6.5 Science and Politics

Science as I understand it is a global adventure. It comprises all mankind. Science is less political than other issues (Mohr 2003). I preach to my students that science is a bridge for peace. Moreover, science is a moral adventure based on values and on mutual trust among the members of a scientific community.

The aim of science is reliable knowledge. Some call it 'truth'. Irrespective of semantics, it is agreed, that knowledge is a supreme value, the only means to overcome ignorance, superstition and poverty (Mohr 1999). Truth, reliable knowledge, can only be achieved if two functional values are obeyed: (1) the scientific method and (2) the ethics of science. Both the search for truth and ethical autonomy are alien to politics, which is necessarily ruled by doctrine and power, public opinions and compromise. Thus, science and politics are distinct and basically unrelated segments of social reality. However, science and politics are connected by two fluxes:

- Science needs alimentation. This implies that we must make the public aware of the value that basic science contributes to society.
- Politics needs knowledge. There is not the slightest chance to run a modern society without reliable knowledge.

Obviously, problems will arise: the individual scientist is expected to be a loyal member of the global scientific community and – at the same time – he or she is expected to be a loyal citizen of a politically defined community. Tensions and conflicts will arise whenever the two loyalties are not compatible.

When I wrote the present chapter, a story from the late 1960s came to mind. James Shapiro, who by that time belonged to the left wing liberals on the Harvard campus, declared his intention of renouncing his scientific career at the AAAS meetings in Boston in 1969. He had been part of the Harvard team that isolated the lac operon. Shapiro argued "that so long as men like Nixon and Agnew prescribe what is going to be done with scientific knowledge, we scientists should quit giving them the materials for their bad decisions." Shapiro could hardly be blamed as long as he spoke as an individual. However, could we fellow researchers still follow him when he tried to make his conviction compulsory for the scientific community? Could we be convinced that he had thought enough about the relationship between the scientific community, society and political power in different parts of the world? Or about the relationship between knowledge, wealth, security and power?

As for myself I reached the following conclusion: no scientist should be forced to do research he or she does not want to do for ethical reasons. If a scientist feels that society would inevitably and immediately use the knowledge originating from his/her research for purposes he/she considers to be evil, he/she must have the right to give up at any time. However, the scientist may leave no doubt that his/her decision is based on his/her personal value system, which is extrinsic to science.

This distinction is essential, since allegiance to any particular ideology may blind even a scientist. As an example, in the 1930s partisans of the

Marxist 'social relations of science' movement in England such as Bernal and Haldane openly used their high prestige as scientists to communicate and spread their ideological convictions even though the horrific excesses of Stalinism were already known. During the same period, some German physicists, among them two Nobel Prize laureates, created what they called German Physics, a direct support of Nazi ideology. Obviously, the outstanding analytical powers of the Nobelists had fallen sway under political prejudice.

6.5.1 First Case Study: Science and War

The utmost problem for the integrity of science is war among nations. Let us briefly consider the matter of war and law in view of the present tensions, in particular in view of biological weapons. A majority of the world's leading scientists stands behind the call – recently issued by John Polanyi – for replacement of war by international law. It is a call, an impressive document indeed, not to arms, but to disarm the sources of major tensions in the present world. Polanyi justifies the call: "I don't think politics can afford to discount the thinking of scientists in an age of science." Others such as Mario Molina from the Massachusetts Institute of Technology (MIT) are more sceptical about the political power of the scientific community: "Science alone, technology alone, is not sufficient to deal with these issues, we need strong commitments and signals from society that science and technology are put to good use." This, of course, is the crucial point: what does good use of knowledge mean in view of terror and warfare?

Let me briefly look back to 1940. World War II made science the most powerful political institution humankind has yet devised. Radar, pioneered in Britain, had the greatest effect on the war. "Radar won the war", US scientists used to say, "but the atomic bomb ended it." During World War II, science became subordinated to government to reach a superior goal, to win the war against Hitler and the Japanese generals.

However, the nuclear bombing of Hiroshima, which was intended to be and conceived of as an ethical act (to save the lives, both American and Japanese, which would be lost in a full-scale invasion), was later classified as unethical, even by brilliant thinkers such as the fabulous British physicist and Nobel Prize winner of 1948, Blackett, one of the inventors of airborne radar. His book of 1949 – *Fear, War and the Bomb* – has had a strong impact on the political scene as well as on the scientific community. Moreover, the apparent subordination of science to government in military matters became a major cause of public distrust of scientists. In any case, the use of the atomic bomb in World War II illustrates the obvious in the starkest terms, namely that even the most considered application of knowledge as well as moral calculus do not lead to unambiguous answers. In any case, Hitler and his allies as well as the Japanese generals no longer had a chance once science entered the war scene in England and America in 1940/1941. Moreover, the ultimate triumph of Western democracy over Marxist doctrine in the Cold War is to be attributed to superior science.

The excelling performance of Western science over German and Soviet science is to be attributed to freedom of thought and freedom from censorship. The catastrophic consequences of constraining freedom of inquiry for ideological reasons are well illustrated by the cases of Galileo – the loss of Italy's leading position in the rise of science during the Renaissance – and Lysenko – the rapid loss of genetic knowledge and competence and concomitantly of agricultural potential and productivity in the Soviet Union during the Lysenko–Stalin era. Millions of people died because an obsolete political doctrine prevented the use of scientific progress in Soviet agriculture and medicine.

6.5.2 Second Case Study: Robust Knowledge

This case study is based on my mixed experience as a science adviser to the government. Under the prevailing contract between science and society, science has been expected to follow sound scientific practice and to produce 'reliable' knowledge, provided merely that it communicates its progress to society. Some social scientists (Michael Gibbons, Helga Nowotny, Peter Scott) have argued that a fresh approach – virtually a complete rethinking of science's relationship with the rest of society – is needed (Nowotny et al. 2001). The argument is that we are currently witnessing a significant shift from 'reliable' to 'socially robust' knowledge. The latter characterization is intended to embrace the growing contextualization and socialization of knowledge. For knowledge to be 'socially robust' three aspects are decisive:

1. It is valid not only inside by also outside the laboratory.
2. This validity is achieved through involving an extended group of experts, including lay 'experts'.
3. Because 'society' has participated in its genesis, such knowledge is less likely to be contested than that which is merely 'reliable'.

This means, as my colleagues believe, that science can no longer be validated as reliable by conventional discipline-bound norms; while remaining robust, science must now be sensitive to a much wider range of social implications.

An example is the current debate surrounding genetically modified organisms (GMOs). Here, specialist peer groups have been challenged not only by parapolitical forces such as Greenpeace but also by ordinary consumers, for whom the research process is far from transparent and who are demanding that it be more so. To emphasize the decisive point again, knowledge of the health and ecological implications of GMOs may be 'reliable' in the conventional scientific sense; but it is not socially robust, and will not become so until the peer group is broadened to take into account the perspectives and concerns of a much wider section of society.

This novel type of scientific activity has been established in the USA and here in Germany under the term 'technology assessment'. I summarize my experience in this new discipline as follows (Mohr 1998): in the modern world at least some first-class experts must respond to issues and questions that are never merely scientific and technical, and must address audiences that never consist only of other experts. The limits of competence of the individual expert call for the involvement of a wide base of expertise that has to be carefully orchestrated if it is to speak in unison.

The suggestions of the advocates of 'robust knowledge' extend further, however. My colleagues from the social sciences demand that the process of knowledge *production* – not only knowledge application – become, what they call, transparent and participative. To quote Mike Gibbons (Gibbons 1999): "Under the prevailing contract, science was left to make discoveries and then make them available to society. The new contract will be based upon the joint production of knowledge by society and science."

Most natural scientists insist that the rigor of the scientific method/ethics and the robustness of the scientific practice may not be weakened. More than ever the modern world depends on 'reliable knowledge'. The advocates of socially 'robust knowledge' concede, of course, that "reliable and/or objective knowledge continues to provide the foundations on which our knowledge of the natural world depends" (Helga Nowotny) (Nowotny et al. 2001). However, the inherent problems have not been solved yet: to what extent does the engagement with extrascientific forces during the process of research undermine science's capacity to produce reliable knowledge? How can scientific quality be maintained?

While the proponents of 'robust knowledge' argue that there is no suggestion that scientific objectivity, consensibility and consensuality cannot be practiced in such a context, my (and others) practical experience in the field of technology assessment has been sobering. In fact, as far as I am aware, participation of lay 'experts' in the process of research has been a disaster – wherever it was tried, including my own efforts in the field of gene technology. Most citizens are neither willing nor prepared to accept this role. Moreover, most scientists are equally unprepared to assess the significance of their scientific results in the context of public opinion or policy guidance. The distinction between science and policy advice cannot simply be suspended by a strong dose of good will.

A particular confusion will be produced when well-intended but politically naive scientists enter the political fray as individuals or groups, as in the case of the Intergovernmental Panel on Climate Change: science as a whole becomes politicized and compromised. This is my major concern at present: how can we prevent the practice of science from being compromised? (Mohr 1996). A realistic strategy is to avoid any merging of science and politics and instead to strengthen the trust between science and the public, in particular between science and the political decision-makers.

What does 'trust' mean in this context? The politician must be sure that our factual statements are as 'reliable' as possible by any means. Scientists, on the other hand, must be sure that political practice is based on factual knowledge, Aristotelian logic and a coherent and consistent value system.

To emphasize again the decisive point: irrespective of the degree of 'public understanding of science', the modern world is totally dependent on *reliable* knowledge, on scientific truth. Scientific truth is the greatest treasure mankind has. It is the privilege and the obligation of the established institutions of science to foster and to preserve this knowledge and to guarantee the cultural boundary conditions for scientific truth and human welfare. This requires the autonomy of science *and* a positive coaction with the political institutions on the basis of mutual trust. In fact, our most urgent task is to restore the confidence of the public – including the political sector – in the trustworthiness of the natural sciences. The scientific enterprise – objective knowledge as the supreme good – must prove its integrity beyond any serious doubt or it will lose its status as the prime cultural force of our age (Mohr 1996).

7 Retrospect and on to the Future

Why did I write the present essay? What kind of message did I want to communicate to the inclined reader? Let me summarize my intentions by referring to a recent interview.

I was asked by a TV-master whether or not – under the present circumstances – I would try again to become a professor? My answer: Given my propensity I would certainly try anew to become a researcher but not necessarily strive for the position of a university professor. The politically overregulated German university of today is no longer the best place to pursue a scientific career.

Do you regret, the moderator continued his inquiry, that in the late 1950s you did not venture a career in the USA? My answer: Even though I liked and admired America, my wife and myself preferred in the end to live closer to the genuine European culture.

Next question: Do you feel that your long-lasting flirtation with Queen Philosophy has affected your scientific achievements in an adverse sense? My answer: Yes and no! Undoubtedly, I have neglected at times my scientific work (and my family!) in favor of urgent philosophical (and political) challenges. On the other hand, the fact that my scientific studies were embedded in a philosophical (or at least epistemological) framework made me sure of my scientific aims and fostered my purposeful behavior in the laboratory and in the lecture hall.

Towards the end of our conversation the moderator asked me four very distinct questions and requested a brief answer to each of them:

- Do you believe in the future of science? – Of course, I responded quickly, since mankind cannot survive without knowledge. Science is the only source of genuine, reliable knowledge (and consistent ethics). Moreover, like it or not, science has become the prime cultural force of our age, and in practice an ever-more pervasive way of life for all people on this planet. In brief: the future of man is linked irreversibly to the future of science.
- Will there be any future for philosophy? – The power of philosophical thinking, mainly an advanced form of hermeneutics, will become an integral force in the ripening processes of the great naturalistic theories. Critical realism will be the common epistemological basis. As A.J. Ayer, the ingenious philosopher, pointed out decades ago (Ayer 1936), philosophy will flourish as the 'logic' or metatheory of the modern sciences. On the other hand, traditional but largely outdated disciplines such as ontology, metaphysics, idealism or phenomenology in its varieties will simply disappear.
- Will your scientific discipline, plant physiology, disappear as well, in this case in favor of molecular biology? – As I have argued repeatedly (Mohr 1989), there are limits to reductionism in the sciences. I am indeed afraid of 'the molecular collapse of quantitative physiology'. An important challenge in the next decade will be constructing an interface between genomics and whole plant and animal physiology.
- Science and technology often encounter scepticism and wariness in academia as well as in the media and general public. How can the scientific community promote a much more positive view of science in society? – A traditional response of the scientific community to what it views as a lack of appreciation by the media and the general public has

been to mount public understanding campaigns to 'enlighten' the populace about specific issues in question. I have participated ex officio in quite a number of these well-aimed campaigns (Mohr 1998). The results have been sobering: simply trying to educate the public about specific science-based issues is not working; what is needed is a broadening of science education in general far beyond the present scope. Science in the modern world must definitely become an integral part of liberal education and cultivated life.

References

Ayer AJ (1936) Language, truth and logic. Gollancz, London
Blackett PMS (1949) Fear, war and the bomb: military and political consequences of atomic energy. Whittlesey House, New York
Bünning E (1949) Theoretische Grundfragen der Physiologie. Piscator, Stuttgart
Dittes H, Rissland I, Mohr H (1971) On the regulation of enzyme levels in different organs of a plant (*Sinapis alba* L.). Z Naturforsch 26b:1175–1180
Durst F, Mohr H (1966) Phytochrome-mediated induction of enzyme synthesis in mustard seedlings (*Sinapis alba* L.). Naturwissenschaften 53:531–532
Frankhauser C, Staiger D (2002) Photoreceptors in *Arabidopsis thaliana*: light perception, signal transduction and entrainment of the endogenous clock. Planta 216:1–16
Gibbons M (1999) Science's new social contract with society. Nature 402 (Suppl 2):C81–C84
Hull DL (1975) Philosophy of reductionism in biology. Nature 257:429–432
Karow H, Mohr H (1967) Aktivitätsänderungen der Isocitratlyase (EC 4.1.3.1) während der Photomorphogenese beim Senfkeimling (*Sinapis alba* L.). Planta 72:170–186
Magee B (1971) Modern British philosophy. St. Martin's Press, New York (conversation with A.J. Ayer)
Melvin T (2003) Wallenberg prize acceptance speech. The future of biology: reason for concern? Plant Physiol 131:3–5
Mohr H (1956) Die Beeinflussung der Keimung von Farnsporen durch Licht und andere Faktoren. Planta 46:534–551
Mohr H (1969) Lehrbuch der Pflanzenphysiologie. Springer, Berlin Heidelberg New York
Mohr H (1972) Lectures on photomorphogenesis. Springer, Berlin Heidelberg New York
Mohr H (1977a) Structure and significance of science. Springer, Berlin Heidelberg New York
Mohr H (1977b) Epistemology and evolution. In: Mohr H (ed) Structure and significance of science, Chap 15. Springer, Berlin Heidelberg New York
Mohr H (1979a) Erbgut und Umwelt. In: Hassenstein B (ed) Freiburger Vorlesungen zur Biologie des Menschen. Quelle and Meyer, Heidelberg
Mohr H (1979b) The ethics of science. Interdisciplin Sci Rev 4:45–53
Mohr H (1987) Mode of coaction between blue/UV light and light absorbed by phytochrome in higher plants. In: Senger H (ed) Blue light responses: phenomena and occurrence in plants and microorganisms, vol 1. CRC Press, Boca Raton
Mohr H (1988) Control of plant development: signals from without - signals from within. Bot Mag (Tokyo) 101:79–101
Mohr H (1989) Is the program of molecular biology reductionistic? In: Hoyningen-Huene P, Wuketits FM (eds) Reductionism and systems theory in the life sciences. Kluwer, Dordrecht

Mohr H (1990a) Der Stickstoff – ein kritisches Element der Biosphäre. Sitzungsberichte der Heidelberger Akademie der Wissenschaften, Jg. 1990, 5. Abhandlung. Springer, Berlin Heidelberg New York, pp 293–330
Mohr H (1990b) Biologie und Ökonomik – Chance für eine Interdisziplinarität. In: Witt K (ed) Studien zur Evolutorischen Ökonomik I. Duncker and Humblot, Berlin
Mohr H (1995) Qualitatives Wachstum. Weitbrecht, Stuttgart
Mohr H (1996) Das Expertendilemma, eine Pilotstudie. In: Nennen HU, Garbe D (eds) Das Expertendilemma. Zur Rolle wissenschaftlicher Gutachter in der öffentlichen Meinungsbildung. Springer, Berlin Heidelberg New York
Mohr H (1998) Technikfolgenabschätzung in Theorie und Praxis. Springer, Berlin Heidelberg New York
Mohr H (1999) Wissen – Prinzip und Ressource. Springer, Berlin Heidelberg New York
Mohr H (2003) On the precarious relationship between science and politics. Nova Acta Leopoldina NF 87(325):177–181
Mohr H, Neininger A (1994) Schwachstellen der Nitrat- und Ammoniumassimilation – eine Chance für die Gentechnik. Sitzungsberichte der Heidelberger Akademie der Wissenschaften, Jg. 1993/94, 6. Abhandlung: 7–71. Springer, Berlin Heidelberg New York
Mohr H, Schopfer P (1995) Plant physiology. Springer, Berlin Heidelberg New York
Nowotny H, Scott P, Gibbons M (2001) Re-thinking science: knowledge production and the public in an age of uncertainty. Polity Press, Oxford
Oelmüller R, Mohr H (1986) Photooxidative destruction of chloroplasts and its consequences for expression of nuclear genes. Planta 167:106–113
Oelze-Karow H, Schopfer P, Mohr H (1970) Phytochrome-mediated repression of enzyme synthesis (lipoxygenase): a threshold phenomenon. Proc Natl Acad Sci USA 65:51–57
Rajasekhar VK, Mohr H (1986) Appearance of nitrite reductase in cotyledons of the mustard (*Sinapis alba* L.) seedling as affected by nitrate, phytochrome and photooxidative damage of plastids. Planta 168:369–376
Sage LC (1992) Pigment of the imagination. A history of phytochrome research. Academic Press, New York
Schuster C, Mohr H (1990) Appearance of nitrite reductase mRNA in mustard seedling cotyledons is regulated by phytochrome. Planta 181:327–334
Stent GS (2002) Paradoxes of free will. American Philosophical Society, Philadelphia
Vollmer G (1975) Evolutionäre Erkenntnistheorie. Hirzel, Stuttgart

Prof. Dr. Dr. h.c. mult. Hans Mohr
Biological Institute II, University of Freiburg
Schaenzlestrasse 1
79104 Freiburg, Germany

Genetics

Recombination: RNA – A Powerful Tool for Recombination and Regulated Expression of Genes*

Dirk Müller and Ulf Stahl

1 Introduction

The natural frequency of any base mutating to another is about 10^{-7} to 10^{-8}. The process of screening for a special phenotype is extremely time-consuming when one just relies on this natural mutation. Thus, a broad spectrum of methods have been developed to speed up the process and to obtain mutated or recombinant organisms.

Random methods such as UV mutagenesis and chemical mutagenesis yield clones with one or more unknown mutations. Although these methods can provide a desired phenotype, e.g. a high producing strain, several additional and undesired mutations can counteract the improvement by slowing growth or introduce other unwanted abilities. These side effects can have severe consequences for the strains concerned, particularly as the whole metabolic spectrum can be thrown out of balance which then results in reduced yield of certain metabolites.

Targeted mutations can only produce the desired phenotype (e.g. homologous recombination, loss-of-function/gain-of-function, gene disruption, and gene replacement) and are thus used when one wants to avoid unwanted mutations. A major disadvantage of both random and targeted mutageneses is that it targets and/or recombines the DNA of the organism. The mutations introduced therefore cannot be reversed. The choice of targets in this case is limited to those which, after mutagenesis, are non-lethal for the organism.

Recombination techniques that do not target the DNA but rather the messenger RNA have become more and more reliable over recent years. Even genes that are essential for the organism can specifically be targeted when the remaining level of the messenger is sufficient to guarantee survival of the organism. The application of these RNA techniques enables scientists to analyse gene function, even in certain developmental phases.

* This chapter is dedicated to Prof. Dr. Karl Esser on the occasion of his 80th birthday

Progress in Botany, Vol. 66

RNA technologies are also a promising tool for the production of metabolites as cells can be grown to the desired growth phase. The use of the appropriate RNA technology can lead to product formation without necessarily having to shift to an inducing medium. Products that are growth inhibiting or, in the end, lethal for the cell, can even be synthesized when the fermentation process is broken down into separate growth and production phases.

The antisense technique which utilizes RNA molecules in an antisense orientation to the target RNA was the first RNA technique used for this purpose. The discovery of the catalytic properties of some kinds of RNAs called 'ribozymes' caused a sensation among researchers which was only topped by elucidation of RNAi mechanisms. Apart from being promising therapeutic agents, these RNA methods also give us insight into RNA chemistry and evolution.

This chapter provides an overview of the progress in RNA-based recombination methods and their application.

2 RNA Methods

Basically, RNA-based methods for recombination do not change the genetic identity of an organism as they do not attack the DNA, but rather the transcribed RNA. This has the advantage over mutating the target DNA, which can be lethal for the organism. In addition, it is a better system when the mutation takes effect at a particular stage of development. Some genes even have open reading frames on the complementary strand which can be affected when the target gene is silenced by classical means.

2.1 Antisense RNA

Antisense oligonucleotides bind specifically to a complementary mRNA. In order to be able to bind specifically to a given target RNA, the size of the molecule can range from anything as small as 12 to several hundred nucleotides; however, the probability for unspecific binding with additional undesired side effects increases with very small antisense molecules. After binding to the target site, antisense molecules lead to digestion by RNase H, impede splicing, or to translational arrest.

The size of antisense molecules determines the probability for specific or unspecific binding. Given an antisense molecule with a size of 8 nt, we can find this binding motive statistically every 65 kb. The human genome consists of about 2,900,000 kb which would result in nearly 45,000 possible binding sites for this small antisense molecule. In contrast, an antisense molecule of 20 nt would bind every 1,099,511,627 kb, thereby ensuring the required specificity.

Although the antisense approach is universal and specific, many antisense molecules show little or no antisense activity in practice. It is therefore crucial to find the optimal binding site for an efficient antisense molecule. RNA molecules fold into complex secondary structures which can exclude accessibility of some regions on the target RNA. Both a high number of external nucleotides and global flexibility of the antisense RNA can improve the formation of the RNA–RNA duplex. In practice, this dependency on unpredictable binding requires that whole libraries of antisense constructs need to be tested in order to obtain only a few functional antisense molecules. Despite the relatively simple theory behind antisense technology, the need for a large setup can make the antisense approach very expensive and time consuming with no guarantee of success.

Binding site selection could, in principle, be based on secondary structure prediction (Lehmann et al. 2000). However, as our knowledge of RNA folding, duplex stability, and the quality of computer programs for structure prediction is rather limited, a combined approach using computer prediction and experimental methods (either based on hybridization techniques or on ribonuclease H, i.e. RNase H activity) are most likely to succeed at constructing efficient antisense RNAs (Sczakiel 2000; Sohail and Southern 2000).

Various chemical modifications can be introduced to improve stability and affinity for the target. Common modifications of antisense molecules include the exchange of phosphodiester internucleotide linkages by phosphorothioate, polyamides (see Braasch and Corey 2002), replacement of the phosphate-sugar backbone with uncharged N-(2-aminoethyl)glycine (PNA, see Fig. 1; Doyle et al. 2001), replacement of the sugar backbone with locked nucleic acids (LNA; Wahlestedt et al. 2000), or addition of functional groups such as 2-methoxyethyl (Chen et al. 2001). Although chemical modification of RNA leads to increased stability, it must be balanced with effects on activity and potential toxicity.

The application of RNA antisense molecules is always accompanied by the risk of degradation by RNases. The use of oligonucleotide analogues is one way of protecting an antisense molecule as there are no natural enzymes inside the cell that can use these synthetic molecules as a substrate. The introduction of peptide nucleic acids (PNA, Fig. 1) has attracted both scientific and commercial interest. The negatively charged sugar-phosphate backbone of DNA or RNA in PNA has been replaced by a polypeptide backbone, leading to enhanced stability and the formation of stronger hybrids with complementary RNA and DNA.

A number of antisense constructs which are used as pharmaceutical agents or as a tool for crop improvement have been developed and tested. However, only a few of these antisense products have got past the test phase to be used commercially (Table 1). Generally, this seems to be the major obstacle in antisense technology. Although the antisense theory seems to

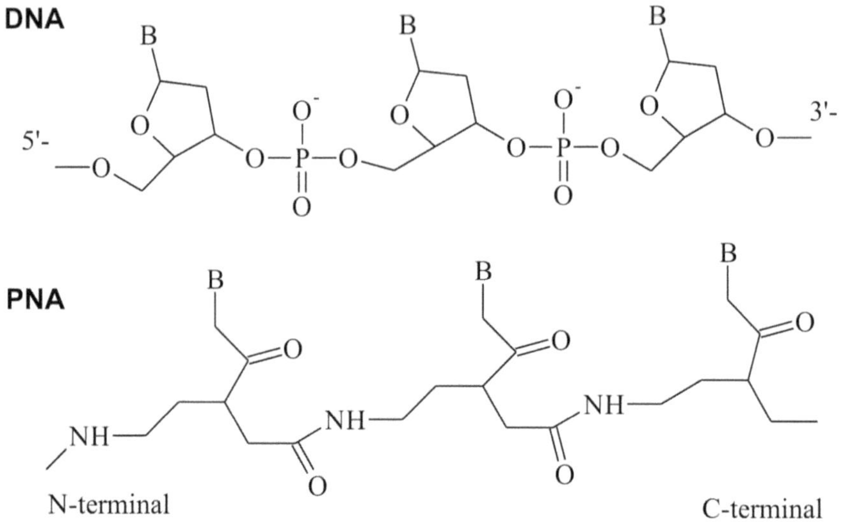

Fig. 1. General structure of a DNA chain and the PNA analogue. *B* Nucleobase

be very elegant in principle, it is extremely complicated when it comes down to its complex in vivo application.

Large amounts of antisense constructs still need to be evaluated to find a functional molecule. This disadvantage requires a large number of transformants, and an easy and speedy selection system. Therefore, antisense technology has been restricted to organisms that are easily transformed and can produce the desired amounts of transformants. There are obviously several factors that play a role in using this technology – last, but not least, unpredictable secondary structures in the target as well as the antisense molecule itself – an area that still requires considerable research.

2.2 Ribozymes

Ribozymes were first discovered in 1982 when Kruger et al. (1982) found an RNA in the ciliated protozoan *Tetrahymena thermophila*. This RNA could cleave and splice itself without any external protein or energy source. Shortly afterwards, a similarly independent RNA in the *Escherichia coli* enzyme ribonuclease P (Guerrier-Takada et al. 1983) was detected.

What exactly are ribozymes? They are RNA molecules that perform a chemical reaction and leave the reaction unmodified. Ribozymes can be divided into groups based on their size and mechanism (Warashina et al. 2000; Doudna and Cech 2002). The first group (Table 2) contains small ribozymes, such as the well-known 'hammerhead' ribozyme, most of which

Table 1. Examples of available antisense products. *Asterisks* indicate product in clinical test phases

Antisense product	Target	Producer
Flavr Savr tomato	Reduction of polygalacturonidase	Calgene, Davis, California, USA (now Monsanto Co.)
AP 12009*	Treatment of malignant brain tumours	Antisense Pharma, Regensburg, Germany
Vitravene*	Treatment of cytomegalovirus (CMV)	Isis Pharmaceuticals, Carlsbad, California, USA
Genasense*	Bcl-2 (protein made by cancer cells that is thought to block chemotherapy-induced cell death)	Genta, Berkeley Heights, New Jersey, USA
Zeaxanthin potato	Zeaxanthinepoxidase	TU München, Munich, Germany

have been derived from ssRNA viruses. The second group (Table 3) consisting of large ribozymes comprises group I and group II introns, RNase P, and the 70S ribosome.

2.2.1 Small Ribozymes

Small ribozymes range from about 40 to 154 nucleotides in length. They catalyze the endonucleolytic cleavage of a RNA phosphodiester backbone, resulting in products with a 5-OH and a 2,3-cyclic phosphate terminus. The smallest and best characterized ribozyme is the *hammerhead ribozyme* (see Table 2). This name is derived from its secondary structure which resembles a carpenter's hammer.

The hammerhead ribozyme was discovered in plant RNA viruses that contain single-strand, circular RNA as genetic material (Buzayan et al. 1986; Rubino et al. 1990). These viruses replicate using the rolling circle mechanism (Fig. 2). The hammerhead ribozyme functions as a scissor in this process, snipping the large mRNA chain into single-genome-length pieces that undergo subsequent ligation into circles.

The *'hairpin' ribozyme* seems to act in much the same way as the *varkud satellite (VS) ribozyme* found in mitochondria of the fungus *Neurospora* and the *hepatitis delta virus (HDV) ribozyme*. Hammerhead, hairpin, and HDV ribozymes require linear RNA as a target, whereas the VS ribozyme requires a stem-loop structure as minimal substrate (Guo and Collins 1995).

Table 2. *Small ribozymes* are involved in replication and result in 2, 3 cyclic phosphate and a 5-OH reaction products. See text for further information

Ribozyme	Size (nt)	Source
Hammerhead (Buzayan et al. 1986; *Rubino et al. 1990)*	ca. 40	Viral satellite and viroids in plants, newt satellite, *Schistosoma*
Hairpin (Fedor 2000)	50–60	Tobacco ringspot virus
Hepatitis delta virus (HDV) (Perrotta and Been 1990, 1991; Shih and Been 2002)	ca. 80	HDV
Varkud satellite (VS) (Collins 2002)	154	VS in *Neurospora crassa* mitochondria

Table 3. *Large ribozymes* are housekeeping RNAs necessary for cellular processes

Ribozyme	Size (nt)	Source	Reaction products	Function
Group I intron (Cech 1990)	200–1,500	Eukaryotes (organelles, nucleus), prokaryotes	5'/3' Ligated exons, intron with 5'guanosin and 3'OH	Splicing
Group II intron (Lehmann and Schmidt 2003)	300–3,000	Eukaryotic organelles, prokaryotes	5'/3' Ligated exons, intron with 2'-5' lariat and 3'OH	Splicing
RNase P (Schon 1999; Xiao et al. 2002)	140–490	Eukaryotes (organelles, nucleus), prokaryotes	5' Phosphate and 3'-OH	tRNA processing
70S ribosome (Moore and Steitz 2003; Steitz and Moore 2003)		Prokaryotes	Peptide bond	Peptidyl transferase

The *Neurospora* varkud-1c strain contains two extremely abundant RNAs of approximately 0.9 kb. These transcripts were found to be the circular and the linear form of a new plasmid named VS. The VS plasmid does not contain an open reading frame; however, it has a sequence that matches the consensus sequence of the minimal promoter on the mitochondrial genome. It is thought that the VS RNA's role is to process multimeric transcripts into monomeric form by site-specific self-cleavage and subsequent circularizing into monomers (Collins 2002).

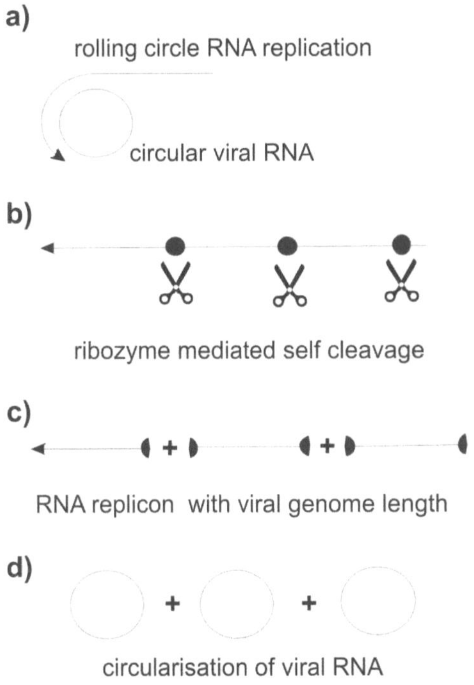

Fig. 2a–d. Ribozyme function in rolling circle replication. **a** The circular viral RNA is transcribed multiple times into a long polycistronic RNA. **b** Internal ribozymes separate the virus genomes by self cleavage. **c** Single-length viral genomes. **d** Single viral genomes then circulate

2.2.2 Large Ribozymes

2.2.2.1 Group I and II Self-Splicing Introns

Group I and group II introns (Table 3; see Cech 1990; Lehmann and Schmidt 2003) catalyze splicing by means of two sequential transesterifications. Group I ribozymes include the *Tetrahymena* ribozyme which was originally discovered by Cech (1990). The introns range in size from a few hundred to just over 2,500 nucleotides in length and catalyze their own removal and the stitching together of the adjacent exons in a two-step transesterification reaction (Schmidt et al. 1992).

2.2.2.2 Ribonuclease P

Ribonuclease P (see Schon 1999; Kirsebom 2002) is a RNA-protein complex found in all cells. RNase P in bacteria is composed of catalytic RNA (~ 400 nt) and a single protein, whereas 9–12 proteins in the eukaryotic nucleus are associated with the RNA section. The main function of this riboprotein involves the maturation of tRNAs.

All tRNAs are transcribed as precursors and need subsequent processing to generate mature 5' and 3' ends. Ribonuclease P removes a 5' leader sequence from the precursor tRNA leaving a 5'-phosphate group. In contrast to eukaryotic RNase P, the RNA part in bacteria and archaea is able to recognize and cleave pre-tRNA without the protein compounds marking them as ribozymes.

2.2.2.3 The Ribosome Is a Ribozyme

Ribosomes are huge aggregates containing three (four in eukaryotes) rRNA molecules and scores of protein molecules. The ribosome catalyzes the formation of peptide bonds by linking amino acids to a growing polypeptide chain. Examination of the crystal structure shows that the formation of the peptide bond is solely catalyzed by the 23S RNA molecule in the large subunit (Nissen et al. 2000). Ribosomal proteins are not directly involved in the catalytic function of the ribosome, but presumably in the formation of the tertiary structure of the rRNA. The finding that only the rRNA part is catalytically active led to the conclusion that the ribosome is, in fact, a ribozyme.

The restriction activity of RNA (such as found in the ribozyme) is mainly regarded as an artificial development. We look at another type of chemical reaction performed by a ribozyme, the ligation of covalent bonds between two amino acids by the RNA part of the ribosome, which indicates that ribozymes can carry out diverse chemical reactions. The fact that an important riboprotein, such as the ribosome, is a ribozyme emphasizes the outstanding properties of RNA within the molecular world (see the following section for more about chemical reactions performed by ribozymes).

2.2.3 Ribozymes as Molecular Tools

The discovery of catalytic activity in RNAs was inherently revolutionary. However, how does this make ribozymes a scientific tool? We will look at the progress made in ribozymes technology in both basic research and commercial applications, using the hammerhead motive as an example.

The analysis of the hammerhead ribozyme provides us with significant insights into the chemistry of RNA, particularly RNA cleavage mechanisms, folding of RNA, and the function of involved cations, respectively. The native form of the hammerhead ribozyme is part of a large viral RNA molecule. Three helixes surround a single-stranded catalytic core region allowing a *cis*-cleavage (Fig. 3a) of the molecule. Removal of the loops in helixes I and III separates the RNA into ribozyme and substrate, thereby allowing multiple turnover *trans*-cleavage (Fig. 3b) of virtually any substrate RNA (Birikh et al. 1997). This modification enables the hammerhead

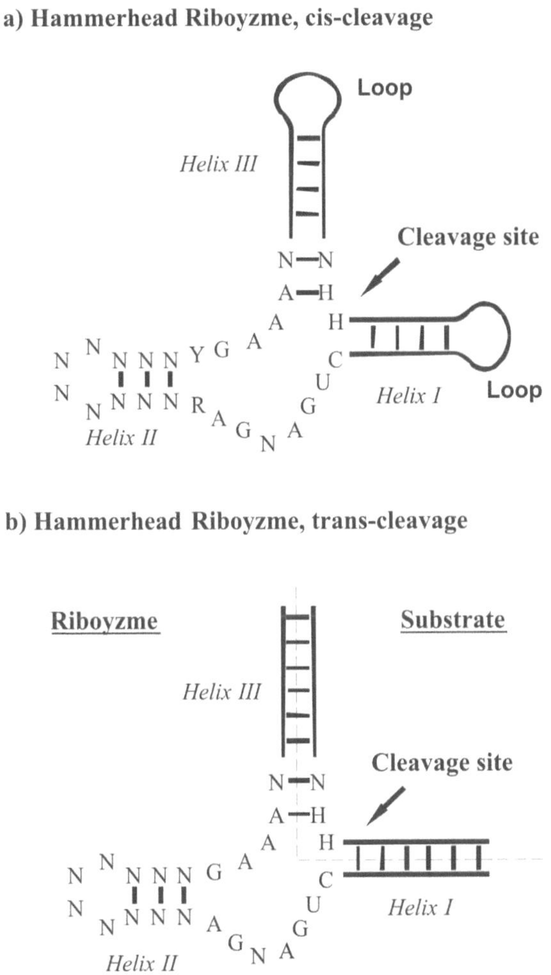

Fig. 3a, b. Minimal consensus structure of the hammerhead ribozyme (N: A, C, G, or U; H: C, A, or U; R: purine; Y: pyrimidine) (Symons 1997). **a** *cis*-cleavage. The ribozyme is a single molecule which can cleave itself. **b** *trans*-cleavage. Both loops in part **a** are removed, thereby dividing the molecule into the ribozyme on the *left-hand side of the dashed line* and the substrate on the *right-hand side*

motive to act as 'real' enzymes: (1) attachment to the target sequence; (2) cleavage; and (3) release of the reaction products to (4) find a new target. In this manner, one ribozyme can catalyze the cleavage of numerous substrate RNAs (Fig. 4).

Natural ribozymes are generally encoded within the sequence they are intended to cleave. As both are located on the same RNA strand, they are named *cis*-acting ribozymes (self-cleaving RNA). In contrast, a *trans*-acting ribozyme is transcribed from a different locus or cistron, respectively. In this case, ribozyme and substrate are located on different transcripts.

As ribozymes have very few requirements, they are a versatile tool. The recognition sequence for the hammerhead ribozyme is simply an 'NHH' triplet [N: A, U, G, or C; H: A, U, or C (Shimayama et al. 1995; Kore et al.

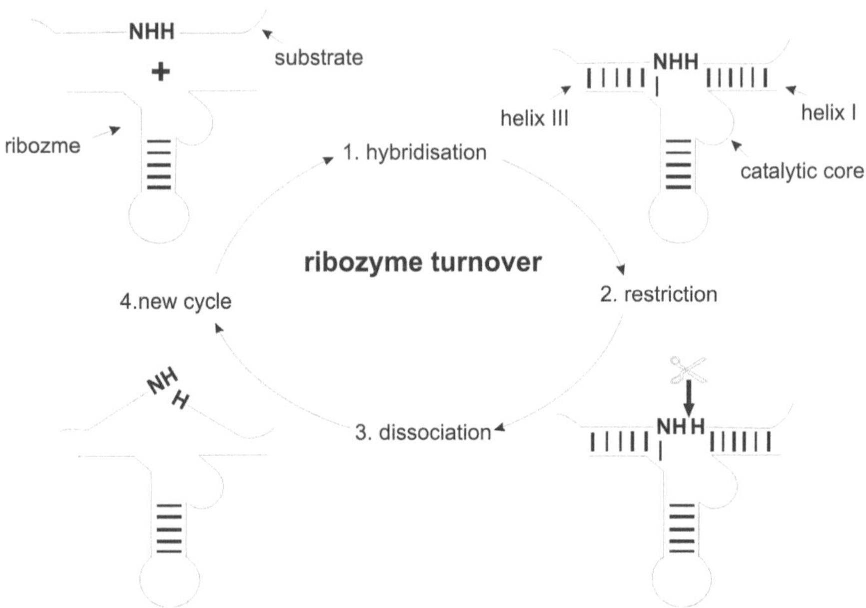

Fig. 4. Ribozyme turnover: *1* The catalytic cycle is initiated when the ribozyme forms a hybrid through the Watson–Crick base pairing with a suitable substrate RNA; *2* the ribozyme cleaves substrate RNA at the cleavage site; *3* reaction products are released; *4* the ribozyme can initiate a new cycle

1998)]. The nucleotides of helix I and III are not conserved and can be designed to perfectly match the target sequence (Fig. 3).

Although the structure of the hammerhead ribozyme is well characterized, the detailed cleavage mechanism still remains unsolved. The hammerhead motive is often regarded as a metalloenzyme, as metal ions are required for cleavage reaction. Divalent metal ions, preferably Mg^{2+}, have to be added for an in vitro reaction. However, cleavage has also been observed under conditions with high molar monovalent cations, although this is not as efficient (Murray et al. 1998).

There are currently numerous applications for hammerhead and other ribozymes in use in both basic research and commerce. One example is to elucidate the polyadenylation machinery responsible for cleaving and polyadenylation of polymerase II-transcribed mRNA. This multiprotein complex consists of several factors which construct a complex network which are thought to function in cooperation.

In this connection, tRNA and rDNA genes were constructed in yeast with *cis*-acting hammerhead motives that cleave transcripts in the 3'-region (Duvel et al. 2003). The resulting 5'-products were efficiently polyadenylated, showing that polyadenylation in yeast is uncoupled from cleavage and from transcription by RNA polymerase II, respectively.

Ribozymes are extremely promising as pharmaceutical agents as they can directly target the genetic origin of a disease and not only the symptoms. Several ribozymes have been tested as anti-cancer agents and some even made it to the clinical phase, e.g. Herazyme and Aagiozyme (Sirna Therapheutics).

Other important pharmaceutical applications are the treatment of virus infections. Ribozymes are especially attractive as agents against retroviruses such as HIV (Chang et al. 2002) and the hepatitis virus (Hoeprich et al. 2003; Sriram et al. 2003) as both genomic RNA and mRNA are susceptible targets. Another example of extensive ribozyme application are allosteric ribozymes (Breaker 2002). These complex ribozymes are controllable RNA catalysts which act as molecular switches and cellular biosensors (Rajendran and Ellington 2002). Allosteric ribozymes are composed of two independently functioning domains, one a specific receptor and the other a self-cleaving ribozyme domain. Once bound to a specific effector molecule, conformational changes occur that either increase or decrease the catalytic activity of the ribozyme (Burke et al. 2002; Vaish et al. 2002).

Ribozymes are capable of more than simply cleaving a substrate molecule. Extensive research has produced novel ribozymes catalyzing different reactions. Most of this work can be associated with the 'RNA world' theory of a precellular life where RNA in a very primitive self-replicating system carries out all catalytic activities and functions as storage for genetic information (Burke 2003).

Using in vitro evolution Johnston et al. (2001) found an RNA polymerase enzyme that catalyzes a primer extension reaction. Tsukiji et al. (2003) developed a ribozyme capable of oxidizing benzyl alcohol coupled with the regeneration of NAD^+, thereby providing support for the existence of an RNA-based metabolic system. Another approach followed by Baskerville and Bartel (2002) was to develop a ribozyme that ligates RNA to a protein through formation of a phosphoamide bond by a combination of rational design and in vitro selection.

Further support for the 'RNA world' hypothesis is the production of a nucleotide synthase ribozyme in vitro which forms a glycosidic link between a tethered sugar and a pyrimidine (Chapple et al. 2003). Kawasaki and Taira (2002) overcame the problem of finding an accessible binding site in the substrate RNA for the ribozyme be adding a 60-nt poly(A) sequence to the ribozyme. This naturally occurring RNA motive interacts with the RNA helicase eIF4AI via both the poly(A)-binding protein and the PABP-interacting protein-1, thereby conferring to the new hybrid ribozyme the ability to unwind and cleave. The new hybrid ribozyme is able to cleave target sites that are not cleaved by the pure ribozyme.

2.3 RNAi/siRNA – The New and Promising Candidate

RNA interference (RNAi) has been a conserved cellular defence mechanism to protect against RNA intruders or aberrant RNAs throughout evolution. Endogene RNA is mediated to sequence specific destruction by cellular machinery directed by short interfering RNAs (siRNA). First evidence for this mechanism was found as post-transcriptional gene silencing (PTGS) (Vaucheret et al. 2001) in transgene plants and as 'quelling' (Pickford et al. 2002) in filamentous fungi (*Neurospora*). The range of organisms showing RNAi continues to grow. Most basic research was done in *Caenorhabditis* (Grishok and Mello 2002) and *Drosophila* (Kao and Megraw 2004); however, mammalian cells (Paddison and Hannon 2003), protozoa (Ullu et al. 2002), and fission yeast (Raponi and Arndt 2003) are also targets for RNAi-mediated gene silencing.

The RNAi technique is based on double-stranded siRNA molecules which are 21–26 nucleotides long.

Figure 5 schematically represents RNAi's mode of action. The small dsRNAs are produced by degrading long dsRNA molecules by an RNase III-related nuclease termed Dicer. The siRNA is then bound by the RNA-induced silencing complex (RISC, see below), a multiprotein complex which first unwinds the double strand, resulting in an ssRNA that is subsequently used by RISC to detect and degrade homologous RNA molecules. In addition, the hybridized single-stranded siRNA can act as a primer for RNA-dependent RNA polymerase (RdRP), thereby amplifying a target mRNA double strand which is in turn substrate for the Dicer (Hammond et al. 2000). This results in a reduction of the target mRNA level and leads to subsequent reduction of or abolished translation.

The *Dicer* is a dsRNA-specific endonuclease. It degrades the long dsRNA (trigger) into siRNA with a defined length.

The *RISC* (RNA-induced silencing complex) complex is a multicomponent nuclease that degrades mRNAs which are homologous to the silencing trigger. RISC binds short siRNAs derived from the double-stranded RNA trigger. A current model suggests that Dicer facilitates the incorporation of siRNAs into RISC as part of the RISC complex. This then dissociates from the Dicer and targets homologue mRNAs for destruction (Hammond et al. 2001).

The length of the initial dsRNA is of particular importance. Hammond et al. (2000) showed a direct connection between length of the dsRNA and the RNAi effect by means of in-vitro experiments with *Drosophila* cell extract. Most RNAi phenotypes were obtained with dsRNA of 500 bp or more. Smaller dsRNA drastically reduced the RNAi effect, whereas no effect was detected with dsRNA of about 100 bp or less.

Recent reports (Hamilton et al. 2002; Tang et al. 2003) show that plants produce two classes of siRNAs which are heterogeneous in size and function. The long class RNAi (24–26 nt) has been linked to transcriptional silencing and to the systemic spread of silencing, whereas the short class RNAi (21–23 nt) seems to act as a guide for RISC, thereby leading to mRNA destruction. Biochemical studies explain the distinct lengths of the two plant siRNA classes

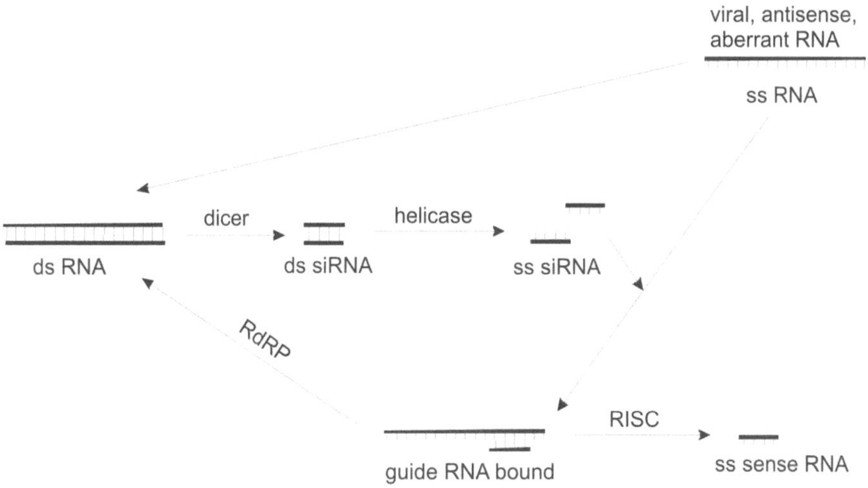

Fig. 5. Schematic model of RNAi. *RISC* Random induced silencing complex; *RdRP* RNA-dependent RNA polymerase

in that they are produced by different members of the Dicer family. Complexity in RNAi gene silencing does not stop at the cellular border. It could be shown that RNAi in *Caenorhabditis elegans* systemically inhibits gene expression throughout the organism (Winston et al. 2002).

RNA interference technology has a growing impact on functional genomics. RNAi has already been used to elucidate the function of many genes in *Caenorhabditis*, *Drosophila*, several plants, and in mammals (see McManus and Sharp 2002; Kittler and Buchholz 2003). Kamath et al. (2003) used a genome-wide approach to identify phenotypes for 1,722 genes in *Caenorhabditis elegans*. It is very difficult and laborious to produce recessive, loss-of-function alleles for genetic analyses in polyploid species. Lawrence and Pikaard (2003) used a single RNAi-inducing transgene to dominantly repress multiple orthologs in *Arabidopsis*. A chemical-inducible Cre/loxP recombination system that triggers the expression of an intron-containing inverted-repeat RNA (RNAi) was developed in *Arabidopsis thaliana* and *Nicotiana benthamiana* (Guo et al. 2003). They could induce silencing of both transgenes and endogenous genes at different stages of development without detectable secondary effects.

In addition to gene function analyses, RNAi shows great promise for use in therapeutic strategies designed to suppress the expression of pathogenic genes (Dave and Pomerantz 2003). The chances of achieving success in single or multiple knock-down gene expression using RNAi technology is much greater than with conventional methods.

3 Conclusion

RNA technologies have several advantages over classical gene regulation methods. However, there are still open questions and problems when it comes to practice. The antisense and ribozyme approach always faces the problem of target-site accessibility. Reduced specificity can become a problem, particularly when using very short antisense molecules. This may lead to a dramatic increase of resources invested and requires extensive preparatory work until gene expression is efficiently reduced. RNAi, however, as an evolutionary conserved mechanism taking advantage of a natural cellular process and thereby circumvents these problems. As a result, RNAi has rapidly become the new method of choice for a wide range of applications, complementing or replacing conventional antisense and ribozyme technologies.

RNA methods as a scientific tool can have diverse functions. They provide important insights into the role of RNA in evolution. The 'RNA-world' hypothesis states that RNA was the first molecule that carried out replication, data storage, and diverse catalytically functions before the appearance of proteins or DNA. This hypothesis has been reinforced by the fact that several cellular functions are indeed implemented by RNA and not by DNA or a protein. One of the most basic mechanisms in the cell, the production of polypeptide chains at the ribosome, is now attributed to the RNA part of the ribosome. Recent results show that short RNA molecules play an important role in the regulation of gene expression.

Drug target discovery and validation are major fields of application in the pharmaceutical industry. The development of drugs involves substantial economic burdens. Therefore, precise and accurate target selection is required. RNA technologies offer the necessary specificity. The vast amount of proteins which play a potential role in disease in addition to the size of sequence data provided by the Human Genome Project require high throughput screening to evaluate gene function. RNA methods are an ideal base for such screens as they are highly specific. A broad range of available modifications facilitate the adaptation to particular needs. Furthermore, the components are easy to synthesize.

As the world population continues to increase, food supplies must also grow to meet nutritional requirements. Prevention of crop loss caused by plant pathogens, on the one hand, and improvement of productivity, on the other, are of major interest for crop scientists. RNA methods can complement classical methods such as breeding. Initial RNA techniques have optimized products (e.g. Flavr Savr tomato) which are already on the market. RNA silencing mechanisms are common in plants as a defence against viruses and can be further developed to confer resistance to other plant pathogens.

What is the main advantage of RNA methods? All DNA-based methods used for regulation of gene expression are irreversible. However, RNA methods – because they do not chance the genetic background of the organism – allow gene expression to be switched on or off. In addition, expression can be controlled using a regulable promoter gene even in a short time frame. Thereby, cells can first grow to a desired growth phase; gene expression is then reduced to a certain level or switched off by inducing the RNA method. This is of particular importance for the biotechnological production of metabolites. In this case the process is often divided into two phases, the growth phase and the production phase. Secondary metabolites such as fungal antibiotics in particular (e.g. penicillin) are produced in the stationary growth phase. Once the stationary phase is reached RNA techniques can be switched on to influence metabolic flow and improve product formation. This high flexibility and specificity facilitated by modern RNA technology is advantageous in comparison to classical DNA-based methods for gene regulation.

Acknowledgements. We would like to thank Roslin Bensmann for checking the English.

References

Baskerville S, Bartel DP (2002) A ribozyme that ligates RNA to protein. Proc Natl Acad Sci USA 99:9154–9159
Bell MA, Johnson AK, Testa SM (2002) Ribozyme-catalyzed excision of targeted sequences from within RNAs. Biochemistry 41:15327–15333
Birikh KR, Heaton PA, Eckstein F (1997) The structure, function and application of the hammerhead ribozyme. Eur J Biochem 245:1–16
Blount KF, Uhlenbeck OC (2002) The hammerhead ribozyme. Biochem Soc Trans 30:1119–1122
Braasch DA, Corey DR (2002) Novel antisense and peptide nucleic acid strategies for controlling gene expression. Biochemistry 41:4503–4510
Brantl S (2002) Antisense RNAs in plasmids: control of replication and maintenance. Plasmid 48:165–173
Breaker RR (2002) Engineered allosteric ribozymes as biosensor components. Curr Opin Biotechnol 13:31–39
Burke DH (2003) Ribozyme-catalyzed genetics. In: Pouplana LRD (ed) The genetic code and the origin of life. Eurekah.com, Houston
Burke DH, Ozerova ND, Nilsen-Hamilton M (2002) Allosteric hammerhead ribozyme TRAPs. Biochemistry 41:6588–6594
Buzayan JM, Hampel A, Bruening G (1986) Nucleotide sequence and newly formed phosphodiester bond of spontaneously ligated satellite tobacco ringspot virus RNA. Nucleic Acids Res 14:9729–9743
Byun J, Lan N, Long M, Sullenger BA (2003) Efficient and specific repair of sickle beta-globin RNA by trans-splicing ribozymes. RNA 9:1254–1263
Cech TR (1990) Self-splicing of group I introns. Annu Rev Biochem 59:543–568

Chang Z, Westaway S, Li S, Zaia JA, Rossi JJ, Scherer LJ (2002) Enhanced expression and HIV-1 inhibition of chimeric tRNA(Lys3)-ribozymes under dual U6 snRNA and tRNA promoters. Mol Ther 6:481–489

Chapple KE, Bartel DP, Unrau PJ (2003) Combinatorial minimization and secondary structure determination of a nucleotide synthase ribozyme. RNA 9:1208–1220

Chen W, Bennett CF, Condon TP, Stecker K, Tian L, Kahan BD, Stepkowski SM (2001) Methoxyethyl modification of phosphorothioate ICAM-1 antisense oligonucleotides improves prevention of ischemic/reperfusion injury. Transplant Proc 33:854

Chu J, Dolnick BJ (2002) Natural antisense (rTSalpha) RNA induces site-specific cleavage of thymidylate synthase mRNA. Biochim Biophys Acta 1587:183–193

Collins RA (2002) The *Neurospora* varkud satellite ribozyme. Biochem Soc Trans 30:1122–1126

Dave RS, Pomerantz RJ (2003) RNA interference: on the road to an alternate therapeutic strategy! Rev Med Virol 13:373–385

DeRose VJ (2003) Metal ion binding to catalytic RNA molecules. Curr Opin Struct Biol 13:317–324

Doudna JA, Cech TR (2002) The chemical repertoire of natural ribozymes. Nature 418:222–228

Doyle DF, Braasch DA, Simmons CG, Janowski BA, Corey DR (2001) Inhibition of gene expression inside cells by peptide nucleic acids: effect of mRNA target sequence, mismatched bases, and PNA length. Biochemistry 40:53–64

Dunham CM, Murray JB, Scott WG (2003) A helical twist-induced conformational switch activates cleavage in the hammerhead ribozyme. J Mol Biol 332:327–336

Duvel K, Pries R, Braus GH (2003) Polyadenylation of rRNA- and tRNA-based yeast transcripts cleaved by internal ribozyme activity. Curr Genet 43:255–262

Emilsson GM, Nakamura S, Roth A, Breaker RR (2003) Ribozyme speed limits. RNA 9:907–918

Fedor MJ (2000) Structure and function of the hairpin ribozyme. J Mol Biol 297:269–291

Grassi G, Grassi M, Kuhn A, Kandolf R (2002) Determination of hammerhead ribozyme kinetic constants at high molar ratio ribozyme-substrate. J Math Biol 45:261–277

Grishok A, Mello CC (2002) RNAi (nematodes: *Caenorhabditis elegans*). Adv Genet 46:339–360

Guerrier-Takada C, Gardiner K, Marsh T, Pace N, Altman S (1983) The RNA moiety of ribonuclease P is the catalytic subunit of the enzyme. Cell 35:849–857

Guo F, Cech TR (2002) Evolution of *Tetrahymena* ribozyme mutants with increased structural stability. Nat Struct Biol 9:855–861

Guo HC, Collins RA (1995) Efficient trans-cleavage of a stem-loop RNA substrate by a ribozyme derived from *Neurospora* VS RNA. EMBO J 14:368–376

Guo HS, Fei JF, Xie Q, Chua NH (2003) A chemical-regulated inducible RNAi system in plants. Plant J 34:383–392

Hamilton A, Voinnet O, Chappell L, Baulcombe D (2002) Two classes of short interfering RNA in RNA silencing. EMBO J 21:4671–4679

Hammond SM, Bernstein E, Beach D, Hannon GJ (2000) An RNA-directed nuclease mediates post-transcriptional gene silencing in *Drosophila* cells. Nature 404:293–296

Hammond SM, Boettcher S, Caudy AA, Kobayashi R, Hannon GJ (2001) Argonaute 2, a link between genetic and biochemical analyses of RNAi. Science 293:1146–1150

Hampel KJ, Burke JM (2003) Solvent protection of the hammerhead ribozyme in the ground state: evidence for a cation-assisted conformational change leading to catalysis. Biochemistry 42:4421–4429

Hatanaka H, Abe Y, Naruke M, Asai S, Miyachi H, Kawakami T, Nagata J, Kamochi J, Kijima H, Yamazaki H, Scanlon KJ, Ueyama Y, Nakamura M (2001) Modulation of multidrug resistance in a cancer cell line by anti-multidrug resistance-associated protein (MRP) ribozyme. Anticancer Res 21:879–885

Hertel KJ, Uhlenbeck OC (1995) The internal equilibrium of the hammerhead ribozyme reaction. Biochemistry 34:1744–1749

Hoeprich S, Zhou Q, Guo S, Shu D, Qi G, Wang Y, Guo P (2003) Bacterial virus phi29 pRNA as a hammerhead ribozyme escort to destroy hepatitis B virus. Gene Ther 10:1258–1267

Johnston WK, Unrau PJ, Lawrence MS, Glasner ME, Bartel DP (2001) RNA-catalyzed RNA polymerization: accurate and general RNA-templated primer extension. Science 292:1319–1325

Kamath RS, Fraser AG, Dong Y, Poulin G, Durbin R, Gotta M, Kanapin A, Le Bot N, Moreno S, Sohrmann M, Welchman DP, Zipperlen P, Ahringer J (2003) Systematic functional analysis of the *Caenorhabditis elegans* genome using RNAi. Nature 421:231–237

Kao LR, Megraw TL (2004) RNAi in cultured *Drosophila* cells. Methods Mol Biol 247:443–458

Kasai Y, Shizuku H, Takagi Y, Warashina M, Taira K (2002) Measurements of weak interactions between truncated substrates and a hammerhead ribozyme by competitive kinetic analyses: implications for the design of new and efficient ribozymes with high sequence specificity. Nucleic Acids Res 30:2383–2389

Kawasaki H, Taira K (2002) Identification of genes by hybrid ribozymes that couple cleavage activity with the unwinding activity of an endogenous RNA helicase. EMBO Rep 3:443–450

Kirsebom LA (2002) RNase P RNA-mediated catalysis. Biochem Soc Trans 30:1153–1158

Kittler R, Buchholz F (2003) RNA interference: gene silencing in the fast lane. Semin Cancer Biol 13:259–265

Kore AR, Vaish NK, Kutzke U, Eckstein F (1998) Sequence specificity of the hammerhead ribozyme revisited; the NHH rule. Nucleic Acids Res 26:4116–4120

Kruger K, Grabowski PJ, Zaug AJ, Sands J, Gottschling DE, Cech TR (1982) Self-splicing RNA: autoexcision and autocyclization of the ribosomal RNA intervening sequence of *Tetrahymena*. Cell 31:147–157

Lawrence RJ, Pikaard CS (2003) Transgene-induced RNA interference: a strategy for overcoming gene redundancy in polyploids to generate loss-of-function mutations. Plant J 36:114–121

Lehmann K, Schmidt U (2003) Group II introns: structure and catalytic versatility of large natural ribozymes. Crit Rev Biochem Mol Biol 38:249–303

Lehmann MJ, Patzel V, Sczakiel G (2000) Theoretical design of antisense genes with statistically increased efficacy. Nucleic Acids Res 28:2597–2604

McManus MT, Sharp PA (2002) Gene silencing in mammals by small interfering RNAs. Nat Rev Genet 3:737–747

Misra VK, Draper DE (2002) The linkage between magnesium binding and RNA folding. J Mol Biol 317:507–521

Mochizuki K, Fine NA, Fujisawa T, Gorovsky MA (2002) Analysis of a piwi-related gene implicates small RNAs in genome rearrangement in *Tetrahymena*. Cell 110:689–699

Moore PB, Steitz TA (2003) After the ribosome structures: how does peptidyl transferase work? RNA 9:155–159

Murray JB, Seyhan AA, Walter NG, Burke JM, Scott WG (1998) The hammerhead, hairpin and VS ribozymes are catalytically proficient in monovalent cations alone. Chem Biol 5:587–595

Nagata J, Kijima H, Hatanaka H, Asai S, Miyachi H, Abe Y, Yamazaki H, Nakamura M, Watanabe N, Mine T, Kondo T, Scanlon KJ, Ueyama Y (2002) Reversal of drug resistance using hammerhead ribozymes against multidrug resistance-associated protein and multidrug resistance 1 gene. Int J Oncol 21:1021–1026

Narlikar GJ, Herschlag D (1997) Mechanistic aspects of enzymatic catalysis: lessons from comparison of RNA and protein enzymes. Annu Rev Biochem 66:19–59

Nishikura K (2001) A short primer on RNAi: RNA-directed RNA polymerase acts as a key catalyst. Cell 107:415–418

Nissen P, Hansen J, Ban N, Moore PB, Steitz TA (2000) The structural basis of ribosome activity in peptide bond synthesis. Science 289:920–930

Osada H, Tokunaga T, Abe Y, Asai S, Miyachi H, Hatanaka H, Tsugu A, Kijima H, Yamazaki H, Shima K, Ueyama Y, Osamura Y, Nakamura M (2003) Reversal of drug resistance mediated by hammerhead ribozyme against multidrug resistance-associated protein 1 in a human glioma cell line. Int J Oncol 22:823–827

Paddison PJ, Hannon GJ (2003) siRNAs and shRNAs: skeleton keys to the human genome. Curr Opin Mol Ther 5:217–224

Perrotta AT, Been MD (1990) The self-cleaving domain from the genomic RNA of hepatitis delta virus: sequence requirements and the effects of denaturant. Nucleic Acids Res 18:6821–6827

Perrotta AT, Been MD (1991) A pseudoknot-like structure required for efficient self-cleavage of hepatitis delta virus RNA. Nature 350:434–436

Pickford AS, Catalanotto C, Cogoni C, Macino G (2002) Quelling in *Neurospora crassa*. Adv Genet 46:277–303

Pyle AM (2002) Metal ions in the structure and function of RNA. J Biol Inorg Chem 7:679–690

Raines RT (1998) Ribonuclease A. Chem Rev 98:1045–1066

Rajendran M, Ellington AD (2002) Selecting nucleic acids for biosensor applications. Comb Chem High Throughput Screen 5:263–270

Raponi M, Arndt GM (2003) Double-stranded RNA-mediated gene silencing in fission yeast. Nucleic Acids Res 31:4481–4489

Rubino L, Tousignant ME, Steger G, Kaper JM (1990) Nucleotide sequence and structural analysis of two satellite RNAs associated with chicory yellow mottle virus. J Gen Virol 71 (Part 9):1897–1903

Schmidt U, Budde E, Stahl U (1992) Self-splicing of a mitochondrial group I intron from the cytochrome b gene of the ascomycete *Podospora anserina*. Mol Gen Genet 233:71–80

Schon A (1999) Ribonuclease P: the diversity of a ubiquitous RNA processing enzyme. FEMS Microbiol Rev 23:391–406

Sczakiel G (2000) Theoretical and experimental approaches to design effective antisense oligonucleotides. Front Biosci 5:D194–D201

Shih IH, Been MD (2002) Catalytic strategies of the hepatitis delta virus ribozymes. Annu Rev Biochem 71:887–917

Shimayama T, Nishikawa S, Taira K (1995) Generality of the NUX rule: kinetic analysis of the results of systematic mutations in the trinucleotide at the cleavage site of hammerhead ribozymes. Biochemistry 34:3649–3654

Sledz CA, Holko M, de Veer MJ, Silverman RH, Williams BR (2003) Activation of the interferon system by short-interfering RNAs. Nat Cell Biol 5:834–839

Sohail M, Southern EM (2000) Selecting optimal antisense reagents. Adv Drug Deliv Rev 44:23–34

Sriram B, Thakral D, Panda SK (2003) Targeted cleavage of hepatitis E virus 3 end RNA mediated by hammerhead ribozymes inhibits viral RNA replication. Virology 312:350–358

Stage-Zimmermann TK, Uhlenbeck OC (1998) Circular substrates of the hammerhead ribozyme shift the internal equilibrium further toward cleavage. Biochemistry 37:9386–9393

Stage-Zimmermann TK, Uhlenbeck OC (2001) A covalent crosslink converts the hammerhead ribozyme from a ribonuclease to an RNA ligase. Nat Struct Biol 8:863–867

Steitz TA, Moore PB (2003) RNA, the first macromolecular catalyst: the ribosome is a ribozyme. Trends Biochem Sci 28:411–418

Symons RH (1997) Plant pathogenic RNAs and RNA catalysis. Nucleic Acids Res 25:2683–2689

Takagi Y, Warashina M, Stec WJ, Yoshinari K, Taira K (2001) Recent advances in the elucidation of the mechanisms of action of ribozymes. Nucleic Acids Res 29:1815–1834

Tang G, Reinhart BJ, Bartel DP, Zamore PD (2003) A biochemical framework for RNA silencing in plants. Genes Dev 17:49–63

Tsukiji S, Pattnaik SB, Suga H (2003) An alcohol dehydrogenase ribozyme. Nat Struct Biol 10:713–717
Ullu E, Djikeng A, Shi H, Tschudi C (2002) RNA interference: advances and questions. Philos Trans R Soc Lond B Biol Sci 357:65–70
Vaish NK, Dong F, Andrews L, Schweppe RE, Ahn NG, Blatt L, Seiwert SD (2002) Monitoring post-translational modification of proteins with allosteric ribozymes. Nat Biotechnol 20:810–815
Vaucheret H, Beclin C, Fagard M (2001) Post-transcriptional gene silencing in plants. J Cell Sci 114:3083–3091
Volpe TA, Kidner C, Hall IM, Teng G, Grewal SI, Martienssen RA (2002) Regulation of heterochromatic silencing and histone H3 lysine-9 methylation by RNAi. Science 297:1833–1837
Wagner EG, Altuvia S, Romby P (2002) Antisense RNAs in bacteria and their genetic elements. Adv Genet 46:361–398
Wagner EGH, Vogel J (2003) Noncoding RNAs encoded by bacterial chromosomes. In: Barciszewski J, Erdmann VA (eds) Noncoding RNAs. Landes Bioscience, Georgetown, USA, pp 243–259
Wahlestedt C, Salmi P, Good L, Kela J, Johnsson T, Hokfelt T, Broberger C, Porreca F, Lai J, Ren K, Ossipov M, Koshkin A, Jakobsen N, Skouv J, Oerum H, Jacobsen MH, Wengel J (2000) Potent and nontoxic antisense oligonucleotides containing locked nucleic acids. Proc Natl Acad Sci USA 97:5633–5638
Warashina M, Takagi Y, Stec WJ, Taira K (2000) Differences among mechanisms of ribozyme-catalyzed reactions. Curr Opin Biotechnol 11:354–362
Watanabe T, Sullenger BA (2000) Induction of wild-type p53 activity in human cancer cells by ribozymes that repair mutant p53 transcripts. Proc Natl Acad Sci USA 97:8490–8494
Winston WM, Molodowitch C, Hunter CP (2002) Systemic RNAi in *C. elegans* requires the putative transmembrane protein SID-1. Science 295:2456–2459
Xiao S, Scott F, Fierke CA, Engelke DR (2002) Eukaryotic ribonuclease P: a plurality of ribonucleoprotein enzymes. Annu Rev Biochem 71:165–189
Yelin R, Dahary D, Sorek R, Levanon EY, Goldstein O, Shoshan A, Diber A, Biton S, Tamir Y, Khosravi R, Nemzer S, Pinner E, Walach S, Bernstein J, Savitsky K, Rotman G (2003) Widespread occurrence of antisense transcription in the human genome. Nat Biotechnol 21:379–386

Dirk Müller
Prof. Dr. Ulf Stahl
Department of Microbiology and Genetics
University of Technology Berlin
13355 Berlin, Germany
Tel.: +49-30-31472271
Fax: +49-30-31472922
e-mail: ulf.stahl@lb.tu-berlin.de

Mutation: Sugar Signaling Mutants in *Arabidopsis*

Christer Jansson

1 Introduction

Sugar signaling cascades are important components of regulatory networks in cells. Compared to the situation in bacteria, yeast and animals, participants in the sugar signaling pathways in plants are poorly understood. In plants, sugar sensing and signaling play pivotal roles in controlling many aspects of growth, metabolism and development throughout the whole plant life cycle. Through photosynthesis, plants convert atmospheric carbon dioxide to sugars, which are transported as sucrose from sugar-exporting (source) organs, such as leaves, to sugar-importing (sink) organs, such as seeds. The coordinated modulation of gene expression in source and sink organs is to a large extent choreographed by the sugar status in the cells. In general, low sugar levels promote photosynthesis and mobilization of energy reserves, whereas high sugar levels stimulate growth and storage of starch and other carbohydrates.

Sugar signaling can be dissected into three steps: sugar sensing, signal transduction and target gene expression. The picture is clouded by the dual function of sugars as nutrients and signaling molecules, and by the interaction (in plants and animals) between sugar signaling and hormonal networks. In plants the complexity is further increased by the vital role of sugar production through photosynthesis. Hexoses, sucrose and trehalose might serve as elicitors of plant sugar signaling. Hexokinase, sucrose and glucose transporters, and various sugar receptors, have been proposed as components of the sugar sensing machinery.

This chapter presents the current status in our understanding of plant sugar signaling and the nature and employment of the large number of sugar signaling mutants in *Arabidopsis* that are now available. Information has been extracted from a large number of articles, including several excellent reviews (Smeekens and Rook 1997; Smeekens 2000; Finkelstein and Gibson 2001; Loreti et al. 2001; Paul and Foyer 2001; Rolland et al. 2002; León and Sheen 2003; Salerno and Curatti 2003).

Progress in Botany, Vol. 66

2 Sugars Sensed in Sugar Signaling Cascades

Hexoses such as glucose and fructose, and disaccharides such as sucrose and, to some extent, maltose predominate among plant carbohydrates. It is not difficult to determine which of these sugars are sensed in different sugar signaling pathways since sugars added exogenously to test systems are readily interconverted; sucrose is hydrolyzed to glucose and fructose, and hexoses are combined to sucrose. Another relevant issue is whether the sensing event is intracellular or extracellular. The use of various sugar analogues has been instrumental in addressing these questions. Some of the sugars used in sugar signaling studies are described below.

2.1 Disaccharides

Sucrose (glucose-1,2-fructose) is known to cause a wide array of transcriptional signaling, such as repression of photosynthesis genes, e.g. the *PC*, *CAB* and *RBCS* genes, encoding, respectively, plastocyanin, chlorophyll a,b-binding proteins and the small subunit of rubisco (Jang and Sheen 1994, and references therein; Smeekens 2000; Rolland et al. 2002), and induction of starch synthesis genes (Sun et al. 2003, and references therein) and the patatin (*PAT*) and phloem-specific promoters (Smeekens 2000). Since sucrose is the major transport form of photoassimilate from source to sink organs, sucrose signaling serves to provide information about the energy status in the plant; high sucrose concentrations slow down photosynthesis and promote starch synthesis, and vice versa. During germination and early seedling development, α-amylase gene expression (Loreti et al. 2000) and nutrient mobilization in general (Rolland et al. 2002) can be repressed by sucrose. This is also illustrated by the repression of the glyoxylate cycle genes for malate synthase (*MS*) and isocitrate lyase (*ICL*) in cucumber cell culture systems (Graham et al. 1994a,b). In higher plants, the glyoxylate cycle plays a crucial role in the mobilization of storage lipids in the early stages after germination (Graham et al. 1994b). Other postgerminative processes that are subject to sucrose repression include greening, leaf formation, cotyledon expansion and hypocotyl and root elongation (Zhou et al. 1998; Rolland et al. 2002). The physiological rationale for sugar repression during and shortly after germination may be that high sugar accumulation reflects suboptimal growth conditions (Lopez-Molina et al. 2001) and, therefore, arrest of developmental programs is a protective measure (Rolland et al. 2002).

Arguably, many of the effects observed for sucrose are likely to be triggered by glucose or fructose following hydrolysis, but sucrose-specific signaling has also been demonstrated (Jefferson et al. 1990; Yokoyama et al. 1994; Chiou and Bush 1998; Rook et al. 1998). It has been suggested that in the developing seed, sucrose controls processes involved in differentiation and storage, whereas hexoses regulate growth and metabolism (Weber et al. 1997; Wobus and Weber 1999).

Sucrose analogs such as lactulose (galactose-1,3-fructose), palatinose (glucose-1,6-fructose) and turanose (glucose-1,3-fructose) were found to repress -amylase gene expression, although these sugars apparently are not metabolized by the plant cells (Loreti et al. 2000, 2001). Melbiose (galactose-1,6-glucose), which is also not metabolized, had no effect on -amylase gene expression (Loreti et al. 2001). Furthermore, palatinose and turanose sensing was demonstrated in tobacco leaves (Sonnewald and Herbers 1999). The sucrose derivative raffinose (galactose-1,6-glucose-1,2-fructose), which naturally occurs in sugar beet, was as effective as glucose or sucrose in downregulating the *MS* and *ICL* genes in cucumber culture systems (Graham et al. 1994b).

Trehalose (glucose-1,1-glucose), which is metabolized by plant cells, was able to induce the *APL3* gene (encoding a subunit of ADP-glucose pyrophosphorylase) in *Arabidopsis* (Wingler et al. 2000; Fritzius et al. 2001). Trehalose also affects plant metabolism and development in other ways. For example, inhibiting the trehalase enzyme leads to trehalose accumulation and a strong reduction in starch and sucrose content, indicating a role for trehalase and/or trehalose in carbon allocation (Müller et al. 2001).

2.2 Hexoses

Glucose is most likely the predominant hexose signal in gene regulation in plants and other organisms. Just like sucrose, glucose has been shown to repress photosynthesis and germinative and postgerminative developmental programs in different plant systems (Smeekens 2000; Rolland et al. 2002). For example, a powerful mutant screening strategy relies on the arrest in greening, leaf formation, cotyledon expansion and hypocotyl and root elongation observed with light-grown *Arabidopsis* seedlings on high (6%) glucose concentrations (Zhou et al. 1998). Glucose repression of photosynthesis genes provides a strong metabolic signal that overrides light activation (Sheen 1990). Glucose is phosphorylated in plant cells by both unspecific hexokinases (HXK) and glucose-specific HXKs (glucokinases).

The glucose analog 2-deoxyglucose and the glucose epimer mannose are transported into the plant cells and phosphorylated by HXK to 2-deoxyglucose-6P and mannose-6P, respectively (Loreti et al. 2001). Two other glucose analogs, 6-deoxyglucose and 3-O-methylglucose, are transported into the cells but not phosphorylated by HXK (Smeekens 2000; Loreti et al. 2001; Rolland et al. 2002). Generally, glucose-induced repression of photosynthesis and seed germination can be mimicked by 2-deoxyglucose and mannose but not by 6-deoxyglucose or 3-O-methylglucose, demonstrating that hexose transport as such does not suffice for gene repression, but that the sensor is intracellular and that hexose phosphorylation is essential. This conclusion is corroborated by the finding that direct delivery of glucose or 2-deoxyglucose to maize protoplasts by electroporation resulted in the same degree of photosynthesis gene repression as that caused by incubation with the sugars (Jang and Sheen 1994). Since both 2-deoxyglucose-6P and mannose-6P are poorly metabolized in plant cells (Loreti et al. 2001), it has further been concluded that metabolism of hexoses beyond the phosphorylation step is not needed to trigger repression of sugar-regulated genes (Graham et al. 1994b; Jang and Sheen 1994; Loreti et al. 2001). Glucose-6P and other sugar phosphates delivered into protoplasts by electroporation did not elicit the same repression as glucose (Loreti et al. 2001; Rolland et al. 2002). This implies that it is the phosphorylation event as such, and not the accumulation of hexose phosphates, that is important for repression of sugar-regulated genes. The alternative explanation that depletion of P_i and ATP is the reason for reduced gene activity has been discounted since addition of P_i and ATP failed to relieve repression (Graham et al. 1994b; Jang and Sheen 1994).

Although the non-phosphorylatable sugars 6-deoxyglucose and 3-O-methylglucose are ineffective in triggering sugar repression of gene activity, they have been shown to be capable of inducing several genes, e.g. those for apoplastic invertase (Roitsch et al. 1995), sucrose synthase (Godt et al. 1995) and class I of the *PAT* promoter (Martin et al. 1997). Thus in these instances, hexose signaling proceeds via a phosphorylation-independent route.

Fructose and galactose are also phosphorylated by HXK and it is likely that they mediate signals via the same pathway(s) as glucose. Repression of photosynthesis genes by fructose or galactose (Jang and Sheen 1994) and glyoxylate genes by fructose (Graham et al. 1994b) was similar to that obtained by glucose. The presence of specific HXKs for fructose and galactose (fructokinases and galactokinases, respectively) suggests that hexose sensing mechanisms independent of those for glucose might also exist.

3 Sugar sensors

3.1 HXK as a Hexose Sensor

There is ample evidence that HXK is an important hexose sensor in sugar signaling in yeast and animal systems (Matschinsky et al. 1993; Grupe et al. 1995; Carlson 1998; Gancedo 1998; Johnston 1999). In particular, a large body of literature has accumulated regarding the role of yeast HXK (YHXK) in glucose repression of gene expression.

The participation of HXK as a sugar sensor also in plants (Fig. 1) is supported by several lines of investigation. (1) The requirement for phosphorylation in glucose repression of gene expression points to the involvement of HXK (see Sect. 2.2 above). (2) Glucose repression of gene activity was lifted by antisense inhibition of HXK activity in transgenic *Arabidopsis* plants, and, conversely, was exaggerated by overexpression of HXK activity (Jang et al. 1997). Also, overexpression of the gene for *Arabidopsis* HXK (*AtHXK*) in transgenic tomato resulted in tomato plants exhibiting a glucose-hypersensitivity phenotype (Dai et al. 1999). In contrast, overexpressing the gene for a heterologous YHXK in *Arabidopsis* did not yield the hypersensitivity phenotype as with the *AtHXK* but rather resulted in glucose hyposensitivity (Jang et al. 1997). The observation that *YHXK*-expressing lines of *Arabidopsis* showed reduced glucose sensitivity has been ascribed to the competition of YHXK with AtHXK for sugars and, hence, to a diminished AtHXK phosphorylation activity (Jang et al. 1997). It has been concluded that YHXK in transgenic plants supplies hexose phosphorylation capacity (Jang et al. 1997) but lacks signaling effects (Smeekens 2000). Thus, it can be deduced that the sugar-sensing capacity of HXK in plants is separate from its metabolic function as a hexose kinase in the glycolytic pathway. (3) HXK inhibitors such as mannoheptulose and glucosamine can reverse glucose repression of gene activity (Jang and Sheen 1994; Umemura et al. 1998; Pego et al. 1999).

Just like the situation in yeast and animals, plants contain several HXKs. The *Arabidopsis* genome contains six *HXK* and *HXK*-like genes, three fructokinase genes, several fructokinase-like genes, plus genes for galactokinase and arabinose kinase (Kaplan et al. 1997; Sherson et al. 1999; Pego and Smeekens 2000; Rolland et al. 2002). Zymogram analyses of rice embryos have resolved three glucose-specific, one fructose-specific and two unspecific HXKs (Guglielminetti et al. 2000). Whether fructokinases, galactokinases or arabinose kinases are involved in sugar signaling is not known, although there are indications, at least for fructokinase, that this might occasionally be the case (Pego and Smeekens 2000).

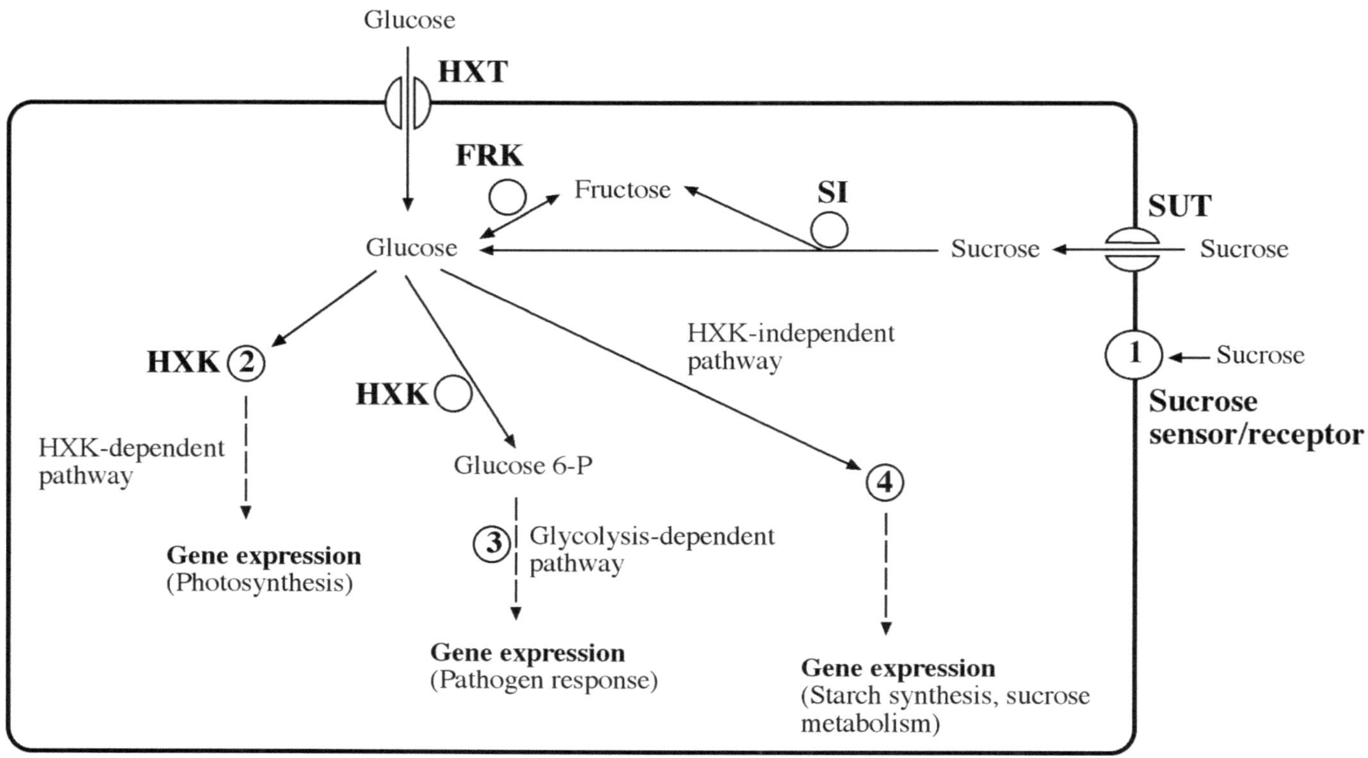

Fig. 1. Sugar sensing in plants. Sucrose is sensed by an unknown disacharide sensor/receptor (1); Glucose is sensed by HXK (2) via a circuit distinct from glycolysis, via glycolysis by an unknown sensor (3), or via a HXK-independent route by an unknown sensor (4). Abbreviations: HXT, hexose transporter; SUT, sucrose transporter; FRK, fructokinase; HXK, hexokinase; SI, sucrose invertase

The function of HXK as a sugar sensor remains unknown. However, it is now clear that its sphere of action is not confined to a cytosolic enzyme involved in glycolysis (Rolland et al. 2002). It has been proposed that one of the YHXKs (YHXK2) is translocated to the nucleus as part of a DNA-binding repressor complex (Herrero et al. 1998; Randez-Gil et al. 1998). YHXK2 has also been demonstrated to link up with the phosphatase complex that modifies Snf1 (see Section 4 below) kinase activity (Alms et al. 1999; Sanz et al. 2000). Mammalian glucokinase (HXKIV) can be translocated to the nucleus (Farrelly et al. 1999; Munoz-Alonso et al. 2000; Shiraishi et al. 2001), and human glucokinase can complement yeast *hxk2* mutants (Mayordomo and Sanz 2001). In plants, the association of HXK with chloroplast, amyloplast or mitochondrial membranes, or with the Golgi apparatus, has been reported (Rolland et al. 2002).

3.2 Other Hexose Sensors

The ability of the non-HXK substrates 6-deoxyglucose and 3-*O*-methylglucose to activate certain genes (see Sect. 2.2) is suggestive of hexose-signaling pathways independent from those with HXK as a sensor (Fig. 1). Furthermore, dissecting the responses for several glucose-controlled genes in transgenic *Arabidopsis* plants revealed that sensing could be classified in three distinct groups (Xiao et al. 2000). (1) Glucose repression of seedling development and photosynthesis gene expression was increased by overexpression of *AtHXK1*, whereas no repression was seen in *antiAtHXK1* plants. These data indicate that glucose sensing occurs at the level of HXK as discussed in Section 3.1. (2) The pathogenesis-related (PR) genes PR1 and PR5 were induced by glucose. The induction was enhanced in plants overexpressing *AtHXK1* and lost in *antiAtHXK1* plants. However, as opposed to the situation for group 1 (see Sect. 3.1), the glucose-triggered effects for these genes were exaggerated also in *Arabidopsis* lines overexpressing YHXK2. In this case, regulation is dependent on the catalytic activity of HXK and glycolysis, but with a sensor downstream of HXK. (3) Certain genes, including *APL3* and *CHS* (encoding chalcone synthase), are upregulated by glucose, but the effect did not differ between control plants and *AtHXK1* or *antiAtHXK1* plants. Here, glucose control of gene expression is mediated via an HXK-independent pathway.

3.3 Disaccharide Sensors

The interconvertibility between sucrose and hexoses makes it difficult to study direct sensing of sucrose as a specific signaling agent. One helpful approach has been to investigate the effects on genes whose expression is affected by sucrose but not by hexoses (Loreti et al. 2001), e.g. the repression of the gene encoding a proton-sucrose symporter (Chiou and Bush 1998). Another strategy is the employment of disaccharides that function as signal elicitors but are not metabolized by the plant cells (see Sect. 2.1). For example, repression of *α-amylase* gene expression was observed both with sucrose and sucrose analogs like palatinose and turanose (Loreti et al. 2000). This indicates the existence of a specific sucrose sensor (Fig. 1). Structure-function analyses of disaccharides that were able or not able to trigger repression suggest that the fructose moiety is crucial for disaccharide sensing (Loreti et al. 2000, 2001). Since trehalose but not palatinose or turanose were capable of inducing the *APL3* gene (Loreti et al. 2001), it is plausible that trehalose and sucrose utilize different sensors.

The nature of the sucrose sensor is elusive. Sucrose is the predominant form of photoassimilate that is imported to heterotrophic organs such as seeds and tubers (see Lalonde et al. 1999 for a recent review). Thus it is tempting to assume that a sucrose transporter (SUT) also serves a dual function as sensor, and such a scenario has been discussed (Smeekens and Rook 1997). However, there is one compelling argument against this notion; the sucrose analogs palatinose, turanose and lactulose are all able to repress *α-amylase* gene expression although none of the disaccharides is recognized by SUTs and they do not compete for sucrose transport (Loreti et al. 2000). In addition, at least palatinose does not seem to be transported across the plant plasma membrane at all (Bouteau et al. 1999). This is indicative of a sucrose sensor in the plasma membrane separate from the SUTs. An interesting solution to this dilemma is that sucrose sensors have evolved from SUTs that have lost their translocation activity and become dedicated sensors. One such example might be SUT2, which has been implicated as a sucrose sensor in sieve elements (Barker et al. 2000). In contrast to the sucrose transporters SUT1 and SUT4, SUT2 lacks detectable transport capacity. Also, as opposed to SUT1 and SUT4, SUT2 contains an extensive cytoplasmic domain (allowing for interaction with cytosolic proteins) and exhibits structural similarities to the yeast sugar sensors SNF3 and RGT2.

4 Downstream Regulators

The signal transduction pathway between sugar sensors and the target genes is poorly understood. In yeast sugar signaling, the Ser/Thr protein kinase Snf1 is a central participant (Carlson 1999). Snf1 phosphorylates downstream components and is also itself activated by phosphorylation. Snf-related protein kinases (SnRKs) are found in yeast, mammals and plants and are involved in a large number of regulatory functions (Halford and Hardie 1998; Hardie et al. 1998). Some plant SnRKs are activated by sugars and partake in sugar-activated gene expression, although the exact nature of their responses to sugars remains to be clarified (Rolland et al. 2002). Other players implicated in sugar signaling transduction pathways in plants are sugar metabolites, 14-3-3 proteins and Ca^{2+} (Smeekens 2000; Rolland et al. 2002).

Little is known about the *cis* and *trans* factors mediating the final steps in plant sugar signaling. To date, five different types of *cis* elements have been identified in sugar-regulated plant promoters: sugar-responsive elements (SURE; Grierson et al. 1994; Sun et al. 2003), SP8 (Ishiguro and Nakamura 1994), TGGACGG (Maeo et al. 2001), G box (Giuliano et al. 1988) and B box (Grierson et al. 1994; Zourelidou et al. 2002) elements. Recently, the SURE-binding transcription factor SUSIBA2 was described (Sun et al. 2003). SUSIBA2 binds as an activator to SURE elements in starch synthesis genes. Two other transcription factors, SPF1 and STK, with relevance to plant sugar signaling have been isolated. SPF1 binds to the SP8 sequence where it functions as a repressor (Ishiguro and Nakamura 1994) and STK binds to the B box as an activator (Zourelidou et al. 2002).

5 Cross-Talk Between Sugar Signaling and Other Regulatory Pathways

Sugar signaling does not operate in splendid isolation but, rather, is integrated in cellular regulatory networks. Most notably, the molecular characterization of sugar signaling mutants (see Sec. 6) has revealed tight and extensive interactions between sugar and hormonal signaling (see León and Sheen 2003 for a recent review), particularly for abscisic acid (ABA) and ethylene. Glucose activates ABA biosynthesis and signaling and both glucose and ABA signaling are antagonistic to ethylene signaling (Rolland et al. 2002; León and Sheen 2003). It is plausible that ABA mediates glucose signaling in many regulatory pathways. There are also indications for connections between sugar signaling and auxin and cytokinin signaling (Rolland et al. 2002).

Induction of stress-related genes, such as the *PR* genes, can be triggered by elevated sugar levels (Roitsch 1999). It was also shown that increased sugar concentrations during fruit ripening induced synthesis of antifungal proteins (Salzman et al. 1998). Apart from biotic stresses, many abiotic stress factors, such as cold and drought, cause major alterations in carbohydrate metabolism (Tomashow 1999; Wanner and Junttila 1999), and it is reasonable to assume that sugar signaling networks connect with stress pathways to modulate metabolism (Smeekens 2000). Finally, in this context, it is interesting to note that the SUSIBA transcription factor family that mediates sugar control of starch synthesis (Sun et al. 2003) belongs to the WRKY family. The WRKY proteins seem to be plant-specific transcription factors and are best known for their participation in various stress responses and senescence (Eulgem et al. 2000).

6 Sugar Signaling Mutants

Mutants offer powerful means for the elucidation of complex signal transduction pathways. *Arabidopsis* has been used extensively in the studies on sugar sensing and signaling, and different mutant screens have been developed. One effective screen rests on the sugar-induced repression of photosynthesis and seedling development (see Sect. 2). Molecular analyses of such mutants have grouped them into two categories: those that are insensitive to high sugar levels, such as *glucose insensitive* (*gin*), *carbohydrate insensitive* (*cai*), *sucrose uncoupled* (*sun*) and *mannose-insensitive-germination* (*mig*) mutants (Table 1), and those that are oversensitive to sugar, such as *glucose oversensitive* (*glo*), *glucose supersensitive* (*gss*) and *sucrose supersensitive* (*sss*) mutants (Table 2).

Genetic screens of *Arabidopsis* sugar signaling mutants have revealed important clues about the signal transduction network that controls plant metabolism. Many of the mutated genes have been isolated and the corresponding protein product identified (Tables 1 and 2). The identification of *gin2* as a mutation in the *AtHXK1* gene (Table 1) corroborates the interpretations from other studies (see Sect. 3.1). The observation that several of the sugar signaling mutants are allelic to ABA or ethylene signaling or biosynthesis mutants has yielded valuable insight into the sugar-hormone regulatory network. For example, *gin1*, *isi4* and *sis4* are allelic to each other and to the ABA-deficient *aba2* mutant (Table 1; Rolland et al. 2002). The *ABA2* gene encodes a short-chain dehydrogenase/reductase (SDR1) that catalyzes one of the steps in the ABA biosynthesis pathways (Rolland et al. 2002; León and Sheen 2003). Another example is provided by *gin6*, *isi3*, *sis5* and *sun6*, which are allelic to each other and to the ABA-insensitive mutant *abi4*

Table 1. *Arabidopsis* mutants with reduced sugar sensitivity. Mutants: *gin* glucose insensitive; *cai* carbohydrate insensitive; *isi* impaired sucrose induction; *ctr* constitutive ethylene triple response; *sis* sucrose insensitive; *sun* sucrose uncoupled; *sig* sucrose insensitive growth; *mig* mannose insensitive germination; *rsr* reduced sucrose response; *lba* low-level β-amylase; *aba* abscisic acid; *abi* abscisic acid insensitive; *eto* ethylene overproduction

Mutant	Phenotype	Gene	Protein	References
gin1	Insensitive to glucose repression of growth	*ABA2*	SDR1	Zhou et al. (1998); Cheng et al. (2002)
gin2	Insensitive to glucose repression of growth	*GIN2/HXK1*	HXK1	Rolland et al. (2002)
gin4	Insensitive to glucose repression of growth	*CTR1*	CTR1	Cheng et al. (2002)
gin5	Insensitive to glucose repression of growth	*ABA3/LOS5*	MCSU	Arenas-Huertero et al. (2000)
gin6	Insensitive to glucose and ABA repression of growth	*ABI4*	ABI4	Arenas-Huertero et al. (2000)
cai	Insensitive to sucrose repression of growth			Boxall et al. (1997); Rolland et al. (2002)
ctr	Insensitive to glucose repression of growth	*CTR*	CTR	Zhou et al. (1998)
isi1	Reduced sucrose induction of the *APL3* promoter	*ISI1*	ISI1	Rook et al. (2001); Rook and Bevan (2003)
isi2	Reduced sucrose induction of the *APL3* promoter			Rook et al. (2001)
isi3	Reduced sucrose induction of the *APL3* promoter	*ABI4*	ABI4	Rook et al. (2001)
isi4	Reduced sucrose induction of the *APL3* promoter; insensitive to glucose repression of growth	*ABA2*	SDR1	Rook et al. (2001)
sis1	Insensitive to glucose and mannose repression of growth	*CTR1*	CTR1	Gibson et al. (2001)
sis4	Insensitive to sucrose and glucose repression of growth	*ABA2*	SDR1	Laby et al. (2000
sis5	Insensitive to sucrose, glucose and mannose repression of growth	*ABI4*	ABI4	Laby et al. (2000)

Table 1. *Continued*

Mutant	Phenotype	Gene	Protein	References
sun6	Insensitive to sucrose, glucose, mannose and ABA repression of the *PC* promoter	*ABI4*	ABI4	Dijkwel et al. (1997)
sig	Insensitive to sucrose repression of growth			Pego et al. (2000)
mig	Insensitive to mannose repression of germination			Pego et al. (1999)
rsr1	Reduced sucrose induction of the *pat* promoter			Martin et al. (1997)
rsr4	Reduced sucrose induction of the *pat* promoter			Martin et al. (1997)
lba1	Reduced sucrose, glucose and fructose induction of the β-*AMY* promoter			Mita et al. (1997b)
lba2	Reduced sucrose, glucose and fructose induction of the β-*AMY* promoter			Mita et al. (1997b)
aba1	Insensitive to sucrose and glucose repression of growth	*ABA1*	ZEP1	Arenas-Huertero et al. (2000); Huijser et al. (2000); Laby et al. (2000)
abi5	Insensitive to glucose repression of growth	*ABI5*	ABI5	Arenas-Huertero et al. (2000); Laby et al. (2000)
eto1	Insensitive to glucose repression of growth	*ETO1*	ETO1	Zhou et al. (1998); Wang et al. (2002))

(Table 1; Rolland et al. 2002). The *ABI4* gene encodes a transcription factor involved in ABA signaling (Arenas-Huertero et al. 2000; Rolland et al. 2002; León and Sheen 2003). Both sugar-hypersensitive and -hyposensitive mutants were found to be allelic to ethylene synthesis or signaling mutants, e.g. *eto1* (Table 1) and *ein2*, *ein3*, *ein6* and *etr1* (Table 2).

Some mutant screens have focused on aberrations in expression of genes that are known to be sugar regulated. Using reporter gene constructs, mutants have been selected that are defective in sugar repression of the *PC* gene (*sun* mutants; Dijkwel et al. 1997) or the *PAT* gene (*rsr* mutants; Martin

Table 2. *Arabidopsis* mutants with enhanced sugar sensitivity. Mutants: *glo* glucose oversensitive; *gss* glucose supersensitive; *fus* fusca; *core* conditional root expansion; *prl* pleiotrophic regulatory locus; *hba* high-level β-amylase; *hsr* high sugar response; *hys* hypersenescence; *ein* ethylene insensitive; *etr* ethylene response

Mutant	Phenotype	Gene	Protein	References
glo	Hypersensitive to glucose repression of growth			Sheen et al. (1999); Rolland et al. (2002)
gss	Hypersensitive to glucose repression of growth			Pego et al. (2000)
sss	Hypersensitive to sucrose repression of growth			Pego et al. (2000)
fus	Hypersensitive to sucrose repression of growth			Castle and Meinke (1994)
core	Hypersensitive to sucrose repression of growth			Hauser et al. (1995)
prl1	Hypersensitive to sucrose, glucose and ABA repression of growth	*PRL1*	PRL1	Németh et al. (1998)
hba	Increased sucrose, glucose and fructose induction of the β-*AMY* promoter			Mita et al. (1997a)
hsr	Increased sucrose induction of the *APL3* promoter			Hadingham et al. (2002); Rook and Bevan (2003)
hys1	Hypersensitive to glucose repression of growth			Yoshida et al. (2002)
ein2	Hypersensitive to glucose repression of growth; insensitive to ethylene induction of growth	*EIN2*	EIN2	Alonso et al. (1999); Cheng et al. (2002)
ein3	Hypersensitive to glucose repression of growth; insensitive to ethylene induction of growth	*EIN3*	EIN3	Wang et al. (2002)
ein6	Hypersensitive to glucose repression of growth; insensitive to ethylene induction of growth			Wang et al. (2002)
etr1	Hypersensitive to glucose repression of growth; insensitive to ethylene induction of growth	*ETR1*	ETR1	Zhou et al. (1998); Wang et al. (2002)

et al. 1997), or that showed either decreased (*lba* mutants; Mita et al. 1997b) or increased (*hba* mutants; Mita et al. 1997a) sugar induction of the -*AMY* gene.

7 Conclusions and Prospects

Sugar signaling impacts plants in all aspects of their life cycle, from germination to senescence. Our understanding of plant sugar signaling is in its infancy, but classical biochemical studies with sugar analogs, together with analyses of transgenic and mutant plants, both primarily in *Arabidopsis*, have shed light on some of the steps in this very important regulatory pathway. It now seems clear that plants contain at least four different sugar sensors: one for sucrose, and one HXK-dependent and two HXK-independent for glucose and other hexoses. Apart from HXK, none of the sensors has been identified. Our knowledge of the processes downstream of the sensors is also very poor and fragmented. The involvement of SnRKs as mediators seems likely and a few transcription factors and *cis* elements have been identified. Some of the immediate tasks in our continued work on sugar metabolism and signaling in plants are to identify the different sugar sensors, other than HXK, and to clarify the relationship between sucrose transporters and sensors.

The phenotypic and genotypic characterization of sugar signaling mutants in *Arabidopsis* has provided crucial information about the participants in the sugar-signaling pathway. Perhaps the most notable outcome of the mutant studies is the realization of the highly integrated nature of sugar, ABA and ethylene signaling routes. Future investigations of sugar signaling mutants in *Arabidopsis*, as well as in plants such as rice and maize, will undoubtedly produce new and valuable clues to our understanding of the different regulators and elements that constitute plant sugar signaling and its interactions with other signaling pathways. To untangle this regulatory web will be a major challenge in plant biology for years to come.

References

Alms GR, Sanz P, Carlson M, Haystead TA (1999) Reg1p targets protein phosphatase 1 to dephosphorylate hexokinase II in Saccharomyces *cerevisiae*: characterizing the effects of a phosphatase subunit in the yeast proteome. EMBO J 18:4157–4168

Alonso JM, Hirayama T, Roman G, Nourizadeh S, Ecker JR (1999) EIN2, a bifunctional transducer of ethylene and stress responses in *Arabidopsis*. Science 284:2148–2152

Arenas-Huertero F, Arroyo A, Zhou L, Sheen J, León P (2000) Analysis of *Arabidopsis* glucose insensitive mutants, *gin5* and *gin6*, reveals a central role of the plant hormone ABA in the regulation of plant vegetative development by sugar. Genes Dev 14:2085–2096

Barker L, Kühn C, Weise A, Schultz A, Gebhardt C, Hirner B, Hellman H, Schultze W, Ward JM, Frommer WB (2000) SUT2, a putative sucrose sensor in sieve elements. Plant Cell 12:1153–1164

Bouteau F, Dellis O, Bousquet U, Rona JP (1999) Evidence of multiple sugar uptake across the plasma membrane of lacticifer protoplasts from *Hevea*. Bioelectrochem Bioenerg 4:135–139

Boxall SF, Martin TR, Graham IA (1997) *Arabidopsis thaliana* mutants that are carbohydrate insensitive (abstract no. 1266). Plant Physiol 114:S247

Carlson M (1998) Regulation of glucose utilization in yeast. Curr Opin Genet Dev 8:650–664

Carlson M (1999) Glucose repression in yeast. Curr Opin Microbiol 2:202–207

Castle LA, Meinke DW (1994) A FUSCA gene of *Arabidopsis* encodes a novel protein essential for plant development. Plant Cell 6:25–41

Cheng WH, Endo A, Zhou L, Penney J, Chen H-C, Arroyo A, León P, Nambara E, Asamif T, Seo M, Koshiba T, Sheen J (2002) A unique short-chain dehydrogenase/reductase in *Arabidopsis* glucose signaling and abscisic acid biosynthesis and functions. Plant Cell 14:2723–2743

Chiou T-J, Bush DR (1998) Sucrose is a signal molecule in assimilate partitioning. Proc Natl Acad Sci USA 95:4784–4788

Dai N, Schaffer A, Petreikov M, Shahak Y, Giller Y, Ratner K, Levine A, Granot D (1999) Overexpression of *Arabidopsis* hexokinase in tomato plants inhibits growth, reduces photosynthesis, and induces rapid senescence. Plant Cell 11:1253–1266

Dijkwel PP, Huijser C, Weisbeek PJ, Chua N-H, Smeekens SCM (1997) Sucrose control of phytochrome A signaling in *Arabidopsis*. Plant Cell 9:583–595

Eulgem T, Rushton PJ, Robatzek S, Somssich IE (2000) The WRKY superfamily of plant transcription factors. Trends Plant Sci 5:199–206

Farrelly D, Brown KS, Tieman A, Ren J, Lira SA, Hagan D, Gregg R, Mookhtiar KA, Hariharan N (1999) Mice mutant for glucokinase regulatory protein exhibit decreased liver glucokinase: a sequestration mechanism in metabolic regulation. Proc Natl Acad Sci USA 96:14511–14516

Finkelstein R, Gibson SI (2001) ABA and sugar interactions regulating development: cross-talk or voices in a crowd? Curr Opin Plant Biol 5:26–32

Fritzius T, Aeschbacher R, Wiemken A, Wingler A (2001) Induction of *APL3* expression by trehalose complements the starch-deficient *Arabidopsis* mutant *adg2-1* lacking APL1, the large subunit of ADP-glucose pyrophosphorylase. Plant Physiol 126:883–889

Gancedo JM (1998) Yeast carbon catabolite repression. Microbiol Mol Biol Rev 62:334–361

Gibson SI, Laby RJ, Kim D (2001) The *sugar-insensitive1* (*sis1*) mutant of *Arabidopsis* is allelic to ctr1. Biochem Biophys Res Commun 280:196–203

Giuliano G, Pichersky E, Malik VS, Timko MP, Scolnic PA, Cashmore AR (1988) An evolutionary conserved protein binding sequence upstream of a plant light-regulated gene. Proc Natl Acad Sci USA 85:7089–7093

Godt DE, Riegel A, Roitsch T (1995) Regulation of sucrose synthase expression in *Chenopodium rubrum*: characterization of sugar-induced expression in photoautotrophic suspension cultures and sink tissue-specific expression in plants. J Plant Physiol 146:231–238

Graham IA, Baker CJ, Leaver CJ (1994a) Analysis of the cucumber malate synthase gene promoter by transient expression and gel retardation assays. Plant J 6:893–902

Graham IA, Denby KJ, Leaver CJ (1994b) Carbon catabolite repression regulates glyoxylate gene-expression in cucumber. Plant Cell 6:761–772

Grierson C, Du J-S, de Torres Zabala M, Beggs K, Smith C, Holdsworth M, Bevan MW (1994) Separate *cis* sequences and *trans* factors direct metabolic and developmental regulation of a potato tuber storage protein gene. Plant J 5:815–826

Grupe A, Hultgren B, Ryan A, Ma YH, Bauer M, Stewart TA (1995) Transgenic knockouts reveal a critical requirement for pancreatic cell glucokinase in maintaining glucose homeostasis. Cell 83:69–78

Guglielminetti L, Perata P, Morita A, Loretti E, Yamaguchi J, Alpi A (2000) Characterization of isoforms of hexose kinases in rice embryo. Phytochem 53:195–200

Hadingham SA, Hemmann G, Smith TC, Corke F, Holman R, Baier M, Rook F, Bevan MW (2002) Sugar-response mutants with altered expression of the starch biosynthetic gene, *APL3*. Poster Abstracts from the 13th Int Conf on *Arabidopsis* Research, Abstract 12–04

Halford NG, Hardie DG (1998) SNF1-related protein kinases: global regulators of carbon metabolism in plants? Plant Mol Biol 37:735–748

Hardie DG, Carling D, Carlson M (1998) The AMP-activated/SNF1 protein kinase subfamily: metabolic sensors of the eukaryotic cell? Annu Rev Biochem 67:821–855

Hauser MT, Morikami A, Benfey PN (1995) Conditional root expansion mutants of *Arabidopsis*. Development 121:1237–1252

Herrero P, Martinez-Campa C, Moreno F (1998) The hexokinase 2 protein participates in regulatory DNA-protein complexes necessary for glucose repression of the *SUC2* gene in *Saccharomyces cerevisiae*. FEBS Lett 434:71–76

Huijser C, Kortsee A, Pego J, Weisbeek P, Wisman E, Smeekens S (2000) The *Arabidopsis* SUCROSE UNCOUPLED-6 gene is identical to ABSCISIC ACID INSENSITIVE-4: involvement of abscisic acid in sugar responses. Plant J 23:577–585

Ishiguro S, Nakamura K (1994) Characterization of a cDNA encoding a novel DNA-binding protein, SPF1, that recognizes SP8 sequences in the 5 upstream regions of genes coding for sporamine and ß-amylase from sweet potato. Mol Gen Genet 244:563–571

Jang J-C, Sheen J (1994) Sugar sensing in higher plants. Plant Cell 6:1665–1679

Jang J-C, León P, Zhou L, Sheen J (1997) Hexokinase as a sugar sensor in higher plants. Plant Cell 9:5–19

Jefferson R, Goldsbrough A, Bevan M (1990) Transcriptional regulation of patatin-1 gene in potato. Plant Mol Biol 14:995–1006

Johnston M (1999) Feasting, fasting and fermenting. Glucose sensing in yeast and other cells. Trends Genet 15:29–33

Kaplan CP, Tugal HB, Baker A (1997) Isolation of a cDNA encoding an *Arabidopsis* galactokinase by functional expression in yeast. Plant Mol Biol 34:497–506

Laby RJ, Kincaid MS, Kim D, Gibson SI (2000) The *Arabidopsis* sugar-insensitive mutants *sis4* and *sis5* are defective in abscisic acid synthesis and response. Plant J 23:587–596

Lalonde S, Boles E, Hellman H, Barker L, Patrick JW, Frommer WB, Ward JM (1999) The dual function of sugar carriers: transport and sugar sensing. Plant Cell 11:707–726

León P, Sheen J (2003) Sugar and hormone connections. Trends Plant Sci 8:110–116

Lopez-Molina L, Mongrand S, Chua N-H (2001) A post-germination developmental arrest checkpoint is mediated by abscisic acid and requires the AB15 transcription factor in *Arabidopsis*. Proc Natl Acad Sci USA 98:4782–4787

Loreti E, Alpi A, Perata P (2000) Glucose and disaccharide-sensing mechanisms modulate the expression of -amylase in barley embryos. Plant J 123:939–948

Loreti E, de Bellis L, Alpi A, Perata P (2001) Why and how do plants sense sugars? Ann Bot 88:803–812

Maeo K, Tomiya T, Hayashi K, Akaike M, Morikami A, Ishiguro S, Nakamura K (2001) Sugar-responsible elements in the promoter of a gene for ß-amylase of sweet potato. Plant Mol Biol 46:627–637

Martin T, Hellman H, Schmidt R, Willmitzer L, Frommer WB (1997) Identification of mutants in metabolically regulated gene expression. Plant J 11:53–62

Matschinsky F, Liang Y, Kesavan P, Wang L (1993) Glucokinase as pancreatic cell glucose sensor and diabetes gene. J Clin Invest 92:2092–2098

Mayordomo I, Sanz P (2001) Human pancreatic glucokinase (GlkB) complements the glucose signaling defect of *Saccharomyces cerevisiae hxk2* mutants. Yeast 18:1309–1316

Mita S, Hirano H, Nakamura K (1997a) Negative regulation in the expression of a sugar-inducible gene in *Arabidopsis thaliana*: a recessive mutation causing enhanced expression of a gene for -amylase. Plant Physiol 114:2575–2582
Mita S, Murano N, Akaike M, Nakamura K (1997b) Mutants of *Arabidopsis thaliana* with pleiotropic effects on the expression of the gene for -amylase and on the accumulation of anthocyanin that are inducible by sugars. Plant J 11:841–851
Munoz-Alonso MJ, Guillemain G, Kassis N, Girard J, Burnol AF, Leturque A (2000) A novel and cytosolic dual specificity for phosphatase, interacting with glucokinase, increases glucose phosphorylation rate. J Biol Chem 275:32406–32412
Müller J, Aeschbacher RA, Wingler A, Boller T, Wiemken A (2001) Trehalose and trehalase in *Arabidopsis*. Plant Physiol 125:1086–1093
Németh K, Salchert K, Putnoky P, Bhalerao R, Konc-Kalman Z, Stankovic-Stangeland B, Bakó L, Mathur J, Ökrész L, Stabel S, Geigenberger P, Stitt M, Rédei GP, Schell J, Koncz C (1998) Pleiotropic control of glucose and hormone responses by PRL1, a nuclear WD protein, in *Arabidopsis*. Genes Dev 12:3059–3073
Paul MJ, Foyer CH (2001) Sink regulation of photosynthesis. J Exp Bot 52:1383–1400
Pego JV, Smeekens SC (2000) Plant fructokinases: a sweet family get-together. Trends Plant Sci 5:531–536
Pego JV, Weisbeek PJ, Smeekens SCM (1999) Mannose inhibits *Arabidopsis* germination via a hexokinase-mediated step. Plant Physiol 119:1017–1023
Pego JV, Kortsee AJ, Huijser C, Smeekens SC (2000) Photosynthesis, sugars and the regulation of gene expression. J Exp Bot 51:407–416
Randez-Gil F, Herrero P, Sanz P, Prieto J, Moreno F (1998) Hexokinase PII has a double cytosolic-nuclear localization in *Saccharomyces cervisiae*. FEBS Lett 425:475–478
Rolland F, Moore B, Sheen J (2002) Sugar sensing and signaling in higher plants. Plant Cell 14:S185–S205
Roitsch T (1999) Source-sink regulation by sugar and stress. Curr Opin Plant Biol 2:198–206
Roitsch T, Bittner M, Godt DE (19995) Induction of apoplastic invertase of *Chenopodium rubrum* by D-glucose and a glucose analog and tissue-specific expression suggest a role in sink-source regulation. Plant Physiol 108:285–294
Rook F, Bevan MW (2003) Genetic approaches to understanding sugar-response pathways. J Exp Bot 54:495–501
Rook F, Gerrits N, Kortstee A, van Kampe M, Borrias M, Weisbeek P, Smeekens S (1998) Sucrose-specific signaling represses translation of the *Arabidopsis ATB2* bZIP transcription factor gene. Plant J 15:253–263
Rook F, Corke F, Card R, Munz G, Smith C, Bevan MW (2001) Impaired sucrose-induction mutants reveal the modulation of sugar-induced starch biosynthetic gene expression by abscisic acid signaling. Plant J 26:421–433
Salerno GL, Curatti L (2003) Origin of sucrose metabolism in higher plants: when, how and why? Trends Plant Sci 8:63–69
Salzman RA, Tikhonova I, Bordelon BP, Hasegawa PM, Bressan RA (1998) Coordinate accumulation of antifungal proteins and hexoses constitutes a developmentally controlled defense response during fruit ripening in grape. Plant Physiol 117:465–472
Sanz P, Alms G, Haystead T, Carlson M (2000) Regulatory interactions between the Reg1-Glc7 protein phosphatase and the Snf1 protein kinase. Mol Cell Biol 20:1321–1328
Sheen J (1990) Metabolic repression of transcription in higher plants. Plant Cell 2:1027–1038
Sheen J, Zhou L, Jang JC (1999) Sugars as signaling molecules. Curr Opin Plant Biol 2:410–418
Sherson S, Gy I, Medd J, Schmidt R, Dean C, Kreis M, Lecharny A, Cobbett C (1999) The arabinose kinase, ARA1, gene of *Arabidopsis* is a novel member of the galactose kinase gene family. Plant Mol Biol 39:1003–1012
Shiraishi A, Yamada Y, Tsuura Y, Fijimoto S, Tsukiyama K, Mukai E, Toyoda Y, Miwa L, Seino Y (2001) A novel glucokinase regulator in pancreatic cells: precursor of propionyl-CoA carboxylase subunit interacts with glucokinase and augments its activity. J Biol Chem 276:2325–2328

Smeekens S (2000) Sugar-induced signal transduction in plants. Annu Rev Plant Physiol Plant Mol Biol 51:49–81
Smeekens S, Rook F (1997) Sugar sensing and sugar-mediated signal transduction in plants. Plant Physiol 115:7–13
Sonnewald U, Herbers K (1999) Sugars, far more than just fuel for plant growth. In: Bryant JA, Burrel MM, Kruger NJ (eds) Plant carbohydrate chemistry. Bios Scientific Publishers, Oxford, pp 69–78
Sun C, Ahlandsberg S, Palmqvist S, Ohlsson H, Borén M, Jansson C (2003) A novel WRKY transcription factor, SUSIBA2, participates in sugar signaling in barley by binding to the SURE element of the *iso1* promoter. Plant Cell 15:2076–2092
Tomashow MF (1999) Plant cold acclimation: freezing tolerance genes and regulatory mechanisms. Annu Rev Plant Physiol Plant Mol Biol 50:571–599
Umemura T, Perata P, Futsuhara Y, Yamaguchi J (1998) Sugar sensing and -amylase gene repression in rice embryos. Planta 204:420–428
Wang KLC, Hai L, Ecker JR (2002) Ethylene biosynthesis and signaling networks. Plant Cell 14:S131–S151
Wanner LA, Junttila O (1999) Cold-induced freezing tolerance in *Arabidopsis*. Plant Physiol 120:391–399
Weber H, Borisjuk L, Wobus U (1997) Sugar import and metabolism during seed development. Trends Plant Sci 2:169–174
Wingler A, Fritzius T, Wiemken A, Boller T, Aeschbacher RA (2000) Trehalose induces the ADP-glucose pyrophosphorylase gene, *APL3*, and starch synthesis in *Arabiopsis*. Plant Physiol 124:105–114
Wobus U, Weber H (1999) Sugars as signal molecules in plant seed development. Biol Chem 380:937–944
Xiao W, Sheen J, Jang J-C (2000) The role of hexokinase in plant sugar signal transduction and growth and development. Plant Mol Biol 44:451–461
Yoloyama R, Hirose T, Fujii N, Aspuria ET, Kato A, Uchimiya H (1994) The rolC promoter of *Agrobacterium rhizogenes* Ri plasmid is activated by sucrose in transgenic tobacco plants. Mol Gene Genet 244:15–22
Yoshida S, Ito M, Nishida I, Watanabe A (2002) Identification of a novel gene HYS/CPR5 that has a repressive role in the induction of leaf senescence and pathogen-defense responses in *Arabidopsis thaliana*. Plant J 29:427–437
Zhou L, Jang J-C, Jones TL, Sheen J (1998) Glucose and ethylene signal transduction crosstalk revealed by an *Arabidopsis* glucose-insensitive mutant. Proc Natl Acad Sci USA 95:10294–10299
Zourelidou M, de Torres-Zabala M. Smith C, Bevan MW (2002) Storekeeper defines a new class of plant-specific DNA-binding proteins and is a putative regulator of patatin expression. Plant J 30:489–497

Christer Jansson
Department of Plant Biology and Forest Genetics
Swedish University of Agricultural Sciences
Uppsala, Sweden
e-mail: christer.jansson@vbsg.slu.se

Cell Biology: The Green Alga *Chlamydomonas reinhardtii* – A Genetic Model Organism

Jörg Nickelsen

1 Introduction

Over the last decades, a handful of photoautotrophically growing organisms have developed into principal genetic model organisms, each reflecting a specific evolutionary level and, thus, exhibiting distinct morphological and functional characteristics. These model organisms include the prokaryotic cyanobacterium *Synechocystis* ssp. PCC 6803, the eukaryotic unicellular green alga *Chlamydomonas reinhardtii* and the multicellular angiosperm *Arabidopsis thaliana*. *Chlamydomonas* has just recently become part of this group when its complete genome sequence – similar to that of *Synechocystis* and *Arabidopsis* – became available, a feature that, nowadays, is expected from a 'modern' model organism.

The moss *Physcomitrella patens* is not yet included in this list since the determination of its genome sequence still awaits completion. However, *Physcomitrella*, certainly, will play an important role in plant genetics in the near future due to the presence of an efficient system for homologous genetic recombination within its nuclear compartment (Holtorf et al. 2002).

However, besides a complete genomic sequence, several additional molecular prerequisites must be fulfilled before an organism might be considered as a genetic model organism. First, classical genetic techniques like crossing of strains should be established. It should be easy to generate and maintain mutants in physiological processes of interest. Furthermore, the genetic material should be accessible for targeted manipulation via transformation or, in other words, a molecular toolkit should be available for the respective organism. The aim of this chapter is to give a brief overview of the genetics of *Chlamydomonas* and the recent new molecular techniques that have been established for this green alga, and to discuss whether this alga deserves to be called a genetic model organism.

Progress in Botany, Vol. 66

2 Cell Architecture and Actual Scientific Interests

Figure 1 presents the structure of a wild-type *Chlamydomonas* cell which is approximately 10 μm in diameter and surrounded by a cell wall consisting mainly of hydroxyprolin-rich glycoproteins. *Chlamydomonas* has two anterior flagella, 10–12 μm in length, which show the typical 9+2 microtubular structure and arise from a pair of basal bodies. These are also involved in the microtubular organization of the entire cell as well as the formation of the mitotic spindle. In the centre of the cell, the nucleus is situated with the nucleolus surrounded by the single cup-shaped chloroplast. Within the chloroplast, which makes up approximately 40% of the cell volume, grana

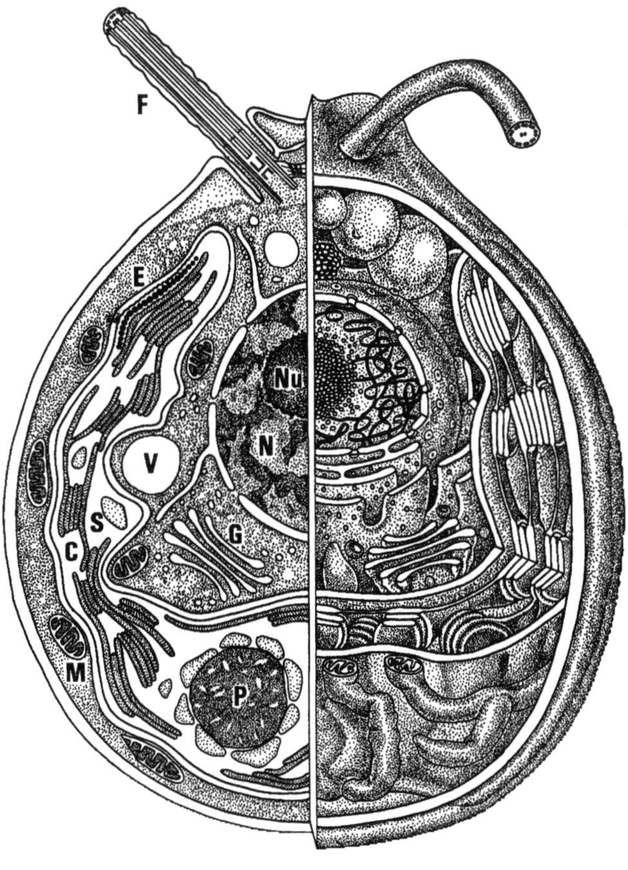

Fig. 1. Scheme of *Chlamydomonas* cell structure. The two-/three-dimensional view of the cell shows two flagellae (*F*), the nucleus (*N*) with nucleolus (*Nu*), the chloroplast (*C*) with eyespot (*E*) and pyrenoid (*P*), mitochondria (*M*) as well as Golgi vesicles (*G*), starch grains (*S*) and vacuoles (*V*). (Adapted from Nickelsen and Kück 2000)

and stromal thylakoid membranes, smaller starch grains and the starch-containing pyrenoid can be distinguished. The pyrenoid, which is also the site of CO_2 fixation, represents an important species-specific characteristic in traditional taxonomy. Furthermore, the bright orange eyespot appears at the inner chloroplast membrane which forms part of the vision apparatus that directs phototactic responses and leads to an inherent asymmetry of the *Chlamydomonas* cell.

In addition, the cytoplasm contains contractile vacuoles, Golgi vesicles and several mitochondria which, in contrast to the chloroplast, occupy only 1–3% of the cell volume. Throughout the cell cycle, the morphology of these mitochondria continuously changes due to fusions and divisions of single organelles (Harris 1989).

In the laboratory, *Chlamydomonas*, as a microorganism, can easily be maintained on agar plates. Growth in liquid cultures is not size-limited and, thus, allows the easy production of large quantities of cells. When grown photoautotrophically, cultures should be bubbled with air containing 5% CO_2 in order to obtain maximum rates. The ease of laboratory handling makes *Chlamydomonas* ideally suited for biochemical/proteomic approaches and several projects addressing the content of various subproteomes have been initiated. For instance, a set of light-harvesting proteins have recently been identified as part of a proteomic analysis of the chloroplast compartment (Stauber et al. 2003).

Consequently, the biological processes that are preferentially studied in this simple eukaryotic organism comprise the analysis of flagellar as well as basal body structure and function, the biogenesis of chloroplasts and photosynthetic functions, light perception and phototactic responses, the control of the cell cycle in physiological processes like nutrient uptake, and circadian rhythms (for recent reviews see Dent et al. 2001 and Harris 2001).

3 Classical Haploid Genetics

Genetics of *Chlamydomonas* started about 50 years ago (Harris 1989) and was facilitated by the fact that *Chlamydomonas* has a simple life cycle. Normally, vegetative cells are haploid and grow rapidly, with an average doubling time of 6–8 h under optimal conditions. Nevertheless, in the laboratory, vegetative diploids can easily be generated and maintained by mating of strains that contain complementary auxotrophic markers. This allows one to determine immediately the dominant/recessive character of a mutation within the nuclear genome. The onset of the sexual cycle, i.e. gametogenesis, is induced by minimal growth conditions. Cells, which are either mating type + (mt+) or mating type – (mt–), then differentiate into mating competent isogametes that pair along their flagella and, finally, fuse and form the diploid zygospore.

The reproductive process includes a series of complex events which start with flagellar adhesion between mt+ and mt– gametes mediated by sex-specific agglutinins. Agglutination, then, induces downstream events via a cAMP signal transduction pathway. Flagellar tips change their morphology and the cell wall is lysed by gamete-specific autolysins. Furthermore, so-called mating structures are formed at the anterior ends of the cells, forming a tubular connection between the gametes. Subsequently, cells fuse to the posterior pole resulting in a quadriflagellate, motile cell. Afterwards, the zygospore is surrounded by a newly synthesized, hard and impermeable cell wall. In a few days, then, zygospore maturation and meiosis take place, leading to the release of a tetrad of four haploid progeny cells.

3.1 Genetic Crossings

In the laboratory, the sexual life cycle is induced by nitrogen deprivation and the resulting meiotic products can easily be separated. Therefore, tetrad analysis – similar to the situation in yeast – has became a simple and useful tool to initially characterize *Chlamydomonas* mutations by following their inheritance during crosses. Nuclear genes are inherited in a Mendelian fashion and segregate 2:2 in a cross. If two nuclear markers are present, the tetrads can be classified as parental ditypes (PD), nonparental ditypes (NPD) or tetratypes (T). In contrast, both chloroplast and mitochondrial genes exhibit a uniparental inheritance with chloroplast genomes being transmitted from the mt+ and mitochondrial genomes from the mt– parent. The molecular basis for this inheritance pattern still awaits clarification; however, it appears that, at least in the case of chloroplasts, a targeted degradation of DNA from the mt– parent is mediated by an mt+ gamete-specific nuclease (Nishimura et al. 2002).

'Forward genetic' approaches are based on the availability of mutants with distinct phenotypes. In *Chlamydomonas*, mutations have been induced in the different genomes of the cell by UV-light treatment or chemical mutagenesis.

The most frequently used chemicals for inducing nuclear mutations include nitrosoguanidine (MNNG), methanesulfonate (MMS) and ethyl methanesulfonate (EMS), whilst chloroplast mutations are effectively obtained after treatment with the thymidine analogue 5-fluorodeoxyuridine (FdUrd).

More recently, transposon-mediated insertion mutations have been produced (Silflow 1998). However, with the development of efficient nuclear transformation systems, nowadays insertional mutagenesis with exogenously added DNA which randomly integrates into the nuclear genome is the most popular technique for generating DNA-tagged mutants (see below).

Since *Chlamydomonas* is haploid, the phenotypic effects of induced mutations can readily be followed without the need to generate homozygotes by self-fertilization as in *Arabidopsis*. To date, a set of more than 200 genetic loci have been mapped by mutation. Mutant strains are provided by the Chlamydomonas Genetics Center at Duke University (http://www.biology.duke.edu/chlamy) and include those that are affected in the above-mentioned essential biological functions which are preferentially studied in this alga, such as photosynthesis, motility, phototaxis, prototrophic growth and resistance to antibiotics.

The generation and maintenance of nonphotosynthetic mutants was significantly facilitated by the fact that *Chlamydomonas* can grow heterotrophically on a medium that provides acetate as a reduced carbon source to the cell. These photosynthetic mutants are viable throughout their whole life cycle and, in addition, are fertile. In contrast, homozygous *Arabidopsis* mutants can be grown heterotrophically only at the seedling stage and, thus, cannot be further analyzed by genetic crossings. Instead, they have to be propagated as heterozygotes. As far as photosynthesis is concerned, another important feature of a *Chlamydomonas* cell, which distinguishes it from *Arabidopsis*, is its capability to synthesize chlorophyll in the dark, resulting in the accurate assembly of photosynthetic complexes even under these conditions. Therefore, light-induced stress effects during the growth and analysis of mutant strains can be minimized.

3.2 Suppressor Genetics

Similar to the budding yeast *Saccharomyces cerevisiae*, *Chlamydomonas* is amenable to suppressor genetics. This allows the application of screening strategies which are based on the selection of 'gain-of-function' instead of 'loss-of-function' mutants. For instance, suppressor mutations define loci involved in flagellar function, basal body assembly or transgene silencing (Porter et al. 1992; Asleson and Lefebvre 1998; Preble et al. 2001; Jeong et al. 2002). In particular, the identification of mutations restoring photoautotrophic growth has been shown to be a powerful tool in elucidating molecular details of chloroplast biogenesis. Defects in chloroplast gene expression have been shown to be suppressed by both intragenic and extragenic mutations.

An inversion in the chloroplast genome of *Chlamydomonas* restores partial *petD* gene expression by creating chimeric but functional *petD* mRNAs (Higgs et al. 1998), and a 5-bp duplication within a mutated translational element of the *psbD* 5' UTR (untranslated region) restored binding of a translation factor (RBP40) to this region and, thus, allowed D2 synthesis (Ossenbühl and Nickelsen 2000). Furthermore, mutations that alter the secondary structure of the chloroplast *rps7* mRNA and which affect either its stability or translation are suppressed by structure-restoring *cis*-acting suppressor mutations (Fargo et al. 1999, 2000). A similar situation has been observed for a secondary structure element which is involved in *psbC* mRNA translation (Rochaix et al. 1989).

In contrast to these intragenic suppressors, several extragenic suppressors identifying novel nuclear loci involved in the regulation of chloroplast gene expression have been characterized. Nucleus-encoded suppressors affect the splicing of group I introns of the 16S rRNA and the *psbA* mRNA (Li et al. 2002). They act on the stability/processing of various chloroplast mRNAs (Levy et al. 1997; Nickelsen 2000; Esposito et al. 2001; Komine et al. 2002) or might be involved in translational control mechanisms (Wu and Kuchka 1995).

After the development of an efficient chloroplast transformation system (see below), current approaches to identify *trans*-acting factors interacting with chloroplast encoded *cis*-acting regulatory elements are based on the introduction of site-directed mutations within the *cis*-acting elements and subsequent screening for suppressors, usually, after a mild UV-light treatment of mutant strains in order to increase mutation rates. Isolated suppressors are further genetically characterized to clarify whether the suppressor mutation, indeed, is tagging a *trans*-acting factor. As depicted in Fig. 2, the suppressor strain with mt+ is crossed with a wild-type strain bearing mt–. Because chloroplasts are uniparentally inherited from the mt+ parent, a 4:0 segregation of photosynthetic to nonphotosynthetic growth would indicate a chloroplast-encoded suppressor mutation. On the other hand, a nuclear suppressor would lead to a Mendelian 2:2 segregation. This nuclear suppressor then can be further characterized by genetic techniques, for instance, crossing of the suppressor mt– with the wild-type mt+ would give rise to a progeny that contains exclusively wild-type chloroplasts, allowing us to test whether the suppressor mutation already has a selectable phenotype. This would facilitate the cloning of the corresponding gene by a complementation approach (see below). If this is not the case, the dominant/recessive character of the suppressor mutation is determined by the generation and analysis of vegetative diploids between the chloroplast mutant and the nuclear suppressor strain. Given a dominant or semidominant character of the suppressor mutation, then the affected gene might be isolated by complementation cloning using an indexed library generated from the genomic DNA of the suppressor strain.

4 Genome Analysis

As a eukaryotic alga, *Chlamydomonas* contains three autonomous genetic systems which include the nuclear (nc), the mitochondrial (mt) and the chloroplast (cp) genomes. Whereas the structure and sequence of the mitochondrial genome have been known for some time, the complete sequences of the chloroplast and the nuclear DNA have been determined only recently by the Department of Energy Joint Genome Institute (JGI).

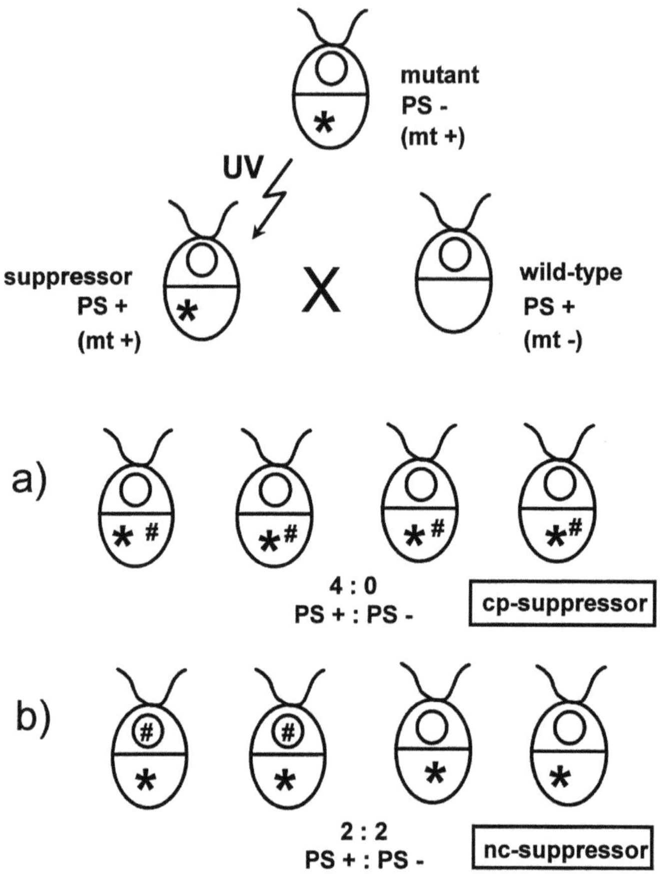

Fig. 2. Crossing scheme for the localization of suppressor mutations. Schematic *Chlamydomonas* cells contain the chloroplast in the *bottom part* and nucleus in the *top part*. * denotes the initial, site-directed chloroplast mutation; # represents the suppressor mutation. For further explanation see text

This information has dramatically enhanced the utilization of *Chlamydomonas* as a model organism.

4.1 Nuclear Genome

The nc-genome of *Chlamydomonas* consists of at least 17 genetic linkage groups. Recently, a detailed molecular map has been established containing the positions of 264 molecular markers which were determined by a combination of RFLP- and PCR-based markers and aligned with a genetic map of 200 phenotypic markers (Kathir et al. 2003).This map was designed such that any position on the *Chlamydomonas* genome is within 2 cM (corresponding to 150–200 kbp) of a mapped molecular marker. Overall, the size of the complete genome is 100–110 million bp, which is in the range of the *Arabidopsis* genome of 120 million bp. Since 2003, a rough draft of the near-complete, partially annotated genome sequence has been made accessible to the scientific community at the JGI Chlamydomonas Genome Portal (http://www.jgi.doe.gov/chlamy). Moreover, a high-quality draft is anticipated by the autumn of 2004 (Grossman et al. 2003).

To date, three *Chlamydomonas* databases exist. First, ChlamyDB contains descriptions of strains, information on genetic loci and mutant alleles, sequenced genes and a platform for information exchange of community members. Second, sequence data of ESTs, gene annotations and specialized data sets containing sequences from other Volvocales are available at ChlamyEST. Third, the genomic sequence is accessible at the above-mentioned JGI Portal. In the future, these data sets are planned to be integrated into a single database representing a version of the Generic Model Organism Database which has already been used for the creation of Flybase from *Drosophila melanogaster* (Stein et al. 2002; Grossman et al. 2003).

Moreover, a BAC library of 15,000 clones is available (http://www.genome.clemson.edu) which covers the nc-genome eightfold. Each of these clones has been sequenced from both ends and, together with the molecular map as well as the genomic sequence, they are being placed into BAC contigs that, to date, cover 25% of the genome. With the proceedings in the analysis of the genome, this will finally result in a complete set of overlapping BAC clones representing the entire nuclear DNA (Grossman et al. 2003). This will significantly facilitate the map-based cloning of genes from *Chlamydomonas*, a strategy that has already been successfully applied (Dutcher et al. 2002).

Besides the analysis of the genome sequence, a few years ago two EST sequencing projects were initiated by the Kazusa DNA Research Institute in Japan and the Carnegie Institution of Washington/Genome Technology Center at Stanford in the USA to identify transcribed regions of the *Chlamydomonas* genome. Starting from a series of different cDNA libraries - which were constructed from cells grown under various environmental conditions - to date, 200,000 clones have been sequenced from their ends and full-length sequences are in the process of being produced (Shrager et al. 2003). These cDNAs were used to generate a first generation of microarrays containing approximately 3,000 distinct gene fragments. An array representing the entire set of *Chlamydomonas* genes is scheduled for 2005 (Grossman et al. 2003). Initial experiments using these arrays have already revealed genes that are regulated by CO_2 or light (Miura et al. 2001; Im et al. 2003).

4.2 Chloroplast Genome

The cp-genome of *Chlamydomonas* principally resembles the one of vascular plants. It is present in multiple copies and contains an inverted repeat (21.2 kbp) which harbors the ribosomal RNA genes and which separates two similar-sized single copy regions of approximately 80 kbp. Recently, the complete cp-genome sequence of 203,395 bp (GenBank accession number BK000554) has been established (Maul et al. 2002) and is available on the web (http://www.biology.duke.edu/chlamy_genome/chloro.html) and, in addition, an interactive site can be visited at http://bti.cornell.edu/bti2/chlamyweb. In general, the *Chlamydomonas* cp-genome possesses a conventional set of cp-genes with, however, some remarkable features: (1) *ndh* genes which are ubiquitous on cp-DNA from vascular plants are absent from the genome, while it contains (2) *tufA*, encoding elongation factor Ef-Tu, (3) *tscA*, which encodes a small RNA required for *psaA trans*-splicing, (4) two *Chlamydomonas*-specific essential large open reading frames (ORF1995 and ORF2971), and (5) atypical RNA polymerase genes. In total, *Chlamydomonas* harbors 99 cp genes including 30 tRNA genes, surprisingly representing the smallest plastid gene set within the photosynthetic Viridiplantae (Maul et al. 2002).

An additional interesting feature is the presence of ca. 19,500 small repeated sequences which are located in intergenic regions and cover = 20% of the genome. These elements – named SDRs (short dispersed repeats) – comprise, on average, 30 bp and, based on their sequences, can be grouped into at least ten different most prevalent classes (Maul et al. 2002). Amongst the chlorophytes, including vascular plants and green algae, similar numerous SDRs have only been found for the green alga *Chlorella* sp.; however, no sequence similarities are shared by SDRs from *Chlorella* and *Chlamydomonas*. Although the origin and function of these SDRs are still unclear, it has been hypothesized that they might play a role during (1) light-dependent conformational DNA changes affecting chloroplast transcription or (2) recombination events (Maul et al. 2002).

With the completion of the cp-genome sequence from *Chlamydomonas* and the subsequent generation of plastome-wide microarrays, a thorough investigation of transcribed cp-DNA regions under different environmental/experimental conditions was performed. The data indicate that, besides the known effects of different light conditions, nutrient stress (especially phosphate and sulfur) and UV-light stress had substantial effects on plastid RNA accumulation (Lilly et al. 2002).

4.3 Mitochondrial Genome

The sequence of the linear mt-genome of 15.7 kbp was already established in 1993 (Vahrenholz et al. 1993; GenBank accession number U03843). The mt-DNA contains 13 genes including those for three tRNAs and the rRNAs for the large and small ribosomal subunits which are scattered in multiple pieces. Furthermore, five subunits of NADH:ubiquinone oxidoreductase, *cox1* and the *cob* gene are encoded by the mt-genome. Transcription, probably, starts at a bidirectional promoter nearly in the middle part of the genome and proceeds towards its ends, resulting in two transcription units (for a review see Remacle and Matagne 1998).

5 Molecular Tools

One particular advantage of using *Chlamydomonas* as a model system is that it represents the only photosynthetic organism that is amenable to the transformation of all three genetic compartments, i.e. the nuclear, the chloroplast and the mitochondrial genome. While the transformation of the nc- and the cp-genomes represents a 'standard' technique in many laboratories, successful transformation of mitochondria has been reported only once (Randolph-Anderson et al. 1993) and, thus, needs to be further optimized. However, mutated versions of the mitochondrial *cob* gene have been isolated which confer resistance to the respiration inhibitors myxothiazol and mucidin (Bennoun et al. 1992). These versions, hopefully, may prove to be used as new selectable markers for mitochondrial transformation in *Chlamydomonas*.

5.1 Transformation and Manipulation of the Nuclear Genome

Transformation of the nc-genome from *Chlamydomonas* has been achieved by various techniques such as biolistic particle bombardment, agitation with glass beads and electroporation (Debuchy et al. 1989; Kindle 1990; Shimogawara et al. 1998).

Each of the nuclear transformation methods has certain advantages/disadvantages. Particle bombardment yields only a low number of transformants and typically results in high numbers of integrated gene copies, thereby complicating subsequent molecular analysis (Kindle 1990). To date, the method of choice is the so-called glass beads method which requires no specialized equipment, is fast and efficient (transformation rate: 10^{-4} to 10^{-5}/cell) and, hence, is sufficient for most routine applications. Otherwise, in contrast to particle bombardment, the method is dependent on the availability of cell-wall-deficient recipient strains, either a *cw* mutant or cells that have been treated with the lytic *Chlamydomonas* enzyme autolysin. Even higher transformation rates (2.5×10^{-3}/cell) can be achieved by using electroporation of cell-wall-deficient strains (Shimogawara et al. 1998). Independent of the used method, in contrast to the chloroplast, DNA integrates ectopically into the genome via heterologous recombination, once the DNA reaches the nucleus.

Initially, the first selection strategies for nuclear *Chlamydomonas* transformants were based on the complementation of auxotrophic mutants by wild-type *Chlamydomonas* genes. The widely used *Arg7* gene, for instance, encodes an argininosuccinate lyase which restores growth of transformants on medium lacking arginine, or the *Nit1* gene encoding a nitrate reductase allows one to select for transformants growing with nitrate as the sole nitrogen source (Debuchy et al. 1989; Kindle et al. 1989). A major restriction on the use of these markers is their dependence on a genetic background containing the mutated version of the marker gene itself. This weakness has been overcome by the development of new dominant selectable markers which confer drug resistance to transformed cells.

At present, several nuclear markers and reporters are available including, e.g., the *Chlamydomonas CRY-1* gene conferring resistance to emetine and cryptopleurine (Nelson et al. 1994), the bacterial *ble* gene conferring resistance to zeocin (Stevens et al. 1996) and the *aphVIII* gene from *Streptomyces rimosus* which enables the selection of transformants on media containing kanamycin (Sizova et al. 2001). For the analysis of nuclear gene expression, the *Chlamydomonas Ars* gene encoding a periplasmatic arysulfatase has often been used as a reporter gene. For a comprehensive overview on nuclear markers and reporters, the interested reader is referred to a recent review by Fuhrmann (2002).

The development of nuclear markers has been hampered for a long time by the fact that foreign genes are not at all or only poorly expressed when integrated into the nc-genome of *Chlamydomonas*. Apparently, a combination of many different features of the introduced DNA can affect the level of transgene expression, i.e. the chosen promoters and promoter structures (Schroda et al. 2000), the presence of introns (Lumbreras et al. 1998; Fischer and Rochaix 2001) and the codon usage of transgenes. The latter problem is caused by the strong codon bias of nuclear genes which share a significant predominance for G or C residues at the wobble position of most codons. It has now been shown that, by using engineered, codon-adapted versions of foreign genes, their high-level expression in the nucleus of *Chlamydomonas* can be achieved.

The first codon-adapted gene to be expressed in *Chlamydomonas* was the frequently used reporter gene *gfp* encoding the green fluorescent protein from jellyfish (Fuhrmann et al. 1999). The availability of this versatile reporter now allows us to easily analyze regulatory elements for nuclear gene expression as well as cellular localization of single gene products fused to *cgfp* (Lechtreck et al. 2002; Ruiz-Binder et al. 2002; Koblenz et al. 2003). Recently, two additional marker/reporter genes have been expressed in *Chlamydomonas* as codon-adapted versions, namely the *aph7* gene from *Streptomyces hygroscopicus* conferring hygromycin B resistance to transformants and a luciferase from the coral *Renilla reniformis* offering the possibility of applying sensitive luminescence assays to measure reporter gene activities (Berthold et al. 2002; Fuhrmann 2002).

As mentioned above, 'forward genetics' in *Chlamydomonas* is facilitated by the easy generation and maintenance of mutant strains which, initially, were obtained during classical physical/chemical mutagenesis programs. With the establishment of an easy-to-use transformation system of the nc-genome, now insertional mutagenesis with nuclear selectable marker genes is most commonly used for this purpose, since non-homologous DNA recombination within the nucleus occurs at random sites.

During DNA integration of marker DNA into the genome, often, rearrangements or even significant deletions of parts of the genome are observed and, thus, cosegregation of the marker and the mutant phenotype has to be tested by genetic crossings to confirm that the mutant gene is tagged by marker sequences. Independently, random DNA integration has led to the development of promoter trap approaches (Haring and Beck 1997).

Subsequent isolation of affected genes can be achieved by different strategies. One involves so-called plasmid rescue experiments in which – after restriction and ligation of mutant DNA – flanking regions of inserted mutagenic sequences containing additional bacterial markers are cloned in *E. coli* (Adam and Loppes 1998; Boudreau et al. 2000). Another approach is to complement mutants by transforming them with an indexed genomic library from the wild type (Zhang et al. 1994).

Similar to the situation in *Arabidopsis*, 'reverse genetic' approaches, aimed at elucidating nuclear gene functions, are difficult to perform in *Chlamydomonas* due to the absence of an efficient homologous recombination system. However, directed manipulation of gene expression levels can give first insights into the function of the corresponding gene products. For instance, the over- and underexpression of a chloroplast-targeted heat shock protein (HSP70) by using sense and antisense constructs of the *HSP70* gene revealed an unexpected contribution of this chaperone to photoprotection and photosystem II repair (Schroda et al. 1999). Moreover, the ubiquitous eukaryotic phenomenon of gene silencing also occurs in *Chlamydomonas* (Cerutti et al. 1997) and the first components constituting the underlying machinery have recently been identified (Wu-Scharf et al. 2000; Jeong et al. 2002; Zhang et al. 2002). Especially, the common RNA interference (RNAi) technique has been successfully adapted to *Chlamydomonas*.

Most efficient induction of RNAi in *Chlamydomonas* is obtained with constructs that contain the genomic gene fragment of interest and, downstream of it, the corresponding cDNA in inverse orientation (Fuhrmann et al. 2001). Usually, expression rates can be reduced to less than 5% of the wild-type level in resulting transformants, leading to severe phenotypes, as has been shown for the opsin-related COP protein (Fuhrmann et al. 2001), the phototaxis receptors CSRA and CSRB (Sineshchekov et al. 2002), SF-assemblin (Lechtreck et al. 2002) and the blue-light receptor phototropin (Huang and Beck 2003).

As discussed by Fuhrmann (2002), an essential weakness of RNAi-mediated knockdown of gene expression is the lack of subsequent complementation experiments which would allow us to introduce site-directed mutations into genes of interest. On the contrary, an obvious advantage includes the analysis of essential genes whose complete inactivation by gene knockouts would prevent cell viability.

5.2 Transformation and Manipulation of the Chloroplast Genome

With the development of the biolistic transformation technique, which uses a particle gun for the delivery of DNA into the chloroplast, it became feasible to genetically manipulate this organelle. It should be noted, however, that transformation of *Chlamydomonas* chloroplasts by agitating cell-wall-deficient cells with glass beads has also been achieved (Kindle et al. 1991). However, the pioneer work was performed by using *Chlamydomonas* as a recipient for DNA transformation (Boynton et al. 1988) or, later on, *Nicotiana tabacum* (Svab et al. 1990); and, still, *Chlamydomonas* and tobacco represent the only organisms in which chloroplast transformation can be routinely carried out. Nevertheless, the successful transformation of the chloroplasts from *Euglena gracilis*, the unicellular red alga *Porphyridium* sp. and tomato has also been reported (Doetsch et al. 2001; Ruf et al. 2001; Lapidot et al. 2002).

Initially, biolistic transformation was performed by accelerating and delivering DNA-coated tungsten particles via powder explosion. Later on, the development of a particle inflow gun improved the transformation process (Finer et al. 1992; Nickelsen and Kück 2000). Nowadays, chloroplast transformation rates are in the range of hundreds of transformants per microgram DNA, depending on the *Chlamydomonas* strains and, thus, selection strategies for transformants in use. Once the tungsten particles have entered the chloroplast, DNA integrates into the cp-genome via homologous recombination. Afterwards, the introduced DNA rapidly segregates, resulting in a homoplasmic state where all cp-DNA copies are identical. However, if the transforming DNA affects an essential gene for cell viability only a heteroplasmic state can be reached which still contains recipient cp-DNA copies.

Selection strategies for *Chlamydomonas* chloroplast transformants include (1) the restoration of chloroplast defects in photosynthetic functions by complementing, for instance, *atpB*, *tscA* or *psbA* deletion mutants, (2) the introduction of drug resistances including those to antibiotics such as spectinomycin or erythromycin or herbicides like metribuzin or DCMU (for an overview see Goldschmidt-Clermont 1998) and, currently, (3) the expression of foreign selectable marker and reporter genes.

The first foreign, dominant selectable marker which was used in *Chlamydomonas* was the bacterial *aadA* gene encoding an aminoglycoside 3" adenyl transferase which confers resistance to spectinomycin and streptomycin (Goldschmidt-Clermont 1991; Table 1). More recently, a second marker, based on the bacterial *aphA-6* gene encoding an aminoglycoside phosphotransferase, has been successfully used to select for *Chlamydomonas* chloroplast transformants on kanamycin- or amikacin-containing medium (Bateman and Purton 2000). This now offers the possibility of successive transformations of the cp-genome without the need for marker recycling strategies (Goldschmidt-Clermont 1998).

In addition, several reporter genes have been designed to allow the analysis of chloroplast gene expression. These include the bacterial *uidA* gene for the application of β-glucuronidase (GUS) assays (Sakamoto et al. 1993) and the above-mentioned gene for a luciferase from *Renilla reniformis*, allowing us to measure luminescence and, hence, gene expression

Table 1. Expression of foreign genes in *Chlamydomonas* chloroplast

Gene name	Gene product	Origin	Reference
aadA	Aminoglcoside 3" adenyl transferase (selection marker)	*Escherichia coli*	Goldschmidt-Clermont (1991)
aphA-6	Aminoglycoside phosphotransferase (selection marker)	*Acinetobacter baumannii*	Bateman and Purton (2000)
uidA	β-Glucuronidase (reporter)	*Escherichia coli*	Sakamoto et al. (1993)
rluc	Luciferase (reporter)	*Renilla reniformis*	Minko et al. (1999)
gfp	Green fluorescent protein (reporter)	*Aequorea aequorea*	Franklin et al. (2002)
recA	RecA	*Escherichia coli*	Cerutti et al. (1995)
VP1+CTB	Fusion of the protein VP1 and the cholera toxin B subunit	Foot-and-mouth disease virus (FMDV) and *Vibrio cholerae*	Sun et al. (2003)
lsc	Large single-chain antibody	*Homo sapiens*	Mayfield et al. (2003)

levels (Minko et al. 1999). Recently, also a synthetic, codon-adapted version of the *gfp* reporter gene encoding the green fluorescent protein has been found to be highly expressed in *Chlamydomonas* chloroplasts, adding another potent chloroplast reporter gene to this list (Franklin et al. 2002; Komine et al. 2002; Table 1).

As mentioned above, the chloroplast – like its cyanobacterial ancestor – contains a homologous recombination system which allows one to apply 'reverse genetic' approaches to elucidate the function of chloroplast genes after they have been inactivated by targeted mutagenesis, for instance, by insertion of the *aadA* cassette. To date, a total of 35 protein-coding cp-genes have been disrupted (for a list see Grossman et al. 2003 and references herein). Six out of these, including genes for three subunits of the plastid-encoded RNA polymerase, a ribosomal protein, ORF1995 of unknown function and the *clpP* gene, are essential as suggested by the finding that they cannot be brought to homoplasmicity.

Although the *clpP* gene encoding the chloroplast ClpP protease was found to be essential in *Chlamydomonas*, preventing its functional analysis (Huang et al. 1994), recently a novel mutagenesis approach – which might work also for other essential cp-genes – has been successfully applied to solve this problem. This strategy, named translational attenuation, is based on the observation that the alteration of AUG initiation codons does not necessarily switch off translation but, in some cases, just reduces protein synthesis rates (Chen et al. 1993). Consequently, the *clpP* initiation codon was changed into AUU resulting in reduced ClpP levels of 25–45% compared to the wild-type. In these mutant strains, the degradation of the cytochrome b_6f complex was affected, suggesting that ClpP is involved in the proteolytic disposal of its fully or partially assembled subunits (Majeran et al. 2000).

The remaining inactivated genes have photosynthetic functions and, as such, are not essential. They include subunits of all four thylakoid membrane complexes which mediate photosynthetic electron flow, i.e. photosystems I and II, the cytochrome b_6f complex and the ATP synthase. Furthermore, genes for chlorophyll synthesis in the dark (*chlL* and *chlN*), heme attachment (*ccsA*), carbon uptake (*cemA*) and the large subunit of Rubisco (*rbcL*) have been inactivated (Spreitzer et al. 1985; Choquet et al. 1992; Suzuki and Bauer 1992; Xie and Merchant 1996; Rolland et al. 1997).

Besides the analysis of cp-ORFs, especially the dissection of regulatory *cis*-acting elements for chloroplast gene expression has attracted much attention during recent years. By cointegration of the selectable marker or by its cotransformation, principally, any single position of the cp-genome can be changed by site-directed mutagenesis. This enabled the precise mapping of numerous elements that are essential for various gene expression steps, i.e. transcription, RNA maturation and stabilization as well as translation initiation (for a recent overview see a forthcoming special issue of *Photosynthesis Research*). The combination of both site-directed chloroplast mutations and nuclear mutations affecting *trans*-acting factors involved in these steps via classical genetic crossings has been proven to be a fruitful approach to determine the chloroplast target sites of the regulatory functions (for a review see Nickelsen 1998). Moreover, site-directed chloroplast mutations have revealed a number of structure/function relationships from subunits of most photosynthetic complexes.

As exemplified in Table 1, most recent work has entered the field of biotechnological application using chloroplast transformation in *Chlamydomonas*, similar to the situation in higher plants (Maliga 2003). High-yield expression (3% of total soluble protein) of a fusion protein consisting of the VP1 protein from the foot-and-mouth disease virus and the cholera toxin B subunit with retained antigenicity supports the possibility that transplastomic *Chlamydomonas* cells might be used as a mucosal vaccine source (Sun et al. 2003). In addition, Mayfield and coworkers (2003) succeeded in expressing a fully active human antibody directed against glycoprotein D of the herpes simplex virus in chloroplasts of *Chlamydomonas*, offering the opportunity of cost-savings in mass production of pharmaceutical drugs from this green alga.

6 Conclusions and Perspectives

The green alga *Chlamydomonas reinhardtii* has frequently been named 'the green yeast' to emphasize its advantages as a model organism (Goodenough 1992; Rochaix 1995). Obviously, *Chlamydomonas* shares many experimental and organic properties with *Saccharomyces cerevisiae*, e.g. its unicellular morphology, its haploid genetics and its accessibility to genetic manipulation of both the nuclear and organellar genomes. However, a major difference is the lack of a nuclear homologous recombination system, which would allow one to easily perform 'reverse genetic' approaches to elucidate gene functions similar to the situation in the chloroplast. As such, *Chlamydomonas* resembles the other well-accepted phototrophic model organism, *Arabidopsis thaliana*, which suffers from the same problem. In both eukaryotic systems, this obstacle has now been overcome by the application of RNAi- and antisense RNA-mediated knockdowns of nuclear gene expression.

Another prerequisite for a comprehensive molecular analysis is the availability of the complete *Chlamydomonas* genome sequence, which is planned to be finished by the end of 2004. This information will have an enormous impact on the molecular analysis of the biology of this alga. The determination of flanking regions from insertional mutants will readily identify the affected genes that are responsible for specific phenotypes. BAC clones covering the respective regions, then, will be used for complementation of the mutants. Global gene expression at the transcript level will be analyzed by using genome-wide microarrays, and proteomic approaches will provide information on the qualitative and quantitative accumulation of proteins. Finally, all the information will be integrated in a database which is currently being established.

Another aspect of future work with *Chlamydomonas* concerns its use in biotechnical applications which are becoming apparent and which will be facilitated by the further improvement of the molecular toolkit for the manipulation of both nuclear and chloroplast gene expression levels. Taken together, the latest developments in *Chlamydomonas* research clearly place this alga in line with the established phototrophic model organisms *Synechocystis* ssp. PCC 6803 and *Arabidopsis thaliana*. Due to its unique advantages, *Chlamydomonas* faces a bright future as a model system, especially for the analysis of processes such as phototaxis, motility, photosynthesis and chloroplast biogenesis.

Acknowledgements. I thank U. Kück for basic support and H.J. Rathke for the excellent artwork. The experimental work of the author was funded by the Deutsche Forschungsgemeinschaft (SFB480-B8).

References

Adam M, Loppes R (1998) Use of the *Arg7* gene as an insertional mutagen to clone *PHON24*, a gene required for derepressible neutral phosphatase activity in *Chlamydomonas reinhardtii*. Mol Gen Genet 258:123–132

Asleson CM, Lefebvre PA (1998) Genetic analysis of flagellar length control in *Chlamydomonas reinhardtii*, a new long-flagella locus and extragenic suppressor mutations. Genetics 148:693–702

Bateman JM, Purton S (2000) Tools for chloroplast transformation in *Chlamydomonas*: expression vectors and new dominant selectable marker. Mol Gen Genet 263:404–410

Bennoun P, Delosme M, Kück U (1992) New tools for mitochondrial genetics of *Chlamydomonas reinhardtii*: manganese mutagenesis and cytoduction. Mol Gen Genet 234:147–154

Berthold P, Schmitt R, Mages W (2002) An engineered *Streptomyces hygroscopicus aph7* gene mediates dominant resistance against hygromycin B in *Chlamydomonas reinhardtii*. Protist 153:401–412

Boudreau E, Nickelsen J, Lemaire S, Ossenbühl F, Rochaix J-D (2000) The *Nac2* gene of *Chlamydomonas reinhardtii* encodes a chloroplast TPR protein involved in *psbD* mRNA stability, processing and/or translation. EMBO J 19:3366–3376

Boynton JE, Gillham NW, Harris EH, Hosler JP, Johnston AR, Jones BL, Randolph-Anderson D, Robertson TM, Klein KB, Shark B, Sanford J (1988) Chloroplast transformation in *Chlamydomonas* with high velocity microprojectiles. Science 240:1534–1538

Cerutti H, Johnson AM, Boynton JE, Gillham NW (1995) Inhibition of chloroplast DNA recombination and repair by dominant negative mutants of *Escherichia coli* RecA. Mol Cell Biol 15:3003–3011

Cerutti H, Johnson AM, Gillham NW, Boynton JE (1997) Epigenetic silencing of a foreign gene in nuclear transformants of *Chlamydomonas.* Plant Cell 9:925–945

Chen X, Kindle KL, Stern DB (1993) Initiation codon mutations in the *Chlamydomonas* chloroplast *petD* gene result in temperature-sensitive photosynthetic growth. EMBO J 12:3627–3635

Choquet Y, Rahire M, Girard-Bascou J, Erickson J, Rochaix J-D (1992) A chloroplast gene is required for the light-independent accumulation of chlorophyll in *Chlamydomonas reinhardtii.* EMBO J 11:1697–1704

Debuchy R, Purton S, Rochaix J-D (1989) The argininosuccinate lyase gene of *Chlamydomonas reinhardtii*: an important tool for nuclear transformation and for correlating the genetic and molecular maps of the *ARG7* locus. EMBO J 8:2803–2809

Dent RM, Han M, Niyogi KK (2001) Functional genomics of plant photosynthesis in the fast lane using *Chlamydomonas reinhardtii.* Trends Plant Sci 6:364–371

Doetsch N, Favreau M, Kuscuoglu N, Thompson M, Hallick RB (2001) Chloroplast transformation in *Euglena gracilis*: splicing of a group III twintron transcribed from a transgenic *psbK* operon. Curr Genet 39:49–60

Dutcher SK, Morrissette NS, Preble AM, Rackley C, Stanga J (2002) ε-Tubulin is an essential component of the centriole. Mol Biol Cell 13:3859–3869

Esposito D, Higgs DC, Drager RG, Stern DB, Girard-Bascou J (2001) A nucleus-encoded suppressor defines a new factor which can promote *petD* mRNA stability in the chloroplast of *Chlamydomonas reinhardtii.* Curr Genet 39:40–48

Fargo DC, Boynton JE, Gilham NW (1999) Mutations altering the predicted secondary structure of a chloroplast 5' untranslated region affect its physical and biochemical properties as well as its ability to promote translation of reporter mRNAs both in the *Chlamydomonas reinhardtii* chloroplast and in *Escherichia coli.* Mol Cell Biol 19:6980–6990

Fargo DC, Hu E, Boynton JE, Gillham NW (2000) Mutations that alter the higher-order structure of its 5' untranslated region affect the stability of chloroplast *rps7* mRNA. Mol Gen Genet 264:291–299

Finer JJ, Vain P, Jones MW, McMullen MD (1992) Development of the particle inflow gun for DNA delivery to plant cells. Plant Cell Rep 11:323–328

Fischer N, Rochaix J-D (2001) The flanking regions of *PsaD* drive efficient gene expression in the nucleus of the green alga *Chlamydomonas reinhardtii.* Mol Genet Genom 265:888–894

Franklin S, Ngo B, Efuet E, Mayfield SP (2002) Development of a GFP reporter gene for *Chlamydomonas reinhardtii* chloroplast. Plant J 30:733–744

Fuhrmann M (2002) Expanding the molecular toolkit for *Chlamydomonas reinhardtii* – from history to new frontiers. Protist 153:357–364

Fuhrmann M, Oertel W, Hegemann P (1999) A synthetic gene coding for the green fluorescent protein (GFP) is a versatile reporter in *Chlamydomonas reinhardtii.* Plant J 19:353–361

Fuhrmann M, Stahlberg A, Govorunova E, Rank S, Hegemann P (2001) The abundant retinal protein of the *Chlamydomonas* eye is not the photoreceptor for phototaxis and photophobic responses. J Cell Sci 114:3857–3863

Goldschmidt-Clermont M (1991) Transgenic expression of aminoglycoside adenine transferase in the chloroplast: a selectable marker for site-directed transformation of *Chlamydomonas.* Nucleic Acids Res 19:4083–4089

Goldschmidt-Clermont M (1998) Chloroplast transformation. In: Rochaix J-D, Goldschmidt-Clermont M, Merchant S (eds) The molecular biology of chloroplasts and mitochondria in *Chlamydomonas.* Kluwer, Dordrecht, pp 139–149

Goodenough UW (1992) Green yeast. Cell 70:533–538

Grossman AR, Harris EH, Hauser C, Lefebvre PA, Martinez D, Rokhsar D, Shrager J, Silflow CD, Stern D, Vallon O, Zhang Z (2003) *Chlamydomonas reinhardtii* at the crossroads of genomics. Eukaryot Cell 2:1137–1150

Haring MA, Beck CF (1997) A promoter trap for *Chlamydomonas reinhardtii*: development of a gene cloning method using 5' RACE-based probes. Plant J 11:1341–1348

Harris EH (1989) The *Chlamydomonas* sourcebook. Academic Press, San Diego

Harris EH (2001) *Chlamydomonas* as a model organism. Annu Rev Plant Physiol Plant Mol Biol 52:363–406

Higgs DC, Kuras R, Kindle KL, Wollman FA, Stern DB (1998) Inversions in the *Chlamydomonas* chloroplast genome suppress a *petD* 5' untranslated region deletion by creating functional chimeric mRNAs. Plant J 14:663–671

Holtorf H, Guitton MC, Reski R (2002) Plant functional genomics. Naturwissenschaften 89:235–249

Huang C, Wang S, Chen L, Lemieux C, Otis C, Turmel M, Liu XQ (1994) The *Chlamydomonas clpP* gene contains translated large insertion sequences and is essential for cell growth. Mol Gen Genet 244:151–159

Huang K, Beck CF (2003) Phototropin is the blue-light receptor that controls multiple steps in the sexual life cycle of the green alga *Chlamydomonas reinhardtii*. Proc Natl Acad Sci USA 100:6269–6274

Im CS, Zhang Z, Shrager J, Chang CW, Grossman AR (2003) Analysis of light and CO_2 regulation in *Chlamydomonas reinhardtii* using genome-wide approaches. Photosynth Res 75:111–125

Jeong BR, Wu-Scharf D, Zhang C, Cerutti H (2002) Suppressors of transcriptional transgenic silencing in *Chlamydomonas* are sensitive to DNA-damaging agents and reactive transposable elements. Proc Natl Acad Sci USA 99:1076–1081

Kathir P, LaVoie M, Brazelton WJ, Haas NA, Lefebvre PA, Silflow CD (2003) Molecular map of the *Chlamydomonas reinhardtii* nuclear genome. Eukaryot Cell 2:362–379

Kindle KL (1990) High-frequency nuclear transformation of *Chlamydomonas reinhardtii*. Proc Natl Acad Sci USA 87:1228–1232

Kindle KL, Schnell RA, Fernandez E, Lefebvre PA (1989) Stable nuclear transformation of *Chlamydomonas* using the *Chlamydomonas* gene for nitrate reductase. J Cell Biol 109:2589–2601

Kindle KL, Richards KL, Stern DB (1991) Engineering the chloroplast genome: techniques and capabilities for chloroplast transformation in *Chlamydomonas reinhardtii*. Proc Natl Acad Sci USA 88:1721–1725

Koblenz B, Schoppmeier J, Grunow A, Lechtreck KF (2003) Centrin deficiency in *Chlamydomonas reinhardtii* causes defects in basal body replication, segregation and maturation. J Cell Sci 116:2635–2646

Komine Y, Kikis E, Schuster G, Stern DB (2002) Evidence for in vivo modulation of chloroplast RNA stability by 3' UTR homopolymeric tails in *Chlamydomonas reinhardtii*. Proc Natl Acad Sci USA 99:4085–4090

Lapidot M, Raveh D, Sivan A, Shoshana A, Shapira M (2002) Stable chloroplast transformation of the unicellular red alga *Porphyridium* spec. Plant Physiol 129:7–12

Lechtreck KF, Rostmann J, Grunow A (2002) Analysis of *Chlamydomonas* SF-assemblin by GFP tagging expression of antisense constructs. J Cell Sci 115:1511–1522

Levy H, Kindle KL, Stern DB (1997) A nuclear mutation that affects the 3' processing of several mRNAs in *Chlamydomonas* chloroplasts. Plant Cell 9:825–836

Li F, Holloway SP, Lee J, Herrin DL (2002) Nuclear genes that promote splicing of group I introns in the chloroplast 23S rRNA and *psbA* genes in *Chlamydomonas reinhardtii*. Plant J 32:467–480

Lilly JW, Maul JE, Stern DB (2002) The *Chlamydomonas reinhardtii* organellar genomes respond transcriptionally and post-transcriptionally to abiotic stimuli. Plant Cell 14:2681–2706

Lumbreras V, Stevens DR, Purton S (1998) Efficient foreign gene expression in *Chlamydomonas reinhardtii* mediated by an endogenous intron. Plant J 14:441–447

Majeran W, Wollman F-A, Vallon O (2000) Evidence for a role of ClpP in the degradation of the chloroplast cytochrome b_6f complex. Plant Cell 12:137–149

Maliga P (2003) Progress towards commercialization of plastid transformation technology. Trends Biotechnol 21:20–28

Maul JE, Lilly JW, Cui L, dePamphilis CW, Harris EH, Stern DB (2002) The *Chlamydomonas reinhardtii* plastid chromosome: islands of genes in a sea of repeats. Plant Cell 14:2659–2679

Mayfield SP, Franklin SE, Lerner RA (2003) Expression and assembly of a fully active antibody in algae. Proc Natl Acad Sci USA 100:438–442

Minko I, Holloway SP, Nikaido S, Carter M, Odom OW, Johnson C, Herrin DL (1999) *Renilla* luciferase as a vital reporter for chloroplast gene expression in *Chlamydomonas*. Mol Gen Genet 262:421–425

Miura K, Kucho K, Inoue Y, Taniguchi F, Kohinata T, Yoshioka S, Asamizu E, Nakamura Y, Tabata S, Ohyama K, Fukuzawa H (2001) The structure of a regulatory gene *CCM1* and screening of its target genes by using *Chlamydomonas* cDNA macroarray. Plant Cell Physiol 42 (Suppl):116

Nelson JA, Savereide PB, Lefebvre PA (1994) The *CRY1* gene in *Chlamydomonas reinhardtii*: structure and use as a dominant selectable marker for nuclear transformation. Mol Cell Biol 14:4011–4019

Nickelsen J (1998) Chloroplast RNA stability. In: Rochaix J-D, Goldschmidt-Clermont M, Merchant S (eds) The molecular biology of chloroplasts and mitochondria in *Chlamydomonas*. Kluwer, Dordrecht, pp 151–163

Nickelsen J (2000) Mutations at three different nuclear loci of *Chlamydomonas* suppress a defect in chloroplast *psbD* mRNA accumulation. Curr Genet 37:136–142

Nickelsen J, Kück U (2000) The unicellular green alga *Chlamydomonas reinhardtii* – an experimental system to study chloroplast RNA metabolism. Naturwissenschaften 87:97–107

Nishimura Y, Misumi O, Kato K, Inada N, Higashiyama T, Momoyama Y, Kuroiwa T (2002) An mt(+) gamete-specific nuclease that targets mt(–) chloroplasts during sexual reproduction. Genes Dev 16:1116–1128

Ossenbühl F, Nickelsen J (2000) *cis*- and *trans*-acting determinants for translation of *psbD* mRNA in *Chlamydomonas reinhardtii*. Mol Cell Biol 20:8134–8142

Porter ME, Power J, Dutcher SK (1992) Extragenic suppressors of paralyzed flagellar mutations in *Chlamydomonas reinhardtii* identify loci that alter the inner dynein arms. J Cell Biol 118:1163–1176

Preble AM, Giddings TH, Dutcher SK (2001) Extragenic bypass suppressors of mutations in the essential gene *BLD2* promote assembly of basal bodies with abnormal microtubules in *Chlamydomonas reinhardtii*. Genetics 157:163–181

Randolph-Anderson BL, Boynton JE, Gilham NW, Harris EH, Johnson AM, Dorthu MP, Matagne RF (1993) Further characterization of the respiratory deficient *dum-1* mutation of *Chlamydomonas reinhardtii* and its use as a recipient for mitochondrial transformation. Mol Gen Genet 236:235–244

Remacle C, Matagne RF (1998) Mitochondrial genetics. In: Rochaix JD, Goldschmidt-Clermont M, Merchant S (eds) The molecular biology of chloroplasts and mitochondria in *Chlamydomonas*. Kluwer, Dordrecht, pp 183–195

Rochaix J-D (1995) *Chlamydomonas reinhardtii* as the photosynthetic yeast. Annu Rev Genet 29:209–230

Rochaix J-D, Kuchka M, Mayfield SP, Rahire M, Girard-Bascou J, Bennoun P (1989) Nuclear and chloroplast mutations affect the synthesis or stability of the chloroplast *psbC* gene product in *Chlamydomonas reinhardtii*. EMBO J 8:1013–1021

Rolland N, Dome AJ, Amoroso G, Sultemeyer DF, Joyard J, Rochaix J-D (1997) Disruption of the plastid *ycf10* open reading frame affects uptake of inorganic carbon in the chloroplast of *Chlamydomonas*. EMBO J 16:6713–6726

Ruiz-Binder NE, Geimer S, Melkonian M (2002) In vivo localization of centrin in the green alga *Chlamydomonas reinhardtii*. Cell Motil Cytoskeleton 52:43–55

Ruf S, Hermann M, Berger IJ, Carrer H, Boch R (2001) Stable genetic transformation of tomato plastids and expression of a foreign protein in fruit. Nat Biotechnol 19:870–875
Sakamoto W, Kindle K, Stern DB (1993) In vivo analysis of *Chlamydomonas* chloroplast *petD* gene expression using stable transformation of beta-glucuronidase translational fusions. Proc Natl Acad Sci USA 90:497–501
Schroda M, Vallon O, Wollman F-A, Beck CF (1999) A chloroplast-targeted heat shock protein 70 (HSP70) contributes to the photoprotection and repair of photosystem II during and after photoinhibition. Plant Cell 11:1165–1178
Schroda M, Blocker D, Beck CF (2000) The HSP70A promoter as a tool for the improved expression of transgenes in *Chlamydomonas*. Plant J 21:121–131
Shimogawara K, Fujiwara S, Grossman A, Usuda H (1998) High-efficiency transformation of *Chlamydomonas reinhardtii* by electroporation. Genetics 148:1821–1828
Shrager J, Hauser C, Chang CW, Harris EH, Davies J, McDermott J, Tamse R, Zhang Z, Grossman AR (2003) *Chlamydomonas reinhardtii* genome project. A guide to the generation and use of the cDNA information. Plant Physiol 131:401–408
Silflow CD (1998) Organization of the nuclear genome. In: Rochaix J-D, Goldschmidt-Clermont M, Merchant S (eds) The molecular biology of chloroplasts and mitochondria in *Chlamydomonas*. Kluwer, Dordrecht, pp 25–40
Sineshchekov OA, Jung KH, Spudich JL (2002) Two rhodopsins mediate phototaxis to low- and high-intensity light in *Chlamydomonas reinhardtii*. Proc Natl Acad Sci USA 99:8689–8694
Sizova I, Fuhrmann M, Hegemann P (2001) A *Streptomyces rimosus aphVIII* gene coding for a new type phosphotransferase provides stable antibiotic resistance to *Chlamydomonas reinhardtii*. Gene 277:221–229
Spreitzer RJ, Goldschmidt-Clermont M, Rahire M, Rochaix J-D (1985) Nonsense mutations in the *Chlamydomonas* chloroplast gene that codes for the large subunit of ribulose biphosphate carboxylase/oxygenase. Proc Natl Acad Sci USA 82:5460–5464
Stauber EJ, Fink A, Markert C, Kruse O, Johanningmeier U, Hippler M (2003) Proteomics of *Chlamydomonas reinhardtii* light-harvesting proteins. Eukaryot Cell 2:978–994
Stein LD, Mungall C, Shu S, Caudy M, Mangone M, Day A, Nickerson E, Stajich JE, Harris TW, Arva A, Lewis S (2002) The generic genome browser: a building block for a model organism system database. Genome Res 12:1599–1610
Stevens DR, Rochaix J-D, Purton S (1996) The bacterial phleomycin resistance gene *ble* as a dominant selectable marker in *Chlamydomonas*. Mol Gen Genet 251:23–30
Sun M, Qian K, Su N, Chang H, Liu J, Chen G (2003) Foot-and-mouth disease virus VP1 protein fused with cholera toxin B subunit expressed in *Chlamydomonas reinhardtii* chloroplast. Biotechnol Lett 25:1087–1092
Suzuki JY, Bauer CE (1992) Light-independent chlorophyll biosynthesis: involvement of chloroplast gene *chlL* (*frxC*). Plant Cell 4:929–940
Svab Z, Hajdukiewicz P, Maliga P (1990) Stable transformation of plastids in higher plants. Proc Natl Acad Sci USA 87:8526–8530
Vahrenholz C, Riemen G, Pratje E, Dujon B, Michaelis G (1993) Mitochondrial DNA of *Chlamydomonas reinhardtii*: the structure of the ends of the linear 15.8 kbp genome suggests mechanisms for DNA replication. Curr Genet 24:241–247
Wu HY, Kuchka MR (1995) A nuclear suppressor overcomes defects in the synthesis of the chloroplast *psbD* gene product caused by mutations in two distinct nuclear genes of *Chlamydomonas*. Curr Genet 27:263–269
Wu-Scharf D, Jeong BR, Zhang C, Cerutti H (2000) Transgene and transposon silencing in *Chlamydomonas reinhardtii* by a DEAH-box RNA helicase. Science 290:1159–1162
Xie Z, Merchant S (1996) The plastid-encoded *ccsA* gene is required for heme attachment to chloroplast c-type cytochromes. J Biol Chem 271:4632–4639
Zhang C, Wu-Scharf D, Jeong BR, Cerutti H (2002) A WD40-repeat containing protein, similar to a fungal co-repressor, is required for transcriptional gene silencing in *Chlamydomonas*. Plant J 31:25–36

Zhang H, Herman P, Weeks DP (1994) Gene isolation through genomic complementation using an indexed library of *Chlamydomonas reinhardtii* DNA. Plant Mol Biol 24:663–672

Jörg Nickelsen
Lehrstuhl Allgemeine und Molekulare Botanik
Ruhr-Universität Bochum
Universitätsstr. 150
44780 Bochum, Germany
e-mail: joerg.nickelsen@ruhr-uni-bochum.de

Extranuclear Inheritance: Mitochondrial Genetics and Biogenesis

Karlheinz Esser, Georg Michaelis, and Elke Pratje

1 Introduction

The sixth International Congress on 'Plant Mitochondria' took place in Perth, Western Australia, in July 2002. The ten symposia dealt with respiration and environment, mitochondrial biogenesis, mitochondrial enzymes and their regulation, nucleic acids in plant mitochondrial function, electron transport, mitochondrial genomics and proteomics, stress physiology, nuclear gene expression, oxidative stress and programmed cell death, and mutations affecting mitochondrial function in plants. Abstracts of the contributions are included in the respective congress book.

'Chloroplasts and mitochondria: functional genomics and evolution' is the title of the January 2003 issue of the *Philosophical Transactions of the Royal Society, London, Biological Sciences* (vol. 358, issue 1429), which summarizes lectures and discussions of a meeting on this subject. Finally, a special issue (May 2003) of the *Biochemical and Biophysical Research Communications* (vol. 304, issue 3) is devoted to mitochondrial control of apoptosis.

The present chapter reviews some selected topics on mitochondrial genomics and proteomics, import of nucleic acids into mitochondria, cytochrome c maturation in plants and mitochondrial fusion.

Of the many reviews that have been published, several are pointed out here to the reader: mitochondrial genomes (Burger et al. 2003), mitochondrial DNA repair (Mandavilli et al. 2002), genes for mitochondrial function (Chinnery 2003), transcriptional and post-transcriptional processes (Binder and Brennicke 2003), protein import into mitochondria (Rehling et al. 2003; Truscott et al. 2003), RNA import (Entelis et al. 2002), cytochrome c oxidase (Richter and Ludwig 2003), respiratory chain supercomplexes (Schagger 2002), programmed cell death (Hoeberichts and Woltering 2003; Huettenbrenner et al. 2003; Newmeyer and Ferguson-Miller 2003; Nieminen 2003; Priault et al. 2003), division of mitochondria and chloroplasts (Miyagishima et al. 2003; Osteryoung and Nunnari 2003), fusion of mitochondria (Mozdy and Shaw 2003), and evolutionary aspects of mitochondria (Berry 2003; Emelyanov 2003; Karlberg and Andersson 2003; Searcy 2003).

Progress in Botany, Vol. 66

2 Mitochondrial Genomes and Proteomes

2.1 Mitochondrial Genomes

The increasing information on mitochondrial genomes and functions is updated in the GOBASE database containing 408 complete mitochondrial genomes in release 6 (O'Brien et al. 2003; http://megasun.bch.umontreal.ca/gobase/). GOBASE contains 130,780 mitochondrial sequences of which 50,948 are proteins; in June 2003 the entries of plant mitochondria listed sequences of 2,858 genes and 1,431 proteins. The OGRe database is a relational database of complete mitochondrial genome sequences for over 250 Metazoan species (Jameson et al. 2003; http://bioinf.man.ac.uk/ogre/).

The size of mitochondrial DNA completely sequenced varies from 6 kb (*Plasmodium*) to 490 kb (*Oryza sativa*). The mitochondrial genomes of some *Curcubitaceae* are larger than 2,000 kb (Ward et al. 1981), but yet unsequenced. The size differences are mainly due to repeated sequences, introns, intergenic spacers, mobile elements or foreign DNA of nuclear, plastid or viral origin. The coding capacity varies from five mitochondrial genes in *Plasmodium* to about 100 genes in *Reclinomonas americana* (see *Progress in Botany*, vol. 60), but in general there is no correlation between the size and the gene content. Mitochondrial genes are involved in respiration, oxidative phosphorylation and mitochondrial protein synthesis. In addition, mitochondrial DNA of some organisms like *Reclinomonas americana* contains genes affecting transcription, RNA maturation or protein import.

The diversity in genes for proteins reflects mainly the transfer of mitochondrial genes to the nucleus. Transfer of ribosomal protein and succinate dehydrogenase genes to the nucleus has been extensively studied and potential mechanisms for activation of the transferred genes are discussed by Adams and Palmer (2003).

The cox2 gene, encoding subunit 2 of cytochrome c oxidase, is a mitochondrial gene in most organisms, but it is transferred to the nucleus in legumes, the chlamydomonad green algae and apicomplexan parasites (Adams et al. 1999; Perez-Martinez et al. 2001; Funes et al. 2002; Gardner et al. 2002). Nine legume genera out of 392 tested contain a mitochondrial and a nuclear copy of the cox2 gene (Adams et al. 1999). Only in *Dumasia* are both gene copies transcribed in significant amounts. In all other species one of the two copies is inactivated. The mitochondrial and nuclear copies seem to be inactivated with similar frequencies. For example, in *Pseudovigna* only the mitochondrial gene copy is active, whereas in soybean, *Glycine max*, the nuclear copy is expressed and the mitochondrial copy is silent. A specific mitochondrial targeting sequence is required to import the hydrophobic Cox2 protein with two membrane-spanning domains into soybean mitochondria. This targeting sequence of 124 amino acid residues is longer than any other mitochondrial targeting presequence reported to date (Covello and Gray 1992; Daley et al. 2002a). When the nuclear encoded presequence is added to the mitochondrial encoded Cox2 protein the in vitro synthesized chimeric protein is not imported into isolated soybean mitochondria (Daley et al. 2002b). However, the chimeric protein is imported when the hydrophobicity of the first transmembrane region of Cox2 is reduced by replacing two amino acid residues. The first transmembrane region of the nuclear encoded soybean protein is less hydrophobic than that of the mitochondrial encoded protein. Thus a decrease in hydrophobicity of the first transmembrane region was required for mitochondrial import of the nuclear encoded protein. In soybean the targeting sequence is cleaved off in a unique three-step process (Daley et al. 2002a). The import mechanism of the soybean Cox2 protein was also studied in yeast as a model system (Qualmann et al. 2003). In Chlamydomonad algae the cox2 gene is split in the nucleus and the proteins are probably imported as separate polypeptides (Perez-Martinez et al. 2001).

Mitochondrial genes have been transferred to mitochondrial genomes of other organisms as well. Horizontal gene transfers seem to be widespread and to occur frequently in flowering plants (Bergthorsson et al. 2003), resulting in gene duplication or re-acquisition of genes previously transferred to the nucleus. The mechanism of how genes move from one plant to another is unknown. DNA transfer into mitochondrial genomes of angiosperms includes transfer from monocots to dicots. The rps11 gene of *Sanguinaria canadensis* is a chimeric gene, its 5 half is of dicot and its 3 half of monocot origin.

Horizontal gene transfer from an angiosperm (asterid) to an Asian clade of *Gnetum* (gymnosperm) has been reported by Won and Renner (2003). The authors suggested an age of 2–5 million years that *Gnetum* received the nad1 intron2 and the two adjacent exons b and c from a flowering plant.

2.2 Mitochondrial Proteomes

The mitochondrial proteomes of man (Taylor et al. 2003), *Saccharomyces cerevisiae* (Sickmann et al. 2003) and *Arabidopsis thaliana* (Heazlewood et al. 2004) were analysed in detail. The mitochondrial and nuclear encoded proteins of purified mitochondria were separated and identified by various methods as two-dimensional polyacrylamid gelelectrophoresis (2D-PAGE), matrix-assisted laser desorption/ionization-time of flight mass spectrometry (MALDI-Tof MS), multidimensional peptide separation and/or tandem MS.

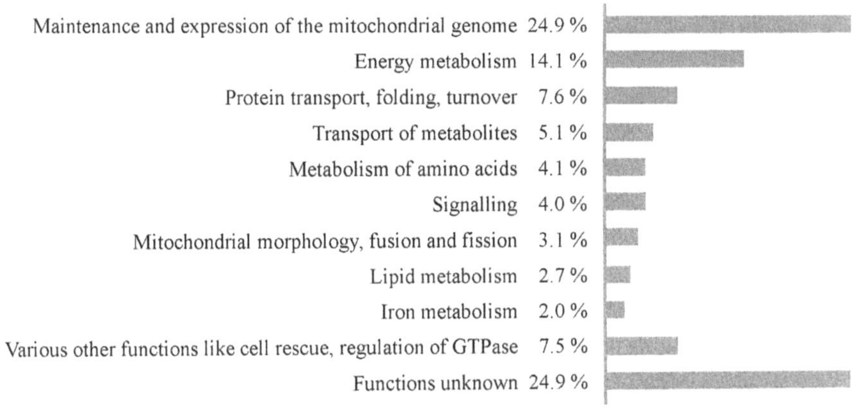

Fig. 1. Functional classification of the proteins in the mitochondria of *Saccharomyces cerevisiae*. (Data taken from Sickmann et al. 2003)

In yeast a total of 750 different proteins were identified which represent about 90% of all putative mitochondrial proteins (Sickmann et al. 2003). A classical mitochondrial presequence is found in 43% of the proteins only. Surprisingly, only 14% of the proteins within mitochondria are involved in energy metabolism including oxidative phosphorylation, the tricarboxylic acid cycle and pyruvate dehydrogenase (Fig. 1). Twice as many proteins (24.9%) act in maintaining and expressing the mitochondrial genome, which means that more than 180 proteins are required to express the eight major proteins encoded by the yeast mitochondrial DNA. Approximately 13% of the proteins function in transport of metabolites and proteins, including protein folding and turnover, and approximately 9% in amino acid-, lipid- and iron metabolism. The function of 25% of the proteins is still unknown.

Studies of the *Arabidopsis* mitochondrial proteome identified 416 proteins (Millar et al. 2001; Heazlewood et al. 2004). As in yeast, the function of about 20% of these proteins remains unknown and a mitochondrial localization was predicted for only half of the proteins by the various targeting prediction programs.

In recent years several phosphoproteins in plant mitochondria have been described, which are involved in the tricarboxylic acid cycle, respiratory complexes, defence against oxidative stress and heat shock (Struglics and Hakansson 1999; Struglics et al. 2000; Lund et al. 2001; Mooney et al. 2002; Bykova et al. 2003). In the *Arabidopsis* mitochondrial proteome ten protein kinases have been identified, which may function in phosphorylation signalling.

2.3 The PPR Family

Sequencing projects of plant nuclear genomes have revealed a huge novel protein family involved in organellar gene expression. This family of plant proteins contains tandem repeats of a 35 amino acid motif, called pentatricopeptide repeat (PPR) (Small and Peeters 2000). More than 450 members of this PPR family were identified in the *Arabidopsis* genome. The majority of these proteins seem to be targeted to mitochondria and comprise up to 15% of the predicted mitochondrial proteins in *Arabidopsis*. Small and Peeters (2000) suggested that PPR proteins bind RNA, thus facilitating RNA editing and RNA processing. The genes are expressed at very low levels, confirming their function in regulation of gene expression. Nevertheless some PPR proteins have been directly identified by two-dimensional gel electrophoresis and reverse-phase high performance liquid chromatography separation of the rice and *Arabidopsis* mitochondrial proteome (Heazlewood et al. 2003, 2004).

In an effort to identify the nuclear Rf restorer gene for cytoplasmic male-sterile (CMS) *Petunia* plants, an amplified fragment length polymorphism (AFLP) marker that cosegregated with the restorer was cloned and used to screen a *Petunia* artificial chromosome library. The *Petunia* Rf locus was identified and found to consist of duplicated genes containing 14 repeats of the PPR motif (Bentolila et al. 2002). It is assumed that the nuclear restorer modifies processing of the mRNA transcribed from the chimeric CMS gene.

Recently, the nuclear restorer genes of cytoplasmic male sterility in radish (Brown et al. 2003; Desloire et al. 2003; Koizuka et al. 2003) and rice (Kazama and Toriyama 2003) have also been identified. These restorer genes encode proteins with 16 or 18 PPR motifs, respectively.

PPR proteins are not restricted to higher plants. Similar repeats are found in a few mitochondrial proteins of *Saccharomyces cerevisiae*, *Neurospora crassa*, *Schizosaccharomyces pombe* (Small and Peeters 2000) and man (Mili and Pinol-Roma 2003). These proteins seem to function in RNA processing or translation and affect cytochrome c oxidase activity.

3 Import of Nucleic Acids into Mitochondria

Little is known about the in vivo mechanism of how nucleic acids are transported into mitochondria, but several in vitro systems have been developed for DNA and RNA import.

3.1 DNA Import

D'Souza et al. (2003) and Flierl et al.(2003) used the protein import machinery to deliver DNA into mitochondria. Mitochondriotropic cationic liposomes (DQAsomes) containing a plasmid DNA-signal peptide conjugate selectively release this conjugate when in contact with mitochondrial membranes, and the conjugate enters the mitochondria via the protein import machinery (D'Souza et al. 2003). Similarly, oligonucleotides annealed to peptide nucleic acids and coupled to mitochondrial targeting peptides could be introduced into mitochondria of mammalian cells (Flierl et al. 2003).

The protein import machinery is not required to import DNA into plant mitochondria. Koulintchenko et al. (2003) demonstrated the DNA import via a permeability transition pore complex which involves the voltage-dependent anion channel and the adenine nucleotide translocator. This process is restricted to double-stranded DNA and is most efficient with small linear DNA fragments. Imported sequences are transcribed within the organelle.

3.2 RNA Import

The mitochondrial genomes of various species contain an insufficient number of tRNA genes to translate mitochondrial mRNAs. Thus nuclear encoded tRNAs have to be imported into the organelle. Several papers describe the import of tRNAs into mitochondria of kinetoplastid protozoa, yeast, plants, marsupials and even man, in which tRNAs normally are not imported.

In *Leishmania tarentolae* and *Trypanosoma brucei* all mitochondrial tRNAs are encoded in the nucleus and most tRNAs function in both the cytosol and the mitochondria. Yermovsky-Kammerer and Hajduk (1999) transcribed the tRNA(Ser) and tRNA(Leu) genes in tandem with a 59-nt intergenic region in vitro. This tandem tRNA precursor was rapidly imported into isolated mitochondria of *Trypanosoma brucei*, while the mature-size tRNA(Leu) failed to be imported. Recently, the sequence motif YGG(C/A)RRC upstream of the genes encoding the mature tRNAs has been identified on the precursor tRNA that influences tRNA trafficking in trypanosomes (Sherrer et al. 2003). A membrane potential, ATP and a protein associated with the outer membrane were essential for the import of the tandem tRNA precursor, suggesting that the import into mitochondria is a receptor-mediated process. For *Leishmania*, it has been shown that mature, fully processed tRNAs are imported. To identify the sequences recognized by the RNA-specific import receptor, *Leishmania* mitochondria were incubated with a combinatorial RNA library (Bhattacharyya et al. 2002). The analysis of the imported sequences identified sequence motifs present in the anticodon arm, the D arm, the V-T region or the acceptor stem of tRNAs as import signals. These signals were grouped into two types: type I are A and D arm homologues, type II the V-T homologues. Subnanomolar concentrations of type I RNAs stimulated the entry of type II RNAs into the mitochondrial matrix, whereas type II RNAs inhibited inner membrane transfer of type I RNAs (Goswami et al. 2003). A 640-kDa multisubunit complex from the *Leishmania* inner mitochondrial membrane was isolated, which binds tRNAs and type I and type II RNAs (Bhattacharyya et al. 2003).

The structural differences between the cytosolic and the mitochondrial tRNAs for glutamate and glutamine from *Leishmania tarentolae* were investigated by mass spectrometry (Kaneko et al. 2003). A unique modification difference in both tRNAs was identified at the anticodon wobble position: the cytosolic tRNAs have 5-methoxycarbonylmethyl-2-thiouridine, whereas the mitochondrial tRNAs have 5-methoxycarbonylmethyl-2-*O*-methyluridine.

Experimental in vitro approaches to import tRNAs or 5S rRNA into yeast and human mitochondria for a possible gene therapy application are discussed by Kolesnikova et al. (2000) and Entelis et al. (2001a,b, 2002). The idea is to suppress pathological mutations in mitochondrial DNA. It has been shown that the import of tRNA(Lys) requires amino acylation, the precursor of the tRNA synthetase and the mitochondrial protein import machinery.

Import of a tRNA has been described in marsupials in vivo. Analysing the mitochondrial DNA of different marsupials revealed unusual features for the lysine tRNA genes (Dorner et al. 2001). For example, the anticodon was lacking in 8 species out of 11 analysed, and this feature was not corrected by RNA editing. Thus the mitochondrial lysyl tRNA genes must be pseudogenes. The demonstration of nuclear encoded tRNA(Lys) within marsupial mitochondria is evidence for tRNA import in vivo.

In higher plants up to one half of the tRNAs inside mitochondria are encoded in the nucleus and imported into the organelle. Transgenic *Arabidopsis thaliana* plants import tRNA(Ala) like wild-type plants, but a mutated tRNA(Ala) that cannot be aminoacylated by the alanyl-tRNA synthetase is not imported. These experiments suggest that aminoacyl-tRNA synthetases are implicated in this import process (Dietrich et al. 1996). The import selectivity depends in addition on the anticodon and the D-domain of the tRNA as shown by mutations in the respective sequences (Delage et al. 2003b). However, in vitro import experiments of tRNAs into isolated mitochondria from *Solanum tuberosum* have shown that the import proceeds in the absence of any added cytosolic protein fraction but requires a protein receptor on the surface of the organelle, ATP and a membrane potential (Delage et al. 2003a). In other words, in these in vitro import experiments, tRNA synthetases do not seem to be required. To explain this discrepancy, it has been speculated that tRNA synthetases may be involved in carrying the tRNAs to the surface of the mitochondrial outer membrane or to stabilize the tRNAs after entry into the organelle rather than to catalyze the import through the protein import channel. Further experiments will show whether in higher plants tRNAs use the protein import machinery to cross mitochondrial membranes.

4 Cytochrome c Maturation

4.1 Three Different Systems of Cytochrome c Maturation

C-type cytochromes are important components of electron transfer chains. Cytochrome c (Cyt c) maturation takes place in bacteria, and in mitochondria and chloroplasts of eukaryotic cells, and generally involves steps of heme and protein transport, in some cases proteolytic processing of an N-terminal pre-sequence, attachment of the heme cofactor to the apoprotein and correct folding of holo-cytochrome c (see Fig. 2). Covalent heme attachment occurs by thioether bonds to the cysteine residues of the heme binding motifs (CXXCH) that are present in c-type cytochromes. In bacteria, maturation takes place in the periplasm, thus heme and the apoprotein have to be translocated across the plasma membrane, in fact separately. The apoprotein is synthesized as precursor, translocated by a sec-dependent pathway, proteolytically processed, and finally heme attachment is exerted by a membrane-bound enzyme complex. Mitochondria harbour cytochrome c_1 in complex III (cytochrome c reductase) and cytochrome c between complex III and complex IV (cytochrome c oxidase). Both the cytochrome apoprotein and the heme are synthesized apart from their final localization and have to be translocated into the inner membrane or intermembrane space, where attachment of heme to the apoprotein occurs.

Although c-type cytochromes represent an evolutionary ancient type of electron transfer proteins that are present in all respiring organisms, their maturation pathway varies in diverse organisms. Three different systems of post-translational cytochrome c maturation have been described (Kranz et al. 1998; Page et al. 1998):

- System I is known to occur in plant (and protozoan) mitochondria, gram-negative bacteria such as *Escherichia coli* and diverse prokaryotes including *Pseudomonas* and *Rhizobium*.
- System II is found in plant chloroplasts and in gram-positive bacteria.

These two systems have common features and the involved genes (for example CCM genes for cytochrome c maturation) mainly encode conserved components of the complex cytochrome c maturation machinery.

- System III is characteristic for mitochondria of yeast, *Neurospora*, *Caenorhabditis* and Mammalia. It represents the most simple system and involves the intermembrane space enzyme cytochrome c heme lyase

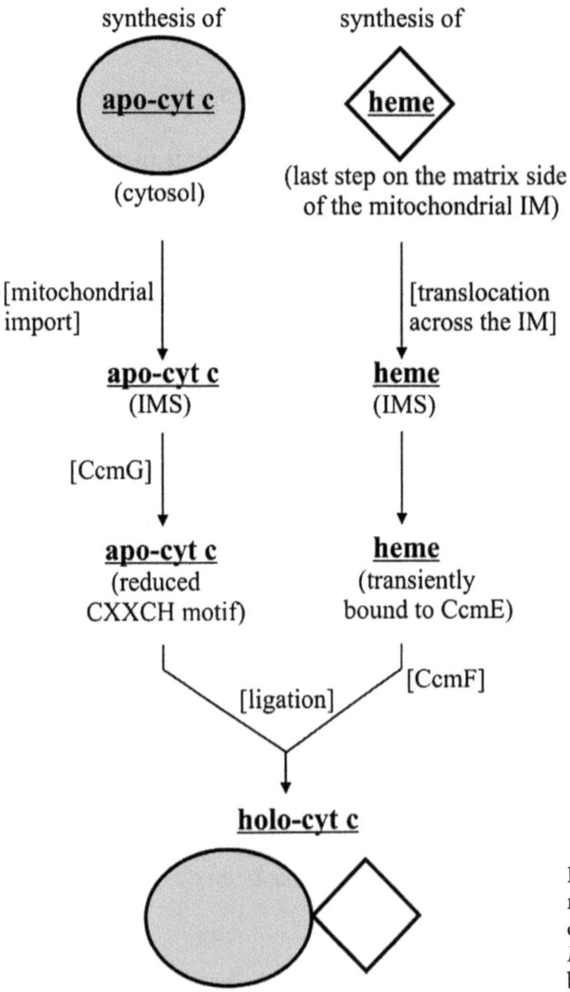

Fig. 2. Steps of cytochrome c maturation. *IM* Mitochondrial inner membrane; *IMS* mitochondrial intermembrane space

(CCHL). By heterologous expression in *E. coli* it was shown recently that this enzyme can also exert heme binding to the bacterial apocytochrome c in the cytoplasm (Sanders and Lill 2000), demonstrating similarity of the basic mechanism regardless of the different systems used.

4.2 Genes Involved in Cytochrome c Maturation in Plant Mitochondria

System I represents the most complex pathway analysed so far, and a number of factors appear conserved among the bacterial and the plant mitochondrial system. However, no homologues of the fungal cytochrome c heme lyases were found in plants.

In *E. coli* at least 12 genes are involved in Cyt c maturation: genes encoding components of an ABC-transporter complex (CcmA, CcmB), heme transfer proteins (CcmC, D, E, F), a thioredoxine system (CcmG, CcmH) to reduce the heme binding cysteine motif of the apoprotein, additional disulfide bond reducing components (DsbA, B, D) and the thioredoxine TrxA (Allen et al. 2003).

Plant nuclear genomes contain numerous homologues of the bacterial ccm genes; however, a precise assignment to the mitochondrial pathway (system I) is often difficult, because the plastids (system II) use Ccm proteins, too. Nevertheless, in recent studies, some Ccm proteins have been described actually acting at specific points of the mitochondrial pathway. An example is the *Arabidopsis thaliana* heme chaperone CcmE (Spielewoy et al. 2001). It has been shown that CcmE is an integral mitochondrial inner membrane (IM) protein with its C-terminal hydrophilic part protruding into the mitochondrial intermembrane space (IMS). It probably transiently binds heme similar to its *E. coli* homologue (Thöny-Meyer 2002) and interacts with CcmC (Ren and Thöny-Meyer 2001).

In higher plant mitochondria at least CcmA, CcmB, CcmC, CcmE, CcmF and possibly CcmH and others have been identified (Spielewoy et al. 2001). CcmH and possibly CcmG homologues might exert thioredoxin-associated functions in the Cyt c maturation system. Unlike in the bacterial system, it is not yet clear whether the apo-cytochrome disulfide bond must really be reduced in the mitochondrial IMS before heme can be attached. It has been discussed that this reduction, which is thought to be exerted by CcmG and/or CcmH, might not be necessary in the plant mitochondrial IMS (Allen et al. 2003), which could explain the lack of certain reducing factors in plant mitochondria. On the other hand, reducing conditions are thought to be required for cytochrome assembly in the mitochondrial IMS (Daltrop et al. 2002).

The ABC transporter CcmB and the heme transfer proteins CcmC and CcmF are encoded in plant mitochondrial genomes and in the mitochondrial genome of *Reclinomonas americana* (Lang et al. 1997; Unseld et al. 1997; Allen et al. 2003). In wheat, four mitochondrially encoded proteins are thought to be involved in cytochrome c maturation: using antibodies Faivre-Nitschke et al. (2001) investigated the localization of the proteins CcmB, an ABC-transporter of the mitochondrial IM, and CcmC, another integral IM protein which has been shown to be part of a high molecular weight complex by blue-native gel electrophoresis. Two other proteins, CcmFn and CcmFc, are homologous to the N- or C-terminal parts of the *E. coli* CcmF, respectively, which seems to be involved in the release of heme transiently bound to CcmE.

Additional Ccm homologues and other genes probably involved in mitochondrial cytochrome c maturation have been analysed in different plant species. Examples are TaCcmB of *Triticum aestivum* (Faivre-Nitschke et al. 2001), and the *Arabidopsis* genes AtNap10, a CcmA homologue (Rayapuram et al., unpubl.), AtTrx-o1, AtTrx-o2 encoding mitochondrial thioredoxins, and AtTra, an NADPH-dependent thioredoxin reductase (Laloi et al. 2001).

4.3 Conclusions and Perspectives

Recent studies on the maturation of c-type cytochromes have again demonstrated significant differences of plant mitochondria when compared with mammalian and fungal systems. The different systems might reflect distinct evolutionary origins: the two pathways I and II are of prokaryotic type and have been retained in plant mitochondria or chloroplasts, respectively. The third (fungal and animal) system is of yet unknown origin. All systems, however, execute the similar basic steps of cytochrome c maturation.

To date, system I is best characterized in *E. coli*. More detailed analysis will surely identify additional factors in plant mitochondria. Sequencing of plant genomes, identification and functional characterization of genes and heterologous studies will further increase our knowledge on composition and structure of the involved protein complexes.

5 Mitochondrial Fusion

5.1 Identification of Factors Involved in Mitochondrial Dynamics

In many species and cell types mitochondria rather build up a dynamic intracellular network than a number of separate organelles (Fig. 3; Egner et al. 2002; Jakobs et al. 2003). Mitochondrial movement, division and fusion are important processes required for correct distribution, number and size of mitochondria, maintenance of functional mitochondria and mitochondrial DNA, compensation of oxidative damages, developmental processes, e.g. apoptosis, intracellular signalling, or accumulation of mitochondria in cellular regions with enhanced requirement of mitochondrial activity. Therefore, these processes seem to be important for vegetative growth and adaption to physiological conditions and developmental challenges (Köhler et al. 1997; Jensen et al. 2000; Logan 2003; Westermann 2003).

A mitochondrial fusion must ensure function and the correct connection (Fritz et al. 2001) and integrity of the four mitochondrial compartments (outer membrane, intermembrane space, inner membrane and matrix space), and therefore seems to be a rather complicated process. From studies in yeast it has been estimated that every 2 min one fusion event per cell occurs (Nunnari et al. 1997). Various components involved in this process have been identified. Mutations of components of mitochondrial dynamics usually result in more or less severely affected mitochondrial morphology and often in loss of respiration, defective inheritance of mitochondrial DNA and developmental blocks.

The FZO ('fuzzy onions') gene probably represents the first identified component of the mitochondrial fusion machinery, which has originally been identified in *Drosophila melanogaster* (Hales and Fuller 1997). The designation 'fuzzy onions' describes the mitochondrial morphology in spermatids of mutants. Members of the FZO family (also termed mitofusins) are widespread among diverse organisms and are thought to represent

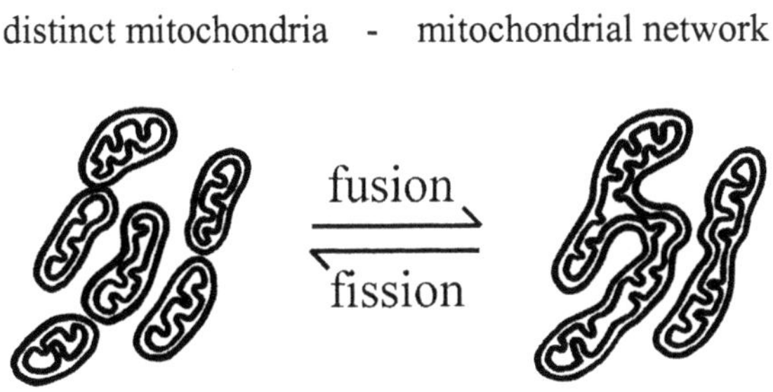

Fig. 3. Mitochondrial dynamics regulating distribution, size, shape and number of organelles

the main component of the mitochondrial fusion machinery in the outer membrane (Hermann et al. 1998). Mitochondrial fusion in plants, however, seems rather dependent on genes that are not related to the FZO family (see below). Most work on mitochondrial dynamics has been done in the yeast system. Systematic analyses of mutants lead to the identification of genes responsible for a changed mitochondrial morphology, resulting from defective distribution and division (see below).

The large number of involved proteins, identification of homologues in diverse organisms, and studies on functional characterization of some genes suggest a complex mechanism for regulation and modulation of mitochondrial dynamics.

5.2 Fzo Homologues: Key Components of the Mitochondrial Fusion Machinery

In yeast and animal systems, Fzo proteins have been identified to represent main components of the mitochondrial fusion machinery in the outer membrane (Hales and Fuller 1997; Rapaport et al. 1998; Fritz et al. 2001; Hwa et al. 2002). Experiments using different mitochondrial fluorescent markers in yeast have shown that the fusion of mitochondria after mating is blocked in conditional fzo1 mutants (Hermann et al. 1998). In metazoans two mitofusin proteins designated Mfn1 and Mfn2 have been found, which may exert similar functions (Legros et al. 2002; Rojo et al. 2002; Chen et al. 2003). Inactivation of each of these two genes in the mouse proved to be lethal in early embryo development and lead to fragmentation of mitochondria in cell cultures.

The mitofusins have a characteristic domain structure and a topology, with two membrane-spanning domains near the C terminus (Hermann et al. 1998; Fritz et al. 2001; Rojo et al. 2002). These transmembrane domains and a short loop separating them, which is rich in positively charged amino acid residues, seem to be important for targeting the protein to the outer membrane, for localization of the Fzo protein to contact sites of the outer and inner membrane, and possibly for interaction with inner membrane components (Fritz et al. 2001; Santel and Fuller 2001; Rojo et al. 2002). The main parts of Fzo, a GTPase domain near the N terminus and two or three coiled coil structures are exposed to the cytosol. Mutational analyses suggest that a functional GTPase domain exerts a regulatory or energy-providing function for mitochondrial fusion, while the coiled coil regions seem to be involved in mitochondrial interactions mediated by inter- and intramolecular events (Hales and Fuller 1997; Hermann et al. 1998; Santel and Fuller 2001; Rojo et al. 2002). Fzo1 is part of a 800-kDa complex in the outer membrane (Rapaport et al. 1998), consisting of yet unknown components. The complex is possibly localized at contact sites between the outer and inner membrane, suggesting a function in coordinated fusion of both membranes.

Mitochondrial fusion and division are controlled antagonistic processes. Ongoing mitochondrial division in fzo1 knock-out mutants results in fragmented mitochondria, a phenotype that can be functionally complemented by inactivation of the mitochondrial division factor Dnm1 (Sesaki and Jensen 1999).

5.3 Additional Components Involved in Mitochondrial Dynamics

The outer membrane protein Ugo1 (Sesaki and Jensen 2001; ugo is the Japanese word for fusion) is another component required for mitochondrial fusion. Its precise function is not yet clear; unlike Fzo1, it is not enriched in contact sites of the outer and inner mitochondrial membrane, but null mutants of ugo1 and fzo1 display similar phenotypes. As in fzo1 mutants, the mitochondrial fragmentation in ugo1 mutants can be restored by deletion of DNM1.

In yeast the MDM genes (mitochondrial distribution and morphology) represent a large group of possibly functionally related genes, and respective mutants are always associated with aberrant mitochondrial morphology (Dimmer et al. 2002). Mdm30 is described as regulating the level of the Fzo1 protein in the mitochondrial outer membrane (Fritz et al. 2003), and consequently mdm30 or fzo1 deletions result in similar phenotypes. The PCP1/MDM 37 gene (processing of cytochrome c peroxidase) encodes a rhomboid-type intramembrane protease involved in proteolytic processing of certain intermembrane space proteins (Esser et al. 2002; Herlan et al. 2003; McQuibban et al. 2003; Sesaki et al. 2003a). Rhomboid proteases represent a nearly ubiquitous family of serine proteases with functions in the release of protein signals, growth factors and other initially membrane tethered proteins (Urban et al. 2002; Koonin et al. 2003). The rhomboid substrate Mgm1 (mitochondrial genome maintenance) released from the inner membrane by proteolytic cleavage appears to be a key factor for remodelling and dynamics of mitochondrial membranes, and the Mgm1 homologue OPA1 in humans was identified as an important factor in dominant optical atrophy (DOA), a mitochondrial disease. Deletions of pcp1 and mgm1 have been described as displaying phenotypes similar to that of fzo1 mutants.

A large number of proteins involved in mitochondrial dynamics represent dynamin-like or similar mechanoenzymes. Mgm1 is involved in mitochondrial fusion (Sesaki et al. 2003b), while the two dynamin-related proteins Dnm1 and Drp1 (Bleazard et al. 1999) are important components of mitochondrial division/fission. The latter two are homologous to the GTPase protein FtsZ, a key component of bacterial division. Other factors, such as Mdv1/Fis2, Mdv2/Fis1, Gag3 and Net2, interact with the GTPase Dnm1 for mitochondrial fission (Fekkes et al. 2000; Mozdy et al. 2000; Tieu and Nunnari 2000; Cerveny et al. 2001; Tieu et al. 2002).

5.4 Components Identified in Higher Plant Mitochondrial Dynamics

As in other organisms, plant mitochondria display various types of dynamics, changing number, size, shape and subcellular localization of the organelle. These changes are often cell or tissue specific (Logan and Leaver 2000). However, morphology and distribution of plant mitochondria differ from yeast mitochondria. Only a few factors with a possible function in mitochondrial dynamics have been identified in the plant system, and homologues of FZO and some other genes are completely absent (Logan et al. 2003). Most data result from studies based on sequencing of plant genomes. As an example, a homologue of FIS1/MDV2 is present in the *Arabidopsis* genome (Logan 2003). Homologues of most MDM genes from yeast do not seem to occur in plants, with the exception of PCP1 and MDM 38, which are both evolutionary conserved in diverse organisms, and possibly MDM 33. At least one homologue of the conserved rhomboid-type protease PCP1 seems to occur in *Arabidopsis* and may be mitochondrially localized as predicted by computer analysis of the targeting sequence. The *Arabidopsis* Mdm38 homologue might be involved in mitochondrial calcium dynamics and morphology, and carries a mitochondrial targeting sequence. The third protein, Mdm33, shows weak homology to an *Arabidopsis* myosine homologue.

In general, dynamin-related proteins are involved in mitochondrial division and dynamics in various organisms, including plants (Danino and Hinshaw 2001). Recently, in plants the ADL2b gene (*Arabidopsis* dynamin-like protein) was identified, encoding a mitochondrial protein which is localized in the constriction sites where division occurs (Arimura and Tsutsumi 2002). A second factor termed Adl2a, initially localized to plastids, may be involved in mitochondrial morphology, too (Kang et al. 1998; Logan 2003). In addition, Jin et al. (2003) identified the *Arabidopsis* dynamin-like proteins ADL1C and ADL1E which are suggested to be involved in mitochondrial fission of plant cells.

Logan et al. (2003) analysed new ethyl methane sulphonate (EMS)-induced mutations affecting genes involved in *Arabidopsis* mitochondrial dynamics. Five different loci were identified and termed according to their mitochondrial phenotypes: FMT (friendly mitochondria), MMT1, MMT2 (motley mitochondrial mutant), BMT1 (big mitochondrial mutant) and NMT (network mitochondrial mutant). In the case of FMT the gene was identified and shown to represent a homologue of *Dictyostelium discoideum* CluA and *S. cerevisiae* Clu1. A function in preventing unwanted associations between microtubules and the mitochondrial outer membrane was discussed.

5.5 Conclusions and Perspectives

Generally, mitochondrial dynamics is a process that involves key factors providing or regulating dynamin-like functions. These functions directly influence mitochondrial morphology, and it seems obvious that movements, fusion, fission, inter- and intramitochondrial membrane contacts are regulated by interacting proteins or complexes.

Mitochondrial fusion is executed by a specialized and evolutionary conserved machinery, which is distinct from fusion machineries of the other known cellular structures, and the Fzo proteins exert key functions in a high molecular weight outer membrane protein complex. Studies on interacting factors, mutational analyses, large-scale functional analyses and the important new fluorescent and microscopic imaging methods might provide new insights into the principal mechanisms of these processes. Ongoing research on the respective proteins in metazoan cells and various tissues will be of value for studying expression, distribution, developmental aspects, e.g. apoptotic functions, and mitochondrial diseases.

The unexpected lack of conserved Fzo homologues and diverse other important factors in the mitochondria of higher plants once more demonstrate significant differences compared to animal and fungal systems.

References

Adams KL, Palmer JD (2003) Evolution of mitochondrial gene content: gene loss and transfer to the nucleus. Mol Phylogenet Evol 29:380–395

Adams KL, Song K, Roessler PG, Nugent JM, Doyle JL, Doyle JJ, Palmer JD (1999) Intracellular gene transfer in action: dual transcription and multiple silencings of nuclear and mitochondrial cox2 genes in legumes. Proc Natl Acad Sci USA 96:13863–13868

Allen JW, Barker PD, Ferguson SJ (2003) A cytochrome b562 variant with a c-type cytochrome CXXCH heme-binding motif as a probe of the *Escherichia coli* cytochrome c maturation system. J Biol Chem 278:52075–52083

Arimura S, Tsutsumi N (2002) A dynamin-like protein (ADL2b), rather than FtsZ, is involved in *Arabidopsis* mitochondrial division. Proc Natl Acad Sci USA 99:5727–5731

Bentolila S, Alfonso AA, Hanson MR (2002) A pentatricopeptide repeat-containing gene restores fertility to cytoplasmic male-sterile plants. Proc Natl Acad Sci USA 99:10887–10892
Bentolila S, Alfonso AA, Hanson MR (2002) A pentatricopeptide repeat-containing gene restores fertility to cytoplasmic male-sterile plants. Proc Natl Acad Sci USA 99:10887–10892
Bergthorsson U, Adams KL, Thomason B, Palmer JD (2003) Widespread horizontal transfer of mitochondrial genes in flowering plants. Nature 424:197–201

Berry S (2003) Endosymbiosis and the design of eukaryotic electron transport. Biochim Biophys Acta 1606:57–72

Bhattacharyya SN, Chatterjee S, Adhya S (2002) Mitochondrial RNA import in *Leishmania tropica*: aptamers homologous to multiple tRNA domains that interact cooperatively or antagonistically at the inner membrane. Mol Cell Biol 22:4372–4382

Bhattacharyya SN, Chatterjee S, Goswami S, Tripathi G, Dey SN, Adhya S (2003) 'Ping-pong' interactions between mitochondrial tRNA import receptors within a multiprotein complex. Mol Cell Biol 23:5217–5224

Binder S, Brennicke A (2003) Gene expression in plant mitochondria: transcriptional and post-transcriptional control. Philos Trans R Soc Lond B Bio Sci 358:181–188

Bleazard W, McCaffery JM, King EJ, Bale S, Mozdy A, Tieu Q, Nunnari J, Shaw JM (1999) The dynamin-related GTPase Dnm1 regulates mitochondrial fission in yeast. Nat Cell Biol 1:298–304

Brown GG, Formanova N, Jin H, Wargachuk R, Dendy C, Patil P, Laforest M, Zhang J, Cheung WY, Landry BS (2003) The radish Rfo restorer gene of *Ogura* cytoplasmic male sterility encodes a protein with multiple pentatricopeptide repeats. Plant J 35:262–272

Burger G, Gray MW, Franz Lang B (2003) Mitochondrial genomes: anything goes. Trends Genet 19:709–716

Bykova NV, Egsgaard H, Moller IM (2003) Identification of 14 new phosphoproteins involved in important plant mitochondrial processes. FEBS Lett 540:141–146

Cerveny KL, McCaffery JM, Jensen RE (2001) Division of mitochondria requires a novel DMN1-interacting protein, Net2p. Mol Biol Cell 12:309–321

Chen H, Detmer SA, Ewald AJ, Griffin EE, Fraser SE, Chan DC (2003) Mitofusins Mfn1 and Mfn2 coordinately regulate mitochondrial fusion and are essential for embryonic development. J Cell Biol 160:189–200

Chinnery PF (2003) Searching for nuclear-mitochondrial genes. Trends Genet 19:60–62

Covello PS, Gray MW (1992) Silent mitochondrial and active nuclear genes for subunit 2 of cytochrome c oxidase (cox2) in soybean: evidence for RNA-mediated gene transfer. EMBO J 11:3815–3820

Daley DO, Adams KL, Clifton R, Qualmann S, Millar AH, Palmer JD, Pratje E, Whelan J (2002a) Gene transfer from mitochondrion to nucleus: novel mechanisms for gene activation from Cox2. Plant J 30:11–21

Daley DO, Clifton R, Whelan J (2002b) Intracellular gene transfer: reduced hydrophobicity facilitates gene transfer for subunit 2 of cytochrome c oxidase. Proc Natl Acad Sci USA 99:10510–10515

Daltrop O, Stevens JM, Higham CW, Ferguson SJ (2002) The CcmE protein of the c-type cytochrome biogenesis system: unusual in vitro heme incorporation into apo-CcmE and transfer from holo-CcmE to apocytochrome. Proc Natl Acad Sci USA 99:9703–9708

Danino D, Hinshaw JE (2001) Dynamin family of mechanoenzymes. Curr Opin Cell Biol 13:454–460

Delage L, Dietrich A, Cosset A, Marechal-Drouard L (2003a) In vitro import of a nuclearly encoded tRNA into mitochondria of *Solanum tuberosum*. Mol Cell Biol 23:4000–4012

Delage L, Duchene AM, Zaepfel M, Marechal-Drouard L (2003b) The anticodon and the D-domain sequences are essential determinants for plant cytosolic tRNA(Val) import into mitochondria. Plant J 34:623–633

Desloire S, Gherbi H, Laloui W, Marhadour S, Clouet V, Cattolico L, Falentin C, Giancola S, Renard M, Budar F, Small I, Caboche M, Delourme R, Bendahmane A (2003) Identification of the fertility restoration locus, Rfo, in radish, as a member of the pentatricopeptide-repeat protein family. EMBO Rep 4:588–594

Dietrich A, Marechal-Drouard L, Carneiro V, Cosser A, Small I. (1996) A single base change prevents import of cytosolic tRNA(Ala) into mitochondria in transgenic plants. Plant J 10:913–918

Dimmer KS, Fritz S, Fuchs F, Messerschmitt M, Weinbach N, Neupert W, Westermann B (2002) Genetic basis of mitochondrial function and morphology in *Saccharomyces cerevisiae*. Mol Biol Cell 13:847–853

Dorner M, Altmann M, Paabo S, Morl M (2001) Evidence for import of a lysyl-tRNA into marsupial mitochondria. Mol Biol Cell 12:2688–2698

D'Souza GG, Rammohan R, Cheng SM, Torchilin VP, Weissig V (2003) DQAsome-mediated delivery of plasmid DNA toward mitochondria in living cells. J Control Release 92:189–197

Egner A, Jakobs S, Hell SW (2002) Fast 100-nm resolution three-dimensional microscope reveals structural plasticity of mitochondria in live yeast. Proc Natl Acad Sci USA 99:3370–3375

Emelyanov VV (2003) Mitochondrial connection to the origin of the eukaryotic cell. Eur J Biochem 270:1599–1618

Entelis NS, Kolesnikova OA, Dogan S, Martin RP, Tarassov IA (2001a) 5 S rRNA and tRNA import into human mitochondria. Comparison of in vitro requirements. J Biol Chem 276:45642–45653

Entelis NS, Kolesnikova OA, Martin RP, Tarassov IA (2001b) RNA delivery into mitochondria. Adv Drug Deliv Rev 49:199–215

Entelis N, Kolesnikova O, Kazakova H, Brandina I, Kamenski P, Martin RP, Tarassov I (2002) Import of nuclear encoded RNAs into yeast and human mitochondria: experimental approaches and possible biomedical applications. Genet Eng (NY) 24:191–213

Esser K, Tursun B, Ingenhoven M, Michaelis G, Pratje E (2002) A novel two-step mechanism for removal of a mitochondrial signal sequence involves the mAAA complex and the putative rhomboid protease Pcp1. J Mol Biol 323:835–843

Faivre-Nitschke SE, Nazoa P, Gualberto JM, Grienenberger JM, Bonnard G (2001) Wheat mitochondria ccmB encodes the membrane domain of a putative ABC transporter involved in cytochrome c biogenesis. Biochim Biophys Acta 1519:199–208

Fekkes P, Shepard KA, Yaffe MP (2000) Gag3p, an outer membrane protein required for fission of mitochondrial tubules. J Cell Biol 151:333–340

Flierl A, Jackson C, Cottrell B, Murdock D, Seibel P, Wallace DC (2003) Targeted delivery of DNA to the mitochondrial compartment via import sequence-conjugated peptide nucleic acid. Mol Ther 7:550–557

Fritz S, Rapaport D, Klanner E, Neupert W, Westermann B (2001) Connection of the mitochondrial outer and inner membranes by Fzo1 is critical for organellar fusion. J Cell Biol 152:683–692

Fritz S, Weinbach N, Westermann B (2003) Mdm30 is an F-box protein required for maintenance of fusion-competent mitochondria in yeast. Mol Biol Cell 14:2303–2313

Faivre-Nitschke SE, Nazoa P, Gualberto JM, Grienenberger JM, Bonnard G (2001) Wheat mitochondria ccmB encodes the membrane domain of a putative ABC transporter involved in cytochrome c biogenesis. Biochim Biophys Acta 1519:199–208

Funes S, Davidson E, Reyes-Prieto A, Magallon S, Herion P, King MP, Gonzalez-Halphen D (2002) A green algal apicoplast ancestor. Science 298:2155

Gardner MJ, Hall N, Fung E, White O, Berriman M, Hyman RW, Carlton JM, Pain A, Nelson KE, Bowman S, Paulsen IT, James K, Eisen JA, Rutherford K, Salzberg SL, Craig A, Kyes S, Chan MS, Nene V, Shallom SJ, Suh B, Peterson J, Angiuoli S, Pertea M, Allen J, Selengut J, Haft D, Mather MW, Vaidya AB, Martin DM, Fairlamb AH, Fraunholz MJ, Roos DS, Ralph SA, McFadden GI, Cummings LM, Subramanian GM, Mungall C, Venter JC,

Carucci DJ, Hoffman SL, Newbold C, Davis RW, Fraser CM, Barrell B (2002) Genome sequence of the human malaria parasite *Plasmodium falciparum*. Nature 419:498–511

Goswami S, Chatterjee S, Bhattacharyya SN, Basu S, Adhya S (2003) Allosteric regulation of tRNA import: interactions between tRNA domains at the inner membrane of *Leishmania* mitochondria. Nucleic Acids Res 31:5552–5559

Hales KG, Fuller MT (1997) Developmentally regulated mitochondrial fusion mediated by a conserved, novel, predicted GTPase. Cell 90:121–129

Heazlewood JL, Howell KA, Whelan J, Millar AH (2003) Towards an analysis of the rice mitochondrial proteome. Plant Physiol 132:230–242

Heazlewood JL, Tonti-Filippini JS, Gout AM, Day DA, Whelan J, Millar AH (2004) Experimental analysis of the *Arabidopsis* mitochondrial proteome highlights signaling and regulatory components, provides assessment of targeting prediction programs, and indicates plant-specific mitochondrial proteins. Plant Cell 16:241–256

Herlan M, Vogel F, Bornhövd C, Neupert W, Reichert AS (2003) Processing of Mgm1 by the rhomboid-type protease Pcp1 is required for maintenance of mitochondrial morphology and of mitochondrial DNA. J Biol Chem 278:27781–27788

Hermann GJ, Thatcher JW, Mills JP, Hales KG, Fuller MT, Nunnari J, Shaw JM (1998) Mitochondrial fusion in yeast requires the transmembrane GTPase Fzo1p. J Cell Biol 143:359–373

Hoeberichts FA, Woltering EJ (2003) Multiple mediators of plant programmed cell death: interplay of conserved cell death mechanisms and plant-specific regulators. Bioessay S 25:47–57

Huettenbrenner S, Maier S, Leisser C, Polgar D, Strasser S, Grusch, Krupitza G (2003) The evolution of cell death programs as prerequisites of multicellularity. Mutat Res 543:235–249

Hwa JJ, Hiller MA, Fuller MT, Santel A (2002) Differential expression of the *Drosophila* mitofusin genes fuzzy onions (fzo) and dmfn. Mech Dev 116:213–216

Jakobs S, Martini N, Schauss AC, Egner A, Westermann B, Hell SW (2003) Spatial and temporal dynamics of budding yeast mitochondria lacking the division component Fis1p. J Cell Sci 116:2005–2014

Jameson D, Gibson AP, Hudelot C, Higgs PG (2003) OGRe: a relational database for comparative analysis of mitochondrial genomes. Nucleic Acids Res 31:202–206

Jensen RE, Hobbs AE, Cerveny KL, Sesaki H (2000) Yeast mitochondrial dynamics: fusion, division, segregation, and shape. Microsci Res Tech 51:573–583

Jin JB, Bae H, Kim SJ, Jin YH, Goh C-H, Kim DH, Lee YJ. Tse YC, Jiang L, Hwang I (2003) The *Arabidopsis* dynamin-like proteins ADL1C and ADL1E play a critical role in mitochondrial morphogenesis. Plant Cell 15:2357–2369

Kaneko T, Suzuki T, Kapushoc ST, Rubio MA, Ghazvini J, Watanabe K, Simpson L, Suzuki T (2003) Wobble modification differences and subcellular localization of tRNAs in *Leishmania* tarentolae: implication for tRNA sorting mechanism. EMBO J 22:657–667

Kang SG, Jin JB, Piao HL, Pih KT, Jang HJ, Lim JH, Hwang I (1998) Molecular cloning of an *Arabidopsis* cDNA encoding a dynamin-like protein that is localized to plastids. Plant Mol Biol 38:437–447

Karlberg EO, Andersson SG (2003) Mitochondrial gene history and mRNA localization: is there a correlation? Nat Rev Genet 4:391–397

Kazama T, Toriyama K (2003) A pentatricopeptide repeat-containing gene that promotes the processing of aberrant atp6 RNA of cytoplasmic male-sterile rice. FEBS Lett 544:99–102

Köhler RH, Zipfel WR, Webb WW, Hanson MR (1997) The green fluorescent protein as a marker to visualize plant mitochondria in vivo. Plant J 11:613–621

Koizuka N, Imai R, Fujimoto H, Hayakawa T, Kimura Y, Kohno-Murase J, Sakai T, Kawasaki S, Imamura J (2003) Genetic characterization of a pentatricopeptide repeat protein gene, orf687, that restores fertility in the cytoplasmic male-sterile *Kosena* radish. Plant J 34:407–415

Koonin EV, Makarova KS, Rogozin IB, Davidovic L, Letellier MC, Pellegrini L (2003) The rhomboids: a nearly ubiquitous family of intramembrane serine proteases that probably evolved by multiple ancient horizontal gene transfers. Genome Biol 4:R19

Kolesnikova OA, Entelis NS, Mireau H, Fox TD, Martin RP, Tarassov IA (2000) Suppression of mutations in mitochondrial DNA by tRNAs imported from the cytoplasm. Science 289:1931–1933

Kranz R, Lill R, Goldman B, Bonnard G, Merchant S (1998) Molecular mechanisms of cytochrome c biogenesis: three distinct systems. Mol Microbiol 29:383–396

Koulintchenko M, Konstantinov Y, Dietrich A (2003) Plant mitochondria actively import DNA via the permeability transition pore complex. EMBO J 22:1245–1254

Laloi C, Rayapuram N, Chartier Y, Grienenberger JM, Bonnard G, Meyer Y (2001) Identification and characterization of a mitochondrial thioredoxin system in plants. Proc Natl Acad Sci USA 98:14144–14149

Lang BF, Burger G, O'Kelly CJ, Cedergren R, Golding GB, Lemieux C, Sankoff D, Turmel M, Gray MW (1997) An ancestral mitochondrial DNA resembling a eubacterial genome in miniature. Nature 387:493–497

Legros F, Lombes A, Frachon P, Rojo M (2002) Mitochondrial fusion in human cells is efficient, requires the inner membrane potential, and is mediated by mitofusins. Mol Biol Cell 13:4343–4354

Logan D (2003) Mitochondrial dynamics. New Phytologist 160:463–478

Logan DC, Leaver CJ (2000) Mitochondria-targeted GFP highlights the heterogeneity of mitochondrial shape, size and movement within living plant cells. J Exp Bot 51:865–871

Logan DC, Scott I, Tobin KA (2003) The genetic control of plant mitochondrial morphology and dynamics. Plant J 36:500–509

Lund AA, Rhoads DM, Lund AL, Cerny RL, Elthon TE (2001) In vivo modifications of the maize mitochondrial small heat stress protein, HSP22. J Biol Chem 276:29924–29929

Mandavilli BS, Santos JH, van Hauten B (2002) Mitochondrial DNA repair and aging. Mutat Res 509:127–151

McQuibban GA, Saurya S, Freeman M (2003) Mitochondrial membrane remodelling regulated by a conserved rhomboid protease. Nature 423:537–541

Mili S, Pinol-Roma S (2003) LRP130, a pentatricopeptide motif protein with a noncanonical RNA-binding domain, is bound in vivo to mitochondrial and nuclear RNAs. Mol Cell Biol 23:4972–4982

Millar AH, Sweetlove LJ, Giegé P, Leaver CJ (2001) Analysis of the *Arabidopsis* mitochondrial proteome. Plant Physiol 127:1711–1727

Miyagishima SY, Nishida K, Kuroiwa T (2003) An evolutionary puzzle, chloroplast and mitochondrial division rings. Trends Plant Sci 8:432–438

Mooney BP, Miernyk JA, Randall DD (2002) The complex fate of alpha-ketoacids. Annu Rev Plant Biol 53:357–375

Mozdy AD, Shaw JM (2003) A fuzzy mitochondrial fusion apparatus comes into focus. Nat Rev Mol Cell Biol 4:468–478

Mozdy AD, McCaffery JM, Shaw JM (2000) Dnm1p GTPase-mediated mitochondrial fission is a multi-step process requiring the novel integral membrane component Fis1p. J Cell Biol 151:367–380

Newmeyer DD, Ferguson-Miller S (2003) Mitochondria: releasing power for life and unleashing the machineries of death. Cell 112:481–490

Nieminen AL (2003) Apoptosis and necrosis in health and disease: role of mitochondria. Int Rev Cytol 224:29–55

Nunnari J, Marshall WF, Straight A, Murray A, Sedat JW, Walter P (1997) Mitochondrial transmission during mating in *Saccharomyces cerevisiae* is determined by mitochondrial fusion and fission and the intramitochondrial segregation of mitochondrial DNA. Mol Biol Cell 8:1233–1242

O'Brien EA, Badidi E, Barbasiewicz A, deSousa C, Lang BF, Burger G (2003) GOBASE – a database of mitochondrial and chloroplast information. Nucleic Acids Res 31:176–178

Osteryoung KW, Nunnari J (2003) The division of endosymbiotic organelles. Science 302:1698–1704
Page MD, Sambongi Y, Ferguson SJ (1998) Contrasting routes of c-type cytochrome assembly in mitochondria, chloroplasts and bacteria. Trends Biochem Sci 23:103–108
Perez-Martinez X, Antaramian A, Vazquez-Acevedo M, Funes S, Tolkunova E, d'Alayer J, Claros MG, Davidson E, King MP, Gonzalez-Halphen D (2001) Subunit II of cytochrome c oxidase in *Chlamydomonad* algae is a heterodimer encoded by two independent nuclear genes. J Biol Chem 276:11302–11309
Priault M, Camougrand N, Kinnally KW, Vallette FM, Manon S (2003) Yeast as a tool to study Bax/mitochondrial interactions in cell death. FEMS Yeast Res 4:15–27
Qualmann SR, Daley DO, Whelan J, Pratje E (2003) Import pathway of nuclear encoded cytochrome *c* oxidase subunit 2 using yeast as a model. Plant Biol 5:481–490
Rapaport D, Brunner M, Neupert W, Westermann B (1998) Fzo1p is a mitochondrial outer membrane protein essential for the biogenesis of functional mitochondria in *Saccharomyces cerevisiae*. J Biol Chem 273:20150–20155
Rehling P, Pfanner N, Meisinger C (2003) Insertion of hydrophobic membrane proteins into the inner mitochondrial membrane – a guided tour. J Mol Biol 326:639–657
Ren Q, Thöny-Meyer L (2001) Physical interaction of CcmC with heme and the heme chaperone CcmE during cytochrome c maturation. J Biol Chem 276:32591–32596
Richter OM, Ludwig B (2003) Cytochrome c oxidase – structure, function, and physiology of a redox-driven molecular machine. Rev Physiol Biochem Pharmacol 147:47–74
Rojo M, Legros F, Chateau D, Lombes A (2002) Membrane topology and mitochondrial targeting of mitofusins, ubiquitous mammalian homologs of the transmembrane GTPase Fzo. J Cell Sci 115:1663–1674
Sanders C, Lill H (2000) Expression of prokaryotic and eukaryotic cytochromes c in *Escherichia coli*. Biochim Biophys Acta 1459:131–138
Santel A, Fuller MT (2001) Control of mitochondrial morphology by a human mitofusin. J Cell Sci 114:867–874
Schagger H (2002) Respiratory chain supercomplexes of mitochondria and bacteria. Biochim Biophys Acta 1555:154–159
Searcy DG (2003) Metabolic integration during the evolutionary origin of mitochondria. Cell Res 13:229–238
Sesaki H, Jensen RE (1999) Division versus fusion: Dnm1p and Fzo1p antagonistically regulate mitochondrial shape. J Cell Biol 147:699–706
Sesaki H, Jensen RE (2001) UGO1 encodes an outer membrane protein required for mitochondrial fusion. J Cell Biol 152:1123–1134
Sesaki H, Southard SM, Hobbs AE, Jensen RE (2003a) Cells lacking Pcp1p/Ugo2p, a rhomboid-like protease required for Mgm1p processing, lose mtDNA and mitochondrial structure in a Dnm1p-dependent manner, but remain competent for mitochondrial fusion. Biochem Biophys Res Commun 308:276–283
Sesaki H, Southard SM, Yaffe MP, Jensen RE (2003b) Mgm1p, a dynamin-related GTPase, is essential for fusion of the mitochondrial outer membrane. Mol Biol Cell 14:2342–2356
Sherrer RL, Yermovsky-Kammerer AE, Hajduk SL (2003) A sequence motif within trypanosome precursor tRNAs influences abundance and mitochondrial localization. Mol Cell Biol 23:9061–9072
Sickmann A, Reinders J, Wagner Y, Joppich C, Zahedi R, Meyer HE, Schönfisch B, Perschil I, Chacinska A, Guiard B, Rehling P, Pfanner N, Meisinger C (2003) The proteome of *Saccharomyces cerevisiae* mitochondria. Proc Natl Acad Sci USA 100:13207–13212
Small ID, Peeters N (2000) The PPR motif – a TPR-related motif prevalent in plant organellar proteins. Trends Biochem Sci 25:46–47
Spielewoy N, Schulz H, Grienenberger JM, Thöny-Meyer L, Bonnard G (2001) CCME, a nuclear-encoded heme-binding protein involved in cytochrome c maturation in plant mitochondria. J Biol Chem 276:5491–5497

Struglics A, Hakansson G (1999) Purification of a serine and histidine phosphorylated mitochondrial nucleoside diphosphate kinase from *Pisum sativum*. Eur J Biochem 262:765196773

Struglics A, Fredlund KM, Konstantinov YM, Allen JF, Moller IM (2000) Protein phosphorylation/dephosphorylation in the inner membrane of potato tuber mitochondria. FEBS Lett 475:213–217

Taylor SW, Fahy E, Zhang B, Glenn GM, Warnock DE, Wiley S, Murphy AN, Gaucher SP, Capaldi RA, Gibson BW, Ghosh SS (2003) Characterization of the human heart mitochondrial proteome. Nat Biotechnol 21:281–286

Thöny-Meyer L (2002) Cytochrome c maturation: a complex pathway for a simple task? Biochem Soc Trans 30:633–638

Tieu Q, Nunnari J (2000) Mdv1p is a WD repeat protein that interacts with the dynamin-related GTPase, Dnm1p, to trigger mitochondrial division. J Cell Biol 151:353–365

Tieu Q, Okreglak V, Naylor K, Nunnari J (2002) The WD repeat protein, Mdv1p, functions as a molecular adaptor by interacting with Dnm1p and Fis1p during mitochondrial fission. J Cell Biol 158:445–452

Truscott KN, Brandner K, Pfanner N (2003) Mechanisms of protein import into mitochondria. Curr Biol 13:R326–R337

Unseld M, Marienfeld JR, Brandt P, Brennicke A (1997) The mitochondrial genome of *Arabidopsis thaliana* contains 57 genes in 366,924 nucleotides. Nat Genet 15:57–61

Urban S, Schlieper D, Freeman M (2002) Conservation of intramembrane proteolytic activity and substrate specificity in prokaryotic and eukaryotic rhomboids. Curr Biol 12:1507–1512

Ward BL, Anderson RS, Bendich AJ (1981) The mitochondrial genome is large and variable in a family of plants (Cucurbitaceae). Cell 25:793–803

Westermann B (2003) Mitochondrial membrane fusion. Biochim Biophys Acta 1641:195–202

Won H, Renner SS (2003) Horizontal gene transfer from flowering plants to *Gnetum*. Proc Natl Acad Sci USA 100:10824–10829

Yermovsky-Kammerer AE, Hajduk SL (1999) In vitro import of a nuclearly encoded tRNA into the mitochondrion of *Trypanosoma brucei*. Mol Cell Biol 19:6253–6259

Karlheinz Esser
Georg Michaelis
Botanisches Institut der Universität Düsseldorf
Universitätsstrasse 1
40225 Düsseldorf, Germany

Elke Pratje
Institut für Allgemeine Botanik der Universität Hamburg
Ohnhorststrasse 18
22609 Hamburg, Germany
Tel.: +49-40-42816288
Fax: +49-40-42816254
e-mail: pratje@botanik.uni-hamburg.de

Population Genetics: Biodiversity in Anthropogenic Landscapes – Population Genetics and Ecological Modelling*

Jürgen Tomiuk, Inga A. Roedenbeck, and Wolfgang Köhler

1 Introduction

Biodiversity can hierarchically be structured on a grand scale into species, populations, individuals and genes. The definitions of these categories, however, are far from unambiguous (Mayr 1988). For example, the species' concept can depend on the reproduction mode, i.e. either sexual or asexual, and in some cases criteria to discriminate closely related species are genetically rather vague, leading to classification as sister species or subspecies. Furthermore, in many species, geographically distinct populations that can be discriminated through the phenotypic habit of their individuals are called races or varieties, although this classification is based on only a minor fraction of often unknown genes (Bodmer and Cavalli-Sforza 1976). The sometimes weak boundaries among evolutionary units may imply that biological systems of interacting elements are evolutionarily rather dynamic than static, and our deep understanding of the dynamic processes governing individuals, populations and species in communities is elemental also in order to assess the consequences of human interference on natural systems (Tomiuk and Bachmann 2002). Natural communities are temporarily and spatially open systems and, therefore, predictions about effects of human disturbances are hard to make. The introduction or extinction of species may, but not necessarily, alter community structure. On the one hand, habitat fragmentation can isolate local populations supporting speciation processes, hence increasing species diversity, whereas, on the other hand, fragmentation in most cases increases the risk of extinction for populations through the reduction of population size. Conservation biology, therefore, attempts to identify minimum population sizes and genetic factors that are fundamental to maintaining viable populations (Soulé 1986). Finally, theoretical models combine ecological elements of a

* This paper is dedicated to Prof. Dr. Hartwig H. Geiger on the occasion of his 65th birthday

Progress in Botany, Vol. 66

species' community, e.g. key species, the genetics of species and even financial interests, in order to find optimum conditions that temporarily stabilize ecosystems (Witting et al. 2000).

Natural populations of many species are often genetically adapted to ecological conditions that they have experienced at specific geographic locations. Interestingly, many cultivated plants, especially self-pollinators, have differentiated into local races more rapidly than wild plants (Allard 1999). Thus, when analysing changes in our natural landscape we must also concentrate on the effects caused by human activities that actively or passively change the mobility of endemic species – but also the mobility of domesticated animals and cultivated plants which underlie the same natural laws as wild species. Genetic analyses of such dynamic processes acting in ecological systems need reliable and easily applicable tools. In this respect the use of simple 'genetic markers' has been very instructive in studies of local differentiation. As early as 1960, Schwartz performed genetic studies of enzymes in maize using electrophoresis, and 6 years later protein variability was considered in population genetic studies of humans and *Drosophila* (Harris 1966; Hubby and Lewontin 1966). Various, more advanced molecular marker systems are available today which can be applied pinpointedly in order to solve specific questions.

Recently, Allard (1999) has reviewed plant population genetics with a focus on the genetic foundations of the processes that have led to improvements in cultivated plants since the earliest domestications. Here, we consider methods and population genetic approaches that are used in population studies of genetic differentiation processes, with special focus on the effects resulting from human activities. Population genetics provides a theoretical basis to study and analyse random as well as directional processes, e.g. selection, random genetic drift, inbreeding, gene flow among populations, population differentiation and speciation; all of these processes may determine the fate of populations in an anthropogenic environment. In the two following sections, genetic marker systems that are most commonly used in population genetic studies are briefly introduced. First, we consider the characteristics of these genetic markers and present some statistical tools that are often used to study genetic variation within and between populations. Subsequently, the focus is on temporal and spatial effects that can determine the genetic structure within populations or species (architecture of populations) in order to understand anthropogenic-induced changes in a natural landscape. Finally, we review some theoretical approaches for finding more general principles and key factors that explain the evolution of species communities.

2 Some Characteristics of Genetic Markers

An important aspect in the choice of a genetic marker system is to optimize the conclusive power of a study. In this context, Powell et al. (1996) proposed the concept of the 'marker index' as a measure of statistical power. The degree of resolution of marker systems is decisive, e.g. identification of the relatedness of individuals needs high resolution, whereas phylogenetic studies require resolution at a lower level. The degree of resolution revealed by a marker is governed by the mode of inheritance and its mutation rate. The latter needs careful attention if gene identity by descent is the focus of studies, i.e. high genetic variability is a prerequisite of high resolution, but it is created by frequent mutation events that can obscure relatedness of individuals.

Some reviews discuss the applications of molecular markers in ecology, systematics, breeding, agriculture and conservation biology. Jones et al. (1997) review their use of genetic marker systems in genome mapping. The suitability of various electrophoretic markers for studies in entomology is discussed by Loxdale and Lushai (1998). They also compare the reliability of techniques, cost effectiveness and statistical analyses utilized. Gupta et al. (1999) focus on applications of molecular markers in wheat breeding, and O'Hanlon et al. (2000) on their utility to weed ecology. Recently, O'Neill et al. (2003) discussed the application and reliability of various marker systems for the analysis of genetic diversity within and between populations. They describe the pros and cons of studies using morphological traits and molecular markers such as single locus approaches [e.g., allozyme variation, restriction fragment length polymorphism (RFLP), simple sequence repeats (SSR), single nucleotide polymorphisms (SNP)] and multilocus fingerprints [DNA fingerprints based on RFLP technique, random amplified polymorphic DNA (RAPD), amplified fragment length polymorphism (AFLP)].

Multilocus analyses have the advantage of revealing polymorphism at many loci simultaneously. The band patterns of multilocus techniques provide information about the degree of genetic polymorphism within and between populations, and facilitate individual identification and parental assignment. The classical DNA-fingerprinting based on RFLP techniques, however, is highly time-consuming, whereas technically simple PCR-based fingerprints are now available: RAPDs have been proven useful for many crops, but in some cases results have been found to be irreproducible across laboratories and even across thermocyclers within one laboratory. Conversely, AFLPs have shown stable properties and produce reliable results. Concerning any fingerprint procedure, however, the fragment length of bands cannot easily be related to allelic variation, and bands do not necessarily mark the same loci across individuals. Furthermore, bands behave like dominant characters and a heterozygous state cannot normally be diagnosed reliably. We deal with the problem that bands cannot easily be compared across species and even across genotypes within species. Therefore, estimates of genetic variability within and between populations on an evolutionary scale seem to be affected by large errors.

Using single-locus analyses (proteins, RFLPs, single-locus fingerprints or minisatellites, microsatellites, SNPs), we can escape from the problem of dealing with non-allelic information. Minisatellites [variable number of tandem repeats (VNTRs) with repeat lengths between 15 and 65 bp] or microsatellites (SSR; with repeat lengths between 2 and 7 bp; see Tautz 1993) are good markers at the individual level as well as at the population level. The probes and primers, respectively, used in the procedure are specific for a DNA segment containing the minisatellite or the microsatellite. In the strict case, allelic variation corresponds to the number of repeats present within populations. Genotypes of diploid individuals show only one or two bands per locus, i.e. they are either homozygous or heterozygous. Substantial variation is usually found due to the high number of alleles at each locus. RFLP analyses and single-locus fingerprints, however, are time-consuming and labour-intensive, and ultimately too cumbersome for the rapid evaluation of a large number of individuals commonly used in plant breeding programmes.

Very high mutational pressure at some VNTRs and SSRs can cause problems for population biological studies. Due to insertions and deletions of repeats, the length of tandem arrays can be a poor indicator of phylogenetic or genealogical relationships. Unrelated individuals can share tandem arrays of identical length by forward–backward mutation events, and direct descendants can appear unrelated because of length mutations in the germline (Rand 1994). Consequently, application of minisatellites or microsatellites should be limited to within-species studies. Therefore, in some cases, the use of traditional protein electrophoresis is a better approach to describe the genetic variability within and between species (Nevo 1990).

Nucleotide substitutions are considered in many studies, either to describe differences in the base sequence of DNA fragments or to use them as single locus markers, e.g. a set of SNPs randomly dispersed over the genome can improve our search for genes. The first-mentioned analysis is nowadays a standard procedure in systematics and studies of species' phylogeny, whereas SNP technologies have been developed recently, particularly for the characterization of the human genome. The high density of oligonucleotide arrays on DNA chips (Gupta et al. 1999) and the improved spectrometry combined with efficient data analyses (Ross et al. 1998) allow genotyping of a large number of SNP loci. Furthermore, the identification of quantitative trait loci (QTLs) and the analyses of complex human diseases are facilitated by such modern but still expensive techniques. SNPs show in some evolutionary studies, however, due to their mostly bi-allelic variation a lower level of information than multi-allelic microsatellites.

3 Measures of Genetic Diversity

The population genetic analysis of processes of population divergence depends on the type of genetic marker used in a study. The type of genetic variation considered, e.g. allelic variation of single loci and multilocus fingerprinting, respectively, but also the mutational process creating genetic variation have consequences for our interpretation of evolutionary processes. Obviously, the study of allele variation requires optimal discrimination of the various allelic states and a reliable estimate of their frequencies in the population. On the other hand, multi-locus procedures provide multiple band patterns whose differences among individuals can be explained by mutation, but the identity of bands does not necessarily correspond with genetic similarity.

3.1 Mutation Models

The infinite allele model assumes that mutations create alleles different from all of the other allelic states, whereas the stepwise mutation model (Otha and Kimura 1973) considers potential changes of alleles to *K* different states that depend on their prior allelic state. Many studies in ecology and evolution have demonstrated the use of the infinite allele model when protein variation is analysed. The stepwise mutation model, however, has been suggested to fit better the processes that are assumed to cause evolutionary changes among populations for microsatellites (Goldstein et al. 1995a,b; Slatkin 1995; Pritchard and Feldman 1996). Nevertheless, recent research has indicated that processes creating microsatellite variation are complex and simple stepwise mutation models cannot easily be applied (Colson and Goldstein 1999; for review see Li et al. 2002): The mutation rate of SSRs is obviously positively correlated with the number of repeats and their increased mutational dynamics needs a threshold number of repeats. In humans Lai and Sun (2003) found a minimum number of four repeats for microsatellites with di- and tetranucleotide motifs. Furthermore, a directional bias of mutations for larger numbers of repeats is observed, e.g. Vigouroux et al. (2002) analysed the mutation patterns of 142 microsatellite loci in maize (*Zea mays* subsp. *mays*). They found a higher probability that a microsatellite mutates to an allele of larger size.

3.2 Degree of Polymorphism and Heterozygosity

The most basic, but also uninformative, measures of genetic variation are given by the number of alleles and the degree of polymorphism that estimates the frequency of variable loci in a population (a locus is polymorphic if its most common allele has a frequency less than 95 or 99%). The degree of heterozygosity (the frequency of heterozygous genotypes), however, is the most commonly used measure to describe variation of individuals, within groups and populations. The measures of heterozygosity provide a potential to test several evolutionary hypotheses. We here have to distinguish between the observed (H_{obs}) and expected (H_{exp}) heterozygosity, respectively, and the heterozygosity of a population (H_{equ}) that is in an evolutionary equilibrium between mutation and random drift. The expected heterozygosity of a locus is defined as the frequency of heterozygotes in a population in Hardy-Weinberg equilibrium:

$$H_{exp} = 1 - \sum_{i=1}^{n} p_i^2 \tag{1}$$

where p_i is the frequency of the ith allele of a locus with n alleles. The observed and expected heterozygosity can be given for one locus and for an individual or group averaged over loci. These estimates are of interest for many biological disciplines, e.g. both measures can be informative for conserving genetic variability, and their close relation to population size as well as their sensitivity to inbreeding make them useful also for breeding efforts. Finally, the equilibrium heterozygosity has some importance for studies in evolution biology that focus on the population history, i.e. the distribution and size of populations. Assuming that the population is in an equilibrium between mutation and random drift, we have under ideal conditions (constant population size and constant mutation rates over generations, no selection):

$$H_{\text{equ}} = \frac{4N\mu}{1 + 4N\mu} \tag{2}$$

where N is the effective population size (only reproducing individuals are considered) and μ is the mutation rate (see Nei 1975, p. 117). The differences among the three measures of heterozygosity can, for example, be used to test whether populations have experienced bottlenecks in their recent history (Cornuet and Lukart 1996).

With the application of microsatellite variation in evolutionary biology, statistics have been established for this marker system assuming *K*-allelic state. Coulson et al. (1998) proposed a measure of heterogeneity that is based on the number of repeats:

$$d^2 = \sum_{i=1}^{r} \frac{(i_a - i_b)^2}{r} \tag{3}$$

where i_a and i_b are the lengths in repeat units of alleles *a* and *b* at locus *i*, and *r* is the number of loci at which an individual is scored (Coulson et al. 1998). This can easily be expanded to a grand mean over groups and populations.

Because genetic variability forms the basis for evolutionary processes (i.e. a loss of variability can diminish adaptive responses of populations to new or altered environmental conditions, and genetic information also characterizes individuals and their fitness), we have to question whether the respective genetic markers have any fitness relevance, i.e. the different marker loci can differ in their meaning for individual and/or population fitness. The term *associated overdominance* was coined by the Danish biologist Frydenberg (1963) to designate the positive correlation between fitness and genetic markers. A controversially discussed topic is currently which set of loci represents the genome that determines the fitness of its carrier. Theoretical models predict for any genetic data set a positive correlation between individual heterozygosity and fitness (e.g. Otha 1971; Charlesworth 1991), but experimental studies are contradictory (for a review see Zouros and Foltz 1987).

Allozymes have long been used as standard markers to characterize genomic heterogeneity and its fitness relevance; however, nowadays microsatellites are preferred because of their high variability and larger number of available loci. Microsatellite variation is described by the use of PCR-supported procedures which, however, can bias genotype classification: (1) null alleles, naturally frequently occurring mutations of microsatellites, but also allelic drop-out during PCR mimic homozygosity (Hare et al. 1996), and (2) fragment length analyses also include insertions and deletions in flanking regions besides repetitive variation of microsatellites. Consequently, such genotypic variation cannot easily be applied to evolutionary models and measures of genetic variability suggested for microsatellites (Colson and Goldstein 1999). Indeed, studies using microsatellites revealed different results for correlations between their variation and fitness, e.g. in habour seal pups (Coltman et al. 1998), in red deer (Coulson et al. 1998) and in great reedwarblers (Hansson et al. 2001) fitness was positively correlated with the measure for microsatellite diversity d^2, but Rowe and Beebee (2001) could not find a relation between fitness parameters and microsatellite variation in *Bufo calamita* and *Rana temporaria* using the same measure. Furthermore, accepting the general effect hypothesis, we expect for any set of marker loci an association between fitness and the degree of variability of marker loci. Thelen and Allendorf (2001), however, found a positive correlation between allozyme variation and growth rates in *Oncorhynchus mykiss*, i.e. they found support for the direct effect hypothesis, but this was not supported by their study of microsatellite variation. Conclusions drawn from these studies are based on estimates using the diversity measure d^2, but in a theoretical approach Tsitrone et al. (2001) have shown that the degree of heterozygosity is a more reliable measure for association studies than the diversity measure d^2 that is based on discrete changes of repeat numbers. Thus the diversity measure d^2 seems to be informative for association studies using microsatellite variation only if repetitive patterns in a strict sense are considered.

There is also controversy about the effects of allozyme variation on individual fitness (see, e.g., Hansson and Westerberg 2002). The direct effect hypothesis simply connects allozyme function with fitness. On the other hand, the general effect hypothesis considers that allozyme variability represents average genomic heterogeneity, i.e. average genetic heterogeneity relates to fitness. Differences between allozyme and microsatellite variation in their strength to fitness relation are discussed with respect to gene function, i.e. if allozyme loci tend to be in gene-rich regions, these loci would be more likely to be in linkage disequilibrium with other loci affecting fitness than would microsatellites (Hansson and Westerberg 2002). However, population structures can also create significant associations between allozyme variation and fitness (the local effect hypothesis), e.g. the action of inbreeding (David et al. 1995). To distinguish among direct, local and general effect hypotheses, Hansson and Westerberg (2002) suggest that research should include pedigree and sibling studies, which, however, are not feasible in many investigations.

Finally, the basic statistics for multi-locus fingerprints are cumbersome, but, nevertheless, some basic procedures are available to compare individuals and populations (band sharing indices: Nei and Li 1979; Lynch and Milligan 1994; and for RAPDs: Clark and Lanigan 1993). Analysis of molecular variance (see AMOVA, Excoffier et al. 1992) similar to analysis of variance applied in quantitative characters has been developed, but comparisons of multi-locus patterns may always cause a redundancy in data input through pairwise comparison of genotypes.

3.3 Analyses of Genetic Variability in Substructured Populations

Populations of species can differ genetically, even without mutation, due to local adaptation, modification of their reproductive mode and simply by isolation or reduced migration. Neutral genetic markers (no significant selection acts on such loci) can give information on the degree of genetic differentiation within and among populations and therewith also on the magnitude of migration. For a diploid, sexually reproducing species, Wright (1951, 1965) developed the *F*-statistics for one locus with two alleles which was modified for multi-allelic loci, the so-called *G*-statistics (Nei 1973). First, in the *i*th subpopulation the observed frequency of heterozygotes is denoted

$$H_{I_i}$$

and the Hardy-Weinberg proportion of heterozygotes H_{S_i}. Assuming n subpopulations, we have on average

$$\overline{H_S} = \sum_{i=1}^{n} \frac{H_{S_i}}{n} \text{ and } \overline{H_I} = \sum_{i=1}^{n} \frac{H_{S_i}}{n} .$$

where H_T equals the Hardy-Weinberg proportion of heterozygotes in the total population neglecting substructures. By comparing these three measures of heterozygosity within populations, we are able to define parameters that are related to population structures. F_{IS} measures the deviation from the Hardy-Weinberg equilibrium across subpopulations:

$$-1 \le F_{IS} := \frac{\overline{H_S} - \overline{H_I}}{\overline{H_S}} \le 1 \tag{4}$$

where F_{IS}=0 if individuals mate randomly, $F_{IS} > 0$ indicates a heterozygote deficiency (e.g. in the case of inbreeding) and $F_{IS} < 0$ if negative assortative mating occurs within subpopulations (e.g. heterozygote excess through balancing selection). F_{IT} measures an individual's deviation from the Hardy-Weinberg equilibrium relative to the total population:

$$-1 \le F_{IT} := \frac{\overline{H_T} - \overline{H_I}}{\overline{H_T}} \le 1 \tag{5}$$

Finally, F_{ST} is a measure of genetic differentiation among subpopulations:

$$0 \le F_{ST} := \frac{\overline{H_T} - \overline{H_S}}{\overline{H_T}} \le 1 \tag{6}$$

where F_{ST}=0 if subpopulations do not differ genetically. The three measures are closely related to each other with $(1-F_{IS})(1-F_{ST})=(1-F_{IT})$. Furthermore, F_{ST} increases with a decreasing number of migrants:

$$F_{ST} = \frac{1}{1 + 4Nm} \tag{7}$$

where N equals the effective population size (constant across generations) and m is the migration rate. Similarly,

$$G_{ST} = \frac{1}{1 + 4Nma}, \text{ where } a = \left(\frac{n}{n-1}\right)^2$$

and n is the number of subpopulations (Latter 1973). Interestingly, an estimate of the number of migrants Nm is given that does not need information about the population size. Theoretically Wright's F-statistics can also be used to estimate the number of generations since populations diverged, but only if population size is known and if effects of mutation are negligible (Nei 1987, p. 217). Nevertheless the basic statistics and F- or G-statistics are nowadays widely used in population biological studies. For example, Cheng et al. (2000) studied allozyme variation in populations of the rare *Myrica adenophora* and its widespread congeneric species *M. rubra*. Populations of the rare species had fewer alleles per locus, a lower percentage of polymorphic loci and lower expected heterozygosity than populations of the widespread species, but intrapopulation differentiation (G_{ST}) was similar in both species and the number of migrants was moderate in both species ($Nm \sim 1.5$).

When microsatellites became commonly used in population genetic studies F-statistics were also developed that took into account the presumed characteristics of microsatellites (Michalakis and Excoffier 1996). Compared, for example, with protein variation, microsatellite variability is much more strongly governed by mutation than random drift and, therefore, the original F-statistics cannot be applied per se to microsatellite data. Slatkin (1995) suggested his R_{ST} measure based on a K-allele model and a generalized stepwise mutation process. Consequently, this alternative measure can again only be applied to data that focus exclusively on repeat variation within microsatellite regions, but may not provide reliable estimates of population divergence when variation in regions flanking the microsatellite is included, as is often done by simply analysing fragment length.

3.4 Genetic Distances Among Populations and Species

The precise estimation of the period (years and generations, respectively) since divergence of populations requires the consideration of many influencing variables, e.g. mutation, migration, population size and selection. Thereby mutation rates, population sizes and selection coefficients may vary temporally and may be characteristic even for local populations. This variation over time is impossible to quantify over long evolutionary periods. Therefore, fluctuation of selection and of population size is often ignored. Furthermore, genetic distance measures are based on additional assumptions concerning the variability of marker systems used that are not always met under natural conditions, e.g. models that analyse mutational processes of microsatellites include the possibility of an infinite number of repeat units and others restrict mutational events to single repeat changes.

In the following, we briefly represent some distance measures that are closely related to evolutionary time under specific assumptions. The parameters in Eqs. (8)–(11) are the frequencies x_{ij} and y_{ij} of the ith allele at the jth locus within two populations x and y, the number of loci r, the mutation rate μ, the number of generations t, the maximum repeat score of microsatellite loci R and the effective population size N. Assuming the infinite allele model and allele frequencies per locus are given, Nei (1972) suggested the standard genetic identity (I_S) for describing the genetic identity of two populations. This is now the most commonly used genetic similarity index in evolutionary studies.

$$I_S = \frac{\sum_{i,j} x_{ij} y_{ij}}{\sqrt{\sum_{i,j} x_{ij}^2 \sum_{i,j} y_{ij}^2}} \tag{8}$$

Similarly, Tomiuk and Loeschcke (1991, 1995) proposed a genetic identity measure $I_{TL,ancestral}$ that is based on the estimated frequency of ancestral alleles over all loci:

$$I_{TL,\, ancestral} = \frac{\sum_{i,j} x_{ij}^{y}}{\mathrm{r}} \frac{\sum_{i,j} y_{ij}^{x}}{\mathrm{r}} \tag{9}$$

where

x_{ij}^{y} and y_{ij}^{x}

are the frequencies of the ith allele at the jth locus in the populations x and y, respectively, which are also present in their sister population. If the biased sampling of ancestral alleles within populations is considered, it can be shown that the $I_{TL,ancestral}$-measure has to be transformed to

$$I_{TL} = +\sqrt{I_{TL,ancestral}}\,.$$

Assuming mutations to be neutral and equilibrium conditions to be met, as well as equal and constant mutation rates at all loci in two populations, Nei (1972) showed a close correlation of his genetic identity measure with evolutionary time $I_S \approx e^{-2\mu t}$. Such a correlation with evolutionary time can also be shown for Tomiuk and Loeschcke's (1991, 1995) measure

$$D_{TL} = -\ln(I_{TL}) \approx 2\mu t \tag{10}$$

Differently to Nei's (1972) measure, the I_{TL} measure can be easily modified for species complexes containing sexual and polyploid asexual races (e.g. Tomiuk and Loeschcke 1992, 1994, 1995, 1996).

Obviously, statistics for estimating genetic distances between species have been suggested that can be applied to mutational processes assumed to create microsatellite variability. Goldstein et al. (1995a,b) developed a measure on which Coulsen et al.'s (1998) procedure to estimate individual heterozygosity was based:

$$(\delta\mu)^2 = \frac{\sum_j (\mu_{xj} - \mu_{yj})^2}{r} \tag{11}$$

where

$$\mu_{x_j} = \sum_i i x_{ij} \text{ and } \mu_{y_j} = \sum_i i y_{ij}$$

are the average number of repeats at the locus j and i is the repeat score of allele i.

There is some discussion about the best estimates for genetic distances with respect to various marker systems that are governed by different mutational processes. Kalinowski (2002) noted that the evolutionary history of populations and the evolutionary patterns that produced observed patterns require careful consideration when a specific distance measure is used, in order to obtain reliable estimates of genetic distance among populations. Tomiuk and Loeschcke (2003), however, argued that the underlying processes are even more complex and their previous simulations demonstrated the limitations of the measures of genetic identity in analysing genetic variability (Tomiuk et al. 1998): For very short-term evolutionary changes where genetic differentiation is mainly due to drift, F_{ST}-based measures are useful for the analysis of population structure. If the genetic similarity of more distantly related populations with large population size and mutation rate (i.e. $4N\mu = 1$) is considered, the genetic identity proposed by Tomiuk and Loeschcke (1991, 1995) is superior to Nei's D_S. Furthermore, Tomiuk et al. (1998) pointed to a general problem with the application of D_S under non-equilibrium conditions: the more the ancestral population deviates from its genetic equilibrium the larger is the bias using Nei's D_S. The tree topology constructed on the basis of D_S data may be not only the result of varying evolutionary rates among populations but also an effect of the genetic structure of the common ancestral population and the relative sizes of the sister populations at the beginning of the divergence process. If alleles can be classified more precisely as being ancestral or mutated, however, this information does not improve the D_S estimate, but D_{TL} converges to the expected genetic identity function.

The sampling strategies should also consider the process that most likely has generated the genetic variation under study. For genetic variation following the stepwise mutation model Takezaki and Nei (1996) proposed to increase the number of loci in order to reduce the variance of distance estimates; only with an increasing degree of heterozygosity should the sample size be enlarged. The required number of microsatellite loci, however, depends on the aim of the study, and Feldman et al. (1997) recommended to study more than 15–20 loci. For the infinite allele model where the ancestral state of alleles can be determined, the estimates of genetic identity can be improved only by increasing both the number of loci and the sample size, where the gain of information by increasing the number of loci depends on the variation of genetic variability among loci.

4 Architecture of Populations

Climatic conditions and landscape structures are, without doubt, the basic elements that determine biodiversity, the distribution of individuals and species abundance in geographic areas. The patterns of species and family richness of angiosperms is consistently associated globally with mean annual temperature, annual water deficit and annual potential evapotranspiration (Francis and Currie 2003). Besides adaptation to geographic conditions, adaptability is the fundamental prerequisite for species in order to persist successfully in a multifactorial system with competing species, host–parasite and prey–predator interactions. An organism must respond permanently to its intraspecific competitors and interspecific attackers, otherwise the risk increases that it will become extinct (the classical Red

Queen model proposed by van Valen 1973). Selection optimizes the genetic structures within a species according to the local conditions that its subpopulations experience. A recent impressive example for the same directional genetic changes in different populations is the chromosomal polymorphism in *Drosophila subobscura*. Frequencies of chromosome inversions showed significant latitudinal clines in Spanish populations of *D. subobscura* (Solé et al. 2002). About two decades ago, however, *Drosophila subobscura* was introduced into the New World, and today almost all inversions found in Spain show the same correlation with latitude as in the Old World (Balanyà et al. 2003). Nevertheless, the discussion still continues as to whether observed genetic variation is selectively neutral or fitness-relevant. Baek et al. (2003) analysed microsatellites of wild barley, *Hordeum spontaneum* (C. Koch) Thell., along a southward transect of increasing aridity. Comparisons across habitats showed that the degree of gene diversity may be associated with altitude, suggesting that microsatellite variation may be adaptive and subjected to natural selection.

In the context of genetic differentiation within species, we must also concentrate on mechanisms that control the dispersal of individuals. Obviously, there is a limitation in the distance that individuals can disperse and species have to adapt their dispersal ability to environmental conditions. As a consequence of restricted migration, effects like inbreeding by proximity or genetic isolation by distance must also be considered (Wright 1943).

In the following two sections, we consider genetic changes in populations that are short- and long-term responses, respectively, to environmental conditions. However, this distinction by no means forms two disjunctive classes of evolutionary changes that can occur on different time scales. Our attempt is rather to consider processes that dominate changes in a natural environment and an anthropogenic landscape, respectively.

4.1 Genetic Variation That Evolved During Long Periods

Biodiversity of a natural habitat is maintained in a dynamic equilibrium that depends on all participants in a long-term evolutionary game. If the balance among species is disturbed the system tends towards a new equilibrium state. The availability of resources modifies the strength of competition among species, and local conditions can determine the genotypic composition of populations within species. Isolation and migration determine the magnitude of differentiation among local populations, e.g. speciation processes will be most effective when populations are completely isolated. However, if isolation barriers are removed during early speciation, gene flow might occur among closely related species. The columbine genus *Aquilegia* is an example of the reproductive isolation of closely related species due to the ecological and floral characteristics of its species (Hodges and Arnold 1994). Clinal variation is evident from RAPD loci and some morphological characters along an altitudinal cline, and the genetic changes were closely associated with the changes in habitat type. However, genetic variation indicates the still-lasting presence of introgression among species of the genus.

Over large geographic ranges, plants often show genetically based differentiation in traits as a result of adaptation to specific sites. Only a few studies, however, have examined the degree to which populations of perennial plants exhibit genetically based differentiation in life-history traits over small spatial scales. Kittelson and Maron (2001) studied genetically based differentiation among bush lupine (*Lupinus arboreus*) from nearby dune and grassland sites. The genetic variation of seeds did not indicate significant spatial differentiation (F_{ST}=0.002), but gene differentiation increased considerably among juvenile plants (F_{ST}=0.041). In short, local selection regimes affect life-history traits differently at different stages of the life cycle of lupines. Many other studies have discovered adaptive genetic differentiation, but in some cases little evidence exists for local adaptation. For example, in populations of the annual legume *Chamaecrista fasciculata* with limited gene flow, it is suggested that metapopulation processes and temporal environmental variation act together in reducing local adaptation (Galloway and Fenster 2000).

The quality and diversity of plants can determine the presence or absence of animal species and their population sizes. Here we briefly consider studies that focus on the local adaptation of animal populations to their host plant. For example, de Jong et al. (2000) studied the genetic basis of the ability of a flea beetle (*Phyllotreta nemorum*) to utilize the crucifer *Barbarea vulgaris* ssp. *arcuata*. They found evidence for a genetic control of this ability and genetic basis of host-plant use across local populations. In contrast, there was no local adaptation of the hemiparasitic plant *Rhinanthus serotinus* to its host plant *Agrostis capillaris* (Mutikainen et al. 2000). However, other parameters such as the parasite impact on the host biomass indicate variation of the parasite virulence across populations, suggesting also a genetic basis for this variation. Finally, the influence of herbivores on plant communities must also be considered when we focus on population structures of plant species. Gomez (2003) investigated effects of herbivores on morphological and floral traits of *Erysimum mediohispanicum* (Cruciferae). Results suggest complex interactions between potential pollinators and herbivores, i.e. pollinator-mediated selection can be disrupted by conflicting effects of plant enemies acting during or subsequent to pollination.

Beyond our interest in understanding the ecology and evolution of species, results of studying natural populations can improve our efforts in species conservation and in animal and plant breeding. Zhang et al. (2002), for example, investigated the evolutionary dynamics of defence genes conferring resistance to the fungal pathogen *Cochliobolus carbonum* in maize and its wild ancestor *Zea mays* ssp. *parviglumis*. Obviously, plant defence genes are subjected to non-neutral evolutionary dynamics which can explain the high level of genetic variation retained in cultivated maize relative to its wild ancestor. Therefore, studies of natural populations provide the basis for breeding programmes where genetic variability of wild populations of cereals and other cultivated plants serve as gene resources. In rice, pest and disease resistance genes from worldwide samples, e.g. resistance to blast, bacterial blight and brown plant hopper, have been used to breed new varieties (Nakagahra et al. 1997).

4.2 Short-Term Genetic Changes

During the last centuries, humans have been the most decisive factor in determining changes in the natural landscape. The environment has been changed drastically by our activities. Areas like wetlands and forests that could not be cultivated intensively in the past were reclaimed from the wild (Waldman and Shedvah 2000). The sudden changes in habitat structures and the fragmentation of habitats have forced never before experienced conditions on many species. The inescapable requirement to adapt instantaneously to new environmental conditions can increase the risk of extinction for species, but species can also be genetically prepared to use new resources, for example, i.e. they are preadapted.

The increased fragmentation of natural habitats will also seriously affect population structures. Large populations with fairly high migration rates among subpopulations can maintain large genetic variability and minimize the negative influence of isolation and inbreeding. For example, in the tropical tree *Symphonia globulifera*, significant inbreeding and genetic differentiation among subpopulations were associated with fragmentation (Aldrich et al. 1998). Landscape management practices that alter habitat fragmentation can therefore considerably change the genetic structure of a population, but can also change structures in its interacting species (Hale et al. 2001). Over the past decades, forest fragments in northern England were connected with those in southern Scotland by planting of conifers. The result of this afforestation was an admixture of gene pools of the Scottish and Cumbrian squirrel populations that inhabit these forests. Finally, genetic variation in a plant population can affect arthropod community richness and composition. Dungey et al. (2000) studied arthropod communities on populations of *Eucalyptus amygdalina*, *E. risdonii* and their hybrids. They found that hybrid populations support significantly different arthropod communities than pure species and hybrids act as centres of biodiversity. This example may also demonstrate a dilemma in conservation biology, i.e. are hybrids, as an evolutionary accident, worthy of conservation?

Besides human changes of natural landscape structures, humans have spread former endemic species all over the globe by their worldwide mobility. Some of these species invaded new communities of species and caused dramatic alterations of existing interactions among species. The study of species' invasion into foreign ecosystems is a major concern in many countries whose biodiversity has evolved in strict isolation from that of other areas (see Drake et al. 1989). For example, extensive human impact in southwestern Australia has been reported to be responsible for a high incidence of rarity throughout the highly endemic flora of the region (Rossetto et al. 1995). Furthermore, we must also consider effects that result from an admixture of locally adapted genotypes within species, e.g. seeds of locally adapted plant populations are brought to new sites where invader plants cross with local varieties. In summary, human activities can disturb an ecosystem and its species' interaction that have evolved over a long period. Species are forced to react to the new conditions and evolutionary processes will drive the system into a new equilibrium state: Species richness can, of course, increase, but in most cases only the risk of species extinction increases.

One major concern in conservation biology is the ongoing reduction of natural habitats and their fragmentation; both influence the viability of plant and animal populations living in these habitats. The characteristics of each species we consider must be taken into account, e.g. genome structures, population history and the reproduction mode (selfing, outcrossing, sexual or asexual reproduction). A large population size or extensive geographic distribution is not always accompanied by a large genetic diversity, as historical factors such as speciation processes and population expansion often play more important roles in determining genetic diversity than the number of remnant individuals (Maki 2003). The mating system of a species also affects genetic diversity; predominantly selfing species tend to have smaller genetic diversity than outcrossing congeners (Maki 2003). In order to obtain baseline data for the conservation of species, levels of genetic diversity within and among populations are often considered. Gaudeul et al. (2000) used AFLP fingerprints and scored polymorphic markers in individuals of the endangered Alpine sea holly *Eryngium alpinum* L. Diversity levels within populations were high (H_{exp}=0.198), and a positive correlation was detected between genetic diversity and population size. Furthermore, F_{ST} values indicate high differentiation among populations (F_{ST}=0.40) which has been explained by founder events during post-glacial colonizations and bottlenecks. The study of genetic variation suggests that it is worthwhile to conserve many populations of *E. alpinum*.

Generalizations of results from individual studies are often limited by the restriction to few species studied and specific test conditions used in the experiments. Evolution, however, is creative and can follow several different pathways to organize life. This, of course, reduces the general validity of results and conclusions of many individual studies. Nevertheless, in some cases there are some basic questions as described above for species richness and altitude that are of great interest in our understanding of natural processes. Meta-analyses and modelling might help us to escape from the limitations of individual experimental studies when we are more interested in general natural principles.

One problem concerning meta-analyses, however, is the comparability of data that are adopted from different studies and/or analysed across species. Clauss et al. (2002) studied microsatellite variation in four closely related *Arabidopsis* species – *A. thaliana*, *A. halleri*, *A. lyrata* ssp. *lyrata* and *A. lyrata* ssp. *petraea* – and the distantly related crucifer *Arabis drummondii*. Their analyses revealed significant differences in population genetic parameters among the *Arabidopsis* species. The degree of polymorphism varied from 8–90% among species. For example, in a population of the self-incompatible perennial herb *A. halleri*, an excess of heterozygosity which can be explained by a recent population bottleneck associated with human-mediated founder events has been observed, whereas a population of the self-incompatible perennial herb *A. lyrata* ssp. *petraea* appeared to be at mutation-drift equilibrium. Therefore, even in closely related species, population structures have evolved differently depending on the ecological conditions that the species have experienced, i.e. population structures are highly dependent on the evolutionary history of a species. However, focusing on genetic structures within species, the analysis of a compilation of data from several studies seems to be a reliable approach to find more general patterns. Hoffmann et al. (2003) analysed allelic information of natural populations of *Arabidopsis thaliana* in order to identify geographical population structures. This analysis disclosed worldwide distribution patterns of *A. thaliana* for which the Mediterranean area is assumed to have been its refuge during glacial periods. Accordingly, the highest genetic variability is observed along the Atlantic coast from the western Iberian

Peninsula to southern Great Britain, while the lowest variability is found in central Europe.

Population genetics has made great progress with development of statistical procedures and in applying mathematical modelling in order to understand the temporal dynamics of genes in populations. Real populations, however, are also spatially structured by geographic features, e.g. mountains, rivers and deserts (Epperson 2003), and therefore processes have to be considered that determine geographic patterns of genetic variation. However, so far relatively few analyses have focused on the role of plant genetic variation in studies common to landscape ecology (Jelinski 1997). Knowledge about genetic variation of natural plant populations at different hierarchical levels is important for our understanding of consequences that result from habitat fragmentation and global changes (see Jelinski 1997). Most studies, however, focus on individual species that are endangered and of interest for conservation biology. For example, Jacquemyn et al. (2003) examined ecological, demographic and genetic consequences of agricultural land use on the long-term persistence of the primrose *Primula vulgaris* in Belgium. Their demographic analyses disclosed a reduced viability of many populations due to their small population sizes and limited seedling recruitment. Furthermore, no difference in genetic variability (allele number, H_{obs}, H_{exp}, F_{ST}) among population types was found, indicating that the quality of the surrounding landscape matrix seems to be more important than within-habitat characteristics, demographic or genetic traits in determining population viability. Furthermore, besides landscape fragmentation the history of populations and geographic patterns must be considered in order to interpret correctly genetic variation observed in studies. For example, the lilioid herb, *Anthericum ramosum*, is restricted to grassland habitats and occurs in four geographically isolated regions (Sjaelland, Skane, Oland and Gotland) in Scandinavia (Rosquist and Prentice 2000). High differentiation of populations from Gotland (G_{ST}=25%), however, could not be explained by a recent habitat fragmentation but by the restricted distribution of limestone bedrock on Gotland and different origins of southern and northern populations (Rosquist and Prentice 2000). Nevertheless, there are many determinant factors affecting biodiversity. Calsbeek et al. (2003) argued that physical historical processes, e.g. geographical barriers and climatological events, have resulted in patterns of genetic diversity in California over the past 2–10 million years. Similarly, a recent study of chloroplast DNA haplotypes identified Pleistocene glacial refugia for the Arctic-Alpine *Saxifraga oppositifolia* in the Arctic (Holderegger et al. 2002). Concluding from present geographic patterns of genetic diversity, the lack of consistent overall genetic pattern found in the study suggests large migration by seed and pollen flow during postglacial times.

5 Biodiversity in Dynamic Systems

Genetic variability in natural populations is caused mainly by changes in the structure of their habitat. Species and populations react to such changes by adapting their behaviour and genetics in order to survive. This may cause – given sufficient time – the development of new species. In other words, genetic variability (micro-scale) is caused by spatial variability at a landscape level (macro-scale). Landscapes, considering space and time, are highly dynamic systems. In order to develop species diversity an ecosystem is mainly dependent on its dynamic and progressive character.

Most of the natural ecosystems in Germany are influenced by human activity, since 80% of the landscape is used agriculturally. In rural cultural landscapes dynamic processes are still ongoing, but they tend to proceed much faster than in natural systems. One might call this phenomenon an 'evolutionary leap' of the landscape. Ecosystems influenced by human land use are always exposed to disturbances, and surviving in frequently disturbed ecosystems demands a high degree of adaptability. While natural ecosystems have time to re-naturalize over a long-term scale, species in frequently disturbed habitats are facing fast-changing conditions. This means that there is a temporal difference in time scales between structural change of habitat and the flexibility of the associated organisms, which have to adapt to the new situation. Especially species with long life cycles only have the possibility to withdraw into potential refuge habitats. Because of the global utilization of landscape by humans, potential refuge habitats have either vanished in the past or are incompletely connected by stepping-stone habitats. Changes in intensively used landscapes may therefore lead to the extinction of species and will inevitably reduce genetic variability.

5.1 Modelling Ecological Dynamics in Anthropogenic Landscapes

The worldwide decline in biodiversity is a global problem and preservation of biodiversity is one of the main targets of biological conservation. In order to protect and preserve biodiversity it is necessary to understand the effects of habitat changes over a macro-scale on genetic variability at the micro-scale. A quantification of the interactions has not yet been achieved. In fact the complexity of the interactions in the system complicates the derivation of causalities. Also, the use of two different scales leads to problems within the research methods. For further research and understanding, modelling approaches seem to be useful. Models make it possible to reduce the complexity of a system by factorizing it. Furthermore, on the basis of rules with which reality is reproduced, the models allow us to predict future scenarios. Therefore models are frequently used in advising political decision-makers.

For ecological modelling, analysing ecosystems and biodiversity, high-end methodical standards are required. For one, the spatial levels of genetic variability and landscape modification must be combined. Not only must one include the natural influences on the system considered, but also the anthropogenically influenced variables have to be taken into consideration as parameters of the system. The main anthropogenic influences on biodiversity at the landscape level are land-use change and landscape fragmentation. Both of these are presented as major causes for the decline of biodiversity in Germany by the federal Council of Experts for Environmental Questions (SRU 1985). Whereas the increasing fragmentation of landscape is related to society's aim for global connection, changes in the agricultural use of landscape are often induced by political interests. Farmers are no longer able to choose cultivation methods according to the natural landscape; rather they must adjust to the ever-changing socio-economic conditions. This may be price trends on the world market, adjustment of production methods to the ongoing globalization and shifting of the financial resources. Land-use change and landscape fragmentation are not natural parameters and therefore have to be handled with different rules than ecological causes and effects.

The method of modelling dynamic processes in rural cultural landscapes is rather new and at the beginning of its development. At this point in time, there are five models used in Germany which allow us an ecological assessment of the effects of land-use changes: KUL (*Kriterien umweltverträglicher Landbewirtschaftung*, Criteria of Ecologically Compatible Land Use; see Breitschuh et al. 2000), REPRO (***Repro****duktion der organischen Bodensubstanz*, Reproduction of the organic soil; see Hülsbergen 2003), MODAM (***M****ulti-****O****bjective* ***D****ecision Support Tool for* ***A****groecosystem* ***M****anagement*; see Zander 2003), RAUMIS (*Regional differenziertes Agrar- und UmweltInformations System*, Regional Differentiated Agro-Environmental Information System; see Henrichsmeyer et al. 1996) and the model network ITE^{2}M (*Integrated Tools in Ecological and Economic Modelling*; see Kuhlmann et al. 2002). These models deal with all scales of landscape that might be affected by land-use changes. The basic scale is a parcel of land with varying cultivations, cropped or fallow land. Some models also represent agricultural farms. At this scale the farmer decides on the type and dynamics of land use to be executed on each parcel of land and on reactions to changing political conditions, which may include a change in land use. Other models analyse the ecological effects of land-use changes on a regional scale. Therefore they are adjusted to natural or political units. All these different model approaches attempt to indicate land-use effects on the environment. Species diversity is only one goal alongside other objects of protection like N, P and K circulation, water balance, soil function, soil fertility or the quality of drinking water. Most of these aspects, however, are related in one way or another to the variety of species. For instance, the impairment of soil function by using heavy machinery leads to worse living conditions for soil fauna and degrades its diversity. Increased nutrients and pesticides reduce the habitat quality of natural ecosystems by peripheral input and endanger their species populations, especially those in oligotrophic habitats. Thus an analysis of land-use effects on abiotic targets considers impacts on species diversity as well. However, how can the direct relations between land-use change and species diversity be described? In which way do different model approaches solve methodical challenges and how can qualitative conclusions be deduced from their output? Here we look at the algorithms of three models (KUL, RAUMIS, ITE^{2}M) used to simulate land-use practices as well as land-use changes, and describe how they implement species diversity in their modelling approach.

The method KUL was developed at the Thuringian State Institute for Agriculture (TLL) in Germany as a weak-point analysis tool to locate and eliminate environmental problems on farms (Breitschuh et al. 2000). Up to now the method has been used in more than 200 farms throughout Germany. The KUL method evaluates the environmental compatibility of farms using 17 different criteria, including aspects of material and energy circulation, soil quality and the use of pesticides. Furthermore the diversity of cultivated plants on each farm is evaluated. The criteria 'diversity of cultivated plants' results from the diversity-index according to *Shannon-Wiener* (Zar 1996). For its calculation the number of cultivated plant species per farm and their ratio of abundance to arable farm land are considered (Breitschuh et al. 2000). An environmentally compatible farm in the sense of the KUL guidelines is allowed to have a 'diversity of cultivated plants-index' of 1.25 including the cultivation of at least four species.

While farms can easily enrich their 'diversity of cultivated plants' by enhancing their crop rotation, assessments of wildlife biodiversity at a farm level are not possible with KUL. According to Eckert et al. (1999), an area-wide determination of wildlife diversity referring to the whole-farm demarcation is impossible. KUL provides no methodical solution to this issue apart from analysing the negative effects of cultivation on species diversity.

The model RAUMIS was developed in the context of a cooperation between the Institute for Agricultural Policy (University of Bonn, Germany) and the agro-economic institutes of the Federal Agricultural Research Centre (FAL Braunschweig-Völkenrode, Germany). The aim of the model development was an illustration of the interdependencies between agricultural policy, agricultural production and environment. The field of application of RAUMIS is both forecasting future trends in land use (prediction) and mid-term analysis of alternative agricultural and environmental policy (simulation) (Geier et al. 1999). RAUMIS includes 326 model regions – the entire area of Germany – based on agricultural-statistical data and representing different production locations. Different agro-political scenarios are defined that might influence the farmers' incomes. Subsequently, a mathematical optimization model simulates the adaptation of land use to different political-economical conditions in any model region. The basic assumption of the model is that the farmer's decision is always based on economic rationality, so that the most profitable type of land use is always practised. Under different scenarios the optimization model calculates the most profitable production of each administrative district. Land-use change is simulated by a comparison between actual and future (scenario) state.

Simulating the effects of anthropogenic land-use change on the environment, 12 indicators are involved referring to aspects of material circulations, soil degradation and conservation of species and habitats. Effects on species diversity are described by the indicator 'potential site and habitat outfit'. This indicator does not evaluate wildlife diversity, but rather the diversity of agricultural land use within the scenarios. The diversity of agricultural land use is described via the *Shannon-Wiener* index and reflects the number of land-use types (realized production practices in each scenario) including their proportions per model region (Meudt 1998). The index defines agricultural landscapes cultivated with many land-use types as ecologically more healthy than regions with only one or few land-use types. These concepts of evaluation are based on the fact that monotonous landscapes generally contain less biodiversity than a variety of land-use patterns.

The model network ITE^{2}M is the primary objective of the collaborative research centre 'Land Use Concepts in Peripheral Regions' at the University of Giessen, Germany. The main task of this research centre is to develop an integrated method for developing sustainable land-use concepts and to define valuation criteria (Möller et al. 1998). In the model network ITE^{2}M, four models in the fields of agricultural economics, ecology and hydrology are adapted and linked in a central interface (Möller et al. 2002). The agro-economic simulation model ProLand creates a land-use prognosis and allows the simulation of land-use changes (Kuhlmann et al. 2002). ProLand is intended to predict land-use changes as a result of changes in the general framework, specified by a particular combination of natural, economic and political characteristics on a regional scale based on the hilly midlands of the 'Lahn-Dill-Bergland' in the German Federal State of Hesse. To predict potential future land-use decisions with ProLand, assumptions on the behaviour of the people involved are necessary. As in RAUMIS it is assumed that land users will maximize the land rent and therefore choose the land use with the maximum economic profit. Under this assumption the behaviour of farmers under different political-economic conditions is simulated. As a result ProLand calculates land-use scenarios that represent the economically most optimal choice of land use.

In order to analyse and predict the influence of land-use change on natural biodiversity of landscape, the simulated land-use scenarios are a basic input to the neighbouring model, named ANIMO. The ecological model ANIMO evaluates the landscape by using the aggregation index from He et al. (2000). Furthermore, the degree of landscape fragmentation is calculated as the connectivity between habitats. ANIMO is able to examine the local number of species in a path (alpha-diversity), the dissimilarity between habitat patches regarding species inventory (beta-diversity) and the overall diversity of landscape (gamma-diversity). As a data input ANIMO uses the land-use scenarios of ProLand as well as generating virtual landscapes itself. The model is based on a two-dimensional cellular automaton, which consists of a lattice of 100×100 quadratic cells. Each cell observed is influenced by all eight neighbouring cells and each cell is linked to the neighbouring cells by specific local rules. Different cell types are defined representing the land-use practices fallow land, arable land and meadow. The state of each cell is defined by a specific spectrum of species, whereas any land-use type consists of a determined number of generalists and specialists (Steiner 2002). The model uses appropriate input data from landscape ecology projects done by Waldhard et al. (2000). Finally different scenarios are simulated by varying the proportions of generalists and specialists or by changing the spatial pattern of the land-use types. The transaction of cells from one state to the other is driven by stochastic and deterministic rules. The coherences between the modelled spatio-temporal patterns and the biodiversity of landscape are displayed by calculating the three biodiversity indices (alpha, beta and gamma diversity).

With the presentation of the three models KUL, RAUMIS and ITE^{2}M, we have addressed the methodical problems that arise when modelling dynamics in cultural landscapes. Modelling the two key factors inducing dynamic processes in cultural landscapes – *land-use change* and *landscape fragmentation* – is so sophisticated that the new science of *Land-Use Modelling* has developed. A consequence of this is the adoption of these models for economic simulations of price policy, while the ecological assessment of the model results plays a more secondary role. The examination of biodiversity is oversimplified and abstracted, and only simple indicators are mapped: two models only quantify the diversity of land use. By estimating the effects of land use on abiotic parameters, the biodiversity of wild species is only indirectly mapped. However, the implementation of several stand-alone models instead of only one large model seems to result in advantages when handling and combining ecological and economic questions.

In summary, we conclude that landscape ecology, the science of the interactions between spatio-temporal patterns of landscape and ecological processes, tends to concentrate on studies on a larger scale (Wiens 1992). The current studies of landscape ecology mainly focus on questions about the spatial patterns in cultural landscapes (using methods of Land-Use Modelling). Politically relevant scales are basically addressed (farms, regions) in this context (Roedenbeck 2004). On the other hand, the rapid development of DNA technologies in the past two decades has led to an increase in knowledge about the genetic variability within and between species (Jelinski and Wu 1996; Jelinski 1997). A consolidation of both fields of knowledge, landscape ecology and genetics, has not been achieved to date. Considering the integration of empirical information about the genetic composition of species (*fine-scale information*) into models that deal with the spatial patterns of landscapes (*macro-scale information*) is imperative (*bottom-up-approach*). Furthermore, the generation of spatio-temporal patterns at the landscape level (in cultural landscapes mostly anthropogenically induced) must be correlated with the genetic variability at the species level (*top-down approach*). A multi-scale approach by modelling dynamic ecosystems is an important and essential component for a knowledge-based, target-oriented conservation of biodiversity in cultural landscapes.

6 Summary

Landscape structures determine without any doubt biodiversity. Species and populations react to abiotic and biotic changes of their habitat, attempting to adapt their behaviour and genetics in order to survive. During their history, however, humans have increasingly caused changes that forced species to adapt rapidly to anthropogenic landscapes, otherwise they became extinct. We here introduce experimental and statistical approaches in population genetics to measure genetic variation within and between populations. Recent population biological studies are presented to show exemplarily adaptation processes of species or ecosystems to natural and anthropogenic habitats. Some approaches to find more general principles that determine the dynamics of ecosystems are considered: (1) the use of meta-analyses is briefly considered, and (2) the application and possibilities of ecological modelling for describing factors that control complex natural systems are discussed. Our conclusion is that approaches in ecological modelling need the integration of population genetics in order to obtain a better insight into evolutionary processes that are induced by human activities.

Acknowledgement. We thank Rod Snowdon for his valuable discussion and comments which improved considerably this manuscript.

References

Aldrich PR, Hamrick JL, Chavarriaga P, Kochert G (1998) Microsatellite analysis of demographic genetic structure in fragmented populations of the tropical tree *Symphonia globulifera*. Mol Ecol 7:933–944

Allard RW (1999) History of plant population genetics. Annu Rev Genet 33:1–27

Baek HJ, Beharav A, Nevo E (2003) Ecological-genomic diversity of microsatellites in wild barley, *Hordeum spontaneum*, populations in Jordan. Theor Appl Genet 106:397–410

Balanyà J, Serra L, Gilchrist GW, Huey RB, Pascual M, Mestres F, Solé E (2003) Evolutionary pace of chromosomal polymorphism in colonizing populations of *Drosophila subobscura*: an evolutionary time series. Evolution 57:1837–1845

Bodmer WF, Cavalli-Sforza LL (1976) Genetics, evolution, and man. Freeman, San Francisco

Breitschuh G, Eckert H, Kuhaupt H, Gernand G, Sauerbeck D, Roth S (2000) Erarbeitung von Beurteilungskriterien und Messparametern für nutzungsbezogene Bodenqualitätsziele – Anpassung und Anwendung von Kriterien zur Bewertung nutzungsbedingter Bodengefährdung. UFOPLAN no. 107 01 022/207 01 022. Umweltbundesamt Berlin, Berlin, 129 pp

Cai HW, Gao ZS, Yuyama N, Ogawa N (2003) Identification of AFLP markers closely linked to the rhm gene for resistance to southern corn leaf blight in maize by using bulked segregant analysis. Mol Genet Genom 269:299–303

Calsbeek R, Thompson JN, Richardson JE (2003) Patterns of molecular evolution and diversification in a biodiversity hotspot: the California Floristic Province. Mol Ecol 12:1021–1029

Charlesworth D (1991) The apparent selection on neutral marker loci in partially inbreeding populations. Genet Res 57:159–175

Cheng YP, Chien CT, Lin TP (2000) Population genetics of geographically restricted and widespread species of Myrica (Myricaceae). J Hered 91:61–66

Clark AG, Lanigan CM (1993) Prospects for estimating nucleotide divergence with RAPDs. Mol Biol Evol 10:1096–1111

Clauss MJ, Cobban H, Mitchell-Olds T (2002) Cross-species microsatellite markers for elucidating population genetic structure in *Arabidopsis* and *Arabis* (Brassicaeae). Mol Ecol 11:591–601

Colson I, Goldstein DB (1999) Evidence for complex mutations at microsatellite loci in *Drosophila*. Genetics 152:617–627

Coltman DW, Don Bowen W, Wright JM (1998) Birth weight and neonatal survival of harbour seal pups are positively correlated with genetic variation measured by microsatellites. Proc R Soc Lond B Biol Sci 265:803–809

Condit R, Hubbell SP (1991) Abundance and DNA sequence of two-base repeat regions in tropical tree genomes. Genome 34:66–71

Cornuet JM, Luikart G (1996) Description and power analysis of two tests for detecting recent population bottlenecks from allele frequency data. Genetics 144:2001–2014

Coulson TN, Pemberton JM, Albon SD, Beaumont M, Marshall TC, Slate J, Guiness FE, Clutton-Brock TH (1998) Microsatellites reveal heterosis in red deer. Proc R Soc Lond B Biol Sci 265:489–495

David P, Delay B, Berthou P, Jarne P (1995) Alternative models for allozyme-associated heterosis in the marine bivalve *Spisula ovalis*. Genetics 139:1719–1726

De Jong PW, Frandsen HO, Rasmussen L, Nielsen JK (2000) Genetics of resistance against defences of the host plant *Barbarea vulgaris* in a Danish flea beetle population. Proc R Soc Lond B Biol Sci 267:1663–1670

Dobrowolski MP, Tommerup IC, O'Brien PA (1998) Microsatellites in the mitochondrial genome of *Phytophthora cinnamomi* failed to provide highly polymorphic markers for population genetics. FEMS Microbiol Lett 163:243–248

Drake JA, Mooney HA, di Castri F, Groves RH, Kruger FJ, Rejmanek M, Williamson M (1989) Biological invasions: a global perspective. Wiley, New York

Dungey HS, Potts BM, Whitham TG, Li HF (2000) Plant genetics affects arthropod community richness and composition: evidence from a synthetic eucalypt hybrid population. Evolution 54:1938–1946

Eckert H, Breitschuh G, Sauerbeck D (1999) Kriterien umweltverträglicher Landbewirtschaftung (KUL) - ein Verfahren zur ökologischen Bewertung von Landwirtschaftsbetrieben. Agribiol Res 52:57–76

Epperson BK (2003) Geographical genetics. Princeton University Press, Princeton

Excoffier L, Smouse PE, Quattro JM (1992) Analysis of molecular variance inferred from metric distances among DNA haplotypes: application to human mitochondrial DNA restriction data. Genetics 131:479–491

Feldman MW, Bergman A, Pollock DD, Goldstein DB (1997) Microsatellite genetic distances with range constraints: analytic description and problems of estimation. Genetics 145:207–216

Francis AP, Currie DJ (2003) A globally consistent richness–climate relationship for angiosperms. Am Nat 161:523–536

Frydenberg O (1963) Population studies of a lethal mutant in *Drosophila melanogaster*. I. Behaviour in populations with discrete generations. Hereditas 50:89–116

Galloway LF, Fenster CB (2000) Population differentiation in an annual legume: local adaptation. Evolution 54:1173–1181

Gaudeul M, Taberlet P, Till-Bottraud I (2000) Genetic diversity in an endangered alpine plant, *Eryngium alpinum* L. (Apiaceae), inferred from amplified fragment length polymorphism markers. Mol Ecol 9:1625–1637

Geier U, Meudt M, Rudloff B, Urfei G (1999) Entwicklung von Parametern und Kriterien als Grundlage zur Bewertung ökologischer Leistungen und Lasten der Landwirtschaft – Indikatorensysteme. Umweltforschungsplan des Bundesministeriums für Umwelt, Naturschutz und Reaktorsicherheit, Forschungsbericht, Berlin, 258 pp

Goldstein DB, Ruis Linares A, Cavalli-Sforza LL, Feldman MW (1995a) Genetic absolute dating based on microsatellites and the origin of modern humans. Proc Natl Acad Sci USA 92:6723–6727

Goldstein DB, Ruis Linares A, Cavalli-Sforza LL, Feldman MW (1995b) An evaluation of genetic distances for use with microsatellite loci. Genetics 139:463–471

Gomez JM (2003) Herbivory reduces the strength of pollinator-mediated selection in the Mediterranean herb *Erysimum mediohispanicum*: consequences for plant specialization. Am Nat 162:242–256

Gupta PK, Varshney RK, Sharma PC, Ramesh B (1999) Molecular markers and their applications in wheat breeding. Plant Breed 118:369–390

Hale ML, Lurz PW, Shirley MD, Rushton S, Fuller RM, Wolff K (2001) Impact of landscape management on the genetic structure of red squirrel populations. Science 293:2246–2248

Hansson B, Westerberg L (2002) On the correlation between heterozygosity and fitness in natural populations. Mol Ecol 11:2467–2474

Hansson B, Bensch S, Hasselquist D, Åkesson M (2001) Microsatellite diversity predicts recruitment of sibling great reed warblers. Proc R Lond B Biol Sci 268:1287–1291

Hare MP, Karl SA, Avise JV (1996) Anonymous nuclear DNA markers in the American oyster and their implications for the heterozygote deficiency phenomenon in marine bivalves. Mol Biol Evol 13:334–345

Harris H (1966) Enzyme polymorphism in man. Proc R Soc Lond B Biol Sci 164:298–310

He H, Dezonia BE, Mladenoff D (2000) An aggregation index (AI) to quantify spatial patterns of landscapes. Landscape Ecol 15:591–601

Henrichsmeyer W, Cypris C, Löhe W, Meudt M, Sander R, Sothen F, Isermeyer F, Schefski A, Schleef KH, Neander E, Fasterding F, Helmcke B, Neumann M, Nieberg H, Mangegold D, Meier T (1996) Entwicklung des gesamtdeutschen Agrarsektormodells RAUMIS 96. Endbericht zum Kooperationsprojekt. Forschungsbericht für das BML (94 HS 021), Bonn/Braunschweig, 393 pp

Hodges SA, Arnold ML (1994) Floral and ecological isolation between *Aquilegia formosa* and *Aquilegia pubescens*. Proc Natl Acad Sci USA 91:2493–2496

Hoffmann MH, Glass AS, Tomiuk J, Schmuths H, Fritsch RM, Bachmann K (2003) Analysis of molecular data of *Arabidopsis thaliana* (L.) Heynh. (Brassicaceae) with geographical information systems (GIS). Mol Ecol 12:1007–1019

Holderegger R, Stehlik I, Abbott RJ (2002) Molecular analysis of the Pleistocene history of *Saxifraga oppositifolia* in the Alps. Mol Ecol 11:1409–1418

Hubby JL, Lewontin RC (1966) A molecular approach to the study of genic heterozygosity in natural populations. I. The number of alleles at different loci in *Drosophila pseudoobscura*. Genetics 54:577–594

Hülsbergen KJ (2003) Entwicklung und Anwendung eines Bilanzierungsmodells zur Bewertung der Nachhaltigkeit landwirtschaftlicher Systeme. Shaker, Aachen, Germany

Jacquemyn H, van Rossum F, Brys R, Endels P, Hermy M, Triest L, de Blust G (2003) Effects of agricultural land use and fragmentation on genetics, demography and population persistence of the rare *Primula vulgaris*, and implications for conservation. Belg J Bot 136:5–22

Jeffreys AJ, Wilson V, Thein SL (1985) Hypervariable 'minisatellite' regions in human. Nature 314:67–73

Jelinski DE (1997) On genes and geography: a landscape perspective on genetic variation in natural plant populations. Landscape Urban Plan 39:11–23

Jelinski DE, Wu J (1996) The modifiable areal unit problem (MAUP) and implications for landscape ecology. Landscape Ecol 11:120–140

Jones CJ,. Edwards KJ, Castaglione S, Winfield MO, Sala F, van de Wiel C, Bredemeijer G, Vosman B, Matthes M, Daly A, Brettschneider R, Bettini P, Buiatti M, Maestri E, Malcevschi A, Marmiroli N, Aert R, Volckaert G, Rueda J, Linacero R, Vazquez A, Karp A (1997) Reproducibility testing of RAPD, AFLP and SSR marker in plants by a network of European laboratories. Mol Breed 3:381–390

Kalinowski ST (2002) Evolutionary and statistical properties of three genetic distances. Mol Ecol 11:1263–1273

Kittelson PM, Maron JL (2001) Fine-scale genetically based differentiation of life-history traits in the perennial shrub *Lupinus arboreus*. Evolution 55:2429–2438

Kuhlmann F, Möller D, Weinmann B (2002) Modellierung der Landnutzung – Regionshöfe oder Raster-Landschaft. Ber Landwirtsch Z Agrarpolitik Landwirtsch Themenheft 'Multifunktionalität der Landnutzung im Rahmen des Sonderforschungsbereiches' 299:351–392

Lai Y, Sun F (2003) The relationship between microsatellite slippage mutation rate and the number of repeat units. Mol Biol Evol 20:2123–2131
Latter BDH (1973) Measures of genetic distance between individuals and populations. In: Morton NE (ed) Genetic structure of populations. University Press of Hawaii, Hawaii, pp 27–37
Li YC, Korol AB, Fahima T, Beiles V, Nevo E (2002) Microsatellites: genome distribution, putative functions and mutational mechanisms: a review. Mol Ecol 11:2453–2465
Loxdale HD, Lushai G (1998) Molecular markers in entomology (review). Bull Entomol Res 88:577–600
Lubjuhn T, Schwaiger FW, Epplen JT (1994) The analysis of simple repeat loci as applied in evolutionary and behavioral sciences. In: Schierwater B, Streit B, Wagner GP, DeSalle R (eds) Molecular ecology and evolution: approaches and applications. Birkhäuser, Basel, pp 33–43
Lynch M, Milligan BG (1994) Analysis of population genetic structure with RAPD markers. Mol Ecol 3:91–99
Maki M (2003) Population genetics of threatened wild plants in Japan. J Plant Res 116:169–174
Maughan PJ, Saghai Maroof MA, Buss GR, Huestis GM (1996) Amplified fragment length polymorphism (AFLP) in soybean: species diversity, inheritance, and near-isogenic line analysis. Theor Appl Genet 93:392–401
Mayr E (1988) Toward a new philosophy of biology – observations of an evolutionist. The Belknap Press of Harvard University Press, Cambridge, Massachusetts
Meudt M (1998) Weiterentwicklung und Anwendung eines Umweltindikatoren- und Politikinformationssystems für die Landwirtschaft der Bundesrepublik Deutschland: dargestellt am Beispiel der Treibhausproblematik. PhD Thesis, University of Bonn, Germany
Michalakis Y, Excoffier L (1996) A generic estimation of population subdivision using distances between alleles with special reference for microsatellite loci. Genetics 142:1061–1064
Möller D, Kirschner M, Weinmann B, Kuhlmann F (1998) Regionale Landnutzungsplanung und GIS: Bioökonomische Modellierung zur Unterstützung politischer Entscheidungsprozesse mit ProLand. Ber Ges Informatik Land-, Forst- Ernährungswirtsch 11:98–104
Möller D, Fohrer N, Steiner N (2002) Quantifizierung regionaler Multifunktionalität land- und forstwirtschaftlicher Nutzungssysteme. Ber Landwirtsch Z Agrarpolitik Landwirtsch Themenheft 'Multifunktionalität der Landnutzung im Rahmen des Sonderforschungsbereiches' 299:393–418
Morgante M, Olivieri AM (1993) PCR-amplified microsatellites as markers in plant genetics. Plant J 3:175–182
Mutikainen P, Salonen V, Puustinen S, Koskela T (2000) Local adaptation, resistance, and virulence in a hemiparasitic plant–host plant interaction. Evolution 54:433–440
Nakagahra M, Okuno K, Vaughan D (1997) Rice genetic resources: history, conservation, investigative characterization and use in Japan. Plant Mol Biol 35:69–77
Nei M (1972) Genetic distance between populations. Am Nat 106:283–292
Nei M (1973) Analysis of gene diversity in subdivided populations. Proc Natl Acad Sci USA 70:3321–3323
Nei M (1975) Molecular population genetics and evolution. North-Holland, Amsterdam
Nei M (1987) Molecular evolutionary genetics. Columbia University Press, New York
Nei M, Li WH (1979) Mathematical models for studying genetic variation in terms of restriction endonucleases. Proc Natl Acad Sci USA 76:5269–5273
Nevo E (1990) Molecular evolutionary genetics of isozymes: pattern, theory, and application. Progr Clin Biol Res 344:701–742
O'Hanlon PC, Peakall R, Briese DT (2000) A review of new PCR-based genetic markers and their utility to weed ecology. Weed Res 40:239–254

O'Neill R, Snowdon R, Köhler W (2003) Population genetics: aspects of biodiversity. In: Esser K (ed) Progress in botany, vol 64. Springer, Berlin Heidelberg New York, pp 115–137
Otha T (1971) Associative overdominance caused by linked detrimental mutations. Genet Res 18:277–286
Otha T, Kimura M (1973) A model of mutation appropriate to estimate the number of electrophoretically detectable alleles in a finite population. Genet Res 22:201–204
Powell W, Morgante M, Andre C, McNicol JW, Machray GC, Doyle JJ, Tingey SV, Rafalski JA (1995) Hypervariable microsatellites provide a general source of polymorphic DNA markers for the chloroplast genome. Curr Biol 5:1023–1029
Powell W, Morgante M, McDevitt R, Vendramin GG, Rafalski JA (1996) Polymorphic simple sequence repeat regions in chloroplast genomes: applications to the population genetics of pines. Proc Natl Acad Sci USA 92:7759–7763
Pritchard JK, Feldman MW (1996) Statistics for microsatellite variation based on coalescence. Theor Popul Biol 50:325–344
Rand DM (1994) Concerted evolution and RAPing in mitochondrial VNTRs and the molecular geography of cricket populations. In: Schierwater B, Streit B, Wagner GP, DeSalle R (eds) Molecular ecology and evolution: approaches and applications. Birkhäuser, Basel, pp 227–245
Roedenbeck IAE (2004) Bewertungskonzepte für eine nachhaltige und umweltverträgliche Landwirtschaft – Fünf Verfahren im Vergleich. BIOGUM-Forschungsbericht/BIOGUM-Research Paper, FG Landwirtschaft, Hamburg, no. 8, 154 pp. http://www.ag-change.de/html/downloads.html
Rosquist G, Prentice HC (2000) Habitat fragmentation and the structure of genetic diversity within disjunct isolates of *Anthericum ramosum* L. (Anthericaceae) in Scandinavia. Biol J Linn Soc 69:193–212
Ross P, Hall L, Smirnov I, Haff L (1998) High level multiplex genotyping by MALDI-TOF mass spectrometry. Nature Biotech 16:1347–1351
Rossetto M, Weaver PK, Dixon KW (1995) Use of RAPD analysis in devising conservation strategies for the rare and endangered *Grevillea scapigera* (Proteaceae). Mol Ecol 4:321–329
Rowe G, Beebee TJC (2001) Fitness and microsatellite diversity estimates were not correlated in two outbred anuran populations. Heredity 87:558–565
Schwartz D (1960) Genetic studies of mutant isozymes in maize: synthesis of hybrid enzymes by heterozygotes. Proc Natl Acad Sci USA 57:12020–1206
Slatkin M (1995) A measure of population subdivision based on microsatellite allele frequencies. Genetics 139:457–462
Solé E, Balanyà J, Sperlich D, Serra L (2002) Long-term changes in the chromosomal inversion polymorphism of *Drosophila subobscura*. I. Mediterranean populations from southwestern Europe. Evolution 56:830–835
Soulé ME (1986) Conservation biology – the science of scarcity and diversity. Sinauer, Sunderland, Massachusetts
SRU – Rat von Sachverständigen für Umweltfragen (1985) Umweltprobleme der Landwirtschaft. Sondergutachten. Kohlhammer, Stuttgart
Steiner N (2002) Modellierung der Artendiversität auf verschiedenen Skalenebenen in Abhängigkeit von der Landschaftsstruktur. Interdisziplinäres Expertentreffen im Rahmen des Übereinkommens über die biologische Vielfalt. Treffpunkt Biologische Vielfalt II. Bundesamt für Naturschutz, Bonn-Bad Godesberg, pp 205–208
Takezaki N, Nei M (1996) Genetic distances and reconstruction of phylogenetic trees from microsatellite DNA. Genetics 144:389–399
Tautz D (1993) Notes on the defunction and nomenclature of tandemly repetitive DNA sequences. In: Pena SDJ, Chakraborty R, Epplen JT, Jeffreys AJ (eds) DNA fingerprinting: state of the science. Birkhäuser, Basel, pp 21–28
Thelen GC, Allendorf FW (2001) Heterozygosity-fitness correlations in rainbow trout: effects of allozyme loci or associative overdominance? Evolution 55:1180–1187

Tomiuk J, Bachmann L (2002) Populations, species and communities. In: Contrafatto G, Minelli A (eds) Biological systematics (subject category: biological and medical sciences). Encyclopedia of life support systems (EOLSS), developed under the auspices of the UNESCO. EOLSS, Oxford, UK (http://www.eolss.net)
Tomiuk J, Loeschcke V (1991) A new measure of genetic identity between populations of sexual and asexual species. Evolution 45:1685–1694
Tomiuk J, Loeschcke V (1992) Evolution of parthenogenesis in the *Otiorhynchus scaber* complex. Heredity 68:391–397
Tomiuk J, Loeschcke V (1994) On the origin of polyploid parthenogenetic races in the weevil *Polydrusus mollis* (Coleoptera: Curculionidae). J Theor Biol 167:89–92
Tomiuk J, Loeschcke V (1995) Genetic identity combining mutation and drift. Heredity 74:607–615
Tomiuk J, Loeschcke V (1996) A maximum-likelihood estimator of the genetic identity between polyploid species. J Theor Biol 179:51–54
Tomiuk J, Loeschcke V (2003) Comments on: evolutionary and statistical properties of three genetic distances (Kalinowski 2002). Mol Ecol 12:2275–2277
Tomiuk J, Guldbrandtsen B, Loeschcke V (1998) Population differentiation through mutation and drift – a comparison of genetic identity measures. Genetica 102/103:545–558
Tsitrone A, Rousset F, David P (2001) Heterosis, marker mutational processes and population inbreeding history. Genetics 159:1845–1859
Tullu A, Buchwaldt L, Warkentin T, Taran B, Vandenberg A (2003) Genetics of resistance to anthracnose and identification of AFLP and RAPD markers linked to the resistance gene in PI 320937 germplasm of lentil (*Lens culinaris* Medikus). Theor Appl Genet 106:428–434
Van Valen LM (1973) A new evolutionary law. Evol Theory 1:1–30
Vigouroux Y, Jaqueth JS, Matsuoka Y, Smith OS, Beavis WD, Smith JS, Doebley J (2002) Rate and pattern of mutation at microsatellite loci in maize. Mol Biol Evol 19:1251–1260
Waldhard R, Fuhr-Bossdorf K, Otte A, Schmidt J, Simmering D (2000) Floristisch-vegetationskundliche Diversitäten einer peripheren Kulturlandschaft in Abhängigkeit von Landnutzung, Raum und Zeit. Agrarspektrum 31:121–147
Waldman M, Shevah Y (2000) Biological diversity – an overview. Water Air Soil Pollut 123:299–310
Weinberg W (1908) Über den Nachweis der Vererbung beim Menschen. Jahresh Ver Vaterl Naturkd Württemb 64:368–382
Wiens JA (1992) What is landscape ecology, really? Landscape Ecol 7:149–150
Witting L, Tomiuk J, Loeschcke V (2000) Conservation of interacting species. Ecol Model 125:123–143
Wright S (1943) Isolation by distance. Genetics 28:114–138
Wright S (1951) The genetical structure of populations. Ann Eugen 15:323–354
Wright S (1965) The interpretation of population structure by F-statistics with special regard to systems of mating. Evolution 19:395–420
Zander PM (2003) Agricultural land use and conservation options – a modelling approach. Dissertation, Wageningen University, The Netherlands, 222 pp
Zar JH (1996) Biostatistical analysis, 3rd edn. Prentice Hall, Englewood Cliffs,718 pp
Zhang L, Peek AS, Dunams D, Gaut BS (2002) Population genetics of duplicated disease-defense genes, hm1 and hm2, in maize (*Zea mays* ssp. *mays* L.) and its wild ancestor (*Zea mays* ssp. *parviglumis*). Genetics 162:851–860
Zhao X, Kochert G (1993) Phylogenetic distribution and genetic mapping of a (GGC)n microsatellite from rice (*Oryza sativa* L.). Plant Mol Biol 21:607–614
Zouros E, Foltz DW (1987) The use of allelic isozyme variation for the study of heterosis. Isozymes Curr Top Biol Med Res 17:1–59

Jürgen Tomiuk
Institute of Anthropology and Human Genetics
University of Tübingen, Germany

Inga A. Roedenbeck
Wolfgang Köhler
Institute of Biometry and Population Genetics
University of Giessen, Germany

Plant Breeding: Recent Advances in Molecular Breeding of Oilseed Rape (*Brassica napus* L.)

Rod Snowdon, Katrin Link, Ana Gloria Badani, and Wolfgang Friedt

1 Introduction

Plant genetic research and breeding have made enormous progress in recent decades with the help of DNA marker techniques for genetic and physical mapping. In particular, the complete sequencing of the model crucifer *Arabidopsis thaliana* (Arabidopsis Initiative 2000), to which *Brassica* species represent the closest crop relatives, has opened the way for detailed comparative investigations into the complex structure of *Brassica* genomes (Quiros et al. 2001; Schmidt et al. 2001; Lukens et al. 2003; Parkin et al. 2003). Together with new, highly informative and high-throughput marker technologies, the DNA sequence and gene data now available from *Arabidopsis* have great potential for use in genetic analysis and breeding of *Brassica* crops. There is considerable potential to use molecular markers to develop new varieties where traditional approaches are unable to achieve breeding aims. New techniques for allele–trait association studies have the potential to further increase the role of markers in *Brassica* breeding, because the genomic information that can potentially be gained will no longer be applicable to only isolated crosses segregating for specific traits of interest. This chapter summarises the present situation with regard to genetic mapping in oilseed rape (*Brassica napus* L.), focusing on broadly used marker systems and their applications. Furthermore, we discuss the importance and potential of new developments in marker technologies and physical mapping concepts with a view to oilseed rape breeding and genetics.

Generally, 80–90% homology is found between the exons of putative orthologous genes in *Arabidopsis* and *Brassica* (reviewed by Schmidt 2002), meaning that knowledge from *Arabidopsis* is highly relevant for gene isolation and characterisation in *Brassica* crops. In comparative studies of genome regions flanking known genes, considerable microsyntenic collinearity has been observed between *Arabidopsis* and *Brassica* genome segments (Sadowski et al. 1996; Cavell et al. 1998). However, minor deletions, insertions and translocations are relatively common in regions surrounding *Brassica* orthologues of *Arabidopsis* genes. For example, Quiros et al. (2001) described small-scale collinearity between the region of *A. thaliana* chromosome 4 containing the *ABI1-Rps2-Ck1* gene complex and a homoeolo-

Progress in Botany, Vol. 66

gous segment of *B. oleracea* chromosome 4. Although almost complete microsynteny was observed, the *B. oleracea* regions contained an extra gene with homology to genes located on *Arabidopsis* chromosomes 2 and 5. In other words, even regions with well-preserved collinearity on a microsyntenic scale can be interrupted by translocations. In fact one or several homologues of *Arabidopsis* genes may be missing from any particular triplicated region in *Brassica* (Schmidt 2002). The different orthologous genes in *Brassica* are therefore often comprised of a different set of genes. Nevertheless, due to the large-scale synteny over long chromosome stretches it is still often feasible to utilise sequence information from markers flanking genes or quantitative trait loci (QTL) of interest in *Brassica* crops, in order to identify possible candidate genes from the corresponding chromosome regions in *Arabidopsis*. For example, different homoeologous regions in *B. rapa* and *B. napus*, which contain various flowering time QTL, each show significant collinearity to *Arabidopsis* chromosome sections containing a number of genes relevant to the onset of flowering (Lagercrantz et al. 1996; Osborn et al. 1997).

2 Molecular Marker Techniques and Their Application in *Brassica napus*

During the past two decades, a large number of different DNA marker techniques have been established and used in crop plant genetics for applications ranging from genetic diversity analysis to linkage mapping and physical gene isolation. Most of the work utilising molecular markers in *Brassica* oilseed breeding has to date been based on genetic mapping using various DNA marker systems, in single segregating populations generated for specific investigations of particular traits of interest. As PCR techniques have developed over the last 15 years, a wealth of new marker technologies have arisen, many of which have been adopted or adapted for use in *Brassica* crops. Molecular markers linked to numerous agronomically important traits have been reported, and a number are now successfully integrated in oilseed breeding programs. New and potentially powerful marker systems like single-nucleotide polymorphisms are still beyond the technical capacity of most end-use laboratories, but their usage can be expected to become routine as cheap, high-throughput assays are developed.

2.1 Widespread Marker Systems

In commercial plant breeding, where cost and ease of use are a high priority in small to mid-sized breeding companies, protein and isoenzyme markers still play a role for specific marker-assisted selection strategies or homogeneity testing. On the other hand, DNA marker applications and genome research in *Brassica* have developed continually since the late 1980s when the first restriction fragment length polymorphism (RFLP) linkage maps were developed for *B. oleracea* (Slocum et al. 1990), *B. rapa* (Song et al. 1991) and *B. napus* (Landry et al. 1991). The polymerase chain reaction (PCR; Mullis and Faloona 1987) delivered the potential to greatly increase the marker density in existing genetic maps through amplification of highly polymorphic PCR fragments, first with randomly amplified polymorphic DNA markers (RAPD; Williams et al. 1990) and later with simple sequence repeat markers (SSR; Grist et al. 1993), inter-simple sequence repeats (ISSR; Zietkewitz et al. 1994) and amplified fragment length polymorphisms (AFLP; Vos et al. 1995). The conversion of anonymous PCR markers closely linked to traits of interest into sequence-characterised amplified region (SCAR) or sequence-tagged site (STS) markers has led to simple PCR-based markers that meet the technical and financial constraints of commercial rapeseed breeders. SSR markers, also known as microsatellites, are highly polymorphic and robust and their simple, relatively inexpensive analysis means they are highly suitable for low-cost, high-throughput commercial use. The predominance of such sequences in coding DNA regions has in some cases led to the use of SSR markers for marker-assisted selection (MAS) of simple traits in oilseed rape. The co-dominant nature of microsatellite polymorphisms also makes them particularly useful for map alignment among different crosses. The number of publicly available *Brassica* SSR primers is increasing as a result of publicly funded international initiatives (see www.brassica.info/ssr/SSRinfo.htm); however, in comparison to other important crop species, relatively few markers are freely available, meaning that although some consensus maps have been reported (e.g. Lombard and Delourme 2001), an internationally accepted consensus nomenclature has to date not been agreed. A public set of well-characterised SSR markers would greatly assist in consensus mapping and integration of the numerous non-standardised genetic maps that have been developed worldwide from different *B. napus* mapping populations.

2.2 New Technologies: Single-Nucleotide Polymorphisms

Single-nucleotide polymorphisms (SNPs) resulting from single-base substitutions in the DNA sequence are the most abundant form of DNA polymorphism in most organisms. This abundance means they represent the ultimate tool for extremely fine genetic mapping, to uncover allelic variation directly within expressed sequences, and to develop haplotypes based on gametic phase disequilibrium for analyses of quantitative traits. In contrast to fragment length polymorphisms, SNPs cannot be resolved by conventional gel electrophoresis, and a variety of novel techniques and technologies have been developed or adapted for SNP detection in recent years.

In *B. napus* (genome AACC, 2n=38, approx. 1,200 Mbp) at least two and often more copies – either active or non-transcribed – of any given candidate gene or marker sequence can be expected. For development of allele-specific SNP markers it is therefore necessary that genuine locus-specific SNPs be distinguishable from sequence polymorphisms among homoeologous loci. A promising approach to locus-specific SNP development in homoeologous *B. napus* genes is the use of comprehensive sets of physical functional markers in the genome regions directly upstream and downstream from *A. thaliana* candidate genes. Using detailed sequence data in polymorphic promoter regions, introns or other gene-flanking sequences can lead to more success in unambiguous identification of orthologous gene loci in *B. napus*. Furthermore, SNPs are potentially more frequent in non-coding than in coding regions; hence a high density of markers in linkage disequilibrium to candidate genes for target traits is feasible.

To date most SNP detection protocols have been based on target sequence PCR amplification combined with fluorescent labelling and/or enzymatic assays. Many of the PCR-based methodologies are time-consuming and costly, but they have considerable potential for automation or miniaturisation using microarray techniques (DNA chips). The use of matrix-assisted laser desorption/ionisation time of flight mass spectrometry (MALDI-TOF MS) now allows extremely high-throughput SNP detection and analysis (see Ross et al. 1998). SNP analysis by MALDI-TOF is initiated by multiplex PCR amplification of genomic DNA, using primers flanking up to 20 or more SNPs of interest. A locus-specific primer adjacent to the variable nucleotide of each SNP is then used in a base-extension reaction, with primer-oligonucleotide extension being performed with DNA polymerase in the presence of dideoxynucleotides that terminate the reactions at a specific length determined by the SNP allele. The extended products are then purified and subjected to mass spectrometric analysis. More recent adaptations utilise specific fragmentation procedures rather than chain termination to generate sets of short oligonucleotides, and these are then compared to sets of fragments derived from known homologous DNA sequences. In this way, whole PCR products can be compared for SNPs along their entire length in one assay, without the need to use specific primers for interrogation of each individual nucleotide. Automated MALDI-TOF MS enables the investigation of many tens of thousands of SNP data-points per day, and despite high set-up costs the throughput capacity makes assay costs per sample relatively cheap. In most cases SNP detection still relies on techniques and equipment that are unaffordable for most plant breeders or smaller research laboratories; however, SNP markers will certainly play a major role in *Brassica* genetics and breeding in coming years as high-throughput SNP discovery and genotyping services developed largely for human genetics become more accessible and cost-effective.

2.3 Genetic Diversity of *Brassica napus*

Molecular marker techniques are particularly suitable for in-depth studies of genetic diversity among collections of varieties or wild material from important crop plants. Because the gene pool of oilseed rape breeding material has been considerably narrowed by the emphasis on specific quality traits, genetic variability is restricted with regard to many valuable characters. Hence the construction of genetic pools, as used, for example, in maize hybrid breeding, is not feasible for oilseed rape. On the other hand, it is well known from molecular marker analyses that winter and spring rapeseed genotypes constitute two genetically different groups (Becker et al. 1995; Diers et al. 1996), and as in other crops genetic distance in oilseed rape is generally correlated with heterosis (Diers et al. 1996). In a recent study, Shengwu et al. (2003) found considerable genetic diversity between European and Chinese oilseed rape using RAPD markers. Using SSR markers, Hasan et al. (2003) found remarkable genetic variation in exotic vegetable and fodder rape genotypes compared to the gene pools of conventional spring and winter oilseed material. Furthermore, a small number of winter and spring oilseed rape accessions were identified that showed large genetic diversity compared to the main pools of these types. Extreme genetic variation compared to conventional rapeseed cultivars was also found in resynthesised rapeseed lines analysed by Becker et al. (1995) using allozyme and RFLP markers, and in further resynthesised rapeseed material investigated by Seyis et al. (2003a) using AFLP markers. In the latter study, the genetic differences were correlated to heterotic yield potential in experimental hybrids (Seyis et al. 2003b). Such material is not only genetically but also often phenotypically very divergent from conventional rapeseed genotypes, and hence must be viewed from a long-term perspective with regard to use in oilseed rape breeding. However, this example shows that genetic diversity analyses using molecular markers have the potential to identify novel genetic variation that might assist in future improvement of heterotic potential in *B. napus*. On the other hand, multiplex SSR markers are also an effective tool to test oilseed rape varieties for genetic distinctness, uniformity and stability (Tommasini et al. 2003).

3 Genetic Mapping

3.1 Comparative Mapping Between *Arabidopsis* and *Brassica*

As the genome relationships between *Arabidopsis* and *Brassica* have been unravelled (e.g. Schmidt et al. 2001), the model plant has developed into the most important resource for gene isolation and characterisation in *Brassica* crops. In recent years, it has become increasingly feasible to integrate genetic mapping with a candidate gene approach (Pflieger et al. 2001) using *Arabidopsis* resources and genome tools to identify gene loci involved in both simple and complex traits. However, the use of *Arabidopsis* information in marker development, map-based gene cloning and candidate gene identification in *B. napus* is complicated by the complex arrangement of the (ancestral) polyploid *Brassica* genomes (see Lagercrantz and Lydiate 1996; Sadowski et al. 1996; Lagercrantz 1998). Fourmann et al. (2002) and Chalhoub et al. (2003) described a strategy to use functional PCR markers for physical mapping of *A. thaliana* gene loci in *B. napus*, by using syntenic regions surrounding candidate genes for better characterisation of orthologous locus copies. BAC clones identified by gene-specific filter hybridisation or PCR were separated into locus-specific contigs through the presence or absence of PCR markers amplified using specific primers within or flanking the relevant gene sequence. These markers not only enable the variable intron-exon structure of the orthologous gene copies to be better characterised, but also can be used as a basis for development of locus-specific SNPs. The latter are a promising tool for allele–trait association studies of relevant candidate genes, and simultaneously provide a basis for integration of the loci in physical functional genetic maps.

The growing availability of *Arabidopsis* EST collections and their localisation in *Brassica* genetic maps (e.g. Fourmann et al. 2002; Babula et al. 2003) enables the fine mapping of genome rearrangements in *Brassica* genomes and the delineation of gene-coding regions. This allows a better correlation of traits in *Brassica* crops with *Arabidopsis* candidate genes and development of genetic markers considerably more closely linked to the relevant genes. It is also possible, however, to develop useful anonymous PCR markers that are nevertheless enriched in gene-coding regions. Li and Quiros (2001) described sequence-related amplified polymorphic (SRAP) markers for *Brassica* that preferentially amplify and detect polymorphisms in open reading frames (ORFs), meaning that markers could be developed to saturate regions surrounding a particular gene without knowledge of the gene sequence. SRAP markers are based on primers containing CCGG-motifs, which anneal preferentially in ORFs due to the predominance of GC bases in coding sequences. Exonic sequences generally exhibit few or no length polymorphisms; hence the ORF primers were combined with primers containing an AATT core near their 3' end. These sequences occur more frequently in promoter regions and introns that are prone to sequence variation, meaning that the resulting anonymous exon-intron markers showed relatively high polymorphism. After sequencing the SRAP amplification products it was found that a large proportion of the markers matched known genes, and their utility for gene tagging was demonstrated by localisation of a glucosinolate desaturation gene in *B. oleracea* (Li and Quiros 2001).

3.2 Conventional Linkage Mapping of Qualitative and Quantitative Traits

Numerous examples have demonstrated the power of high-density genetic maps for identification of genetic markers closely linked to agronomically important traits in oilseed rape. For example, many studies have concentrated on mapping of genome regions controlling flowering time and vernalisation requirement (e.g. Ferreira et al. 1995c), and molecular markers are now available for fertility restoration genes for the cytoplasmic male sterility (CMS) systems 'Ogura' (Brown et al. 2003), 'Kosena' (Imai et al. 2003) and 'tournefortii' (Jeneja et al. 2003). Foisset et al. (1995) and Barret et al. (1998) described genetic mapping leading to a sequence-characterised amplified region (SCAR) marker linked to the dwarf gene *Bzh* in oilseed rape, and Fray et al. (1997) mapped loci for apetalous *B. napus* flowers. The development of genetic markers is of particular interest in efforts to select for complex quantitative traits that are strongly influenced by environment, for example the expression of yellow seed colour. We used AFLP and SSR markers to localise QTL associated with the yellow-seed trait in two independent oilseed rape mapping populations (Badani et al. 2003), and by aligning the maps using consensus SSR markers we were able to show that the QTL from the two different yellow-seeded *B. napus* sources probably represent the same genetic loci. The QTL and putative mode of inheritance corresponded with results published by Somers et al. (2001), who identified RAPD markers linked with a putative dominant gene locus and two further epistatic loci for seed colour in a segregating population derived from yet another source of yellow-seeded rape. QTL for winter survival and other related complex, environment-dependent traits were localised by Kole et al. (2002a).

Ecke et al. (1995) localised the two loci controlling erucic acid content in *B. napus* and identified loci contributing to seed oil content. Based on a candidate gene approach, Fourmann et al. (1998) developed polymorphic markers within the respective *B. napus* A- and C-genome *FAE1* gene loci and showed that these genes co-segregated with the erucic acid loci identified by Ecke et al. (1995). Based on this knowledge, Das et al. (2003) cloned the fatty acid elongase genes (*FAE1*) from *B. campestris* and *B. oleracea* and described their effects on erucic acid levels. A large number of QTL affect the overall seed fatty acid composition in oilseed rape. Burns et al. (2003) used marker-assisted backcrossing to develop a set of intervarietal chromosome substitution lines that enabled a more precise localisation of fatty acid QTL and their comparison with the map positions of genes related to fatty acid biosynthesis.

Seed glucosinolate accumulation in oilseed rape, a complex trait controlled by numerous genes (QTL), was studied in detail by Howell et al. (2003) by mapping in different intervarietal backcross populations and comparisons with previously published maps (e.g. Uzunova et al. 1995). Analysis of the QTL-containing genome regions in the two populations showed that the accumulation of glucosinolates was mainly influenced by three QTL in homoeologous regions of the *B. napus* genome, suggesting that duplicate genes are responsible for the trait. In addition, it was found that high-glucosinolate varieties often carried low-glucosinolate alleles at one or more loci, meaning that transgressive segregation for the trait is common.

Somers et al. (1998) identified RAPD markers associated with linoleic acid (C18:2) desaturation in *B. napus* and found that the gene *FAD3* was localised near one of the identified QTL for this trait. Hu et al. (1999) also found RAPD markers linked to linoleic/oleic acid content in rapeseed oil, and furthermore developed sequence-tagged markers associated with low linolenic acid content that appeared to be linked to an omega-3 (linolenic) desaturase gene. A gene determining linolenic acid (C18:3) content was also localised by Hu et al. (1995). Tanhuanpaa et al. (1995) identified a RAPD marker associated with a gene affecting linolenic acid content, whereas Jourdren et al. (1996) identified specific polymorphisms within one of the *B. napus FAD3* delta-desaturase gene copies that were also associated with variations in linolenic acid concentration. Schierholt et al. (2000) mapped a high oleic acid mutation in *B. napus* and showed that linked AFLP markers were localised near a copy of the *FAD2* gene. Previously, Tanhuanpaa et al. (1998) had identified an association between the *FAD2* gene, which encodes 18:1 desaturase, and a QTL for oleic acid (C18:1) in *B. rapa*. Allele-specific PCR markers for this locus were developed, based on a single SNP in the *B. rapa FAD2* gene that resulted in a functionally relevant amino acid substitution.

Genetic characterisation of disease resistances and marker development for resistance breeding are a high priority in oilseed rape breeding due to the importance of major diseases on yield potential and stability. Dion et al. (1995) and Pilet et al. (1998a,b, 2001), respectively, identified QTL for field resistance to blackleg in crosses of different genetic background, while Ferreira et al. (1995a) and Mayerhofer et al. (1997) mapped major loci associated with race-specific *Leptosphaeria maculans* (*Phoma*) resistance genes. RFLP and RAPD markers linked with B-genome *Phoma* (blackleg) resistance were converted into STS markers for use in marker-assisted backcrossing of the introgressed trait in *B. napus* (Plieske and Struss 2001). Wretblad et al. (2003) isolated the cDNA sequence *Lm1* from black mustard (*B. nigra*), which gave enhanced resistance to *Leptosphaeria maculans* when overexpressed in oilseed rape. Ferreira et al. (1995b) and Kole et al. (2002b) mapped loci conferring resistance to white rust (*Albugo candida*) in rapeseed, and Walsh et al. (1999) mapped genes responsible for turnip mosaic virus resistance. More recently, Zhao and Meng (2003a,b) have described loci contributing to resistance against *Sclerotinia sclerotiorum* (stem rot), which among the serious diseases of *B. napus* is one of the few for which to date little progress has been made in oilseed rape resistance breeding. With regard to *Verticillium* wilt (*Verticillium longisporum*), very little genetic variability exists in the *B. napus* gene pool; however, the identification of resistance in *B. oleracea* and its introgression into resynthesised *B. napus* (Happstadius et al. 2003) creates the opportunity to obtain mapping populations for elucidation of the genetic control of this resistance as well. Availability of genetic markers for resistance genes will assist in the combination of such genes, in order to establish more durable resistance against major oilseed rape pathogens.

Microspore cultivation is now used extensively by *Brassica* oilseed breeders for doubled haploid (DH) production; however, a considerable genotype effect is observed in the regenerative embryogenic ability of microspore-derived populations. Zhang et al. (2003) identified RAPD markers associated with improved microspore responsiveness in *B. rapa* and *B. napus*, indicating that specific genetic loci are involved in suppression of microspore embryogenesis.

3.3 Marker-Assisted Selection in Oilseed Rape

For a number of qualitative traits traditional mapping approaches have led to the development of marker-assisted selection strategies in oilseed *Brassica* breeding, and in some cases to map-based cloning of the responsible genes. A good example of this is selection for restorer genes for different cytoplasmic sterility systems using markers developed by conventional mapping techniques (see Delourme et al. 1998; Jean et al. 1998; Brown et al. 2003; Imai et al. 2003; Janeja et al. 2003). With such traits, whose expression can only be detected following the onset of flowering, selection procedures can be considerably accelerated if the necessity for phenotyping can be avoided. In the case of quantitative traits, however, conventional QTL mapping techniques have often not achieved the desired success in terms of commercial trait-marker development. Markers for QTL derived from a given mapping population are not necessarily transferable to other material, and in many cases the genetic distances between markers flanking QTL are physically very large. For this reason traditional QTL mapping is often not sufficient to develop effective markers for trait introgression or for identification and map-based cloning of the responsible genes. Successful marker-assisted selection for quantitative traits has therefore not been achieved to date.

3.4 Allele-Trait Association Mapping

The basis of QTL localisation is gametic phase disequilibrium (GPD) between simply inherited markers and genetic loci affecting the trait of interest. Although it is often referred to as linkage disequilibrium, GPD can also exist among non-linked alleles; however, it is most commonly observed among tightly linked markers (linkage disequilibrium). In general terms GPD refers to the phenomenon whereby particular sets of alleles (referred to as haplotypes) are preferentially inherited together in the same gamete. GDP is highest in populations derived from a low number of founders, where the haplotypes present in the founders will be more common than would be expected under equilibrium. In plant breeding such populations are common, the classic examples being DH or F_2 populations derived from a cross between two inbred lines: Because all segregating individuals are derived from a single F_1 genotype, the associations between loci can be predicted based on their mapping distances. This leads to a generally low resolution of QTL localisation, however; unless the mapping population is extremely large the recombination frequency will tend to be underestimated due to the greater extent of linkage disequilibrium surrounding the QTL. The accuracy of QTL localisation can be considerably improved when a larger number of meiotic events are considered, since these introduce a greater number of potential recombinations between any given marker and the trait locus. For example, recombinant inbred lines (RILs) can be employed for fine mapping of QTL.

Even if the resolution can be improved, however, the localisation of QTL for complex traits in conventional mapping populations is still limited to those loci for which the cross parents segregate. Association mapping, on the other hand, provides an opportunity to detect allele–trait associations

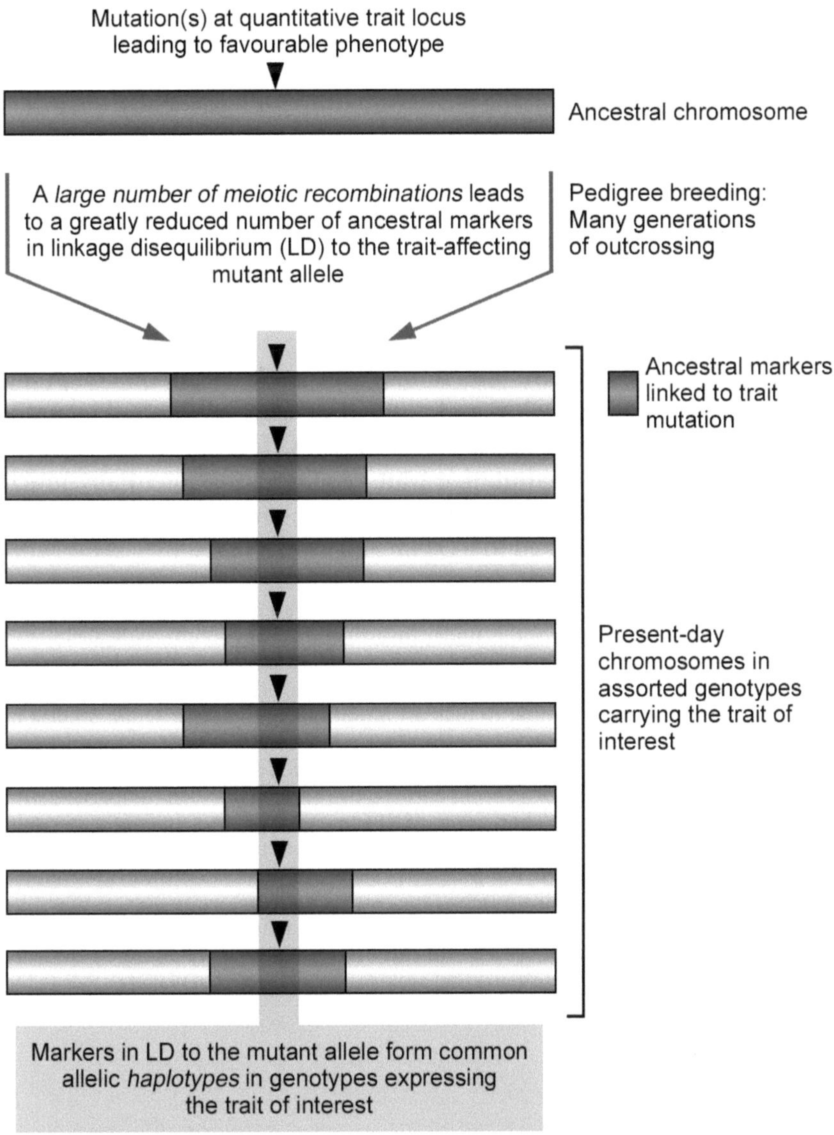

Fig. 1. Allele–trait association mapping techniques are based on gametic phase disequilibrium (normally linkage disequilibrium) between markers surrounding ancestral mutations that contribute allelic diversity to a quantitative trait of interest. After many generations of genetic recombination, for example following pedigree breeding, only extremely tightly linked markers will still be associated with the trait in all genotypes carrying the mutation

in larger sets of genotypes representing a much broader array of the genetic diversity available for a particular trait. Assuming many generations, and therefore meioses, have elapsed since a mutation for a particular quantitative trait arose in a species, recombination will have removed association between a QTL and any marker not tightly linked to it (Fig. 1). This situation occurs in non-structured sets of genetically diverse genotypes derived from pedigree breeding, for example, where ancestral mutations remain associated only with tightly linked markers; spurious associations with non-linked markers are more likely to be removed by recombination. Association mapping thus allows for much finer mapping than standard bi-parental cross approaches, where the number of meioses in the mapping population is generally very low and population sizes are limited by practical considerations. Allelic markers (haplotypes) found to have marker-trait association with a quantitative trait are therefore potentially much more useful for marker-assisted selection than markers found through classical mapping to be linked to a QTL.

The main requirement for successful allele–trait association studies is access to very closely linked markers in the genomic regions under investigation. For genome-wide allele–trait association studies achieving the required marker density is not yet feasible; however, high-throughput genotyping is increasing the availability of the necessary marker data. Particularly SNPs have the potential to play an important role in association studies in *Brassica* in future years.

Because of the extremely high frequency of single nucleotide polymorphisms, meiotic recombination events are less likely to occur between two adjacent SNPs in comparison with other types of markers or genes. Thus, several SNPs in a given DNA sequence may retain complete linkage disequilibrium for a considerable period of time. The degree of linkage disequilibrium among SNPs in *Brassica* genotype collections is still unknown; however, for a relatively young crop species like oilseed rape, with a limited genetic pool, it might be speculated that GDP is relatively well preserved in breeding material. In the meantime, the chance of uncovering allele–trait association can be considerably enhanced by deriving haplotypes from SNPs in candidate genes that are strongly suspected of playing a role in the expression of the trait of interest.

4 Physical Mapping

4.1 Candidate Gene Strategies and Physical Functional Maps

The complete sequencing of the *Arabidopsis thaliana* genome (Arabidopsis Initiative 2000) has provided the catalyst for a huge leap in progress towards understanding the organisation, control and function of the genes responsible for physiological and developmental regulation in the model plant. Understanding of gene function in *Brassica*, the closest crop relatives to *Arabidopsis*, will certainly be considerably accelerated by the information available from the model genome. However, the complex nature of the *Brassica* diploid species, which from their genome makeup are presumed to represent ancestral hexaploids (Lagercrantz and Lydiate 1996; Lagercrantz 1998), means that the ordering of sequences and markers on *Arabidopsis* chromosomes cannot be directly transferred to homoeologous regions on *Brassica* chromosomes. For the amphidiploid *Brassica napus* genome (AACC, 2n=38) the situation is further complicated by the fact that rapeseed is comprised of copies of the extremely similar genomes of *B. rapa* (AA, 2n=20) and *B. oleracea* (CC, 2n=18); however, the chromosomes of the respective A and C genomes are largely conserved in the *B. napus* genome (Parkin et al. 1995; Snowdon et al. 2002). In studies of the local preservation of defined *Arabidopsis* regions in the genomes of *B. rapa* and *B. oleracea*, Conner et al. (1998) and Cavell et al. (1998) were able to show that a certain degree of collinearity is still present, which makes it feasible in some cases to utilise *Arabidopsis* gene and surrounding sequence data for physical mapping in the crop genome. This small-scale interspecific genome alignment, described as microsynteny, is in many cases sufficient to allow limited use of gene, sequence or mapping data from *Arabidopsis* for physical mapping in *B. napus*. Lan et al. (2000) produced a detailed comparative physical map for *Brassica oleracea* by mapping *Arabidopsis* ESTs in a number of different *B. oleracea* and *A. thaliana* mapping populations. Correspondence between *Brassica* and *Arabidopsis* chromosomes accounted for 57% of comparative loci when conserved criteria were used to infer synteny. Based on 186 corresponding loci detected in *B. oleracea* and *A. thaliana*, at least 19 chromosome structural rearrangements were found to differentiate *B. oleracea* and *A. thaliana* orthologs. Chromosomal duplication in the *B. oleracea* genome was strongly suggested by parallel arrangements of duplicated loci on different chromosomes, which accounted for 41% of the loci mapped in *Brassica*.

4.2 Molecular Cytogenetic Techniques

Fluorescence in situ hybridisation (FISH) techniques not only allow more reliable chromosome identification in *Brassica* (Snowdon et al. 2002), but also potentially enable more accurate ordering of molecular markers and measurement of physical distances within marker contigs (Jackson et al. 2000). A combination of genetic and structural chromosome analysis can often assist in physical mapping of genome regions containing large repetitive DNA stretches, or where chromosomal recombination is scarce. Howell et al. (2002) used FISH localisation of mapped BAC clones to integrate an existing *B. oleracea* genetic map with karyotype information. Large-scale physical localisation of mapped large-insert clones will ultimately assist in complete integration of *Brassica* physical and genetic maps. Furthermore, high-resolution FISH can give important information about ordering and physical distances between molecular markers, which are both vital considerations for physical mapping and positional cloning. Kim et al. (2004) demonstrated the use of multi-colour FISH for selection of physically dispersed initial BAC start-points for the complete sequencing of *B. rapa* chromosome 1.

We have developed FISH methods for the accurate localisation of repetitive DNA sequences at chromosomal sub-arm level in *Brassica* species, allowing more reliable chromosome identification and giving new information on genome structure and evolution. In addition we have applied genomic in situ hybridisation (GISH) for identification and characterisation of parental genome components in oilseed rape (*B. napus*) hybrids. For example, GISH on meiotic chromosomes of asymmetric *B. napus* × *Crambe abyssinica* (Abyssinian kale) hybrids gave evidence for intergenomic chromosome recombination that led to novel fatty acid gene patterns in backcross offspring (Wang et al. 2004), and the successful transfer of resistance to *Leptosphaeria maculans* from *Sinapis arvensis* (wild mustard) to oilseed rape lines with a normal *B. napus* karyotype could also be corroborated by GISH (Snowdon et al. 2000). In another approach, a B-genome-specific repetitive sequence isolated from *B. nigra* (Schelfhout et al. 2004) was used as a FISH probe for characterisation of B-genome introgressions in *B. napus* × *B. juncea* (Indian mustard) hybrids containing various positive agronomic traits from the *B. juncea* donor.

4.3 Sequencing of *Brassica* Genomes

Although much progress can be made using *Arabidopsis–Brassica* microsynteny and candidate gene strategies, a complete understanding of gene distribution and function in crop genomes can be achieved only when these genomes themselves are sequenced. A recently grounded international *Brassica* genome initiative (http://www.brassicagenome.org) has initiated efforts for a multinational sequencing of the *B. rapa* genome, and the Korean National Institute of Agricultural Biotechnology began sequencing chromosome 1 of *B. rapa* (Kim et al. 2004). The project aims initially to sequence BAC clones representing the approximately 500-Mb genome of *B. rapa* subspecies *pekinensis*. In order to cut costs a second-phase sequence is foreseen, which although fully oriented and ordered will contain some small sequence gaps and low-quality sequences. The intention is that scientists requiring finished sequence from a specific region will be able to complete it by accessing trace files that will be accessible online. Annotation of the genome sequence will be possible by using the annotated *Arabidopsis* sequence. One important aim of the project is to integrate the completed *B. rapa* physical map with existing genetic maps by anchoring the genomic sequence to a reference genetic map with molecular markers. Initially 1,000 *B. rapa* BAC clones will be unambiguously anchored (via both end sequences) to the *Arabidopsis* genome sequence, and to a *B. rapa* genetic map, via single amplification-product markers (SSRs, SNPs or InDels).

The decision to use the A genome of *B. rapa* (n=10) was taken because this genome is slightly larger than the C genome of *B. oleracea* (n=9). Because the A and C genomes are highly homoeologous it is thought that their size difference arises from a slight excess in non-coding repetitive DNA in the A genome. This may be attributed to additional centromeric heterochromatin on the additional chromosome in *B. rapa*, since to date no evidence has been found to suggest that *B. rapa* possesses more genes than *B. oleracea*. From a functional genomics perspective this additional, presumably repetitive, sequence in *B. rapa* is not as relevant as coding DNA; however, it will certainly give insight into the genome evolution of the two closely related diploid species. Of course, for studies of *B. napus*, it will be advantageous when the genomic sequences of both genomes are available. The Institute for Genome Research (TIGR; www.tigr.org) has already made good progress in a project aimed at partial to whole-genome shotgun sequencing in *B. oleracea*. In this case, the information gained from the crop plant sequence is also to be used to improve the annotation of the model plant genome: Alignment of conserved sequences between *Arabidopsis* and *Brassica* enables identification of conserved regions outside currently annotated gene boundaries. These represent potential novel genes (both protein-coding and structural RNAs). At the time this chapter was written TIGR had produced approximately 400,000 sequences and accumulated around 200,000 small insert clones. The sequences from these are available from NCBI/GenBank, and preliminary contigs can be obtained via a BLAST server.

5 Outlook

The complete sequence of *Arabidopsis* and the increasing availability of data concerning large-scale genome homology and microsynteny between the model plant and *Brassica* species mean that oilseed rape and its close relatives are well-positioned to be among the first major crop species to benefit from the continuing progress in plant molecular genetics and molecular marker technologies. Detailed physical maps and ultimately the complete sequencing of the diploid A and C genomes constituting *B. napus* can be expected within the next 5 years. This will bring information particularly with regard to the genes or gene expression differences that distinguish the *Brassica* crops from the model crucifer weed *Arabidopsis*. Furthermore, the availability of complete gene sequences for *Brassica* will enable rapid advances in detection of single nucleotide polymorphisms for a large number of agronomically relevant candidate genes. By combining SNP haplotype data for *Brassica* candidate genes with pedigree and trait information, it will then be possible to elucidate the genetic control of a large number of important traits via allele–trait association studies. Together with developing DNA-chip and high-throughput SNP genotyping technologies, the potential exists for enormous gains in the knowledge of the molecular basis of trait genetics in *Brassica* crops and a plethora of new molecular markers for agronomic traits. One important consequence of this should be a considerable improvement in marker-assisted breeding for quantitative traits, which has yet to fulfil the practical potential that it promises in theoretical terms.

References

Arabidopsis Initiative (2000) Analysis of the genome sequence of the flowering plant *Arabidopsis thaliana*. Nature 408:796–815

Babula D, Kaczmarek M, Barakat A, Delseny M, Quiros CF, Sadowski J (2003) Chromosomal mapping of *Brassica oleracea* based on ESTs from *Arabidopsis thaliana*: complexity of the comparative map. Mol Gen Genomics 268:656–665

Badani AG, Snowdon R, Baetzel R, Horn R, Friedt W (2003) QTL mapping for yellow seed colour in oilseed rape (*Brassica napus*). In: Proc 11th Int Rapeseed Congr, July 2003, Copenhagen, Denmark, vol 1, pp 85–87

Barret P, Delourme R, Foisset N, Renard M (1998) Development of a SCAR (sequence characterised amplified region) marker for molecular tagging of the dwarf BREIZH (*Bzh*) gene in *Brassica napus* L. Theor Appl Genet 97:828–833

Becker HC, Engqvist GM, Karlsson B (1995) Comparison of rapeseed cultivars and resynthesized lines based on allozyme and RFLP markers. Theor Appl Genet 91:62–67

Brown GG, Formanova N, Jin H, Wargachuk R, Dendy C, Patil P, Laforest M, Zhang JF, Cheung WY, Landry BS (2003) The radish *Rfo* restorer gene of *Ogura* cytoplasmic male sterility encodes a protein with multiple pentatricopeptide repeats. Plant J 35:262–272

Burns MJ, Barnes SR, Bowman JG, Clarke MHE, Werner CP, Kearsey MJ (2003) QTL analysis of an intervarietal set of substitution lines in *Brassica napus*: (1) Seed oil content and fatty acid composition. Heredity 90:39–48

Cavell AC, Lydiate DJ, Parkin IAP, Dean C, Trick M (1998) Collinearity between a 30-centimorgan segment of *Arabidopsis thaliana* chromosome 4 and duplicated regions within the *Brassica* genome. Genome 41:62–69

Chalhoub B, Budin K, Aubourg S, Belcram H, Laffont P, Delourme R, Lecharny A, Renard M, Caboche M (2003) Comparative genomic approaches for high-throughput physical and genetic mapping and gene cloning in oilseed rape (*Brassica napus*) and application to breeding. In: Proc 11th Int Rapeseed Congr, 6–10 July 2003, Copenhagen, vol 1, p 78

Conner JA, Conner P, Nasrallah ME, Nasrallah JB (1998) Comparative mapping of the *Brassica* S locus region and its homeolog in *Arabidopsis*. Implications for the evolution of mating systems in the *Brassicaceae*. Plant Cell 10:801–812
Das S, Roscoe TJ, Delseny M, Srivastava PS, Lakshmikumaran M (2003) Cloning and molecular characterization of the fatty acid elongase 1 (*FAE 1*) gene from high and low erucic acid lines of *Brassica campestris* and *Brassica oleracea*. Plant Sci 162:245–250
Delourme R, Foisset N, Horvais R, Barret P, Champagne G, Cheung WY, Landry BS, Renard M (1998) Characterisation of the radish introgression carrying the *Rfo* restorer gene for the Ogu-INRA cytoplasmic male sterility in rapeseed (*Brassica napus* L.). Theor Appl Genet 97:129–134
Diers BW, McVetty PBE, Osborn TC (1996) Relationship between heterosis and genetic distance based on restriction fragment length polymorphism markers in oilseed rape (*Brassica napus* L.). Crop Sci 36:79–83
Dion Y, Gugel RK, Rakow GFW, Seguin-Swartz G, Landry BS (1995) RFLP mapping of resistance to the blackleg disease [causal agent, *Leptosphaeria maculans* (Desm) Ces et de Not] in canola (*Brassica napus* L). Theor Appl Genet 91:1190–1194
Ecke W, Uzunova M, Weissleder K (1995) Mapping the genome of rapeseed (*Brassica napus* L.). 2: Localization of genes controlling erucic acid synthesis and seed oil content. Theor Appl Genet 91:972–977
Ferreira ME, Rimmer SR, Williams PH, Osborn TC (1995a) Mapping loci controlling *Brassica napus* resistance to *Leptosphaeria maculans* under different screening conditions. Phytopathology 85:213–217
Ferreira ME, Williams PH, Osborn TC (1995b) Mapping of a locus controlling resistance to *Albugo candida* in *Brassica napus* using molecular markers. Phytopathology 85:218–220
Ferreira ME, Satagopan J, Yandell BS, Williams PH, Osborn TC (1995c) Mapping loci controlling vernalisation requirement and flowering time in *Brassica napus*. Theor Appl Genet 90:727–732
Foisset N, Delourme R, Barret P, Renard M (1995) Molecular tagging of the dwarf BREIZH (BZH) gene in *Brassica napus*. Theor Appl Genet 91:756–761
Fourmann M, Barret P, Renard M, Pelletier G, Delourme R, Brunel D (1998) The two genes homologous to *Arabidopsis FAE1* co-segregate with the two loci governing erucic acid content in *Brassica napus*. Theor Appl Genet 96:852–858
Fourmann M, Barret P, Froger N, Baron C, Charlot F, Delourme R, Brunel D (2002) From *Arabidopsis thaliana* to *Brassica napus*: development of amplified consensus genetic markers (ACGM) for construction of a gene map. Theor Appl Genet 105:1196–1206
Fray MJ, Puangsomlee P, Goodrich J, Coupland G, Evans EJ, Arthur AE, Lydiate DJ (1997) The genetics of stamenoid petal production in oilseed rape (*Brassica napus*) and equivalent variation in *Arabidopsis thaliana*. Theor Appl Genet 94:731–736
Grist SA, Firgaira FA, Morley AA (1993) Dinucleotide repeat polymorphisms isolated by the polymerase chain reaction. BioTechniques 15:304–309
Happstadius I, Ljungberg A, Kristiansson B, Dixelius C (2003) Identification of *Brassica oleracea* germplasm with improved resistance to *Verticillium* wilt. Plant Breed 122:30–34
Hasan M, Lühs W, Seyis F, Friedt W, Snowdon R (2003) Surveying genetic diversity in the *Brassica napus* gene pool using SSR markers. In: Proc 11th Int Rapeseed Congr, Copenhagen, July, vol 2, pp 392–394
Howell EC, Barker GC, Jones GH, Kearsey MJ, King GJ, Kop EP, Ryder CD, Teakle GR, Vicente JG, Armstrong SJ (2002) Integration of the cytogenetic and genetic linkage maps of *Brassica oleracea*. Genetics 161:1225–1234
Howell PM, Sharpe AG, Lydiate DJ (2003) Homoeologous loci control the accumulation of seed glucosinolates in oilseed rape (*Brassica napus*). Genome 46:454–460
Hu J, Quiros C, Arus P, Struss D, Röbbelen G (1995) Mapping of a gene determining linolenic acid concentration in rapeseed with DNA based markers. Theor Appl Genet 90:258–262
Hu J, Li D, Struss D, Quiros CF (1999) SCAR and RAPD markers associated with 18-carbon fatty acids in rapeseed, *Brassica napus*. Plant Breed 118:145–150

Imai R, Koizuka N, Fujimoto H, Hayakawa T, Sakai T, Imamura J (2003) Delimitation of the fertility restorer locus Rfk1 to a 43-kb contig in Kosena radish (*Raphanus sativus* L.). Mol Genet Genom 269:388–394

Jackson SA, Cheng Z, Wang ML, Goodman HM, Jiang J (2000) Comparative fluorescence in situ hybridization mapping of a 431-kb *Arabidopsis thaliana* bacterial artificial chromosome reveals the role of chromosomal duplications in the expansion of the *Brassica rapa* genome. Genetics 156:833–838

Janeja HS, Banga SS, Lakshmikumaran M (2003) Identification of AFLP markers linked to fertility restorer genes for *tournefortii* cytoplasmic male-sterility system in *Brassica napus*. Theor Appl Genet 107:148–154

Jean M, Brown GG, Landry BS (1998) Targeted mapping approaches to identify DNA markers linked to the Rfp1 restorer gene for the 'Polima' CMS of canola (*Brassica napus* L.). Theor Appl Genet 97:431–438

Jourdren C, Barret P, Brunel D, Delourme R, Renard M (1996) Specific molecular marker of the genes controlling linolenic acid content in rapeseed. Theor Appl Genet 93:512–518

Kim JS, Lim KB, Park BS, Lee MC, Lee SI, Kim JA, Kim HI (2004) Genome sequencing strategy of chromosome 1 of *Brassica rapa*. Book of abstracts. In: Proc 12th Conf on Plant and Animal Genome, San Diego, California, January, p 521

Kole C, Thormann CE, Karlsson BH, Palta JP, Gaffney P, Yandell B, Osborn TC (2002a) Comparative mapping of loci controlling winter survival and related traits in oilseed *Brassica rapa* and *B. napus*. Mol Breed 9:201–210

Kole C, Williams PH, Rimmer SR, Osborn TC (2002b) Linkage mapping of genes controlling resistance to white rust (*Albugo candida*) in *Brassica rapa* (syn. *campestris*) and comparative mapping to *Brassica napus* and *Arabidopsis thaliana*. Genome 45:22–27

Lagercrantz U (1998) Comparative mapping between *Arabidopsis thaliana* and *Brassica nigra* indicates that *Brassica* genomes have evolved through extensive genome replication accompanied by chromosome fusions and frequent rearrangements. Genetics 150:1217–1228

Lagercrantz U, Lydiate DJ (1996) Comparative genome mapping in *Brassica*. Genetics 144:1903–1910

Lagercrantz U, Putterill J, Coupland G, Lydiate D (1996) Comparative mapping of *Arabidopsis* and *Brassica*, fine scale collinearity and congruence of genes controlling flowering time. Plant J 9:1903–1910

Lan TH, Del Monte TA, Reischmann KP, Hyman J, Kowalski SP, McFerson J, Kresovich S, Paterson AH (2000) An EST-enriched comparative map of *Brassica oleracea* and *Arabidopsis thaliana*. Genome Res 10:776–788

Landry BS, Hubert N, Etoh T, Harada JJ, Lincoln SE (1991) A genetic map for *Brassica napus* based on restriction fragment length polymorphisms detected with expressed DNA sequences. Genome 34:543–552

Li G, Quiros CF (2001) Sequence-related amplified polymorphism (SRAP), a new marker system based on a simple PCR reaction: its application to mapping and gene tagging in *Brassica*. Theor Appl Genet 103:455–461

Lombard V, Delourme R (2001) A consensus linkage map for rapeseed (*Brassica napus* L.): construction and integration of three individual maps from DH populations. Theor Appl Genet 103:491–507

Lukens L, Zou F, Lydiate D, Parkin I, Osborn T (2003) Comparison of a *Brassica oleracea* genetic map with the genome of *Arabidopsis thaliana*. Genetics 164:359–372

Mayerhofer R, Bansal VK, Thiagarajah MR, Stringam GR, Good AG (1997) Molecular mapping of resistance to *Leptosphaeria maculans* in Australian cultivars of *Brassica napus*. Genome 40:294–301

Mullis K, Faloona F (1987) Specific synthesis of DNA in vitro via a polymerase-catalyzed chain reaction. Methods Enzymol 44:224–249

Osborn TC, Kole C, Parkin IA, Sharpe AG, Kuiper M, Lydiate DJ, Trick M (1997) Comparison of flowering time genes in *Brassica rapa, B. napus* and *Arabidopsis thaliana.* Genetics 146:1123–1129

Parkin IAP, Sharpe AG, Keith DJ, Lydiate DJ (1995) Identification of the A and C genomes of amphidiploid *Brassica napus* (oilseed rape). Genome 38:1122–1131

Parkin IAP, Sharpe AG, Lydiate DJ (2003) Patterns of genome duplication within the *Brassica napus* genome. Genome 46:291–303

Pflieger S, Lefebvre V, Causse M (2001) The candidate gene approach in plant genetics: a review. Mol Breed 7:275–291

Pilet ML, Delourme R, Foisset N, Renard M (1998a) Identification of loci contributing to quantitative resistance to blackleg disease, causal agent *Leptosphaeria maculans* (Desm.) Ces. et de Not, in winter rapeseed (*Brassica napus* L.). Theor Appl Genet 96:23–30

Pilet ML, Delourme R, Foisset N, Renard M (1998b) Identification of QTL involved in field resistance to light leaf spot (*Pyrenopeziza brassicae*) and blackleg resistance (*Leptosphaeria maculans*) in winter rapeseed (*Brassica napus* L.). Theor Appl Genet 97:398–406

Pilet ML, Duplan G, Archipiano H, Barret P, Baron C, Horvais R, Tanguy X, Lucas MO, Renard M, Delourme R (2001) Stability of QTL for field resistance to blackleg across two genetic backgrounds in oilseed rape. Crop Sci 41:197–205

Plieske J, Struss D (2001) STS markers linked to *Phoma* resistance genes of the *Brassica* B-genome revealed sequence homology between *Brassica nigra* and *Brassica napus.* Theor Appl Genet 102:483–488

Quiros CF, Grellet F, Sadowski J, Suzuki T, Li G, Wroblewski T (2001) *Arabidopsis* and *Brassica* comparative genomics: sequence, structure and gene content in the *ABI1-Rps2-Ck1* chromosomal segment and related regions. Genetics 157:1321–1330

Ross P, Hall L, Smirnov I, Haff L (1998) High level multiplex genotyping by MALDI-TOF mass spectrometry. Nature Biotech 16:1347–1351

Sadowski J, Gaubier P, Delseny M, Quiros CF (1996) Genetic and physical mapping in *Brassica* diploid species of a gene cluster defined in *Arabidopsis thaliana.* Mol Gen Genet 251:298–306

Schelfhout CJ, Snowdon RJ, Cowling WA, Wroth JM (2004) A PCR based B-genome specific marker in *Brassica* species. Theor Appl Genet (in press)

Schierholt A, Becker HC, Ecke W (2000) Mapping a high oleic acid mutation in winter oilseed rape (*Brassica napus* L.). Theor Appl Genet 101:897–901

Schmidt R (2002) Plant genome evolution: lessons from comparative genomics at the DNA level. Plant Mol Biol 48:21–37

Schmidt R, Acarkan A, Boivin K (2001) Comparative structural genomics in the *Brassicaceae* family. Plant Phys Biochem 39:253–262

Seyis F, Snowdon RJ, Lühs W, Friedt W (2003a) Molecular characterisation of novel resynthesised rapeseed (*B. napus* L.) lines and analysis of their genetic diversity in comparison to spring rapeseed cultivars. Plant Breed 122:473–478

Seyis F, Friedt W, Lühs W (2003b) Resynthesised *Brassica napus* as a resource in rapeseed improvement for quality and agronomic performance. In: Knüpffer H, Ochsmann J (eds) Rudolf Mansfeld and plant genetic resources. ZADI, Bonn. Schriftenr Genet Ressour 19:33–338

Shengwu H, Ovesna J, Kucera L, Kucera V, Vyvadilova M (2003) Evaluation of genetic diversity of *Brassica napus* in germplasm from China and Europe assessed by RAPD markers. Plant Soil Environ 49:106–113

Slocum MK, Figdore SS, Kennard WC, Suzuki JY, Osborn TC (1990) Linkage arrangement of restriction fragment length polymorphisms in *Brassica oleracea.* Theor Appl Genet 80:57–64

Snowdon RJ, Winter H, Diestel A, Sacristan MD (2000) Development and characterisation of *Brassica napus–Sinapis arvensis* addition lines exhibiting resistance to *Leptosphaeria maculans.* Theor Appl Genet 101:1008–1014

Snowdon RJ, Friedrich T, Friedt W, Köhler W (2002) Identifying the chromosomes of the A and C genome diploid *Brassica* species *B. rapa* and *B. oleracea* in their amphidiploid *B. napus*. Theor Appl Genet 104:533–538

Somers DJ, Friesen KRD, Rakow G (1998) Identification of molecular markers associated with linoleic acid desaturation in *Brassica napus*. Theor Appl Genet 96:897–903

Somers DJ, Rakow G, Prabhu VK, Friesen KRD (2001) Identification of a major gene and RAPD markers for yellow seed coat colour in *Brassica napus*. Genome 44:1077–1082

Song KM, Suzuki JY, Slocum MK, Williams PH, Osborn TC (1991) A linkage map of *Brassica rapa* (syn. *campestris*) based on restriction fragment length polymorphisms. Theor Appl Genet 82:296–304

Tanhuanpaa PK, Vilkki JP, Vilkki HJ (1995) Association of a RAPD marker with linolenic acid concentration in the seed oil of rapeseed (*Brassica napus* L.). Genome 38:414–416

Tanhuanpaa P, Vilkki J, Vihinen M (1998) Mapping and cloning of *FAD2* gene to develop allele-specific PCR for oleic acid in spring turnip rape (*Brassica rapa* ssp. *oleifera*). Mol Breed 4:543–550

Tommasini L, Batley J, Arnold GM, Cooke RJ, Donini P, Lee D, Law JR, Lowe C, Moule C, Trick M, Edwards KJ (2003) The development of multiplex simple sequence repeat (SSR) markers to complement distinctness, uniformity and stability testing of rape (*Brassica napus* L.) varieties. Theor Appl Genet 106:1091–1101

Uzunova M, Ecke W, Weißleder K, Röbbelen G (1995) Mapping the genome of rapeseed (*Brassica napus* L.). I. Construction of an RFLP linkage map and localization of QTLs for seed glucosinolate content. Theor Appl Genet 90:194–204

Vos P, Hogers R, Sleeker M, Reijans M, Lee T, Homes M, Freiters A, Pot J, Peleman J, Kuiper M, Zabeau M (1995) AFLP: a new concept for DNA fingerprinting. Nucleic Acids Res 23:4407–4414

Walsh JA, Sharpe AG, Jenner CE, Lydiate DJ (1999) Characterisation of resistance to turnip mosaic virus in oilseed rape (*Brassica napus*) and genetic mapping of TuRB01. Theor Appl Genet 99:1149–1154

Wang YP, Snowdon RJ, Rudloff E, Wehling P, Friedt W, Sonntag K (2004) Cytogenetic characterization and *fae1* gene variation in progenies from asymmetric somatic hybrids between *Brassica napus* and *Crambe abyssinica*. Genome (in press)

Williams JGK, Kubelik AR, Livak KJ, Rafalski JA, Tingey SV (1990) DNA polymorphisms amplified by arbitrary primers are useful as genetic markers. Nucleic Acids Res 18:6531–6535

Wretblad S, Bohman S, Dixelius C (2003) Overexpression of a *Brassica nigra* cDNA gives enhanced resistance to *Leptosphaeria maculans* in *B. napus*. Mol Plant Microbe Interact 16:477–484

Zhang FL, Aoki S, Takahata Y (2003) RAPD markers linked to microspore embryogenic ability in *Brassica* crops. Euphytica 131:207–213

Zhao JW, Meng JL (2003a) Genetic analysis of loci associated with partial resistance to *Sclerotinia sclerotiorum* in rapeseed (*Brassica napus* L.). Theor Appl Genet 106:759–774

Zhao J, Meng J (2003b) Detection of loci controlling seed glucosinolate content and their association with *Sclerotinia* resistance in *Brassica napus*. Plant Breed 122:19–23

Zietkiewicz E, Rafalski A, Labuda D (1994) Genome fingerprinting by simple sequence repeat (SSR)-anchored polymerase chain reaction amplification. Genomics 20:176–183

Rod Snowdon
Katrin Link
Ana Gloria Badani
Wolfgang Friedt
IFZ, Research Centre for Bio Systems,
Land Use and Nutrition
Department of Crop Science and Plant Breeding
Justus-Liebig-University of Giessen
Heinrich-Buff-Ring 26-32
35392 Giessen, Germany

Plant Breeding: Clonality – A Concept for Stability and Variability During Vegetative Propagation

Astrid Forneck

1 Introduction

Clonality means asexual reproduction or propagation without meiosis. This results in offspring that are identical to the ancient progenitor except when mutations occur. It seems that clonality is a dynamic concept. Genetic variation is added by numerous mechanisms to an asexual living strategy to enhance variation and provide an open system for adaptation and selection. For a cultivated woody perennial crop genetic variation has additional meanings since clonality is used to sustain identical genotypes and select for new clones of grapevine cultivars through clonal selection.

Grapevine (*Vitis* ssp. L.) is used as a role model for investigating clonality in a woody perennial since it is a well-investigated cultivated plant that is propagated vegetatively. Grapevine cultivars (*Vitis vinifera* L.) are composed of clones showing homogeneous ampelographic characteristics, and distinguished by minor differences. Asexually derived grapevine lineages afford one of the best systems to study clonality in plants. Estimates of intra-varietal genetic diversity have increased as detection systems with enhanced resolution become available. The most powerful applied marker system for analyzing clonality in grapevine is AFLP-PCR (amplified fragment length polymorphism-polymerase chain reaction). Less resolution has been obtained with RAPD-PCR (random amplified polymorphic DNA) or ISSR-PCR (inter simple sequence repeat). Microsatellites have recently been used to detect clonal variability. The sources of genetic variation in grapevine clones are manifold. Somatic mutations, somaclonal variation, retrotransposition, chimerism and epigenetic changes have all been recently confirmed with molecular marker techniques.

Most perennial plants combine sexual reproduction with some form of clonal propagation. The balance between these forms of reproduction affects propagule size, establishment dynamics and the transmission of genetic variation. This balance is thought to greatly influence population demography, genetic diversity, the accumulation of mutations, metapopulation dynamics and the evolutionary potential of populations (Dorken and

Progress in Botany, Vol. 66

Eckert 2001). Clonal growth (clonality) is known as a mechanism of foraging for resources, competition and compensation of environmental heterogeneity, but also as a low-risk strategy for maintaining local populations and the fittest genotypes within a population (Auge and Brandl 1997). In general, long-lived woody perennials are highly heterozygous and inbreeding results in a decline in vigor and fertility, as in the case with naturally cross-pollinated crops. Vegetative reproduction occurs both naturally in habitats of plant communities and artificially through vegetative propagation of commercially desired plant clones. Clonality is a term that is often vaguely applied in association with asexual reproduction. Here, this term is used to describe the asexual reproduction of an individual resulting in a set of clones. A clone is defined as an individual that descended from a single common ancestor by mitosis. A group of clones originating from the same ancestor (monozygotic) is defined to be genetically identical except for the effect of mutations.

Clonality in woody perennials occurs by means of stolon formation, layering, root suckering, vegetative sprouting, formation of rhizomes and axis splitting. The relative importance of sexual versus clonal recruitment may vary widely among plant species because clonal reproduction allows populations to persist and expand in habitats where, for one reason or another, sexual reproduction cannot occur. The production of asexually derived offspring in woody perennials occurs in many ways. There are species in which new individuals may arise without disturbance; in others, however, asexual reproduction is in direct response to injury or imbalance (Berg and Hamrick 1994). Woody perennials are economically used as vegetatively propagated plants and sold as 'clones' with identical phenotypes. Such cultivated clones consist of single desired genotypes that are multiplicated and spread worldwide as identical individuals with superior combinations of genes. Vegetatively propagated plant material can be used to control growth phases, to shorten time to flowering or conversely to rejuvenate plantings. Furthermore, two superior genotypes may be combined in one plant through grafting techniques.

Clonality has been studied under a variety of perspectives involving morphology, physiology, ecology and evolution of natural asexually reproducing plant populations. The work on clonality of the last decades is relatively limited in the study of woody perennials (e.g. Rajora 1999; Schenk 1999; Chung and Epperson 2000) and has been brought together in several reviews (Eckert 1999). Most approaches to study clonality have focused on plant populations in their natural habitat (de Kroon and van Groenendael 1997) to examine ecological effects and the natural environment for selective forces on the populations. There is increasing interest in analyzing the clonality of cultivated clones to assist in clonal selection processes of superior genotypes. Moreover the need to identify and discriminate among clones has been intensified by legal defenses to breeders' rights. A new area of research focuses on transgenic clones with investigations of performance and identification.

Grapevine (*Vitis vinifera* ssp. L.) is an excellent crop plant for clonality study because of its economic importance and reliance on vegetative propagation. Today's cultivars resulted from the selection of advanced genotypes of ancient origin mostly generated by intentional and spontaneous crosses centuries ago. Each ancient cultivar expresses distinct phenotypes, resulting in sets of morphologically different clones. These clones have spread worldwide adjusting to different environments and cultivation techniques. The causes discussed for clonal variation are virus infection, polyclonality and mutations. For grapevine clones the concept of individuality is straightforward and relies on propagation records and morphological features. Since grapevine is a high-value crop, significant viticultural research in describing and analyzing the phenotypes has been performed since the 19th century. More recently, tissue culture, transformation and molecular genetic techniques have been used in genomic and biotechnological approaches to improving grapevines.

This chapter does not intend to provide a comprehensive literature review on grapevine genetic studies, but to present examples of publications that underline the discussion of clonality.

2 Clonal Variation Assessment

Degrees of clonality in *Vitis* have been assessed through phenotypic and genotypic parameters with differing intentions. Molecular markers in combination with phenotypic traits are used to identify clones. Molecular markers can further be used to quantify neutral genetic variation within and among 'populations' of clones. Several causes have been proposed to account for the variation that occurs among cultivated clones. One is that differing clones are not true clones, but rather resemble a population of very closely related individuals, perhaps siblings, of similar morphology, a concept termed polyclonality. Another explanation considers the interference of pathogens such as viruses on phenotypes with identical genotypes. Both concepts are widely used with respect to woody perennial clones; however, neither proves to measure clonality in its strict sense. The third explanation considers accumulation of the phenotype-altering mutations as the primary factor resulting in true clones (clones identical by descent).

2.1 Polyclonality

Traditionally, phenotypic variation within cultivars has been accepted. Polyclonality can account for the existence of such variation. According to Rives (1961) more than one seedling, all marked with morphological uniformity, gave rise to many grapevine cultivars. Polyclonality in woody perennial cultivars was proposed long ago, but only recently proven by molecular marker techniques. The advent of simple sequence repeat (SSR) techniques permits polyclones to be clearly detected (Vignani et al. 1996; Silvestroni et al. 1997; Filippetti et al. 1999).

2.2 Pathogen-Infected Clones

Virus diseases contribute to increasing the phenotypic variability within grapevine cultivars. Grapevines, in common with other woody perennials, are subject to infection by virus and virus-like pathogens. In addition, there are several other debilitating diseases caused by viroids, phyto- or mycoplasms. In the case of grapevine, plants with recognizable symptoms of virus disease are normally rejected as a source of budwood or cuttings. One problem in viticulture, as in other fruit crops, is that seemingly healthy plants can be symptomless carriers of disease-causing viruses. The extent to which phenotypic variation within a population of clones is due to the genotype or to the presence of viruses has been subject to experimentation with differing results. Studies show that differences among clones exist regardless of their phytosanitary status. Other studies suggest that the overall performance of grapevine clones is improved after sanitation of the plant material (Mannini 2000). According to current knowledge, genomic changes in DNA have not been observed, and thus pathogen-infected clones with altered phenotypes are not true clones.

2.3 Mutation

Clonal variation can be induced through mutations to induce stable genetic changes. These genetic changes are visualized by observed changes in morphology or by applying molecular genetic markers. There is abundant evidence for the occurrence of somatic mutations in plants (e.g. Klekowski and Godrey 1989) and molecular genetic variation has been found among naturally occurring clones in several plant species. Somatic variation might be expected to be most common at loci with high mutation rates, e.g. as reported in some cases for microsatellite DNA sequences (Schlotterer et al.

1998; Udupa and Baum 2001). There is little information about mutation rates during somatic development of plants (Gill et al. 1995).

2.3.1 Phenotype

Grapevine development is controlled by two main factors over the growing season: the pre-growth conditions of the vine, including its size, characteristics of its buds and the amount of stored reserves; and the environmental conditions that regulate and modify shoot and fruit development. Furthermore, how vineyards are managed has a large impact on the phenotypic performance of clones (e.g. Clingeleffer 1988; Cirami et al. 1993). Clones of a cultivar often differ substantially in viticultural performance and in their ability to produce quality wines. Clonal descriptions usually focus on viticultural traits (e.g. yield components: cluster number, cluster weight, numbers of berries/cluster, berry weight, berry size), while traits pertaining to wine quality are very difficult to evaluate. The literature concerning the selection of clones based on viticultural traits is vast. A summary of the German clonal selection program covers the strategies and their success (Schöffling and Stellmach 1993).

Clonal selection focuses on the phenotype and viticultural performance. The selection procedure takes between 20 and 30 years of repeated studies and phytosanitary restoration before a clone will be 'certified'. Differences among the phenotypic plasticity vary among cultivars; examples for diverse *V. vinifera* cultivars are Pinot, Traminer and Nebbiolo, whereas Zinfandel is a good example of very limited genetic diversity. The plasticity of Pinot results in a set of color types (red, grey, white berries), differences in cluster architecture (loose, tight cluster), berry size and growth habit, while Zinfandel, with a very limited genetic base at its point of origin (Maletic et al. 2003), has not developed significant clonal variation even though it has existed for centuries. The phenotypic identification of grapevine clones is an essential means of grapevine improvement due to the reliance on traditional varieties for wine making. Clonality studies, however, require reproducible morphological measurements. Furthermore, because of environmental and pathogen effects on clonal expression, it can be difficult to define true genetic clonal differences.

2.3.2 Genotype

The identification of individuals is challenging in perennial clonal organisms. Grape clones are often poorly defined, and tracking and confirming their identity with worldwide distribution and hundreds or thousands of mothervines is extremely difficult. Molecular markers are now routinely implemented in studies of clonality. There are a range of molecular markers and several reviews discuss the application of different marker classes, their pros and cons in respect of the studied organisms (e.g. Lavi et al. 1994). Most markers applied in clonality studies are PCR-based and generate multi-locus fingerprints. Microsatellites have a unique standing in grapevine genetic research. More than 400 SSR loci have been developed for *Vitis* and are applied for identification and mapping purposes. A brief description of the marker systems follows:

AFLP (amplified fragment length polymorphism): AFLP involves the restriction of DNA with two different endonucleases, followed by a ligation with appropriate adaptors and amplification of DNA fragments in two steps. Various primer combinations with differing selection ends are employed to generate multilocus dominant markers.

SSR (simple sequence repeats): Often referred to as microsatellites, SSR consist of short stretches of tandemly repeated motifs, 2–4 bp in length. Once flanking primer sites are identified, they can be amplified using PCR to generate a locus and scored by size (corresponding to alleles). SSR are inherited in a codominant fashion, allowing an assessment of within-population structure.

RAPD (random amplified polymorphic DNA): Short random primers are employed in PCR reaction to amplify random DNA segments. The presence of a band indicates successful amplification; absence indicates no amplification due to, e.g., mutation in the primer recognition site. RADP in diploid organisms behave in a dominant/recessive fashion.

ISSR (inter simple sequence repeat): Non-anchored ISSR markers are arbitrary, multiloci, PCR-based markers that amplify intermicrosatellite sequences at multiple loci throughout the genome. The marker is PCR-based and dominant.

RFLP (restriction fragment length polymorphism): RFLP analyses involve the restriction of DNA and the production of DNA patterns based on variations occurring in the length of DNA fragments generated by a specific endonuclease. Marker systems applying RFLP may be combined with PCR or hybridization techniques and may vary in the classes of DNA employed.

Some authors have proposed combining different marker systems to enhance resolution of the methods. The combination of markers is relevant in clonality studies for identifying true clones that arise from the accumulation of mutations. In these cases, analysis with allele-specific markers is followed by screening for intra-variety diversity with multilocus dominant markers.

3 Studies Analyzing Clonal Variation in Grape

Many clonal variation studies in grapevine have been performed in recent years in an effort to discriminate among clones and provide techniques for reliable identification. The experimental design and data interpretation vary significantly depending on the study's aims. In the following sections the experimental conditions are summarized.

3.1 Experimental Design and Plant Material

The plant material studied varies among certified variety clones, accessions of particular varieties, which are morphologically similar or dissimilar, color types of clones, sports and their motherplants. The numbers of clones analyzed vary greatly among studies and within a variety. The phytosanitary status of the samples ranged from clones taken from commercial vineyards (virus status unknown), from academic grapevine collections, from tissue culture regenerated plant material, to certified virus-free material (via thermotherapy or shoot tip culture). To prevent the inclusion of different but morphologically similar varieties, some studies test the plant material for polyclonal origin with sets of SSR markers to provide a solid basis to genetically analyze clonality.

3.2 Data Measurement and Interpretation

Analysis of data from these clonal studies varied from the identification of discriminating markers to performing statistical analysis. Computing similarity–dissimilarity or genetic distances matrices with differing coefficients was frequently done together with either constructing dendrograms or principal component analysis. Frequently cophenetic values were calculated or the mantel test performed to test for linkage of genotypic and phenotypic data. The statistical treatment depends on the marker system used. Defining clonal variation via SSR markers is simple because of differing, missing or additional allele sizes, whereas multilocus marker systems need further statistical analysis. Cervera et al. (1998) employed genetic similarity values as an approximate value to characterize clonal genotypes. The value of $= 0.97$ (corresponding to more than 97% similarity in markers) is utilized as a standard. However, a similarity value will depend on the marker system used, the statistical coefficient applied and the number and diversity of samples/clones studied.

3.3 Clonal Variation in Grapevine Clones?

Clonal variation can be repeatedly detected among clones of grapevine cultivars. The degree of genetic variation depends mainly on the molecular marker chosen and the scope and range of plant samples employed. According to the studies performed, AFLP-PCR provides the best resolution of genetic variation deriving from somatic mutations. Table 1 shows published work employing AFLP markers. All of these studies, except one, detected genetic variation among clones of differing grapevine cultivars. The scale of diversity found seems to depend on the number of markers produced and the clonal plant material studied. Cultivated, registered clones seem to exhibit more intra-variety variability than clones derived from mutated shoots from the same vine. The reproducibility of AFLP markers has been challenged (e.g. Goto-Yamamoto 2000; Merdinoglu et al. 2000), but recent studies report no such problems, possibly due to careful scoring procedures and optimization of conditions (e.g. Imazio et al. 2002; Fanizza et al. 2003; Forneck et al. 2003a).

RAPD-PCR was implemented in earlier studies to search for genetic differences among clones. Five studies found clone-specific RAPD markers, whereas three could not detect differences among clones (Table 2). Other studies investigated genetic variation among somaclones (multiple clones arising from the same mothervine through in vitro passages). In general, the genetic variation detected is low and determined by the nature of the marker system.

SSR markers are well suited to grapevine, although few SSR studies have been conducted to search for differences among clones. However, clones, biotypes (clones that show phenotypic divergence) or sports are often studied together while investigating variety identification, parentage and diversity. Early SSR studies did not discover differences in alleles among clones and general opinion confirmed SSR marker to be unsuitable for clonal detection. This was thought because of their codominant inheritance and stability. SSR marker were successfully adopted for establishing pedigree analyses of grapevine cultivars (e.g. Bowers and Meredith 1997) and used for cultivar identification. SSR markers have been employed to study polyclonality of grapevine cultivar accessions (e.g. Filippetti et al. 1999; Kozjak et al. 2003). Given the increasing use of SSR markers in genetic analyses of long-lived plants, it is important to characterize their stability at the inter-variety level. Studies have confirmed genetic variation in SSR sequences and specified variation through missing alleles (null alleles), the addition of one or several alleles (chimerism) and the development of new alleles with differing sizes (Table 3). Recent studies analyzing the chimeric state of grapevine clones have been directed using SSR markers (e.g. Franks et al. 2002). The results of these studies suggest that chimerism occurs frequently in grapevine clones.

ISSR markers were applied by several groups to enhance the resolution and reproducibility of the RAPD technology. Limited levels of genetic

Table 1. Studies implementing AFLP markers to differentiate among clones

Variation	Cultivar and (number of clones or sports)	AFLP technology	Specific comments	Source
Clonal variation	Sangiovese (6), Colorino (4), among others	8 *Mse*I/*Eco*RI combinations with 458 bands	Intra-specific variation	Sensi et al. (1996)
Clonal variation	Pinot gris (1), P. noir (1)	4 *Mse*I/*Eco*RI combinations	Clone-specific patterns	Goto-Yamamoto (2000)
Clonal variation	Napoleon (9), Flame Seedless (5), Italia (5), among others	4 *Mse*I/*Eco*RI combinations with 440 bands	Ten intra-varietal polymorphic bands for F. Seedless and Napoleon	Cervera et al. (2000)
Clonal variation	Traminer (3), Savignin (5), P. blanc (2), P. gris (2), Chasselas (2), Auxerrois (2), Silvaner (2)	8 *Mse*I/*Eco*RI combinations with 197 bands	Clone-specific patterns for all clones	Merdinoglu et al. (2000)
Clonal variation	Flame Seedless (2)	*Mse*I/*Eco*RI combinations with 3,000 bands	Two differentiating bands	Scott et al. (2001)
Clonal variation	Albarino (28 accessions)	2 *Mse*I/*Eco*RI combinations with 199 bands	Distinction among closely related varieties and evidence of intra-varietal variation	Cervera et al. (2001)
Clonal variation	Tempranillo (31 accessions)	2 *Mse*I/*Eco*RI combinations with 206 bands	One to two polymorphic bands among clone groups	Cervera et al. (2002)
Clonal variation	Traminer (24), among others	3 *Mse*I/*Eco*RI combinations with 153 bands	Clones show an average similarity of 97.1%. Eight clones could not be separated	Imazio et al. (2002)
Clonal variation	P. noir (20), P. blanc (10), P. gris (6) among others	10 *Mse*I/*Eco*RI combinations with 422 bands	Clone-specific patterns. Three clones could not be separated	Forneck et al. (2003b)
No variation	Italia (3) (+ 1 sport)	49 *Mse*I/*Eco*RI combinations with 3.880 bands		Fanizza et al. (2003)

Table 2. Studies implementing RAPD marker to differentiate among clones. Studies analyzing somaclonality are not included

Variation	Cultivar and (number of clones or sports)	Specific comments	Source
Clonal variation	Sultania (2)	Genetic differences found in bands produced from 17 of 110 primers, between this pair with a bandsharing ratio of 96.1% (compared to unrelated cultivars with 78.5%)	Striem et al. (1994)
Clonal variation	Pinot, Chardonnay (no. not specified)	Genetic variation among Pinot noir and Pinot gris observed using 118 markers. 'Frequency of polymorphism was not higher between the color types than within one color type'	Regner et al. (2000b)
Clonal variation	Riesling (10)	Genetic variation was found among 9 of 10 clones tested using a 'clone'-specific marker. All clones could be differentiated using primer combinations	Regner et al. (2000a)
Clonal variation	Traminer (3), Savignin (5), Pinot blanc (2), P. gris (2), Chasselas (2), Auxerrois (2), Silvaner (2)	Genetic differences among all clones tested in banding patterns produced from 138 primers	Merdinoglu et al. (2000)
Clonal variation	Traminer (12)	Genetic differences found in bands produced from 20 of 26 primers	Regner and Kaserer (2002)
Clonal variation/ no variation	Chardonnay (3), Pinot noir (+4 sports), Niagara (+3 sports), Concord (+1 sport)	No differences found among 'known' clones of P. noir and Chardonnay using 53 primers. Low genetic variation found among Niagara and Concord	Ye et al. (1998)
Clonal variation/ no variation	Aubin vert (3), Räuschling (2), Aubin blanc (2), Chardonnay (3), Aligoté (3), Pinot (5), St. Laurent (2)	Genetic variation found in banding patterns produced from 20 primers. Clones of Pinot, Aubin vert and Chardonnay could not be differentiated	Tschammer and Zyprian (1994)

Table 3. Studies implementing SSR markers to differentiate among clones. Cultivar expressing clonal variation appears in *italics*

Variation	Variety and (number of clones studied)	SSR loci analyzed	Origin of mutation according to author(s)	Source
Variation	Fortana (5)	VVS1, VVS2, VVS4, VVMD3, VVMD6	Polyclonality	Silvestroni et al. (1997)
Clonal variation	Gewürztraminer (3), Savignin (5), Pinot blanc (2), *P. gris* (2), Chasselas (2), Auxerrois (2), Silvaner (2)	VVS2	Allelic mutation	Merdinoglu et al. (2000)
Clonal variation	Traminer (12)	VRG1, VRG2, VRG3, VRG4, VRG7, VRG9, VRG10, VRG11, VRG15	Allelic mutations (null alleles, additional alleles)	Regner and Kaserer (2002)
Variation	Nebbiolo (15)	VVMD7, VVS5	Polyclonality	Botta et al. (2000)
Clonal variation	Riesling (10)	VRG1, VRG2, VRG3	Allelic mutations (null alleles, additional alleles)	Regner et al. (2000a)
Clonal variation	Pinot noir (25), Chardonnay (22)	15 VMC loci	Chimerism, allelic mutations	Riaz et al. (2002)
Clonal variation	Primitivo (4), Pinot noir (3), P. blanc (3), P. gris (2), P. meunier (4)	VVS1, VVS2, VVS19, VVS29, VVS5, VVS16, VVMD7	Chimerism, allelic mutation	Franks et al. (2002)
Clonal variation	Pinot (145)	VVS2, VVS5, VVMD25, VVMD30, VVMD32, VVMD7, VMC3b12, VMC3c9, VMC8g6, VrZAG25, VrZAG79, VMC5g7	Chimerism, allelic mutation	Hocquigny et al. (2003)
Variation	Refosk (55 accessions)	29 loci	Polyclonality	Kozjak et al. (2003)
Clonal variation	Muscat d'Alsace (5), Grefo di Tufo (6), Primitivo (5), Corvina (3)	VVMD32, VVS2, VVMD7, ISV8, VMC6e4	Chimerism, allelic mutations	Crespan (2004)

Table 3. *Continued*

Variation	Variety and (number of clones studied)	SSR loci analyzed	Origin of mutation according to author(s)	Source
No variation	Pinot noir (3), P. gris (3), P. blanc (5), Cabernet franc (3), C. sauvignon (3), Chardonnay (5), Refosk (4), Ribolla (7)	VVS1, VVS2, VVS3, VVS4, VVS5		Cipriani et al. (1994)
No variation	Sangiovese (12)	VVS2, VVS4, VVS29, VVMD5, VVMD6, VVMD7, VVMD8		Vignani et al. (1996)
No variation	Albarino (18 accessions)	VVS1, VVS2, VVS29, VVMD5, VVMD6, VVMD7		Loureiro et al. (1998)
No variation	Portugieser (2), Silvaner (2), Gutedel (2), Riesling (2), Pinot blanc (1), P. gris (1), P. blanc (1)	VVS1, VVS2, VVS3, VVS4, VVS29, VVMD5, VVMD7, VVMD28, VVMD32, VVMD36		Sefc et al. (1998)
No variation	Pinot clones (number not specified)	34 loci		Regner et al. (2000b)

variation can be revealed by using these techniques. Regner et al. (2000a) found clone-specific ISSR markers among a set of ten *V. vinifera* cv. Riesling clones. Such polymorphic markers were also found by Specht (2002) analyzing Pinot blanc, P. gris, Auxerrois and Chardonnay clones.

Few studies have been published using RFLP-PCR to examine clonal variation and no differences were detected. Gogorcena et al. (1993) tested nine Pinot noir clones together with P. blanc and P. gris, finding no differences. Likewise Bourquin et al. (1995) did not detect differences among rootstock clones.

Marker systems to trace mutations associated with transposition or retrotransposition events in the genome are of great use in clonality studies. Several strategies have been developed and two different strategies were adopted in grapevine to analyze clonal variation. Inverse sequence-tagged repeat (ISTR) analysis (Rohde 1996) makes use of the ubiquitous presence of reverse transcriptase sequences and can be extended into a generally applicable, multiple locus strategy for grapevine. Sensi et al. (1996) found intra-varietal genetic differences among putative clones of *V. vinifera* cv. Sangiovese using ISTR markers. Another technique, sequence-specific amplification polymorphism (S-SAP), uses a combination of AFLP and sequence-specific PCR (based on long terminal repeats). This technique is a powerful method to detect the insertional polymorphism of retrotransposons (Waugh et al. 1997). The discriminating power of S-SAP has been studied by Pelsy et al. (2002, 2003) who found resulting S-SAP markers in good agreement with SSR markers obtained from the same set of samples. S-SAP markers can be employed for clonal studies of *V. vinifera* cv. Pinot, and have shown the potential to discriminate among clones (Forneck and Wedig, in prep.).

3.4 Pitfalls

Experimental errors limiting clonality measures may hamper clonality studies. Potential drawbacks can result due to sampling errors. Misnaming or mixing of grapevine clones has been reported in grapevine research. Although molecular genetic techniques are now routinely used, methodological pitfalls may result in misinterpretation. However such pitfalls may occur, they need to be considered in the data interpretation. A problem with AFLP analysis is the appearance of artifact amplicons. These bands may be caused by reduced specificity of restriction enzymes (Goto-Yamamoto 2000), or when buffer conditions are inadequate. Furthermore, banding patterns can be affected by the DNA extraction method (Konradi et al. 2002). Another critical point is that genetic data generated by AFLP and

RAPD markers may use foreign DNA (e.g. contaminating microbes) as template-producing 'false positives'. RAPD-PCR was employed by several groups to discriminate among clones. However, this technique is not always reproducible and has been dismissed by most authors (e.g. Büscher et al. 1993; Regner et al. 2000b). SSR markers are advantageous since they are species-specific and display typical allele sizes for a particular locus. A quality assessment of different detection systems showed that differences in SSR allele size estimates do occur and can affect the utility of SSR markers. Furthermore, visualization of alleles depends on the resolution of the technique chosen. This may be critical for the correct identification of null or chimeric alleles.

3.5 Somaclones and Protoclones

Another grouping among clones are the soma- and protoclones, which refer to plantlets regenerated through in vitro techniques. The processes underlying somaclonal variation are believed to require multiple genetic and/or epigenetic events that affect patterns of expression or result in gene mutation. For further information a comprehensive review of somaclonal variation has been brought together by various authors and edited by Roubelakis-Angelakis (2001). The first indications of somaclonal variation of regenerated or tissue-culture-propagated plants were gained through phenotypic observations of altered morphology from the donor plant. Genetic variation has been identified using RAPD markers in protoclones of *V. vinifera* cv. Seyval blanc (Schneider et al. 1996) and by applying AFLP markers to characterize somaclones of anther-derived grapevines (Popescu et al. 2002). The latter group also observed changes in methylation patterns in *V. vinifera* cv. Mission. No somaclonal variation of ploidy level was found by Kuksova et al. (1997) in plants regenerated from leaf explants through somatic embryogenesis. Spontaneous somaclonal variation in regenerated Seyval blanc protoclones was confirmed by Reustle and Matt (2000). These authors detected cytogenetic variation (tetraploid regenerates) and clone-specific RAPD-patterns.

3.6 Chimerism

Chimeric grapevines have been observed in the past and used in clonal selection programs. A chimera coexists of at least two different genotypes. By convention a periclinal chimera is a plant with a two-layered tunica above a corpus showing one or more genetically different entire shoot apical cell layers. The underlying mutations can enter these meristems in two ways. Cells can mutate in the initial shoot meristem, or a mutated somatic cell may be incorporated into an adventitious meristem. In both ways the meristem develops into a shoot with the mutant phenotype (sport). The literature on chimerism in woody perennials and grapes is sparse. Descriptions of chimeric grapevine plants were reported in the middle of the 19th century describing red- and white-colored Pinot clusters occurring on one vine. First experimental evidence was provided through cytological studies by Thompson and Olmo (1963), who showed chimeras coexisting with different ploidy levels. Molecular analysis has added proof with recent studies revealing genetic diversity and chimeric state of some P. meunier clones (Franks et al. 2002), P. noir and P. gris clones (Hocquigny et al. 2003), and P. noir and Chardonnay clones (Riaz et al. 2002) with SSR markers.

4 Epigenetic Effects on Clonal Variation

Epigenetic effects cover the study of the mechanisms by which genes bring about their phenotypic effects. Several age-related processes have been described that affect the different stages of growth. In the case of woody perennials the epigenetic 'imprint' on the DNA may persist for several years of culture. There are at least two distinct classes of epigenetic information that can be inherited by chromosomes and thus transmitted via vegetative propagation. One class is DNA methylation, in which a nucleic acid base is modified by a DNA methyltransferase. The other class of epigenetic information involves changes in chromatin proteins (van Steensel and Henikoff 2003). These findings motivated the development of technologies to reveal epigenetic patterns that can be used to elucidate genetic variation. One of the most popular strategies is to use methyl-sensitive restriction endonucleases to map and quantitatively assay the relative abundance of methylated C-residues (e.g. MSAP, methylation sensitive amplified polymorphism). Very few studies have been conducted to reveal methylation patterns among grapevine clones. Imazio et al. (2002) found changes in methylation patterns among clones of *V. vinifera* cv. Traminer. The authors suggest that the phenotypic differences observed are due to synergetic effects of genomic and epigenetic variation. Another study was performed to analyze changes in methylation patterns as potential causes of the recalcitrance of regeneration of *Vitis* ssp. (Harding et al. 1996). This group also found that methylation patterns change during tissue culture passages of *V. vinifera* cv. Sultania.

5 Concluding Remarks

Clonality is a dynamic concept. Genetic variation is sustained by mutation, which provides the basis for the clonal selection of superior clones in woody perennial crops. The old nomenclature that exists for defining clones (offspring produced asexually without recombination events) persists in the light of recent DNA-based studies. The terminology on clonality is expanding and new definitions have been constructed describing clones in all taxa (Loxdale and Lushai 2003). A term describing clones in woody perennial crops could be 'clones are the assemblage of biotypes deriving from a single zygote through somatic mutations. Clones may expose genetic variation. Chimeric clones bearing mutations in a divided genome are thus clones as are polyploid clones.'

Genetic variation is added by numerous mechanisms to enhance variation in asexually reproducing organisms, and allow for adaptation and selection processes. For a cultivated woody perennial crop genetic variation has additional meanings since clonality is used to maintain identical genotypes and allow for the selection of new clones of grapevine cultivars. Research on clonality and grapevine clones will progress. New innovative techniques will facilitate closer examinations of the mechanisms of the genome using microchip and microarray techniques. Expanding research in grapevine clones by incorporating advances from other genomics efforts and techniques will have a large impact on grape clonality research. The position and timing of the mutations will be studied to manipulate variation-inducing events. Furthermore, quantification of such variation will be of great interest.

Acknowledgements. I would like to thank Prof. Dr. Esser for his invitation to put my thoughts together on this interesting topic and Prof. Dr. Blaich, Prof. Dr. Walker and Dr. G. Buffler for their helpful discussions and editorial advice. I am especially grateful to my group in the Department of Viticulture for their ongoing input.

References

Auge H, Brandl R (1997) Seedling recruitment in the invasive clonal shrub, *Mahonia aquifolium* Pursh (Nutt.). Oecologia 110:205–211

Berg EE, Hamrick JL (1994) Spatial and genetic structure of two sandhills oaks: *Quercus laevis* and *Quercus margaretta* (Fagaceae). Am J Bot 81:7–14

Botta R, Schneider A, Akkak A, Scott NS, Thomas MR (2000) Within cultivar grapevine variability studied by morphometrical and molecular marker based techniques. Acta Hortic 528:91–96

Bourquin JC, Otten L, Walter B (1995) PCR-RFLP analysis of *Vitis*, *Ampelopsis* and *Parthenocissus* and its application to the identification of rootstocks. Vitis 34:103–108

Bowers JE, Meredith CP (1997) The parentage of a classic wine grape, Cabernet Sauvignon. Nat Genet 16:84–87

Büscher N, Zyprian E, Blaich R (1933) Identification of grapevine cultivars by DNA analyses: pitfalls of random amplified polymorphism DNA techniques using 10mer primers. Vitis 32:187–188

Caposella A, Silander JA, Jansen RK, Bergen B, Talbot DR (1992) Nuclear ribosomal DNA variation among ramets and genets of white clover. Evolution 46:1240–1247

Cervera MT, Cabezas JA, Sancha, JC, Martinez de Toda, F, Martínez-Zapater JM (1998) Application of AFLPs to the characterization of grapevine *Vitis vinifera* L. genetic resources. A case study with accessions from Rioja (Spain). Theor Appl Genet 97:51–59

Cervera MT, Cabezas JA, Sanches-Escribano E, Cenis JL, Martinez-Japater JM (2000) Characterization of genetic variation within table grape varieties (*Vitis vinifera* L.) based on AFLP markers. Vitis 39:109–114

Cervera MT, Rodriguez I, Cabezas JA, Chavez J, Martinez-Zapater JM, Cabello F (2001) Morphological and molecular characterization of grapevine accessions known as Albillo, one of the oldest varieties grown in Spain. Am J Enol Vitic 52:127–135

Cervera MT, Cabezas JA, Rodríguez-Torres I, Chavez J, Cabello F, Martínez-Zapater JM (2002) Varietal diversity within grapevine accessions of cv. Tempranillo. Vitis 41:33–36

Chung MG, Epperson BK (2000) Clonal and spatial genetic structure in *Eurya emarginata* (Theaceae). Heredity 84:170–177

Cipriani G, Frazza G, Peterlunger E, Testolin R (1994) Grapevine fingerprinting using microsatellite repeats. Vitis 33:211–215

Cirami RM, McCarthy MG, Nicholas PR (1993) Clonal selection and evaluation to improve production of Cabernet Sauvignon grapevines in South Australia. Aust J Exp Agric 33:213–220

Clingeleffer PR (1988) Response of Riesling clones to mechanical hedging and minimal pruning of cordon trained vines (MPTC) - implications for clonal selection. Vitis 27:87–93

Crespan M (2004) Evidence on the evolution of polymorphism of microsatellite markers in varieties of *Vitis vinifera* L. Theor Appl Genet 108:231–237

De Kroon H, van Groenendael J (1997) The ecology and evolution of clonal growth in plants. Backhuys, Leiden, 453 pp

Dorken ME, Eckert CG (2001) Severely reduced sexual reproduction in northern populations of a clonal plant, *Decodon verticillatus* (Lythraceae). J Ecol 89:339–350

Eckert CG (1999) Clonal plant research: proliferation, integration, but not much evolution. Am J Bot 86:1649–1654

Fanizza G, Chaabane R, Ricciardi L, Resta P (2003) Analysis of a spontaneous mutant and selected clones of cv. Italia (*Vitis vinifera*) by AFLP markers. Vitis 42:27–30

Filippetti I, Silvestroni O, Thomas MR, Intrieri C (1999) Diversity assessment of seedlings from self-pollinated Sangiovese grapevines by ampelography and microsatellite DNA analysis. Vitis 38:67–71

Forneck A, Konradi J, Blaich R (2003a) Über die genetische Diversität der Burgunderreben und ihrer Klone. Dtsch Weinbau-Jhrb 54:71–78

Forneck A, Konradi J, Blaich R (2003b) A genetic variation analysis of *V. vinifera* cv. Pinot noir. Acta Hort 603:167–171

Franks T, Botta R, Thomas MR (2002) Chimerism in grapevines: implications for cultivar identity, ancestry and genetic improvement. Theor Appl Genet 109:192–199

Gill DE, Chao L, Perkins SL, Wolf JB (1995) Genetic mosaicism in plants and clonal animals. Annu Rev Ecol Syst 26:423–444

Gogorcena Y, Arulsekar S, Dandekar AM, Parfitt DE (1993) Molecular markers for grape characterization. Vitis 32:183–185

Goto-Yamamoto N (2000) Phenetic clustering of grapes (*Vitis* spp.) by AFLP analysis. Breed Sci 50:53–57

Harding K, Benson EE, Roubelakis-Angelakis KA (1996) Methylated DNA changes associated with the initiation and maintenance of *Vitis vinifera* in vitro shoot and callus cultures: a possible mechanism for age-related changes. Vitis 35:79–85

Hocquigny S, Merdinoglu D, Heloir MC, Relsy F (2003) Chimerism and genetic diversity within the cultivar group of Pinots. Acta Hort 603:535–544

Imazio S, Labra M, Grassi F, Winfield M, Bardini M, Scienza A (2002) Molecular tools for clone identification: the case of the grapevine cultivar 'Traminer'. Plant Breed 121:531–535

Klekowski EJ, Godfrey PJ (1989) Aging and mutation in plants. Nature 340:389–391

Konradi J, Forneck A, Blaich R (2002) Jahrestagung des Forschungsrings des Deutschen Weinbaus (FDW) in Geisenheim 17.-18.4.2002. Jahresbericht 2002. Deutsche Landwirtschafts Gesellschaft, Frankfurt/Main, pp 9–10

Kozjak P, Korosec-Koruza Z, Javornik B (2003) Characterisation of cv. Refosk (*Vitis vinifera* L.) by SSR markers. Vitis 42:83–86

Kuksova VB, Piven NM, Gleba Y (1997) Somaclonal variation and in vitro induced mutagenesis in grapevine. Plant Cell Tissue Org Cult 49:17–27

Lavi U, Cregan P, Schaap T, Hillel J (1994) Application of DNA markers for identification and breeding of perennial fruit crops. Plant Breed Rev 12:195–226

Loureiro MD, Martinez MC, Boursiquot JM, This P (1998) Molecular marker analysis of *Vitis vinifera* 'Albarino' and some similar grapevine cultivars. J Am Soc Hortic Sci 123:842–848

Loxdale HD, Lushai G (2003) Intraclonal genetic variation: ecological and evolutionary aspects. Biol J Linn Soc 279 pp

Maletic E, Pejic I, Kontic JK, Piljac J, Dangl G, Vorkuka A, Lacombe T, Mirosevic B, Meredith C (2003) The identification of Zinfandel on the Dalmatian coast of Croatia. Acta Hort 603:251–254

Mannini F (2000) Clonal selection in grapevine: interactions between genetic and sanitary strategies to improve propagation material. Acta Hortic 528:703–712

Mannini F, Rolle L, Guidoni S (2003) Vineyard management to optimize grape quality in virus-free clones of *Vitis vinifera* L. Acta Hortic 603:121–126

Merdinoglu D, Butterlin G, Baur C, Balthazard J (2000) Comparison of RAPD, AFLP and SSR (microsatellite) markers for genetic diversity analysis in *Vitis vinifera* L. Acta Hortic 528:193–197

Pelsy F, Merdinogly D (2002) Development of grapevine molecular markers based on retrotransposons. In: Proc 10th Int Plant and Animal Genome Conf, 12–16 Jan, San Diego

Pelsy F, Schehrer L, Merdinoglu D (2003) Development of grapevine retrotransposon-based molecular markers (S-SAP). Acta Hort 603:83–87

Popescu CF, Falk A, Glimelius K (2002) Application of AFLPs to characterize somaclonal variation in anther-derived grapevines. Vitis 41:177–182

Rajora OP (1999) Genetic biodiversity impacts of silvicultural practices and phenotypic selection in white spruce. Theor Appl Genet 99:954–961

Regner F, Kaserer H (2002) Investigations into the genetic variability of Traminer clones. Mitt Klosterneuburg 52:177–186

Regner F, Stadelbauer A, Eisenheld C, Kaserer H (2000a) Genetic relationships among Pinots and related cultivars. Am J Enol Vitic 51:7–17

Regner F, Wiedeck E, Stadelbauer A (2000b) Differentiation and identification of White Riesling clones by genetic markers. Vitis 39:103–107

Reustle GM, Matt A (2000) First steps to use the protoplast technique for breeding purposes. Acta Hortic 528:341–347

Riaz S, Garrison KE, Dangl GS, Boursiquot JM, Meredith CP (2002) Genetic divergence and chimerism within ancient asexually propagated winegrape cultivars. J Am Soc HorticSci 127:508–514

Rives M (1961) Bases génétiques des la sélection clonale chez la vigne. Ann Amélior Plant 11:337–348

Rohde J (1996) A novel PCR-based DNA marker technology called inverse sequence tagged repeat (ISTR) analysis. Genet Breed 50:249–261

Roubelakis-Angelakis K (ed) (2001) Molecular biology and biotechnology of grapevine. Kluwer, Dordrecht

Schenk HJ (1999) Clonal splitting in desert shrubs. Plant Ecol 141:41–52

Schlotterer C, Ritter R, Harr B, Brem G (1998) High mutation rate of a long microsatellite allele in *Drosophila melanogaster* provides evidence for allele-specific mutation rates. Mol Biol Evol 15:1269–1274

Schneider S, Reustle G, Zyprian E (1996) Detection of somaclonal variation in grapevine regenerants from protoplasts by RAPD-PCR. Vitis 35:99–100

Schöffling H, Stellmach G (1993) Klon-Züchtung bei Weinreben in Deutschland. Waldkirchner, Waldkirch, 818 pp

Scott KD, Ablett EM, Lee LS, Henry RJ (2001) AFLP markers distinguishing an early mutant of Flame Seedless Grape. Euphytica 113:245–249

Sefc KM, Regner F, Glössl J, Steinkellner H (1998) Genotyping of grapevine and rootstock cultivars using microsatellite markers. Vitis 37:15–20

Sensi E, Vignani R, Rohde W, Biricolti S (1996) Characterization of genetic biodiversity with *Vitis vinifera* L. Sangiovese and Colorino genotypes by AFLP and ISTR DNA marker technology. Vitis 35:183–188

Silvestroni O, di Pietro D, Intrieri C, Vignani R, Filippetti I, del Casino C, Scail M, Cresti M (1997) Detection of genetic diversity among clones of cv. Fortana (*Vitis vinifera* L.) by microsatellite DNA polymorphism analysis. Vitis 36:147–150

Specht C (2002) Examensarbeit. Botanisches Institut, Universität Heidelberg

Striem JM, Ben-Hayyim G, Spiegel-Roy P (1994) Developing molecular genetic markers for grape breeding, using polymerase chain reaction procedures. Vitis 33:53–54

Thompson MM, Olmo HP (1963) Cytohistological studies of cytochimeric and tetraploid grapes. Am J Bot 50:901–906

Tschammer J, Zyprian E (1994) Molecular characterization of grapevine cultivars of Riesling-type and of closely related Burgundies. Vitis 33:249–250

Udupa SM, Baum M (2001) High mutation rate and mutational bias at (TAA)n loci in chickpea (*Cicer arietinum* L). Mol Gen Genet 265:1097–1103

Van Steensel B, Henikoff S (2003) Epigenomic profiling using microarrays. Biotechniques 35:346–357

Vignani R, Bowers JE, Meredith CP (1996) Microsatellite DNA polymorphism analysis of clones of *Vitis vinifera* 'Sangiovese'. Sci Hortic 65:163–169

Waugh R, McLean K, Flavell AJ, Pearce SR, Kumar A, Thomas BB, Powell W (1997) Genetic distribution of Bare-1-like retrotransposable elements in the barley genome revealed by sequence-specific amplification polymorphisms (S-SAP). Mol Gen Genet 53:687–694

Ye GN, Soylemezoglu G, Weeden NF, Laboy WF, Pool RM, Reisch BI (1998) Analysis of the relationship between grapevine cultivars, sports and clones via DNA fingerprinting. Vitis 37:33–38

Astrid Forneck
Institut für Sonderkulturen und Produktionsphysiologie
Universität Hohenheim
Fachgebiet Weinbau (370a)
70593 Stuttgart, Germany

Physiology

Proton Channelling *b*-Type Cytochromes in Plant Plasma Membranes?

Sabine Lüthje, Miachael Böttger, and Olaf Döring

1 Introduction

To maintain an energized status of the cellular components a constant flow of electrons has to be provided. Electron flow goes energetically downhill from reduced molecules (donors) to oxidized molecules (acceptors). This energy, when it comes down to it, mainly stems from light energy captured by photosynthetic processes. The constant flow of electrons keeps cellular (and apoplastic) compounds in a certain redox status, i.e. the ratio between reduced and oxidized compounds. This redox status has to be maintained in order for cellular processes (enzyme activities, transcription, translation, etc.) to function properly. Stress conditions are usually accompanied by change of redox status, thus causing metabolic activities to change, and, as an extreme, after death the redox status inclines towards an oxidized status. In order to keep the redox status in the various cellular compartments and also in the apoplast at the respective level or to transfer electrons to certain molecules, electrons have to be transported from one compartment to another.

Electron transfer is mediated by specialized systems. The electron flow we are going to discuss in this chapter is a flow across membranes; in our case we will concentrate on the plasma membrane (PM) of plant cells. Electron transfer occurs at the PM with electron donors at the cytosolic side, and acceptors that are located in the apoplast or in the cytosol (Lüthje et al. 1997; Bérczi and Møller 2000). One group of components of electron transfer chains – the cytochromes – are proteins containing specialized macrocyclic porphyrin ligands with an iron atom at their center. Electron transfer to and from cytochromes is accomplished by oxidation state alterations of the porphyrin-bound iron. The properties of cytochromes are dependent in part on few variations of the porphyrins – these variations give rise to a broad classification of the cytochromes. Significant variations of measured properties, i.e. redox potential, spectral absorption bands, are caused by the embedding proteins holding the porphyrin.

Progress in Botany, Vol. 66

For electron transfer to take place at any significant rate, donor and acceptor have to be in immediate vicinity. An increase in distance by only 2 nm, a distance corresponding well to distances within membranes or proteins, lowers electron transfer rate by a factor of 10^{12} (Moser et al. 1992). Even then, electrons do not just hop from donor to acceptor, but the wave function of donor and acceptor orbitals has to overlap. This overlap is usually expressed by a coefficient called the Frank-Codon factor.

Cytochromes can be characterized by their absorption spectra. In general, there are four characteristic absorption bands if the cytochromes are measured in their reduced status. The absorption bands have been termed as α-, β-, γ-, and δ-band. The α- and β-bands, lying somewhere around 550 and 520 nm, respectively, completely disappear if the cytochromes are oxidized, while γ- and δ-bands (γ-band somewhat above 400 nm, δ-band usually between 300 and 400 nm) drop to some extent without completely disappearing and also slightly shift the wavelength of their absorption maximum. Mostly the α-band difference spectra (reduced minus oxidized status) are used to characterize and measure the redox status of the cytochromes. Also, the names of some cytochromes have been in part determined by the absorption maximum of the α-band, which is indicated by a numerical add-on or subscript to the cytochrome's name.

One function of membranous cytochrome *b* complexes in all membranes is to transport protons linked to electron transport across the membrane (Wikström et al. 1981; Michel 1998; Mankelow and Henderson 2003). This has been demonstrated for cytochromes involved in photosynthesis and respiration (Rich 1984; Berry and Trumpower 1985; Williams 1985; Joliot and Joliot 1986).

2 *b*-Type Cytochromes in Plant PM

In the past, several reviews especially on PM-bound *b*-type cytochromes have been published (Asard et al. 1994, 1998, 2000). The occurrence of *b*-type cytochromes in plant PMs was demonstrated first by absorbance difference spectra of PM preparations (e.g. Brain and Briggs 1977; Jesaitis et al. 1977; Caubergs et al. 1983; Widell and Larsson 1983; Kurkova and Verhovskaya 1984; Barr et al. 1986; Asard et al. 1989; Askerlund et al. 1989). The total amount of cytochromes in enriched plant PM was reported as between 0.1 and 0.5 nmol mg protein^{-1} (references in Asard et al. 1994, 1998; Lüthje et al. 1997; Bérczi and Møller 2000). There were significant differences between the cytochrome *b* content of material from different plant species and organs. These differences may be highly influenced by the developmental stage of the material used as observed for microsomal cytochrome composition (Rich and Bendall 1975; Ishimamura and Yamazaki 1977; Hendry et al. 1981).

About 10% of the PM-bound cytochrome *b* was found to be reduced after the plants were exposed to blue light (e.g. Caubergs et al. 1978; Widell et al. 1982; Borgeson and Bowman 1985; Asard et al. 1989, 1995a; Short and Briggs 1990). Blue-light-induced absorbance changes (LIAC) in the PM fraction of *Neurospora crassa* were sensitive to 1 mM azide (50% inhibition) and salicylhydroxamic acid (SHAM), suggesting that a heme-containing protein may be involved in this reaction (Borgeson and Bowman 1985). Inhibitors of flavin-mediated electron transfer, such as potassium iodide or phenylacetic acid, were highly effective in preventing this cytochrome *b* reduction (Asard et al. 1995a). These findings were interpreted as a blue-light-mediated electron flow from a flavin chromophore to cytochrome *b* (Asard and Caubergs 1990; Galland and Senger 1991; Rubinstein and Stern 1991a; Asard et al. 1998). Meanwhile three types of photoreceptors were isolated and characterized:

1. Cryptochromes 1 and 2 (Ahmad and Cashmore 1993, 1996; Ahmad et al. 2002);
2. Phototropins 1 and 2 of *Arabidopsis thaliana* (Christie et al. 1998, 1999); and
3. Photoactivated adenylyl cyclase of *Euglena* (Iseki et al. 2002).

A recent study described the purification of a photoreceptor from PM-enriched preparations isolated from plants (Lorenz et al. 2003). Galland and Tölle (2003) suggested a general model, in which the three redox states of the flavin photoreceptor, the oxidized flavin (Fl), the flavosemiquinone (FlH•), and the flavohydroquinone (FlH_2) are acting as chromophores, each with its own characteristic photochemical primary reaction. These reactions consist of the photoreduction of the Fl generating the FlH•, the photoreduction of the FlH• leading to the formation of the FlH_2, and the photooxidation of the FlH_2 regenerating the Fl. The proposed mechanism represents a photocycle with two antagonistic photoreceptor forms. Amounts of Fl and FlH_2 determine the pool size of the biologically active molecule, the FlH•. The redox changes that are associated with the photocycle are mediated by redox partners, namely pterins, which function in the near-UV as secondary chromophores. Evidence for the occurrence of pterins in the plant PM was presented (van Gestelen et al. 1996).

Redox titration and multi-component analysis of highly enriched PM fractions suggest the occurrence of at least two, or perhaps even three, different *b*-type cytochromes in plants (Asard et al. 1989; Askerlund et al. 1989). A low-potential cytochrome *b* (α-band 557-562 nm, E_0'=−65 mV) comparable to that of soluble nitrate reductase (E_0'=−174 to −73 mV, Solomonson and Barber 1990) was identified in maize coleoptile PM isolated by sucrose gradient centrifugation (Jesaitis et al. 1977; Leong et al. 1981) and highly enriched PM from barley roots (Askerlund et al. 1989). Several investigations suggest a function of PM-bound nitrate reductase as a blue light photoreceptor (references in Stöhr 1998). Since PM-bound nitrate reductase may be an inducible redox system, its absence in some material may be explained by the developmental state and different tissues used. At present it is not possible to assign the low potential cytochrome *b* to the nitrate reductase; also the involvement of a cytochrome b_5 has been suggested (Asard et al. 1989; Askerlund et al. 1989).

The major cytochrome found to be ubiquitous in the PM is a *high*-potential cytochrome *b* (E_0' ≈+150 mV) with an α-band maximum between 559 and 562 nm (references in Asard et al. 1998). In some redox titration analyses an additional component was calculated which has a redox potential of E_0' ≈+265 mV.

2.1 Cytochrome b_5

The observation of split α-bands, the identification of low midpoint redox potentials (E_0' ≈–20 to –40 mV) and the relatively high antimycin A-insensitive NADH-cytochrome *c* reductase activity found with PM preparations were taken as indirect evidence for the presence of cytochrome b_5 in plant PMs (Asard et al. 1989; Askerlund et al. 1989). These authors estimate that cytochrome b_5 was about 30% of the total cytochrome content in the PM.

Cytochrome b_5, which has a molecular mass of 16 kDa on SDS-PAGE, consists of two domains and was found in cytosol, microsomal fractions and the outer mitochondrial membrane (references in Ozols 1989; Vergères and Waskell 1995). The domains can be isolated as separate proteins after mild trypsin digestion (Strittmacher and Ozols 1966). The amino acid sequence of the heme-binding domain of microsomal and mitochondrial cytochrome b_5, a hydrophilic globular portion, reveals that only 58% of the amino acids are conserved. The heme group (E_0' ~+20 mV) is capable of accepting and transferring a single electron (Velick and Strittmatter 1956; Iyanagi 1990). The second domain of the protein is hydrophobic and anchors the protein in the membrane (Ito and Sato 1968; Spatz and Strittmatter 1971). The microsomal cytochrome b_5 is located in the endoplasmic reticulum and is involved in lipid biosynthesis. The protein accepts electrons either from cytochrome b_5 reductase or from cytochrome P450 (Vergères and Waskell 1995; Drexler et al. 2003).

A flavin adenine dinucleotide (FAD)-containing cytochrome b_5 reductase activity of 31 kDa was purified from microsomal fractions (Jollie et al.

1987; Sparla et al. 1997). The protein from maize was cloned and shown to have Fe^{3+}-chelate reductase activity (Bagnaresi et al. 1999). Also an NADH dehydrogenase (31 kDa) with capability to reduce quinones and Fe^{3+}-citrate was purified from plant PM (Serrano et al. 1994; van Gestelen et al. 1997; Lüthje et al. 1998). However, antibodies raised against microsomal cytochrome b_5 reductase cross-reacted with a 45-kDa protein in PMs isolated from spinach leaves (Askerlund et al. 1991). Purification and micro sequence analysis of the 45-kDa protein showed homology to ascorbate free radical reductase (Bérczi et al. 1995; Bérczi and Møller 1998a). Ascorbate free radical reductase activity was observed with PMs of several plant materials (references in Buckhout and Luster 1991). Thus evidence for the presence of cytochrome b_5 or cytochrome b_5 reductase in highly enriched PM preparations rather than microsomes appears weak. On the other hand, plant PMs contain sphingolipids (e.g. Bohn et al. 2001) which are modified by Δ8-desaturases (Sperling and Heinz 2001; Sperling et al. 2003). Amino acid sequences of these cytochrome b_5-containing proteins suggest a localization in the PM, although desaturases have not yet been found in enriched PM preparations.

2.2 Cytochrome P450

Multiple forms of cytochrome P450 have been found in plants (Donaldson and Luster 1991). These proteins were characterized and purified from microsomal fractions (Rich et al. 1971; Benveniste et al. 1986; Funk and Croteau 1993; Gerardy and Zenk 1993; Rodgers et al. 1993; Stadler and Zenk 1993; Hansikova et al. 1994; Marabini et al. 1994; Nega and Grunwaldt 1997). They are anchored in the endoplasmic reticulum by one or two transmembrane domains located at the N terminus (Black 1992). The active site is part of a large cytoplasmic domain that may have one or two additional peripheral membrane contacts.

Cytochrome P450 systems are known to have a function in cell detoxification in plants, fungi, and animal cells (Shoun et al. 1991; Durst et al. 1992; Loeper et al. 1998). In yeast PM a cytochrome P450 responsible for metabolizing xenobiotica was reported repeatedly (see Loeper et al. 1998). In plants cytochrome P450 can be induced by wounding (Lesot et al. 1990) or application of herbicide safeners (Barrett and Maxson 1991; Mougin et al. 1991; Persan et al. 2001). The application of such agrochemicals to vegetables might be the reason for the detection of cytochrome P450 in cauliflower inflorescences obtained from local markets (Kjellbom et al. 1985; Askerlund et al. 1989). An attempt to demonstrate the induction of cytochrome P450 in bean PM failed (R.J. Caubergs, pers. comm.) and the search for cytochrome P450/420 (E_0' =−100 mV) in plant PM other than cauliflower inflorescences by CO-difference spectra or redox titrations was unsuccessful (Ramirez et al. 1984; Kjellbom et al. 1985; Asard et al. 1989; Askerlund et al. 1989).

2.3 Peroxidases

Peroxidases may have a part in the low potential cytochrome detected in plant PM. The presence of peroxidase-like activities in the PM of plants has been demonstrated since the beginning of the early 1980s (references in Vianello and Macri 1991; Mika et al. 2004). Recent work has demonstrated the presence of at least two class III peroxidases in the PM of maize roots (Mika and Lüthje 2003a). The corresponding proteins were identified by heme-staining, and absorbance difference spectra of the partial purified proteins showed typical cytochrome *b* spectra with maxima at 425, 528, and 559 nm after reduction by dithionite. The inhibition by iron chelators, such as cyanide or azide, further supports a function of the heme group for the peroxidase activity. The midpoint redox potentials of these *b*-type cytochromes, however, have not been analysed yet.

Several functions were postulated for PM-bound peroxidases (references in Vianello and Macri 1991; Bérczi and Møller 2000; Mika et al. 2004). Although most of these functions were demonstrated for soluble and cell-wall-bound extracellular peroxidases as well (references in Ros Barceló 1997; Hiraga et al. 2001; Kawano 2003), apparently there may be some functions specific for membrane-bound peroxidases, namely a membrane protective function (Bunkelmann and Trelease 1996; Jespersen et al. 1997; Mika et al. 2004). The ascorbate peroxidase activity of PM-bound peroxidase, however, is low (Mika and Lüthje 2003a); thus the mechanism of these enzymes appears to be distinct from that of ascorbate peroxidases in peroxisomes. The increase of flavonoids during oxidative stress (Peters et al. 1986; Liu et al. 1993; Dixon and Paiva 1995; Shirley 1996; Jaakola et al. 2003) suggests a flavonol–peroxidase reaction, as proposed by Yamasaki et al. (1997) for H_2O_2 scavenging (Mika et al. 2004).

2.4 Ascorbate-Reducible Cytochrome *b*

Early work demonstrated that plant microsomes contain an ascorbate-reducible cytochrome b_{559} also named cytochrome b_3 (Martin and Morton 1955, 1957; Lance and Bonner 1968). This cytochrome b_{559} was purified by ion exchange chromatography from mung bean (Shichi et al. 1963) and algae (Sugimura et al. 1980). The protein was purified by anion exchange chromatography, had a molecular weight of 28 kDa, was reducible by ascorbate, cystein, and dithionite, and transfers electrons to ferricyanide (hexacyanoferrate III, HCF III) and molecular oxygen (Shichi et al. 1963). The characterization of highly enriched tonoplast and PM fractions showed the occurrence of ascorbate-reducible *b*-type cytochromes with similar properties in both fractions (Barr et al. 1986; Trost et al. 2000; Bérczi et al. 2003; Griesen et al. 2004).

The ascorbate-reducible cytochrome *b* (E_0'=+150 mV) is a major compound in plant PM with an α-band between 559 and 562 nm (Asard et al. 1989, 1998; Askerlund et al. 1989). This cytochrome was suggested to be glycosylated and to interact with flavins in the blue-light-induced reaction (references in Asard et al. 1994). Studies in vitro suggested the involvement of the bean (*Phaseolus vulgaris* L.) cytochrome b_{561} in transmembrane electron transfer from cytosolic ascorbate as electron donor to apoplastic monodehydroascorbate, thus keeping the apoplastic redox status high (Asard et al. 1992, 1995b; Horemans et al. 1994). This reaction appears to be analogous to the animal cytochrome b_{561} in chromaffin granules (Njus et al. 1983; Kelley and Njus 1986; see also below). The ascorbate regenerating system observed in PM of HL-60 cells or erythrocytes, however, seems to be different compared to the chromaffin system; a cytochrome b_{561} appears not to be involved in this reaction (van Duijin et al. 2001a,b).

3 Cytochrome b_{561}

3.1 Cytochrome b_{561} of Chromaffin Granules

Cytochrome b_{561} in membranes of chromaffin vesicles has a central function in ascorbate regeneration from monodehydroascorbate during catechol biosynthesis (Fig. 1; Njus and Kelley 1993; Njus et al. 2001). Cytochrome b_{561} is reduced by cytosolic ascorbate, which regenerates ascorbate inside the vesicles by a transmembrane electron transfer (references in Njus and Kelley 1993; Rubinstein 1994; Okuyama et al. 1998, Tsubaki et al. 2000). Cytochrome b_{561} is a highly conserved 30-kDa protein with two *heme* prosthetic groups (E_0'=+70 and +170 mV) and five to six transmembrane spanning domains (Flatmark and Terland 1971; Wakefield et al. 1984; Asada et al. 2002). Rates of 950 nmol HCF III (min mg protein)$^{-1}$ were measured for the ascorbate-dependent transmembrane reduction (Kelley and Njus 1986).

The mechanism postulated for ascorbate reduction is (Njus et al. 2001):

1. The ascorbate monoanion binds to an unprotonated site (histidine) on cytochrome b_{561}.
2. This complex donates an electron to reduce the heme.
3. The semidehydroascorbate anion dissociates from the cytochrome, leaving a proton associated with the binding site.
4. The binding site is deprotonated to complete the cycle.

Thus the ascorbate-regenerating system in the chromaffin vesicles is a chain of concerted H^+/e^--transfer reactions (Njus and Kelley 1993). Thermodynamic consideration showed that no steps in this process involve large changes in free energy, i.e. the mechanism is reversible and capable of fulfilling the cytochrome's function of equilibrating ascorbate and semidehydroascorbate.

Recent studies demonstrate the presence of a cytochrome b_{561} homologue (Dcytb) in duodenum, which was suggested to have a function in iron reduction (McKie et al. 2001, 2002; Knöpfel and Solioz 2002).

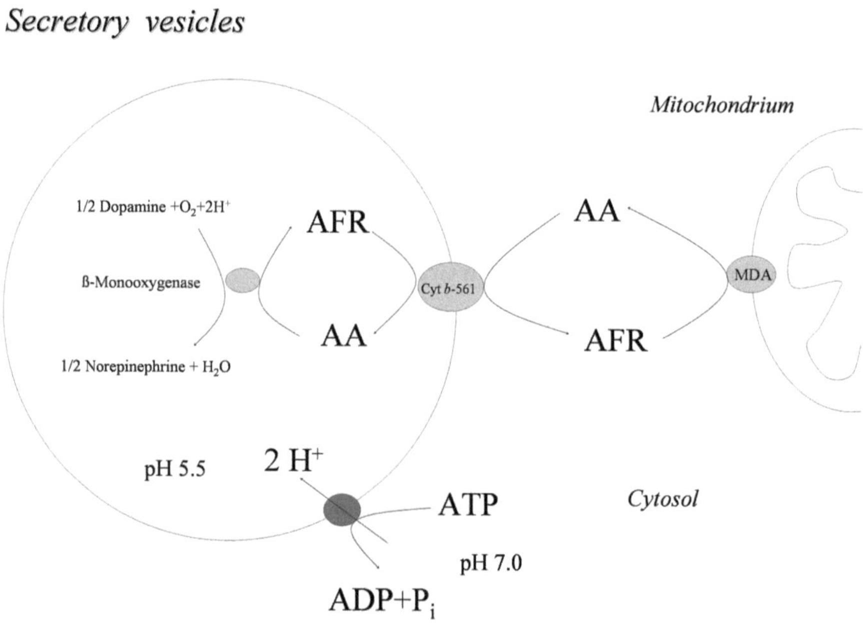

Fig. 1. Ascorbate regeneration in chromaffin vesicles. Monooxygenases inside secretory vesicles oxidize ascorbate (*AA*) to its free radical form, semidehydroascorbate (*AFR*). This intravesicular AFR is reduced back to AA by cytochrome b_{561}. Cytochrome b_{561} is reduced in turn by cytosolic AA, and the cytosolic AFR thus produced is recycled by AFR reductase (*MDA*) in the outer mitochondrial membrane. (After Njus and Kelley 1993)

3.2 Cyt b_{561} Homologues in Plants

At least three genes with a limited homology (30–38%) to the DNA sequence encoding for the mammalian chromaffin granule cytochrome b_{561} have been identified in the genome of *Arabidopsis thaliana* L. (Asard et al. 2000, 2001). Due to the identified sequences, six transmembrane helixes, heme- and ascorbate-binding sites were predicted for the plant homologues (Asard et al. 1998; Bashtovyy et al. 2003). A phylogenetic study identified further homologues in other plant species (Verelst and Asard 2003).

Several attempts were made to solubilize and purify this cytochrome *b* from the PM of higher plants (Bérczi and Møller 1998b; Scagliarini et al. 1998; van Gestelen et al. 1998; Bérczi et al. 2001, 2003; Griesen et al. 2004). Biochemical characteristics (i.e. absorbance difference spectra, redox potential) of the ascorbate-reducible cytochrome were seemingly comparable to the animal cytochrome b_{561} (Scagliarini et al. 1998; Asard et al. 2000; Trost et al. 2000; Bérczi et al. 2003). Furthermore, a transmembrane electron transfer via the cytochrome b_{561} to monodehydroascorbate was shown in ascorbate-loaded right-side-out PM vesicles isolated from bean hooks (Asard et al. 1992). Rates of HCF III reduction, however, were about 100-fold lower (10 nmol HCF III (min mg protein)$^{-1}$) compared to that of the chromaffin system. Fractions with the partially purified bean cytochrome contained two heme groups with standard redox potentials of +135 and +204 mV (Trost et al. 2000). Similar redox potentials (+135 and +180 mV) were found for the ascorbate-reducible cytochrome *b* isolated from *Arabidopsis* leaves (Bérczi et al. 2003). However, relative molecular masses and isoelectric points of the *Arabidopsis* and maize (*Zea mays* L.) cytochromes appeared significantly different compared to the bean cytochrome b_{561} (Bérczi et al. 2001).

The ascorbate-reducible cytochrome *b* isolated from *Arabidopsis* leaves showed up as a 27-kDa protein PAGE band which is in accordance with the predicted molecular mass of the cytochrome b_{561} homologue (Bérczi et al. 2003). The final identification by micro sequence analysis has not been presented so far. Also, the location of further cytochrome b_{561} homologues has yet to be shown. Until then, expression in endoplasmic reticulum and tonoplast may be a feasible alternative possibility (Barr et al. 1986; Verelst 2003; Griesen et al. 2004; Verelst et al. 2004).

The physiological function of the ascorbate-reducible cytochrome in plant PM is not yet clear. Due to the identification of distinct iron reductases (Robinson et al. 1999; Waters et al. 2002), the low rates of HCF III reduction observed in vitro (Asard et al. 1992), and because there is evidence for PM ascorbate transporters (references in Horemanns et al. 2000a,b), a major function of the PM-bound cytochrome b_{561} in apoplastic ascorbate regeneration or iron reduction is debatable.

Grasses have evolved a completely different system for iron uptake from that of dicotyledonous or non-grass monocotyledonous plants (references in Bienfait 1985; Schmidt 1999). Nevertheless, cytochrome b_{561} homologues have been identified in maize and rice as well as in various dicotyledonous plants (Verelst and Asard 2003). Although the localization of these cytochrome b_{561} homologues is not really known today, occurrence of an ascorbate-reducible cytochrome *b* with properties comparable to that purified from *Arabidopsis* PM was reported in maize (Bérczi et al. 2001).

Reduction of the major ascorbate-reducible cytochrome *b* by quinols in vitro suggests a possible function in production of reactive oxygen species (Lüthje et al. 1998; van Gestelen et al. 1998). The physiological relevance of such a reaction involving cytochrome b_{561}, however, has to be further elucidated. In our studies, superoxide anion radical ($O_2^{\bullet-}$) production could be observed only at unphysiologically high quinol concentrations (10–500 μM), whereas physiological relevant concentrations (100–500 nM) rather caused an inhibition of $O_2^{\bullet-}$ production in plant PM (M. Menckhoff and S. Lüthje, unpubl.).

4 Flavocytochrome *b* Family

The name of this protein family originates from the β-subunit of the mammalian phagocyte NADPH oxidase, the gp91^{phox} (*gp* glycosylated protein; *91* rough M_r; *phox* phagocyte oxidase) also named Nox2 (NADPH oxidase). The structure, regulation, and function of the phagocyte NADPH oxidase has been reviewed at regular intervals over the past decades (Rossi 1986; Morel et al. 1991; Babior 1992; Dagher et al. 1995; Henderson and Chappell 1996; Segal et al. 1998; Vignais 2002). Meanwhile several homologues of the gp91^{phox} have been identified (Fig. 2) and recent reviews have discussed possible functions of the flavocytochrome *b* family (Kimball and Sair 2001; Lambeth 2002). Regulation and functions of these homologues, however, are quite different, as detailed below.

4.1 NADPH Oxidase

Among the flavocytochrome *b* family, the phagocyte NADPH oxidase complex is the one of most interest in the literature. Phagocytes are able to actively migrate towards the site of invasion and contribute to defence by phagocytosis of pathogens and by applying a mixture of digestion, poison-

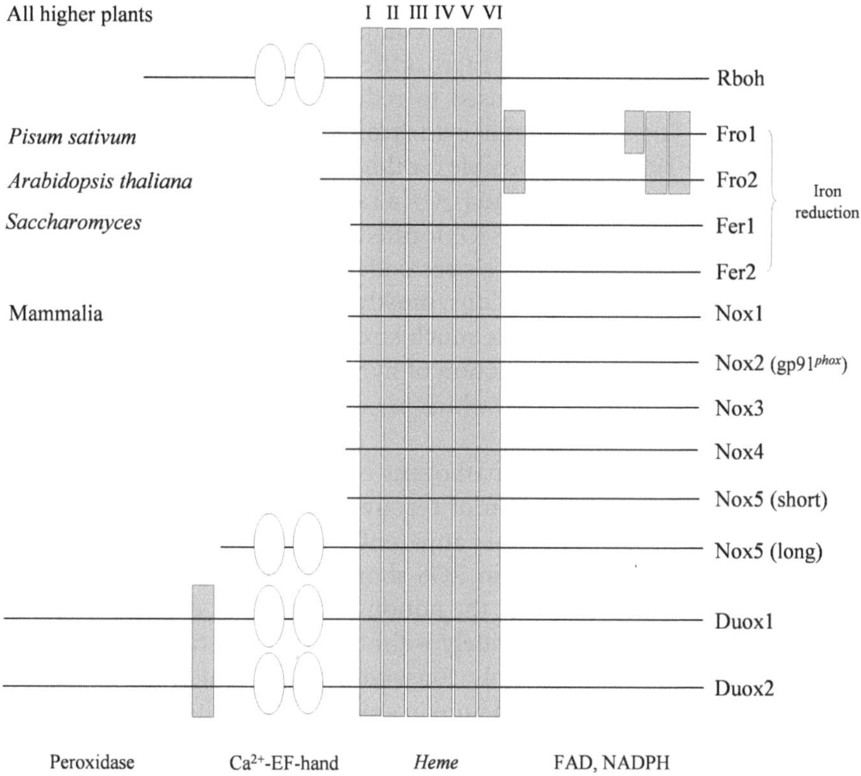

Fig. 2. The flavocytochrome *b* family and homologues in higher plants. The flavocytochrome *b* family is homologous to the β-subunit of the phagocyte NADPH oxidase (gp91^{phox}, Nox2). Nox5, dual oxidases (*Duox*) and plant homologues (*Rboh*) have calcium EF-hands for activation [E and F are the calcium-binding helices identified in calmodulin as defined by Hendrickson and Karle (1973)]. Duox have additional peroxidase-like domains. All homologues seem to contain voltage-dependent proton channels, whereas $O_2^{\bullet -}$ production was demonstrated for Nox2 and the long form of Nox5. Iron reductases (*Fro*) of dicotyledonous plants have additional transmembrane domains and show homology to yeast ferric reductases

ing, and disinfection to the pathogen in the phagolysosome formed this way (references in Henderson and Chappell 1996). Disinfection is mediated by the NADPH oxidase. The NADPH oxidase electron transport chain in the phagocyte PM comprises a *low*-potential cytochrome b_{558} containing heterodimer, with an α-subunit named p22^{phox} and a glycosylated β-subunit the gp91^{phox}. The latter protein has binding-sites for NADPH, FAD, and two heme groups (Segal et al. 1992, 1999; Ravel and Lederer 1993; Doussier et al. 1995; Yu et al. 1998).

The activated NADPH oxidase complex transfers electrons from cytosolic NADPH (E_0'=–320 mV) across the membrane into the phagolysosome. Here the electrons are used to reduce O_2 to form $O_2^{\bullet-}$ ($O_2/O_2^{\bullet-}$; E_0'=–160 mV). The *low*-potential cytochrome *b* (E_0' –245 mV; Cross et al. 1981; Klebanoff 1992) is responsible for this reaction. Two $O_2^{\bullet-}$ can dismutate spontaneously to give O_2 and H_2O_2, a reaction that is greatly enhanced by superoxide dismutase. The H_2O_2 formed this way is a good disinfecting agent. Another enzyme, the myeloperoxidase, uses the produced H_2O_2 as an oxidant to convert halide anions, mostly chloride, to hypohalidic acids (i.e. hypochloric acid) which are much stronger microbicidal agents than H_2O_2 (Weiss 1989). Further reaction of hypochloric acid with amines like taurine give chloroaminic acids which are quite stable and toxic agents by themselves.

Plant cells confronted with pathogens or elicitors exhibit an oxidative burst [i.e. a massive production of reactive oxygen species (ROS, in our context mainly of H_2O_2)] that is apparently so similar to the phagocyte oxidative burst (transient rise in ROS shortly after pathogen attack on a roughly similar time scale) that the notion of homology even down to the molecular level was accepted widely without much questioning (Lüthje et al. 2000). The existence of highly conserved and homologue domains in ferric reductases from yeast and human gp91^{phox} led to the suggestion that these enzymes are representatives of a common transmembrane single electron transport system (Roman et al. 1993; Shatwell et al. 1996). Depending on the cell's requirement this system would be adapted to either iron or oxygen reduction. Based on this hypothesis a large number of publications related to the response of plants to pathogens deal with analogies between the animal and plant systems. This is reflected in a number of hypothetical models which include a transmembrane NADPH oxidase complex in the plant PM (e.g. Dangl et al. 1996; Jabs et al. 1996; Kieffer et al. 1997; Lamb and Dixon 1997; Yoshioka et al. 2001).

The now at least six respiratory burst oxidase homologues (Rboh) found in *Arabidopsis* have considerable homology not only to the phagocyte gp91^{phox}, but also to the yeast ferric reductase genes FRE1 and frp1^{+}, both of which are involved in iron reduction and uptake steps performed at the yeast PM (Dancis et al. 1992; Roman et al. 1993; Groom et al. 1996; Desikan et al. 1998; Torres et al. 1998; see also Cakmak et al. 1987; Robinson et al. 1999; Waters et al. 2002). The similarity between mammalian and plant NADPH oxidase seems so close that even in vitro assays using *Arabidopsis* microsomes and human neutrophil cytosol were able to induce $O_2^{\bullet-}$ generation, albeit, due to the use of microsomes rather than defined membrane preparations, the subcellular location of the activity cannot be concluded from the data presented (Desikan et al. 1996).

Some different evidence for the occurrence of a plant homologue of gp91^{phox} was presented by Keller et al. (1998). These authors detected a protein in plant PM by immunoblot with an anti-L serum directed against a 13 amino acid sequence near the C terminus of the rice Rboh and determined the membrane Rboh is located in. The protein used for raising the antibodies, however, was purified from microsomal fractions that showed a stronger cross reaction than PM preparations. Thus the location and postulated function in the PM have to be further elucidated.

Studies using antibodies directed against the individual regulatory subunits (p47 and p67) of the phagocyte NADPH oxidase showed cross-reactivity with proteins of several plant species (Tenhaken et al. 1995; Desikan et al. 1996; Dwyer et al. 1996; Xing et al. 1997b; Keller et al. 1998). However, this kind of antibody reaction depends on the specificity of the antibody and is only indirect evidence for the presence or absence of plant homologues. For example, several clones of supposed NADPH oxidase components were detected by screening of a soybean *c*DNA expression library with antisera against the human NADPH oxidase subunits (anti-p22^{phox}, anti-p47^{phox}, anti-p67^{phox}). Neither of the corresponding *c*DNAs showed significant homology to any of the NADPH oxidase proteins nor did they have an even distantly related function (Tenhaken and Thulke 1996; Tenhaken and Rübel 1998). In addition to these observations, the *Arabidopsis* genome gave no evidence for the occurrence of any of the cytosolic factors in plant cells (Dangle and Jones 2001). DPI-sensitive NADPH oxidase activity observed with isolated plant PMs appears to be due to peroxidases (Mika et al. 2004). The absence of a cytochrome *b* with a midpoint redox potential similar to that of gp91^{phox} (E_0'=–245 mV, cf. Segal 1995) in plant PM (references in Asard et al. 1998) supports this suggestion.

4.2 Iron Reductases

The first gene shown to be responsible for iron reductase activity of *Saccharomyces cerevisiae* FRE1 (ferric reductase enzyme) was described by Dancis et al. (1990). However, the disruption of FRE1 did not completely abolish the ferric reductase activity of the cells. This fact led to the isolation of FRE2 (Georgatsou and Alexandraki 1994), a second gene involved in ferric reduction activity. Disruption of both genes completely abolished the inducible ferric reductase activity of the cells. Although FRE1, which is a *b*-type cytochrome (Shatwell et al. 1996), and FRE2 are not highly homologous at the amino acid level (24.5% in 693 overlapping amino acids), there are conserved boxes all over the protein which are also found in the ferric reductase enzyme frp1$^+$ of *Schizosaccharomyces pombe* (ferric reductase *S. pombe*) and gp91^{phox} (Roman et al. 1993).

The ferric reductase activity of *S. cerevisiae* and *S. pombe* are inducible at the transcriptional level by iron starvation and FRE1 and FRE2 are active at different times of growth of the yeast culture (Dancis et al. 1992; Roman et al. 1993; Georgatsou and Alexandraki 1994). It has been known for a long time that the reductase activity of *S. cerevisiae* strongly depends on the growth phase and that cells stop reduction of extracellularly added electron acceptors when cells enter the stationary growth phase (Crane et al. 1982).

While FRE1 is active in the early exponential growth phase , FRE2 is active in the later phase (Georgatsou and Alexandraki 1994). Quantification of FRE1 and FRE2 mRNA levels showed that FRE2 is not expressed when cells are cultured in an iron-containing medium (Yamaguchi-Iwai et al. 1995). It can therefore be concluded that the higher reduction activity induced by iron deprivation is due to the induction of FRE1 transcription and the additional expression of FRE2. The extracellularly reduced iron is taken up by the cells independently of FRE1 and FRE2.

The uptake of ferrous iron can be separated into a high and a low affinity uptake system (Dancis et al. 1992; Eide et al. 1992; Askwith et al. 1994; de Silva et al. 1995; Dix et al. 1994). The high affinity uptake system is induced by iron starvation and involves the reoxidation of iron after uptake by a ferrous iron oxidase, thus trapping the iron intracellularly (Fig. 3). The ferrous iron oxidase FET3 (ferrous iron transport; Askwith et al. 1994; Dancis et al. 1994) is a copper protein with high homology to multi-copper oxidases evolutionarily linking the iron to the copper uptake systems of the cells. FET3 is a transmembrane protein with the ferrous oxidase domain pointing to the extracellular space (Lesuisse et al. 1991). As a predicted transmembrane protein, FET3 would span the membrane with only one transmembrane domain and it could therefore comprise an iron transporter only as a homo- or heteromultimer. It has been suggested that the reoxidation of ferrous iron by FET3 is necessary to release the iron from the transporter. In the absence of FET3 activity, the iron would remain protein bound (Askwith et al. 1994). In contrast to FET3, the amino-acid sequence of the FET4 protein, which is involved in the low affinity iron uptake system, suggests that the protein is an integral membrane protein with multiple transmembrane domains. It may be by itself sufficient for ferrous iron uptake (Dix et al. 1994).

It has been concluded that the low and the high affinity iron uptakes are driven by separate systems and that the high affinity uptake system is regulated to compensate alterations in low affinity iron uptake. The observation that deletion of CTR1, the high affinity copper transport protein of *S. cerevisiae*, led to iron starvation of the cells gave first hints to a connection

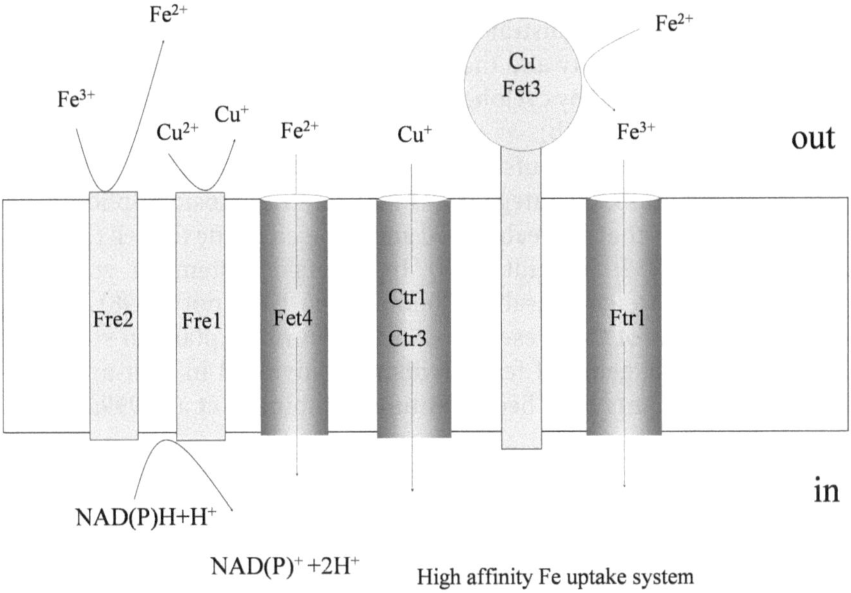

Fig. 3. Iron uptake system of yeast. Fre1 and Fre2 reduce external Fe^{3+}-chelates. Fre1 can also reduce Cu^{2+}, which is essential for subsequent Cu^{+} uptake by Ctr1/Ctr3 and the high-affinity Fe^{3+} uptake system Fet3. Cu^{+} is shuttled to the Golgi compartment and transported by an ATP-dependent copper transporter into the Golgi, where Fet3 is loaded with Cu^{+}. For further details see text. (Modified after Askwith and Kaplan 1998)

between copper and iron metabolism (Dancis et al. 1994). The iron deficiency of CTR1 mutants could be complemented by high amounts of copper in the medium. Moreover, copper is capable of repressing FRE1 transcription and FRE1 accounts for 50–70% of total copper reduction (Hassett et al. 1995). The requirement of copper for the iron oxidase FET3 and the discovery that FRE1, FRE2, and FET3 expression is under the control of one protein, the putative transcription factor AFT1 (Yamaguchi-Iwai 1995), further clarify the coupling of copper and iron metabolism. At the transcriptional level another protein is involved in the regulation of ferric reduction.

Proteomic approaches revealed that the high-affinity iron uptake system of the unicellular green algae *Chlamydomonas rheinhardii* showed similarities to the yeast system (Thimm et al. 2001; Herbik et al. 2002a,b). The high-affinity iron uptake requires an Fe^{3+}-chelate reductase and an iron transporter. Based on the homology of internal peptide sequences to the multi-copper oxidase hephaestin, a 150-kDa protein was proposed as ferroxidase and participation of this ferroxidase in high-affinity iron uptake in *C. reinhardtii* was demonstrated (Herbick et al. 2002a). The induction of the Fe^{3+} reductase activity and the 150-kDa protein under iron deficiency was completely blocked by cycloheximide, indicating transcriptional regulation (Herbick et al. 2002b).

Most plants reduce iron outside of the cell before uptake (strategy I) with the exception of grasses (strategy II), which excrete phytosiderophores and form Fe^{3+}-chelates that are reabsorbed and reduced inside the cell (Bienfait 1985; Römheld 1987; Schmidt 1999). Iron uptake systems of yeast and strategy I plants are comparable. Connolly and Guerinot (1998) give an excellent overview of the present knowledge on iron uptake in strategy I plants. Meanwhile, genes of ferric reductases involved in iron uptake of dicotyledonous plants have been identified (Robinson et al. 1999; Waters et al. 2002).

FRO2 and *IRT1* genes encode the root ferric-chelate reductase and a high-affinity iron transporter, respectively, involved in the iron deficiency-induced uptake system (Vert et al. 2003). The *Arabidopsis FRO2* gene encodes the low-iron-inducible ferric chelate reductase responsible for reduction of iron at the root surface (Connolly et al. 2003). *FRO2* and *IRT1*, the major transporters responsible for high-affinity iron uptake from the soil, are coordinately regulated at both the transcriptional and posttranscriptional levels (Connolly et al. 2003; Vert et al. 2003). Recovery from iron-deficiency and modulation of apoplastic iron pools indicate that iron itself plays a major role in the regulation of root iron deficiency responses at the mRNA and protein levels (Vert et al. 2003).

FRO2 is composed of 725 amino acids, has a predicted isoelectric point of 9.37, and a relative molecular mass of 81.50 kDa, although the presence of glycosylation motifs suggests that the native protein may have a greater molecular mass (Robinson et al. 1999). FRO2 contains sequences identical to the FAD-binding site and the NADPH-binding site of FRE1. The predicted secondary structure indicates six transmembrane domains within the N-terminal regions of FRO2 along with two C-terminal hydrophobic domains. This suggests that an intracellular region of FRO2, containing the deduced cofactor-binding sites, is anchored at both ends by membrane-spanning regions. Cofactor-binding sites and associated C-terminal regions are all predicted to be cytosolic.

Fe^{3+}-EDTA or Fe^{3+}-citrate, used for studies of iron reduction and uptake, have midpoint redox potentials of +120 and +600 mV respectively (Pierre and Fontecave 1999), i.e. the heme involved in these reactions is likely to be a *high*-potential cytochrome *b*. Two ascorbate-reducible *b*-type cytochromes were found in PM isolated from *Arabidopsis* leaves, while PM of bean hooks and maize roots seemingly contain only one ascorbate-reducible cytochrome (Bérczi et al. 2001). Fe^{3+}-citrate reductase activity, however, was also suggested to be involved in intercellular iron transport (references in Schmidt 1999); thus a function in iron reduction may be possible for the second ascorbate-reducible cytochrome in *Arabidopsis*, but this suggestion has to be further investigated.

4.3 Voltage-Gated Proton Channels

While gp91^{phox} and the long form of Nox5 produce an oxidative burst, $O_2^{\bullet-}$ generation by other Nox homologues is low and they appear to be involved in oxygen sensing and signaling rather than generating an oxidative burst (Meier 1998; Hancock et al. 2001; Lambeth 2002). A function of yeast and some plant homologues in iron reduction and uptake has been demonstrated (Shatwell et al. 1996; Robinson et al. 1999; Waters et al. 2002). Redox potentials of the cytochromes (i.e. heme groups) are important for the function of these proteins in electron transfer either to oxygen or ferric chelates. Four conserved histidyl residues (His101, His115, His210, His222) are believed to play roles in heme binding (Mankelow and Henderson 2003). The mutation of only one of these histidines resulted in the total loss of heme from the protein and its function (Finegold et al. 1996; Biberstine-Kinkade et al. 2001).

However, histidine 115 appears to be involved not only in heme binding but also in the proton conduction pathway of gp91^{phox} (Henderson 1998; Mankelow and Henderson 2003), although this conclusion is not unchallenged (DeCoursey et al. 2001; DeCoursey 2003). pH regulation appears to be a major function of the phagocyte NADPH oxidase (Nanda et al. 1993; Henderson et al. 1997; Henderson 2001). Nowadays, it becomes apparent that proton channeling is a common feature of most members in the flavocytochrome *b* protein family identified so far (Kimball and Saier 2002).

5 Proton Transport in Plant PM

The first hypothesis on redox-linked ion transport was presented by Lundegårdh before H^+-ATPases were identified in the plant PM (Lunegårdh 1945). However, the idea of a proton pumping redox system discussed today is quite different from the concepts of Lundegårdh (Barr 1991; Crane et al. 1991; Döring and Lüthje 1996; Kimball and Saier 2002).

During the reduction of HCF III, which was added to the experimental solution, a decrease in medium pH was observed (Novak and Miklashevich 1986; van Beusichem et al. 1988; Barr and Crane 1991; Rubinstein and Stern 1991b). In some experiments acidification of the medium and reduction of

HCF III were tightly coupled (Kurkova and Verhovskaya 1984; Neufeld and Bown 1987; Döring et al. 1990). According to other reports HCF III-induced H^+ extrusion could be detected after a *lag* phase and was sometimes transient, while HCF III reduction could be measured within seconds after its addition (Rubinstein and Stern 1986). Buffering capacities of cell walls and (to a lesser extent) of the incubation solution may partially explain the *lag* phases of H^+ export.

Some evidence for redox-linked proton export arises from investigations using 'specific' inhibitors of either the PM H^+-ATPase or the PM redox system (Elzenga and Prins 1989; Xu 1993). Evidence for ATP-independent H^+ export in the light and in the dark was found in *Asparagus* mesophyll cells (Bown and Crawford 1988). In cultured carrot cells, two sources of H^+ extrusion were found (Elzenga and Prins 1989; Xu 1993). Dependence of trans-PM electrical potential on cellular respiration – limited by low oxygen concentration – did not match the decrease in cytosolic ATP level; thus an ATP-independent membrane energization was proposed (Löppert 1981, 1983). An inhibition of proton secretion at very low concentrations of external acceptor $< 1\ \mu M$ for HCF III has been reported (Böttger and Hilgendorf 1988; Lüthen and Böttger 1988; Döring et al. 1992); however, the significance of this observation was never clarified.

Monitoring the increase in PM H^+-ATPase or PM redox activity by determining the level of reductants or ATP is problematic. If an ATP (NADH, etc.) sink increases activity, the cellular regeneration of the required substrate also increases; thus mainly the turnover of the substrate changes rather than the total level of ATP or electron donor. Therefore changes in reductant or ATP concentration do not quantitatively correspond to H^+ efflux by ATPase or PM redox activity (Qiu et al. 1985; Krüger 1993). Measurements of cytosolic ATP concentration after addition of apoplastic electron acceptor gave contradicting results (Löppert 1981, 1983; Novak and Ivankina 1986; Novak and Miklashevich 1986; Novak et al. 1988; Xu 1993). Despite these limitations, work has been done on steady-state levels of metabolites: In *Elodea* and *Lemna* light-dependent cytosolic ATP concentration did not coincide with the dependency of membrane potential on light (Ivankina and Novak 1980; Löppert 1983). Oxygen and light sensitivity of the membrane potential (Löppert 1983) and proton efflux (Böttger et al. 1985) were also taken as evidence against the PM-H^+-ATPase as the sole source for H^+ extrusion. However, these effects may also be due to a possible involvement of the redox system in cellular regulation and signaling.

The dependence of HCF III reduction and redox-induced H^+ extrusion on light and DCMU (3[3,4-dichlorophenyl]-1,1-dimethylurea) points to at least two different sources of protons extruded: (1) ATPase and (2) redox-dependent activity (Sze 1985). Similar conclusions were drawn from measurements of temperature dependency of ATPase and redox-dependent proton extrusion (Hilgendorf and Böttger 1993).

So far no evidence for proton pumping by plasma membrane-bound redox systems has been found in vitro (Gianinni and Briskin 1988; Hassidim et al. 1987; Klobus and Buckzek 1995). In our studies, a calcium-stimulated NADPH-dependent proton transport – i.e. by a postulated Rboh – could not be observed with inside-out PM vesicles, whereas ATP-driven proton transport was easily observed.

6 Conclusion

Cytochrome b_5 or cytochrome P450/420 both seem not to be present in unstressed plant PM. Thus PM-bound peroxidases, together with other redox systems that occur in the outer permeability barrier of the cell, may be involved in membrane repair mechanisms (references in Döring et al. 1998; Bérczi and Møller 2000; Mika et al. 2004). The absence of low potential cytochrome *b* in plant PM suggests that the major functions of gp91^{phox} homologues in plants may be iron reduction and proton channeling rather than an oxidative burst. The source of $O_2^{\bullet -}$ production in plant PM cannot be assigned to any particular protein found so far (references in Lamb and Dixon 1997; Lüthje et al. 2000; Mika et al. 2004). The expression of Rboh, which often is taken as the source of the plant oxidative burst, in plant PM could not be shown until today. There is no doubt about the occurrence of homologues of the chromaffin granules, cytochrome b_{561} in plants PM, but their possible localizations and functions need further investigation.

Acknowledgements. The authors wish to thank Lydia Henderson (University of Bristol, UK) and Alajos Bérczi, (Hungarian Academy of Sciences, Szeged) for helpful discussions.

References

Ahmad M, Cashmore AR (1993) HY4 gene of *Arabidopsis thaliana* encodes a protein with characteristics of a blue-light photoreceptor. Nature 366:162–166

Ahmad M, Cashmore AR (1996) Seeing blue: the discovery of cryptochrome. Plant Mol Biol 30:851–861

Ahmad M, Heil M, Black RC, Giovani B, Galland P, Lardemer D (2002) Action spectrum for hypocotyl growth inhibition suggests dosage-dependent synergism among cryptochrome photoreceptors of *Arabidopsis thaliana*. Plant Physiol 129:774–785

Asada A, Kusakawa T, Orii H, Agata K, Watanabe K, Tsubaki M (2002) Planarian cytochrome b_{561}: conservation of a six transmembrane structure and localization along central and peripheral nervous system. J Biochem 131:175-183

Asard H, Caubergs R (1990) LIAC activity in higher plants. In: Lenci F, Ghetti F, Colombetti G, Häder DP, Song PS (eds) Biophysics of photoreceptors and photomovement in microorganisms. NATO ASI Ser 211:181–189
Asard H, Venken M, Caubergs R, de Greef JA (1989) *b*-type cytochromes in higher plant plasma membranes. Plant Physiol 90:1077–1083
Asard H, Horemans N, Caubergs RJ (1992) Transmembrane electron transport in ascorbate-loaded plasma membrane vesicles from higher plants involve a *b*-type cytochrome. FEBS Lett 306:143–146
Asard H, Horemans N, Mertens J, Caubergs RJ (1994) The plant plasma membrane *b*-type cytochrome: an overview. Belgien J Bot 127:171–183
Asard H, Horemans N, Briggs WR, Caubergs RJ (1995a) Blue light perception by endogenous redox components of the plant plasma membrane. Photochem Photobiol 61:518–522
Asard H, Horemans N, Caubergs RJ (1995b) Involvement of ascorbic acid and a *b*-type cytochrome in plant plasma membrane redox reactions. Protoplasma 184:36–41
Asard H, Horemans N, Preger V, Trost P (1998) Plasma membrane *b*-type cytochromes. In: Asard H, Bérczi A, Caubergs RJ (eds) Plasma membrane redox systems and their role in biological stress and disease. Kluwer, Dordrecht, pp 1–31
Asard H, Terol-Alcayade J, Preger V, del Favero J, Verelst W, Sparla F, Pérez-Alonso M, Trost P (2000) *Arabidopsis thaliana* sequence analysis confirms the presence of cyt *b*-561 in plants: evidence for a novel protein family. Plant Physiol Biochem 38:905–912
Asard H, Kapila J, Verelst W, Bérczi A (2001) Higher-plant plasma membrane cytochrome b_{561}: a protein in search of a function. Protoplasma 217:77–93
Askerlund P, Larsson C, Widell S (1989) Cytochromes of plant plasma membranes. Characterization by absorbance difference spectrophotometry and redox titration. Physiol Plant 76:123–134
Askerlund P, Laurent P, Nakagawa H, Kadar JC (1991) NADH-ferricyanide reductase of leaf plasma membranes. Partial purification and immunological relation to potato tuber microsomal NADH-ferricyanide reductase and spinach leaf NADH-nitrate reductase. Plant Physiol 95:6–13
Askwith C, Kaplan J (1998) Iron transport in yeast: the involvement of an iron reductase and oxidase. In: Asard H, Bérczi A, Caubergs RJ (eds) Plasma membrane redox systems and their role in biological stress and disease. Kluwer, Dordrecht, pp 157–177
Askwith C, Eide D, van Ho A, Bernhard PS, Li L, Davis-Kaplan S, Sipe DM, Kaplan J (1994) The FET3 gene of *S. cerevisiae* encodes a multicopper oxidase required for ferrous iron uptake. Cell 76:403–410
Babior B (1992) The respiratory burst oxidase. Adv Enzymol Relat Areas Mol Biol 65:49–95
Bagnaresi P, Thoiron S, Mansion M, Rossignol M, Pupillo P, Briat J-F (1999) Cloning and characterization of a maize cytochrome-b_5 reductase with Fe^{3+}-chelate reduction capability. Biochem J 338:499–505
Barr R (1991) The possible role of redox-associated protons in growth of plants. J Bioenerg Biomembr 23:443–467
Barr R, Crane FL (1991) Modulation of plasma membrane redox reactions in plant cells. In: Crane FL, Morre DJ, Low HE (eds) Oxidoreduction at the plasma membrane: relation to growth and transport, vol 2. Plants. CRC Press, Boca Raton, pp 237–279
Barr R, Sandelius AS, Crane FL, Morré DJ (1986) Redox reactions of tonoplast and plasma membranes isolated from soybean hypocotyls by free-flow electrophoresis. Biochim Biophys Acta 852:254–261
Barrett M, Maxson JM (1991) Naphthalic anhydride induces imazethapyr metabolism and cytochrome P-450 activity in maize. Z Naturforsch 46c:897–900
Bashtovyy D, Bérczi A, Asard H, Páli T (2003) Structure prediction for the di-heme cytochrome b_{561} protein family. Protoplasma 221:31–40
Benveniste I, Gabriac B, Durst F (1986) Purification and characterization of the NADPH-cytochrome P-450 (cytochrome c) reductase from higher-plant microsomal fraction. Biochem J 235:365–373

Bérczi A, Møller IM (1998a) NADH-monodehydroascorbate oxidoreductase is one of the redox enzymes in spinach leaf plasma membrane. Plant Physiol 116:1029–1036

Bérczi A, Møller IM (1998b) Characterization and solubilization of residual redox activity in salt-washed and detergent-treated plasma membrane vesicles from spinach leaves. Protoplasma 205:59–65

Bérczi A, Møller IM (2000) Redox enzymes in the plant plasma membrane and their possible roles. Plant Cell Environ 23:1287–1302

Bérczi A, Fredlund KM, Moller IM (1995) Purification and characterization of an NADH-hexacyanoferrate(III) reductase from spinach leaf plasma membrane. Arch Biochem Biophys 320:65–72

Bérczi A, Lüthje S, Asard H (2001) *b*-Type cytochromes in plasma membranes of *Phaseolus vulgaris* hypocotyls, *Arabidopsis thaliana* leaves, and *Zea mays* roots. Protoplasma 217:50–55

Bérczi A, Caubergs RJ, Asard H (2003) Partial purification and characterization of an ascorbate-reducible *b*-type cytochrome from the plasma membrane of *Arabidopsis thaliana* leaves. Protoplasma 221:47–56

Berry EA, Trumpower BL (1985) Pathways of electrons and protons through the cytochrome bc1 complex of the mitochondrial respiratory chain. In: Lenaz G (ed) Coenzyme Q. Wiley, New York, pp 365–389

Biberstine-Kinkade KJ, DeLeo FR, Epstein RI, LeRoy BA, Nauseef WM, Dinauer MC (2001) Heme-ligating histidines in flavocytochrome b558. Identification of specific histidines in $gp91^{phox}$. J Biol Chem 276:31105–31112

Bienfait HF (1985) Regulated redox processes at the plasmalemma of plant root cells and their function in iron uptake. J Bioenerg Biomembr 17:73–83

Black SD (1992) Membrane topology of the mammalian P450 cytochromes. Fed Am Soc Exp Biol J 6:680–685

Bohn M, Heinz E, Lüthje S (2001) Lipid composition and fluidity of plasma membranes isolated from corn (*Zea mays* L.) roots. Arch Biochem Biophys 387:35–40

Borgeson CE, Bowman BJ (1985) Blue light-reducible cytochromes in membrane fractions from *Neurospora crassa*. Plant Physiol 78:433–437

Böttger M, Bigdon M, Soll H-J (1985) Proton translocation in corn coleoptiles: ATPase or redox chain? Planta 163:376–380

Böttger M, Hilgendorf F (1988) Hormone action on transmembrane electron and H^+ transport. Plant Physiol 86:1038–1043

Bown AW, Crawford LA (1988) Evidence that H^+ efflux stimulated by redox activity is independent of plasma membrane ATPase activity. Physiol Plant 73:170–174

Brain RD, Briggs WR (1977) Light-induced cytochrome reduction in *Neurospora crassa* membrane fractions. In: Castellani A (ed) Proc Int Congr on Research in Photobiology. Plenum Press, New York, pp 539–544

Buckhout TJ, Luster DG (1991) Pyridine nucleotide-dependent reductases of the plant plasma membrane. In: Crane FL, Löw H, Morré DJ (eds) Oxidoreduction at the plasma membrane: relation to growth and transport, vol 2. Plants. CRC Press, Boca Raton, pp 61–84

Bunkelmann J, Trelease RN (1996) Ascorbate peroxidase. A prominent membrane protein in oilseed glyoxysomes. Plant Physiol 110:589–598

Cakmak I, van de Wetering DAM, Marschner H, Bienfait HF (1987) Involvement of superoxide radical in extracellular ferric reduction by iron-deficient bean roots. Plant Physiol 85:310–314

Caubergs RJ, Goldsmith MHM, Briggs WR (1978) Effects of inhibitors on light-induced absorbance changes in corn membrane fractions. Car Inst Wash Yearb 77:356–357

Caubergs RJ, Widell S, Larsson C, de Greef JA (1983) Comparison of two methods for the preparation of a membrane fraction of cauliflower inflorescences containing a blue light reducible b-type cytochrome. Physiol Plant 57:291–295

Christie JM, Reymond P, Powell GK, Bernasconi P, Raibekas A, Liscum E, Briggs WR (1998) *Arabidopsis* NPH1: a flavoprotein with the properties of a photoreceptor for phototropism. Science 282:1698–1701
Christie JM, Salomon M, Nozue K, Wada M, Briggs WR (1999) LOV (light, oxygen, or voltage) domains of the blue-light photoreceptor phototropin (nph1): binding sites for the chromophore flavin mononucleotide. Proc Natl Acad Sci USA 96:8779–8783
Connolly EL, Guerinot ML (1998) Reduction and uptake of iron in plants. In: Asard H, Bérczi A, Caubergs RJ (eds) Plasma membrane redox systems and their role in biological stress and disease. Kluwer, Dordrecht, pp 179–192
Connolly EL, Campbell NH, Grotz N, Prichard CL, Guerinot ML (2003) Overexpression of the FRO2 ferric chelate reductase confers tolerance to growth on low iron and uncovers posttranscriptional control. Plant Physiol 133:1102–1110
Crane FL, Roberts H, Linnane AW, Löw H (1982) Transmembrane ferricyanide reduction by cells of the yeast *Saccharomyces cerevisiae*. J Bioenerg Biomembr 14:191–205
Crane FL, Sun IL, Barr R, Löw H (1991) Electron and proton transport across the plasma membrane. J Bioenerg Biomembr 23:773–803
Cross AR, Jones OTG, Harper AM, Segal WA (1981) Oxidation-reduction properties of the cytochrome *b* found in the plasma-membrane fraction of human neutrophils. A possible oxidase in the respiratory burst. Biochem J 194:599–606
Dagher MC, Fuchs A, Bourmeyster N, Jouan A, Vignais PV (1995) Small G proteins and the neutrophil NADPH oxidase. Biochimie 77:651–660
Dancis A, Klausner RD, Hinnebusch AG, Barriocanal JG (1990) Genetic evidence that ferric reductase is required for iron uptake in *Saccharomyces cerevisiae*. Mol Cell Biol 10:2294–2301
Dancis A, Roman DG, Andersen GJ, Hinnenbusch AG, Klaussner RD (1992) Ferric reductase of *Saccharomyces cerevisiae*: molecular characterization, role in iron uptake, and transcriptional control by iron. Proc Natl Acad Sci USA 89:3869–3873
Dancis A, Yuan DS, Haile D, Askwith C, Eide D, Moehle C, Kaplan J, Klausner RD (1994) Molecular characterization of a copper transport protein in *S. cerevisiae*: an unexpected role for copper in iron transport. Cell 76:393–402
Dangl JL, Jones JDG (2001) Plant pathogens and integrated defence responses to infection. Nature 411:826–833
Dangl JL, Dietrich RA, Richberg MH (1996) Death don't have no mercy: cell death programs in plant–microbe interactions. Plant Cell 8:1793–1807
DeCoursey TE (2003) Voltage-gated proton channels and other proton transfer pathways. Physiol Rev 83(2):475–579
DeCoursey TE, Cherny VV, Morgan D, Katz BZ, Dinauer MC (2001) The gp91^{phox} component of NADPH oxidase is not the voltage-gated proton channel in phagocytes, but it helps. J Biol Chem 276:36063–36066
Desikan R, Burnett EC, Hancock JT, Neill SJ (1998) Harpin and hydrogen peroxide induce the expression of a homologue of gp91-phox in *Arabidopsis thaliana* suspension cultures. J Exp Bot 49:1767–1771
Deskian R, Hanncock JT, Coffey MJ, Neill SJ (1996) Generation of active oxygen in elicited cells of *Arabidopsis thaliana* is mediated by a NADPH oxidase-like enzyme. FEBS Lett 382:213–217
De Silva DM, Askwith CC, Eide D, Kaplan J (1995) The *FET3* gene product required for high affinity iron transport in yeast is a cell surface ferroxidase. J Biol Chem 270:1098–1101
Dix DR, Bridgham JT, Broderius MA, Byersdorfer CA, Eide DJ (1994) The FET4 gene encodes the low affinity Fe(II) transport protein of *Saccharomyces cerevisiae*. J Biol Chem 269:26092–26099
Dixon RA, Paiva N (1995) Stress-induced phenylpropanoid metabolism. Plant Cell 7:1085–1097
Donaldson RP, Luster DG (1991) Multiple forms of plant cytochromes P-450. Plant Physiol 96:669–674

Döring O, Lüthje S (1996) Molecular components and biochemistry of electron transport in plant plasma membranes (review). Mol Membr Biol 13:127–142
Döring O, Lüthje S, Hilgendorf F, Böttger M (1990) Membrane depolarization by hexacyanoferrate (III), hexabromoiridate (IV) and hexachloroiridate (IV). J Exp Bot 43:1055–1061
Döring O, Lüthje S, Böttger M (1992) Modification of the activity of the plasma membrane redox system of *Zea mays* L. roots by vitamin K_3 and dicumarol. J Exp Bot 43:175–181
Döring O, Böttger M, Lüthje S (1998) The constitutive electron transport system in plant plasma membranes: a role in defense against oxidative stress in the apoplast and within the plasma membrane? A hypothesis. In: Noga G, Schmitz M (eds) Antioxidants in higher plants: biosynthesis, characteristics, actions and specific functions in stress defense. Shaker, Aachen, pp 19–30
Doussier J, Buzenet G, Vignais PV (1995) Photoaffinity labeling and photoinactivation of the O_2^--generating oxidase of neutrophils by an azido derivative of FAD. Biochem 34:1760–1770
Drexler H, Spiekermann P, Meyer A, Domergue F, Zank T, Sperling P, Abbadi A, Heinz E (2003) Metabolic engineering of fatty acids for breeding of new oilseed crops: strategies, problems and first results. J Plant Physiol 160(7):779–802
Durst F, Beneviste I, Salaün JP, Werck-Reichart D (1992) Function, mechanism and regulation of cytochrome P-450 enzymes in plants. Biochem Soc Trans 20:353–357
Dwyer SC, Legrende L, Low PS, Leto TL (1996) Plant and human neutrophil oxidative burst complexes contain immunologically related proteins. Biochim Biophys Acta 1289:231–237
Eide D, Davis-Kaplan S, Jordan I, Sipes D, Kaplan J (1992) Regulation of iron uptake in *Saccharomyces cerevisiae*. J Biol Chem 267:20774–20781
Elzenga JTM, Prins HBA (1989) Light-induced polar pH changes in leaves of *Elodea canadensis*. I. Effects of carbon concentration and light intensity. Plant Physiol 91:68–72
Finegold AA, Shatwell KP, Segal AW, Klausner RD, Dancis A (1996) Intramembrane bisheme motif for transmembrane electron transport conserved in a yeast reductase and the human NADPH oxidase. J Biol Chem 271:31021–3102
Flatmark T, Terland O (1971) Cytochrome b_{561} of the bovine adrenal chromaffin granules. A high potential *b*-type cytochrome. Biochim Biophys Acta 253:487–491
Funk C, Croteau R (1993) Induction and characterization of a cytochrome-P-450-dependent camphor hydroxylase in tissue cultures of common sage (*Salvia officinalis*). Plant Physiol 101(4):1231–1237
Galland P, Senger H (1991) Flavins as possible blue light photoreceptors. In: Holmes MG (ed) Photoreceptor evolution and function. Academic Press, London, pp 65–124
Galland P, Tölle N (2003) Light-induced fluorescence changes in *Phycomyces*: evidence for blue light-receptor associated flavo-semiquinones. Planta 217:971–982
Georgatsou E, Alexandraki D (1994) Two distinctly regulated genes are required for ferric reduction, the first step of iron uptake in *Saccharomyces cerevisiae*. Mol Cell Biol 14:3065–3073
Gerardy R, Zenk MH (1993) Formation of salutaridine from (R)-reticuline by a membrane-bound cytochrome-P-450 enzyme from *Papaver somniferum*. Phytochemistry 32:79–86
Giannini JL, Briskin DP (1988) Pyridine nucleotide oxidation by a plasma membrane fraction from red beet (*Beta vulgaris* L.) storage tissue. Arch Biochem Biophys 260:653–660
Griesen D, Su D, Bérczi A, Asard H (2004) Localization of an ascorbate-reducible cytochrome b_{561} in the plant tonoplast. Plant Physiol 134:726–734
Groom QJ, Torres MA, Fordham-Skelton AP, Hammond-Kosack KE, Robinson NJ, Jones DJ (1996) RbohA, a rice homologue of the mammalian $gp91^{phox}$ respiratory burst oxidase gene. Plant J 10:515–522
Hancock JT, Deskian R, Neill SJ (2001) Role of reactive oxygen species in cell signalling pathways. Biochem Soc Trans 29:345–350

Hansikova H, Frei E, Anzenbacher P, Stiborova M (1994) Isolation of plant cytochrome P-450 and NADPH-cytochrome P-450 reductase from tulip bulbs (*Tulipa fosteriana* L.) oxidizing xenobiotics. General Physiol Biophys 13:149–169

Hassett R, Kosman DJ (1995) Evidence for Cu(II) reduction as a component of copper uptake by *Saccharomyces cerevisiae*. J Biol Chem 270:128–134

Hassidim M, Rubinstein B, Lerner HR, Reinhold L (1987) Generation of a membrane potential by electron transport in plasmalemma-enriched vesicles of cotton and radish. Plant Physiol 85:872–875

Henderson LM (1998) Role of histidines identified by mutagenesis in the NADPH oxidase-associated H^+ channel. J Biol Chem 273:33216–33223

Henderson LM (2001) NADPH oxidase subunit $gp91^{phox}$: a proton pathway. Protoplasma 217:37–42

Henderson LM, Chappell JB (1996) NADPH oxidase of neutrophils. Biochim Biophys Acta 1273:87–107

Henderson LM, Thomas S, Banting G, Chappell B (1997) The arachidonic-activatable NADPH oxidase-associated H^+ channel is contained with the multi-membrane-spanning N-terminal region of $gp91^{phox}$. Biochem J 325:701–705

Hendrickson WA, Karle J (1973) Carp muscle calcium-binding protein. J Biol Chem 248:3327–3334

Hendry GAF, Houghton JD, Jones OTG (1981) The cytochromes in microsomal fractions of germinating mung beans. Biochem J 194:743–751

Herbik A, Bölling C, Buckhout TJ (2002a) The involvement of a multicopper oxidase in iron uptake by the green algae *Chlamydomonas reinhardtii*. Plant Physiol 130:2039–2048

Herbik A, Haebel A, Buckhout TJ (2002b) Is a ferroxidase involved in the high-affinity iron uptake in *Chlamydomonas reinhardtii*? Plant Soil 241:1–9

Hilgendorf F, Böttger M (1993) Influence of temperature on proton secretion and hexacyanoferrate (III) reduction of *Zea mays* L. roots. Plant Physiol 101:1349–1353

Hiraga S, Sasaki K, Ito H, Ohashi Y, Matsui H (2001) A large family of class III plant peroxidases. Plant Cell Physiol 42:462–468

Horemans N, Asard H, Caubergs RJ (1994) The role of ascorbate free radical as an electron acceptor to cytochrome *b*-mediated trans-plasma membrane electron transport in higher plants. Plant Physiol 104:1455–1458

Horemans N, Briggs WR, Caubergs RJ (1995) Blue light perception by endogenous redox components of the plant plasma membrane. Photochem Photobiol 61:518–522

Horemans N, Foyer C, Asard H (2000a) Transport and action of ascorbate at the plant plasma membrane. Trends Plant Sci 5:63–66

Horemans N, Foyer C, Potters G, Asard H (2000b) Ascorbate function and associated transport systems. Plant Physiol Biochem 38:1–11

Iseki M, Matsunaga S, Muraka A, Ohno K, Shiga K, Yoshida K, Sugai M, Takahashi T, Hori T, Watanabe M (2002) A blue light-activated adenylyl cyclase mediates photoavoidance in *Euglena gracilis*. Nature 415:1047–1051

Ishimamura A, Yamazaki I (1977) The carbon monooxide-binding hemoprotein reducible by hydrogen peroxide in microsomal fractions of pea seeds. J Biol Chem 252:199–204

Ito A, Sato R (1968) Purification by means of detergents and properties of cytochrome b_5 from liver microsomes. Eur J Biochem 243:4922–4923

Ivankina NG, Novak VA (1980) H^+ transport across the plasmalemma. H^+-ATPase or redox chain? In: Lucas RM, Spanswick WJ, Dainty J (eds) Plant membrane transport, current conceptual issues. Elsevier, Amsterdam, pp 503–504

Iyanagi T (1990) On the mechanism of one-electron reduction of quinones by microsomal flavin enzymes: the kinetic analysis between cytochrome b_5 and menadione. Free Rad Res Commun 8:259–268

Jaakola L, Maatta-Riihinen K, Karenlampi S, Hohtola A (2003) Activation of flavonoid biosynthesis by solar radiation in bilberry (*Vaccinium myrtillus* L.) leaves. Planta 218:721–728

Jabs T, Dietrich RA, Dangl JL (1996) Initiation of runaway cell death in an *Arabidopsis* mutant by extracellular superoxide. Science 273:1853–1856

Jesaitis AJ, Heners PR, Hertel R, Briggs WR (1977) Characterization of a membrane fraction containing a *b*-type cytochrome. Plant Physiol 59:941–947

Jespersen HM, Kjærsgård IVH, Østergaard L, Welinder KG (1997) From sequence analysis of three novel ascorbate peroxidases from *Arabidopsis thaliana* to structure, function and evolution of seven types of ascorbate peroxidase. Biochem J 326:305–310

Joliot P, Joliot A (1986) Mechanism of proton pumping in the cytochrome b/f complex. Photosynth Res 9:113–124

Jollie DR, Sligar SG, Schuler M (1987) Purification and characterization of microsomal cytochrome b_5 and NADH cytochrome b_5 reductase from *Pisum sativum*. Plant Physiol 85:457–462

Kawano T (2003) Roles of the reactive oxygen species-generating peroxidase reactions in plant defense and growth induction. Plant Cell Rep 21:829–837

Keller T, Damude HG, Werner D, Doerner P, Dixon RA, Lamb C (1998) A plant homolog of the neutrophil NADPH oxidase $gp91^{phox}$ subunit gene encodes a plasma membrane protein with Ca^{2+} binding motifs. Plant Cell 10:255–266

Kelley PM, Njus D (1986) Cytochrome b_{561} spectral changes associated with electron transfer in cromaffin-vesicle ghosts. J Biol Chem 261:6429–6432

Kieffer F, Simon-Plas F, Maume BF, Blein JP (1997) Tobacco cells contain a protein, immunologically related to the neutrophil small G protein Rac2 and involved in elicitor-induced oxidative burst. FEBS Lett 403:149–153

Kimball RA, Saier MH Jr (2002) Voltage-gated H^+ channels associated with human phagocyte superoxide-generating NADPH oxidase: sequence comparison, structural predictions, and phylogenetic analysis. Mol Membr Biol 19:137–147

Kjellbom P, Larsson C, Askerlund P, Schelin C, Widell S (1985) Cytochrome P-450/420 in plant plasma membranes: a possible component of the blue-light-reducible flavoprotein–cytochrome complex. Photochem Photobiol 42:779–783

Klebanoff SJ (1992) Oxygen metabolites from phagocytes. In: Gallin JG, Goldstein IM, Snydman R (eds) Inflammation: basic principles and clinical correlates, 2nd edn. Raven Press, New York, pp 541–588

Klobus G, Buczek J (1995) The role of plasma membrane oxidoreductase activity in proton transport. J Plant Physiol 146:103–107

Knöpfel M, Solioz M (2002) Characterization of a cytochrome b_{558} ferric/cupric reductase from rabbit duodenal brush border membranes. Biochem Biophys Res Commun 291:220–225

Krüger S (1993) Untersuchungen zur Energetisierung plasmalemmagebundener Redoxprozesse. PhD Thesis, University of Hamburg, Germany

Kurkova EB, Verhovskaya ML (1984) Redox components in the plasmalemma of plant cells. Soviet Plant Physiol 31:386–392

Lamb C, Dixon RA (1997) The oxidative burst in plant disease resistance. Annu Rev Plant Physiol Plant Mol Biol 48:251-275

Lambeth JD (2002) Nox/Duox family of nicotinamide adenine dinucleotide (phosphate) oxidases. Curr Opin Hematol 9:11–17

Lance C, Bonner WD Jr (1968) The respiratory chain components of higher plant mitochondria. Plant Physiol 43:756–766

Leong TY, Vierstra RD, Briggs WR (1981) A blue light-sensitive cytochrome-flavin complex from corn coleoptiles. Further characterization. Photochem Photobiol 34:697–703

Lesot A, Benveniste I, Hasenfratz MP, Durst F (1990) Induction of NADPH-cytochrome P-450 (c) reductase in wounded tissues from *Helianthus tuberosus* tubers. Plant Cell Physiol 31(8):1177–1182

Lesuisse E, Horion B, Labbe P, Hilger F (1991) The plasma membrane ferri reductase activity of *Saccharomyces cerevisiae* is partially controlled by cyclic AMP. Biochem J 280:545–548

Liu Q, Markham KR, Pare PW, Dixon RA, Mabry TJ (1993) Flavonoids from elicitor-treated cell suspension cultures of *Cephalocereus senilis*. Phytochemistry 32:925–928

Loeper J, Louerat-Oriou B, Duport C, Pompon D (1998) Yeast expressed cytochrome P450 2D6 (CYP2D6) exposed on the external face of plasma membrane is functionally competent. Mol Pharmacol 54:8–13

Löppert H (1981) Energy coupling for membrane hyperpolarization in *Lemna*: evidence against an ATP-fueled electrogenic pump as the exclusive mechanism. Planta 151:293–297

Löppert H (1983) Energy coupling for membrane hyperpolarization in *Lemna*: respiration rate, ATP level and membrane potential at low oxygen concentrations. Planta 159:329–335

Lorenz A, Kaldenhoff R, Hertel R (2003) A major integral protein of the plant plasma membrane binds flavin. Protoplasma 221:19–30

Lundegårdh H (1945) Absorption, transport and exudation of inorganic ions by roots. Arkiv Bot 32A(12):1–139

Lundegardh H (1955) Mechanism of absorption, transport, accumulation, and secretion of ions. Annu Rev Plant Physiol 6:1–24

Lüthen H, Böttger M (1988) Hexachloroiridate IV as an electron acceptor for a plasmalemma redox system in maize roots. Plant Physiol 86:1044–1047

Lüthje S, Döring O, Heuer S, Lüthen H, Böttger M (1997) Oxidoreductases in plant plasma membranes. Biochim Biophys Acta 1331:81–102

Lüthje S, van Gestelen P, Córdoba-Pedregosa M-C, González-Reyes JA, Asard H, Villalba JM, Böttger M (1998) Quinones in plant plasma membranes – a missing link? Protoplasma 205:43–51

Lüthje S, Böttger M, Döring O (2000) Are plants stacked neutrophiles? Pathogen-induced oxidative burst in plants and mammals in comparison. Progress in Botany, vol 61. Springer, Berlin Heidelberg New York, pp 187–222

Mankelow TJ, Henderson LM (2003) Proton conduction through full-length gp91phox requires histidin 115. Protoplasma 221:101–108

Marabini L, Radice S, Cipelletti B, Chiesara E (1994) Different amounts of cytochrome P-450 dependent monooxygenases in tulip bulbs, pea seedlings and maize endosperm cells. Plant Sci 99(2):135–140

Martin EM, Morton RK (1955) Cytochrome b_3 of microsomes from plant tissues. Nature 176:113–114

Martin EM, Morton RK (1957) Heam pigments of cytoplasmic particles from non-photosynthetic plant tissues. Biochem J 65:404–413

McKie AT, Barrow D, Latunde-Dada GO, Rolfs A, Sager G, Mudaly E, Mudaly M, Richardson C, Barlow D, Bomford A, Peters TJ, Raja KB, Shirali S, Hediger MA, Farzaneh F, Simpson RJ (2001) An iron-regulated ferric reductase associated with the absorption of dietary iron. Science 291:1755–1759

McKie AT, Latunde-Dada GO, Miret S, McGregor JA, Anderson GJ, Vulpe CD, Wrigglesworth JM, Simpson RJ (2002) Molecular evidence for the role of a ferric reductase in iron transport. Biochem Soc Trans 30:722–724

Meier B (1998) Reactive oxygen intermediates involved in cellular regulation. Protoplasma 217:101–116

Michel H (1998) The mechanism of proton pumping by cytochrome c oxidase. Proc Natl Acad Sci USA 95(22):12819–12824

Mika A, Lüthje S (2003a) Properties of guaiacol peroxidase activities isolated from corn root plasma membranes. Plant Physiol 132:1489–1498

Mika A, Lüthje S (2003b) DPI-sensitive NAD(P)H oxidase activities of plasma membrane-bound peroxidases isolated from corn roots (*Zea mays* L.). Free Rad Res 37 (Suppl 2):41

Mika A, Minibayeva F, Beckett R, Lüthje S (2004) Possible functions of extracellular peroxidases in stress-induced generation and detoxification of active oxygen species. In: Ros

Barceló A, Sottomayor M (eds) Plant peroxidases and phytochemistry, phytochemistry reviews. Kluwer, Dordrecht, in press

Morel F, Doussiére J, Vignais PV (1991) The superoxide-generating oxidase of phagocytic cells. Eur J Biochem 201:523–546

Moser CC, Keske JM, Warncke K, Farid RS, Dutton PL (1992) Nature of biological electron transfer. Nature 355(6363):796–802

Mougin C, Polge N, Scalla R, Cabanne F (1991) Interactions of various agrochemicals with cytochrome P-450-dependent monooxygenases of wheat cells. Pest Biochem Physiol 40:1–11

Nanda A, Grinstein S, Curnutte JT (1993) Abnormal activation of H^+ conductance in NADPH oxidase-defective neutrophils. Proc Natl Acad Sci USA 90:760–764

Nega E, Grunwaldt G (1997) Evidence for and characterization of cytochrome P-450 in *Neurospora crassa.* J Basic Microbiol 37(2):139–145

Neufeld E, Bown AW (1987) A plasma membrane redox system and proton transport in isolated mesophyll cells. Plant Physiol 83:895–899

Njus D, Kelley PM (1993) The secretory-vesicle ascorbate-regenerating system: a chain of concerted H^+/e^- transfer reactions. Biochim Biophys Acta 1144:235–248

Njus D, Knoth J, Cook C, Kelley PM (1983) Electron transfer across the chromaffin granule membrane. Biol Chem 258:27–30

Njus D, Kelley PM, Harnadek GJ, Pacquing YV (1987) Mechanism of ascorbic acid regeneration mediated by cytochrome b_{561}. Ann N Y Acad Sci 493:108–119

Njus D, Wigle M, Kelley PM, Kipp BH, Schlegel HB (2001) Mechanism of ascorbic acid oxidation by cytochrome b_{561}. Biochemistry 40:11905–11911

Novak VA, Ivankina NG (1986) Cellular level of ATP, transport of potassium, and electric characteristics of the *Elodea* plasmalemma in the presence of ferricyanide. Dokl Akad Nauk SSSR 286:498–501

Novak VA, Miklashevich AI (1986) Effect of ferricyanide on the metabolic activity and ion transport of *Chlorella vulgaris.* Zitologia 28:939–947

Novak VA, Ivankina NG, Morokova EA, Miclashevich AI (1988) The link between the electrogenic and redox functions of the plasmalemma and the energy metabolism of the cell. Physiol Plant 73:165–169

Okuyama E, Yamamoto R, Ichikawa Y, Tsubaki M (1998) Structural basis for the electron transfer across the chromaffin vesicle membranes catalyzed by cytochrome b_{561}: analyses of cDNA nucleotide sequences and visible absorption spectra. Biochem Biophys Acta 1383:269–278

Ozols J (1989) Structure of cytochrome b_5 and its topology in the microsomal membrane. Biochim Biophys Acta 997:121–130

Persan MW, Wang J, Schuler MA (2001) Characterization of maize cytochrome P450 monooxygenases induced in response to safeners and bacterial pathogens. Plant Physiol 125:1126–1138

Peters NK, Frost JW, Long S (1986) A plant flavone, luteolin, induces expression of *Rhizobium meliloti* nudolation genes. Science 233:977–980

Pierre JL, Fontecave M (1999) Iron and activated oxygen species in biology: the basic chemistry. BioMetals 12:195–199

Qiu Z-S, Rubinstein B, Stern AI (1985) Evidence for electron transport across the plasma membrane of *Zea mays* root cells. Planta 165:383–391

Ramirez JR, Gallego GG, Serrano R (1984) Electron transfer constituents in plasma membrane fractions of *Avena sativa* and *Saccharomyces cerevisiae.* Plant Sci Lett 34:103–110

Ravel P, Lederer F (1993) Affinity-labeling of an NADPH-binding site on the heavy subunit of flavocytochrome b_{558} in particulate NADPH oxidase from activated human neutrophils. Biochem Biophys Res Commun 196:543–552

Rich PR (1984) Electron and proton transfers through quinones and cytochrome *bc* complexes. Biochem Biophys Acta 768:53–79

Rich PR, Bendall DS (1975) Cytochrome components of plant microsomes. Eur J Biochem 55:333–341
Rich PR, Cammack R, Bendall DS (1971) Electron paramagnetic resonance studies of cytochrome P-450 in plant microsomes. Eur J Biochem 59:281–286
Robinson NC, Tanford C (1975) The binding of deoxycholate, triton X-100, sodium dodecyl sulfate, and phosphatidylcholine vesicles to cytochrome *b*5. Biochemistry 14(2):369–377
Robinson NJ, Procter CM, Connolly EL, Guerinot ML (1999) A ferric-chelate reductase for iron uptake from soils. Nature 397:694–697
Rodgers MW, Zimmerlin A, Werck-Reichart D, Bolwell GP (1993) Microsomally associated heme proteins from French bean: characterization of the cytochrome P450 cinnamate-4-hydroxylase and two peroxidases. Arch Biochem Biophys 304:74–80
Roman DG, Dancis A, Anderson GJ, Klausner RD (1993) The fission yeast ferric reductase gene frp1$^+$ is required for ferric uptake and encodes a protein that is homologous to the gp91-phox subunit of the human NADPH phagocyte oxidoreductase. Mol Cell Biol 13:4342–4350
Römheld V (1987) Existence of two different strategies for the acquisition of iron in higher plants. In: Winkelmann G, van der Helm D, Neilands JB (eds) Iron transport in microbes, plants and animals. VCH, Weinheim, pp 353–374
Ros Barcelü A (1997) Lignification in plant cell walls. Int Rev Cytol 176:87–132
Rossi F (1986) The O_2^- forming NADPH oxidase of the phagocytes: nature, mechanism of activation and function. Biochim Biophys Acta 853:65–89
Rubinstein B (1994) The action of ascorbate in vesicular systems. J Bioenerg Biomembr 26:385–392
Rubinstein B, Stern AI (1986) Relationship of transplasmalemma redox activity to proton and solute transport by roots of *Zea mays*. Plant Physiol 80:805–811
Rubinstein B, Stern AI (1991a) The role of plasma membrane redox activity in light effects in plants. J Bioenerg Biomembr 23:393–408
Rubinstein B, Stern AI (1991b) Proton release and plasmalemma redox in plants. In: Crane FL, Morre DJ, Low HE (eds) Oxidoreduction at the plasma membrane: relation to growth and transport, vol 2. Plants. CRC Press, Boca Raton, pp 167–187
Scagliarini S, Rotino I, Bäurle I, Asard H, Pupillo P, Trost P (1998) Initial purification study of the cytochrome b_{561} of bean hypocotyl plasma membrane. Protoplasma 205:66–73
Schmidt W (1999) Mechanisms and regulation of reduction-based iron uptake in plants. New Phytol 141:1–26
Segal AW (1995) The NADPH oxidase of phagocytic cells is an electron pump that alkalinises the phagocytic vacuole. Protoplasma 184:86–104
Segal AW, West I, Wientjes F, Nugent JHA, Haley B, Garcia RC, Rosen H, Scarce G (1992) Cytochrome *b*-245 is a flavocytochrome containing FAD and the NADPH-binding site of the microbicidal oxidase of phagocytes. Biochem J 284:781–788
Segal AW, Wientjes F, Stockley R, Dekker LV (1998) Components of the NADPH oxidase of phagocytic cells and their abnormality in the molecular pathology of chronic granulomatous disease (CGD). In: Asard H, Bérczi A, Caubergs R (eds) Plasma membrane redox systems and their role in biological stress and disease. Kluwer, Dordrecht, pp 69–101
Segal AW, Wientjes F, Stockley RW, Dekker LV (1999) Components and organization of the NADPH oxidase of phagocytic cells: its role in microbial killing and in the molecular pathology of chronic granulomatous disease. Adv Cell Mol Biol Membr Org 5:441–483
Serrano A, Córdoba F, Gonzáles-Reyes JA, Navas P, Villalba JM (1994) Purification and characterization of two distinct NAD(P)H dehydrogenases from onion (*Allium cepa* L.) root plasma membrane. Plant Physiol 106:87–96
Shatwell KP, Dancis A, Cross AR, Klausner RD, Segal AW (1996) The FRE1 ferric reductase of *Saccharomyces cerevisiae* is a cytochrome b similar to that of NADPH oxidase. J Biol Chem 271:14240–14244

Shichi H, Kasinsky HE, Hackett DP (1963) Studies on the *b*-type cytochromes from mung bean seedlings. IV. Purification and properties of cytochrome b_3 (*b*-559). J Biol Chem 238:1162–1166

Shirley BW (1996) Flavonoid biosynthesis: 'new' functions for an 'old' pathway. Trends Plant Sci 1:377–382

Short TW, Briggs WR (1990) Characterization of a rapid, blue light-mediated change in detectable phosphorylation of a plasma membrane protein from etiolated pea (*Pisum sativum* L.) seedlings. Plant Physiol 92:179–185

Shoun H, Suyama W, Kim D (1991) Unique nitrate/nitrite-inducible cytochrome P-450 in *Fusarium oxysporum* and related fungal species. Agric Biol Chem 55:593–596

Solomonson LP, Barber MJ (1990) Assimilatory nitrate reductase: functional properties and regulation. Annu Rev Plant Physiol Plant Mol Biol 41:225–253

Sparla F, Bagnaresi P, Scagliarini S, Trost P (1997) NADH:Fe(III)-chelate reductase of maize roots is an active cytochrome *b*(5) reductase. FEBS Lett 414:571–575

Spatz L, Strittmatter P (1971) A form of cytochrome b_5 that contains an additional hydrophobic sequence of 40 amino acid residues. Proc Natl Acad Sci USA 68:1042–1046

Sperling P, Heinz E (2001) Desaturases fused to their electron donor. Eur J Lipid Technol 103:158–180

Sperling P, Ternes P, Zank TK, Heinz E (2003) The evolution of desaturases. Prostaglandins Leukotiens Essential Fatty Acids 68:73–95

Stadler R, Zenk MH (1993) The purification and characterization of a unique cytochrome P-450 enzyme from *Berberis–Stolonifera* plant cell cultures. J Biol Chem 268(2):823–831

Stöhr C (1998) Plasma membrane-bound nitrate reductase in algae and higher plants. In: Asard H, Bérczi A, Caubergs RJ (eds) Plasma membrane redox systems and their role in biological stress and disease. Kluwer, Dordrecht, pp 103–119

Strittmatter P, Ozols J (1966) The restricted tryptic cleavage of cytochrome *b*5. J Biol Chem 241(20):4787–4792

Sugimura Y, Yoshizaki F, Katagiri J, Horiuchi C (1980) Studies on algal cytochromes. I. Purification and properties of cytochrome *b*-561 from *Enteromorpha prolifera*. J Biochem 87:541–547

Sze H (1985) H^+-translocating ATPases: advances using membrane vesicles. Annu Rev Plant Physiol 36:175–208

Tenhaken R, Rübel C (1998) Cloning of putative subunits of the soybean plasma membrane NADPH-oxidase involved in the oxidative burst by antibody expression screening. Protoplasma 205:21–28

Tenhaken R, Thulke O (1996) Cloning of an enzyme that synthesizes a key nucleotide-sugar precursor of hemicellulose biosynthesis from soybean: UDP- glucose dehydrogenase. Plant Physiol 112:1127–1134

Tenhaken R, Levine A, Brisson LF, Dixon RA, Lamb C (1995) Function of the oxidative burst in hypersensitive disease resistance. Proc Natl Acad Sci USA 92:4158–4163

Thimm O, Essigmann B, Kloska S, Altmann T, Buckhout TJ (2001) Response of *Arabidopsis* to iron deficiency stress as revealed by microarray analyses. Plant Physiol 127:1030–1043

Torres MA, Onouchi H, Hamada S, Machida C, Hammond-Kosack KE, Jones JDG (1998) Six *Arabidopsis thaliana* homologues of the human respiratory burst oxidase ($gp91^{phox}$). Plant J 14:365–370

Trost P, Bérczi A, Sparla F, Sponza G, Marzadori B, Asard H, Pupillo P (2000) Purification of cytochrome *b*-561 from bean hypocotyls plasma membrane. Evidence of two heme centers. Biochim Biophys Acta 1468:1–5

Tsubaki M, Kobayashi K, Ichise T, Takeuchi F, Tagawa S (2000) Diethyl pyrocarbonate modification abolishes fast electron accepting ability of cytochrome b_{561} from ascorbate but does not influence electron donation to monodehydroascorbate radical: identification of the modification sites by mass spectrometric analysis. Biochemistry 39:3276–3284

Van Beusichem ML, Nelemans JA, Bienfait HF (1988) Interrelationships between transplasma membrane electron/proton transfer stoichiometry, organic acid metabolism, and nitrate reduction in Dwarf bean (*Phaseolus vulgaris*). Plant Physiol 87:269–273

Van Duijn MM, Buijs JT, van der Zee J, van den Broek PJA (2001a) The ascorbate:ascorbate free radical oxidoreductase from the erythrocyte membrane is not cytochrome b_{561}. Protoplasma 217:94–100

Van Duijn MM, van der Zee J, van den Broek PJA (2001b) The ascorbate-driven reduction of extracellular ascorbate free radical by the erythrocyte is an electrogenic process. FEBS Lett 491:67–70

Van Gestelen P, Asard H, Caubergs RJ (1996) Partial purification of a plasma membrane flavoprotein and NAD(P)H-oxidoreductase activity. Physiol Plant 98:389–398

Van Gestelen P, Asard H, Caubergs RJ (1997) Solubilization and separation of a plant plasma membrane NADPH-O-2(-) synthase from other NAD(P)H oxidoreductases. Plant Physiol 115:543–550

Van Gestelen P, Asard H, Horemans N, Caubergs RJ (1998) Superoxide producing NAD(P)H-oxidase in plasma membrane vesicles from elicitor responsive bean plants. Physiol Plant 104:653–660

Velick S, Strittmatter P (1956) The oxidation–reduction stoichiometry and potential of microsomal cytochrome b_5. J Biol Chem 221:265–275

Verelst W (2003) Molecular–biological characterisation and physiological function of cytochromes b_{561} in higher plants. PhD Thesis, University of Antwerp, Belgium

Verelst W, Asard H (2003) A phylogenetic study of cytochrome b_{561} proteins. Gen Biol 4:R38

Verelst W, Kapila J, de Almeida Engler J, Stone JM, Caubergs R, Asard H (2004) Tissue-specific expression and developmental regulation of *Arabidopsis thaliana* cytochrome b_{561} genes. Physiol Plant 120:312–318

Vergères G, Waskell L (1995) Cytochrome b_5, its function, structure, and membrane topology. Biochimie 77:121–130

Vert GA, Franc J, Briat J-F, Curie C (2003) Dual regulation of the *Arabidopsis* high-affinity root iron uptake system by local and long-distance signals. Plant Physiol 132:796–804

Vianello A, Macri F (1991) Generation of superoxide anion and hydrogen peroxide at the surface of plant cells. J Bioenerg Biomembr 23:409–423

Vignais PV (2002) The superoxide-generating NADPH oxidase: structural aspects and activation mechanism. Cell Mol Life Sci 59(9):1428–1459

Wakefield LM, Cass AEG, Radda GK (1984) Isolation of a membrane protein by chromatofocusing: cytochrome *b*-561 of the adrenal chromaffin granule. J Biochem Biophys Meth 9:331–341

Waters BM, Blevins DG, Eide DJ (2002) Characterization of FRO1, a pea ferric-chelate reductase involved in root iron acquisition. Plant Physiol 129:85–94

Weiss SJ (1989) Tissue destruction by neutrophils. N Engl J Med 320:365–376

Widell S, Larsson C (1983) Distribution of cytochrome *b* photoreductions mediated by endogenous photosensitizer or methylene blue in fractions from corn and cauliflower. Physiol Plant 57:196–202

Widell S, Lundborg T, Larsson C (1982) Plasma membranes from oats prepared by partition in an aqueous polymer two-phase system. Plant Physiol 70:1429–1435

Wikström M, Krab K, Saraste M (1981) Proton-translocating cytochrome complexes. Annu Rev Biochem 50:623–655

Williams RJP (1985) Proton diffusion and the bioenergies of enzymes in membranes. In: Martonosi A (ed) The enzymes of biological membranes. Plenum Press, New York, pp 71–110

Xing T, Higgins VJ, Blumwald E (1997b) Race-specific elicitors of *Cladosporium fulvum* promote translocation of cytosolic components of NADPH oxidase to the plasma membrane of tomato cells. Plant Cell 9:249–259

Xu Y (1993) New insight of relationship of transplasma membrane redox activity to proton transport of corn roots. Master Thesis, Beijing University

Yamaguchi-Iwai Y, Dancis A, Klausner RD (1995) AFT1: a mediator of iron regulated transcriptional control in *Saccharomyces cerevisiae*. EMBO J 14:1231–1239
Yamasaki H, Sakihama Y, Ikehara N (1997) Flavonoid-peroxidase reaction as a detoxification mechanism of plant cells against H_2O_2. Plant Physiol 115:1405–1412
Yoshioka H, Sugie K, Park H-J, Maeda H, Tsuda N, Kawakita K, Doke N (2001) Induction of plant gp91 *phox* homolog by fungal cell wall, arachidonic acid, and salicylic acid in potato. Mol Plant Microb Inter 14:725–736
Yu L, Quinn MT, Cross AR, Dinauer MC (1998) Gp91^{phox} is the heme binding subunit of the superoxide-generating NADPH oxidase. Proc Natl Acad Sci USA 95:7993–7998

Sabine Lüthje
Michael Böttger
Olaf Döring
Biozentrum Klein Flottbek und Botanischer Garten
Universität Hamburg
Ohnhorststraße 18
22609 Hamburg, Germany
Fax: +49-40-42816254
e-mail: s.luthje@botanik.uni-hamburg.de

Photosynthesis. Carbon Metabolism: The Calvin Cycle's Golden Jubilee*

Grahame J. Kelly

'A crucial question was how the two molecules of 3PGA might arise from RuBP and CO_2...An ingenious solution to the question soon occurred to Melvin Calvin...I recall him saying that the answer...came to him as he was stopped in traffic waiting for a stoplight to change.'

Bassham (2003)

1 Personal Reflections

With this chapter, I have completed 30 years of reviewing photosynthetic carbon metabolism in *Progress in Botany*. Thirty years ago this series was produced as *Fortschritte der Botanik*, a formidable title for a young man from the Australian bush, freshly arrived in Germany. However, the task of reviewing in those early years was a pleasure: I had Erwin Latzko for a wise counsellor and coauthor, and since many journals were not available in our library at Weihenstephan I was permitted to enjoy a trip on the S-Bahn to the beautiful Botanic Gardens in Munich every now and then to peruse the literature in the magnificent old library there. With the latest research assembled, our task was to connect the newly discovered pathways of C_4 photosynthesis and photorespiration to the Calvin cycle, a challenge that we enthusiastically accepted over coffee in the laboratory, or over beer in the nearby 'oldest brewery in the world'.

Time has moved on, and the science of photosynthetic carbon metabolism has changed; thus with this chapter I will 'call it a day'. Erwin Latzko deserted me in 1998, no doubt because, being wiser than me, he could see that the likelihood of the discovery of new metabolic pathways was slim, while the influence of the emerging era of the gene and molecular biology could only be wondered upon. Indeed, no new pathways have been found,

* This review is dedicated to Professor Dr Erwin Latzko on the occasion of his 80th birthday.

Progress in Botany, Vol. 66

but molecular biology is calling into question the comfortable simplicity of the established pathways by showing us that there are often more forms of most enzymes (and other proteins) than we would have deemed necessary. For example, there are five genes encoding five sucrose transporters in rice (Aoki et al. 2003; Kühn 2003), and in *Arabidopsis* there are four genes encoding four phosphoenolpyruvate carboxylases (Sánchez and Cejudo 2003), six genes encoding six hexokinases (Frommer et al. 2003), six genes encoding eight ADP-glucose pyrophosphorylases (but four are faulty; Crevillén et al. 2003), nine genes encoding nine β-amylases (Laby et al. 2001), and 14 genes encoding 14 carbonic anhydrases (Moroney et al. 2001). The business of retrieving the literature reporting all of this has also changed. Hard copies of journals are being replaced by virtual forms; thus one is more-and-more tied to one's desk and a computer screen rather than walking to the library or enjoying an S-Bahn ride which, it must be admitted, is an opportunity for contemplation and reflection (even better than that provided while waiting in traffic for a stoplight to change?)

2 The Calvin Cycle's Golden Jubilee

This year (2004) can be considered the 50th anniversary of one of the most satisfying discoveries in science: the elucidation of the cycle of biochemical reactions that photosynthetically convert CO_2 to the sugars that support all life on Earth. Properly called the Calvin–Benson–Bassham cycle in recognition of the scientists who contributed so much to its discovery, its details gradually became apparent during several years of research [principally with haploid, clonal unicellular green algae (Raven and Girard-Bascou 2001)], but it was the publication 50 years ago by Bassham et al. (1954) that finalised and established it, as emphasised in the informative and entertaining historical accounts recently published by Benson (2002) and Bassham (2003). The cycle as presented by Bassham (2003), and reproduced here in 'word' form (Fig. 1), emphasises the substance of its discovery: most of the work involved investigations of sugar phosphates (sugar-Ps), while little attention needed to be given to assimilatory power (NADPH and ATP) or enzymes, because the former was known to be available, and for most of the contemplated reactions an appropriate enzyme was already described, or was being described, from research into the carbohydrate biochemistry of all types of organisms in other laboratories. It is remarkable that no alterations have been deemed necessary during the ensuing 50 years, and that the cycle has turned out to be universal in plants, algae and cyanobacteria. Only in a few relatively obscure photosynthetic prokaryotes has evidence for alternative cycles been detected, including a reductive form of the tricar-

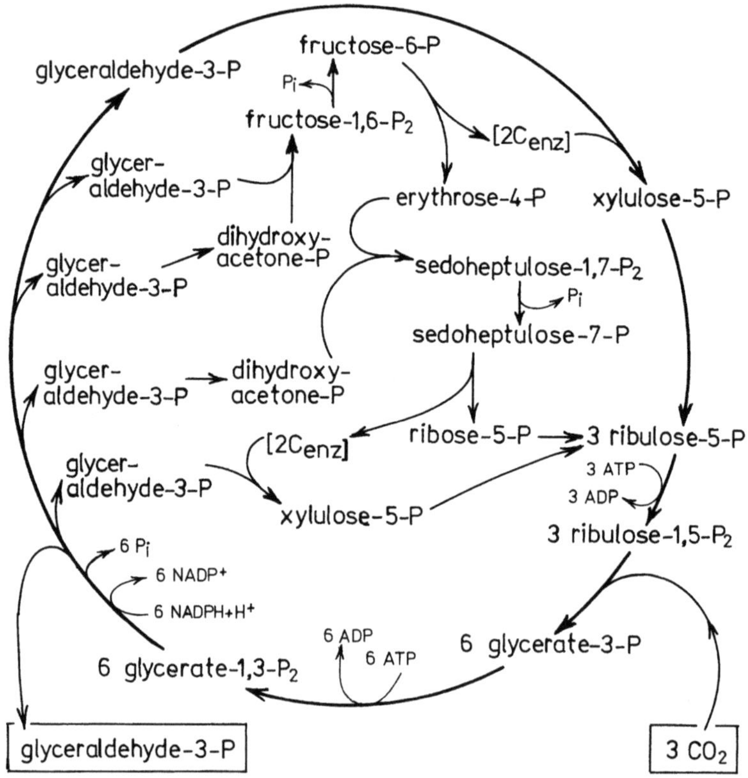

Fig. 1. The Calvin cycle, with emphasis on the sugar-P biochemistry. The sum of the reactions is: $3CO_2 + 6\ (NADPH + H^+) + 9\ ATP \rightarrow$ glyceraldehyde-3-P $+ 6\ NADP^+ + 9\ ADP + 8\ Pi$ [$2C_{enz}$], 2C-piece/enzyme (transketolase) complex. (Adapted from Bassham 2003)

boxylic acid cycle (see our 1978 review) and a bicyclic 3-hydroxypropionate cycle (Herter et al. 2002). Research into the critical carboxylating enzymes of these cycles is progressing (Kanao et al. 2002; Hügler et al. 2003).

3 The Calvin Cycle's Enzymology

3.1 Ribulose-1,5-Bisphosphate Carboxylase/Oxygenase (Rubisco)

Rubisco, the Calvin cycle's CO_2-fixing enzyme, is the protein that helps two molecules of glycerate-3-P (3PGA) to arise from ribulose-1,5-P_2 (RuBP) and CO_2 (see Bassham 2003 quote above). Its discovery was concomitant with that of the Calvin cycle 50 years ago (see Benson 2002). It has been, without doubt, the most studied of the 11 enzymes of the Calvin cycle, and the current perceived potential to manipulate it so that it works better and/or faster (Spreitzer and Salvucci 2002) is maintaining its popularity. Rubisco is not fast. In fact, it is distinctive in that it is a large and cumbersome enzyme. Its catalytic sites struggle to fix more than one molecule of CO_2 per second (von Caemmerer 2003). Thus, to keep global photosynthesis running there needs to be a lot of it. It is the most abundant protein on Earth.

The desire to fully explain the catalytic and regulatory properties of Rubisco has led to more detailed and impressive three-dimensional descriptions of the molecule (Okano et al. 2002; see Spreitzer and Salvucci 2002; Du et al. 2003). Some include its intimate association with a second protein, Rubisco activase, which acts in a three-fold chaperone capacity by dissociating inhibitory sugar-Ps from the enzyme (Portis 2003), by mediating its light activation (Zhang et al. 2002), and by indirectly determining its sensitivity (and therefore the sensitivity of photosynthesis) to high temperature, as recently shown for cotton (Law et al. 2001) and maize (Crafts-Brandner and Salvucci 2002). A realistic picture of Rubisco at sunrise would therefore show not only the 16 subunits of the enzyme (i.e. eight copies of the chloroplast-encoded large subunit and eight copies of the nuclear-encoded small subunit), but also the associated Rubisco activase. An adaptation of such a supramolecular complex shown by Portis (2003) is presented in Fig. 2. Incidentally, there is now stronger evidence that one of the more notable inhibitory sugar-Ps, viz. carboxyarabinitol-P (which binds to Rubisco and prevents it from being active at night), is synthesised from the Calvin cycle intermediate fructose-1,6-P_2 via the sugar hamamelose (Andralojc et al. 2002).

As mentioned above, much of the impetus for current research on Rubisco comes from the desire to apply the tools of modern biotechnology to improve this 'notoriously inefficient' (Spreitzer and Salvucci 2002) enzyme, and thereby increase agricultural production. This latter benefit presupposes that, under good (i.e. not water- or inorganic-nutrient-limited) field conditions, Rubisco activity is the ultimate determinate of photosynthetic rate and production. It certainly is a major (Jiang and Xu 2001; Sage 2002),

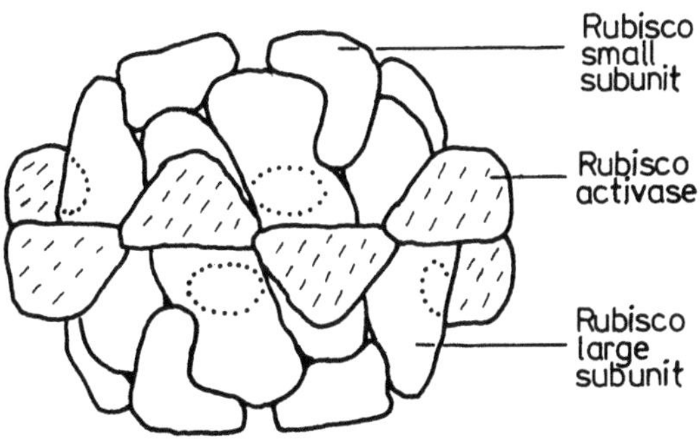

Fig. 2. The Rubisco complex at sunrise. One molecule of Rubisco activase is bound to each of the eight large subunits of Rubisco, facilitating dissociation of any inhibitory sugar-Ps from the enzyme's eight active sites (*dotted ellipses*) which are on these subunits. Having completed this task, the activase molecules themselves dissociate, liberating the active Rubisco for its diurnal chore of CO_2 fixation. Note that not all of the Rubisco subunits and activase molecules are in view in this 3D picture. (Adapted from Portis 2003)

although not exclusive (Bernacchi et al. 2002), determinate of photosynthetic rate in C_3 plants, but whether this translates into any important control function for photosynthesis (Raines 2003) is debatable. Furthermore, as emphasised in our 2000 review and now supported by the studies of Murchie et al. (2002) with rice, there is little evidence that Rubisco activity is a principal determinant of plant production in the field.

Rubisco could be improved by accelerating its activity, and/or by coaxing it to ignore O_2 (Parry et al. 2003). When the enzyme accepts O_2, rather than CO_2, as substrate the 'wasteful' photorespiratory pathway is initiated (see Section 5.5). Spreitzer and Salvucci (2002) are optimistic that some day a Rubisco better than any natural version will be created in a laboratory, but so far this has not happened, and they remind researchers that any alteration to the enzyme that improves one of the above two characteristics will be useless if it causes a depreciation of the other.

Research on Rubiscos from primitive autotrophic prokaryotes is providing intriguing insights. A study by Robinson et al. (2003) reminds us that Rubisco, and indeed the entire Calvin cycle, is not restricted to the sunlit parts of Earth; these authors have isolated Rubisco from chemoautotrophic (as opposed to photoautotrophic) bacteria that exist as endosymbionts in tubeworms colonising hydrothermal vents in the dark depths of the ocean. Rubisco presumably evolved in primitive prokaryotes, and indeed Ashida et al. (2003) have made the startling discovery that the unusually simple two-large-subunit Rubisco from the photosynthetic bacterium *Rhodospirillum rubrum* is so closely related to a similar protein in the common non-photosynthetic bacterium *Bacillus subtilis* that it can perform that protein's function (catalysis of a reaction in a pathway that salvages the amino acid methionine), and thereby rescue cells that have lost the protein but have had the *R. rubrum* Rubisco gene genetically engineered into them. Another impressive event involving *R. rubrum* Rubisco was the creation by Whitney and Andrews (2001) of a tobacco plant lacking its own Rubisco, but transformed to contain *R. rubrum* Rubisco in its chloroplasts. These plants grew well so long as they were treated like *R. rubrum*, i.e. supplied with a relatively high concentration of CO_2 (Whitney and Andrews 2003). Finally, Wang et al. (2001) manufactured prokaryotic/eukaryotic hybrid Rubiscos with cyanobacterial (*Synechococcus*) large subunits and either tobacco, wheat or rice small subunits. None of the hybrids was better than the native *Synechococcus* enzyme.

3.2 The Enzymes That Regenerate RuBP from 3PGA

In the Calvin cycle, 10 enzymes catalyse the 12 reactions that regenerate RuBP from the initial product of CO_2 fixation, 3PGA. The first (which converts 3PGA to glycerate-1,3-P_2) receives scant attention, but the second, NADP-linked glyceraldehyde-3-P dehydrogenase (GAPDH), which reductively forms glyceraldehyde-3-P (GAP) from the glycerate-1,3-P_2, is well researched. The majority of the recent studies on GAPDH also include the cycle's tenth enzyme which finally generates the RuBP, viz. ribulose-5-P kinase (Ru5PK), because these two enzymes come together with a third small protein (named CP12) to form a complex, and it is within this complex that many of the earlier-documented regulatory events, such as aggregation, light-mediated reductive activation and NADP-linked activation, take place and (additionally) interact. Consequences include lowering of the reactions' activation energies (Graciet et al. 2002), indirect reductive activation of GAPDH via intra-complex conformational changes following the reduction of a disulfide bond (to two sulfhydryl groups) in the Ru5PK (Lebreton et al. 2003) and the existence of multiple aggregation states (Scheibe et al. 2002). The plant GAPDH, unlike the algal enzyme, is a heterotetramer of two A-subunits and two B-subunits, and there is evidence that a unique disulfide bond in the C-terminus of the B subunit has critical regulatory functions (Qi et al. 2001; Sparla et al. 2002). While on the topic of molecular complexes, two further items are worth noting. Firstly, the possibility has again been raised (this time from experiments with a cyanobacterium) that the Calvin cycle enzymes may exist as ordered supramolecular complexes in vivo, perhaps with the help of thylakoid membranes (Sainis et al. 2003). Secondly, Giegé et al. (2003) have surprisingly announced that a dollop of the ten enzymes of the classic old pathway of glycolysis is associated with the outside surface of the cell's mitochondria, thereby enhancing the conversion of cytosolic glucose to mitochondrial-matrix pyruvate.

Returning to the Calvin cycle enzymes, four of the remaining seven have received some attention during the past 2 years. Two enzymes, viz. aldolase and triose-P isomerase, were recently observed by Ito et al. (2003) to be susceptible to in vivo modification by the covalent attachment of glutathione. Given the importance of glutathione to life, the significance of this observation will be eagerly awaited. The enzyme that hydrolyses fructose-1,6-P_2 to fructose-6-P [fructose1,6-bisphosphatase (FBPase)] remains popular; it has been the subject of a review (Chueca et al. 2002), and is considered an important factor in explanations of why the photosynthetic capacity of wheat leaves appears better than that of rice leaves (Sudo et al. 2003), and why photosynthesis runs poorly in cooled soybean leaves (van Heerden et al. 2003). Lastly, there is a report addressing transketolase. Maize transketolase has been cloned, crystallised and shown to be a homodimer, similar to its yeast counterpart (Gerhardt et al. 2003).

Regulation of the Calvin cycle and its constituent enzymes is taking some new twists. While it is clear that, upon sunrise, the well-established reductive activation of enzymes such as FBPase (Ocón et al. 2001; Chueca et al. 2002) and Ru5PK (Lebreton et al. 2003) and the increase in chloroplast-stromal Mg^{2+} concentration from 0.5–2 mM (Ishijima et al. 2003) must assist the start-up of photosynthesis, the view is still held that other enzymes such as aldolase and transketolase may be more important than these light-regulated enzymes (and even more important than Rubisco) in controlling steady-state carbon fixation (Raines 2003).

On the other hand, a fourth Calvin cycle enzyme susceptible to reductive activation, viz. sedoheptulose-1,7-bisphosphatase (SBPase), does seem to be an important determinant of the steady-state photosynthetic rate (Raines 2003); thus the reductive activation of enzymes will continue to be closely monitored. It is a burgeoning field. Enzyme reduction is mediated by the ubiquitous small protein thioredoxin (Tdx), and genomics has revealed that our old impression of chloroplasts containing two Tdxs (Tdx *f* and Tdx *m*) must be modified to one in which there are two forms of Tdx *f*, four forms of Tdx *m* and also a Tdx *x* and a Tdx *y* (Collin et al. 2003; Lemaire et al. 2003). Furthermore, many more proteins than the Calvin cycle's GAPDH, FBPase, SBPase and Ru5PK are potentially reduced by Tdx. Several others, including glucose-6-P dehydrogenase and NADP-linked malate dehydrogenase, have been mentioned in our earlier reviews, and now the list can be extended by the addition of two protein kinases [one that phosphorylates maize phosphoenolpyruvate carboxylase (Saze et al. 2001) and the other a thylakoid light-harvesting complex (Martinsuo et al. 2003)], and the initial enzyme of the shikimate pathway [3-deoxy-D-arabino-heptulosonate-P synthase (Entus et al. 2002)]. However, the list could explode, because Balmer et al. (2003) have used genomics to predict that up to 35 chloroplast proteins may be targets for reduced Trx.

4 The Calvin Cycle's Products

When the Calvin cycle is spinning in plant-leaf cells, two main products are generated: starch and sucrose. Starch is synthesised in the vicinity of the cycle, within the chloroplast, by diverting some of the cycle's fructose-6-P into the pathway: fructose-6-P → glucose-6-P → glucose-1-P → ADP-glucose → starch. Sucrose, on the other hand, is synthesised away from the cycle, in the cytosol, using GAP supplied by the cycle but exported into the cytosol via the chloroplast envelope's well-known, and still investigated (Schneider et al. 2002), 'phosphate translocator' protein. Ultimately, the sucrose is translocated to so-called sinks (growing shoots, roots, fruits, seeds and storage organs such as tubers). The starch accumulated in chloroplasts is later hydrolysed to soluble sugars at night, after the Calvin cycle has stopped spinning, and these are moved via chloroplast-envelope transporter proteins into the cytosol and used to continue the process of sucrose biosynthesis and translocation. Aspects of these metabolisms are reviewed below.

4.1 Starch

Research into the biosynthesis and structure of starch, both as a transitory reserve in chloroplasts (see above) and as a storage reserve in sink tissues, continues to flourish. In much of this research, properties of the participating enzymes have been addressed, especially those of the notable regulatory enzyme ADP-glucose pyrophosphorylase (ADPG-PPase) which generates the glucose donor molecule (ADP-glucose) utilised by the starch synthetases. In cyanobacteria ADPG-PPase is a relatively simple, highly regulated (Gómez-Casati et al. 2003) tetrameric protein composed of four identical subunits, whereas in plant chloroplasts it is a tetramer of two pairs of subunits, one slightly larger than the other (Salamone et al. 2002), and furthermore there are four genes encoding four versions of the larger subunit and two genes encoding two versions of the smaller subunit. Differential expression of these genes in different plant tissues possibly fine-tunes the enzyme's regulatory properties (especially relating to the well-established allosteric activation by 3PGA and inhibition by inorganic phosphate) in a way that is appropriate to each tissue (Crevillén et al. 2003). A recent observation indicates that regulation of the enzyme is more complex, in that both light and increased levels of soluble sugars can indirectly activate the enzyme by promoting the reduction of a disulfide bridge that interconnects the pair of smaller subunits when the enzyme is in its less active state (Hendriks et al. 2003). Intriguingly, no prominent mention of thioredoxin is included in this report, and similar regulation of the enzyme in potato tubers (but obviously not involving light!) has also been detected (Tiessen et al. 2002). Finally, Baroja-Fernández et al. (2001) are entertaining the possibility that ADPG-PPase might to some extent be side-stepped when ADP-glucose is synthesised in the cytosol by the enzyme sucrose synthase catalysing a reaction between ADP (instead of the usual UDP) and sucrose, followed by import of the ADP-glucose into the chloroplast. However, these authors have had to propose a complex cycling of carbohydrates between chloroplasts and the cytosol, possibly for control purposes, to explain the very existence of ADPG-PPase. This cycling would entail turnover of starch during the day, but Zeeman et al. (2002) report that this does not occur.

The existence of ADPG-PPase is currently viewed as being most significant, because one of the most successful advances in genetically engineering crop plants to increase productivity has centred on this enzyme. Smidansky et al. (2002, 2003) introduced a mutant form of the gene for the enzyme's larger subunit into wheat and rice. The mutant form of the subunit resulted in the enzyme being less sensitive to inhibition by orthophosphate, and the resultant increased activity effectively increased the rate of starch synthesis such that the wheat yield increased by 38% and the rice yield increased by 20%. Clearly, the plant leaves had the capacity to photosynthesise faster when called upon to do so. A similar success has been reported, rather unexpectedly, for another enzyme: when Regierer et al. (2002) genetically engineered potato to have a reduced activity of adenylate kinase in plastids, the yield of tubers doubled when compared to one plot of wild-type plants, although the increase was only 22% when compared to a second plot of these plants.

Concepts concerning the diurnal biosynthesis of starch granules from ADP-glucose in chloroplasts, and its subsequent nocturnal dissimulation, remain essentially as outlined in our preceding two reviews, but the discovery of two new proteins requires that these concepts be remodelled. The first protein (Ritte et al. 2002) is a granule-bound, enigmatic new enzyme that, during biosynthesis, phosphorylates starch via a reaction in which the central (rather than terminal) phosphate group of ATP is transferred to approximately 0.3% of the glucose moieties in the amylopectin component of starch; the ATP's terminal phosphate is simultaneously transferred to water (i.e. hydrolysed). Entitled 'α-glucan water dikinase' (an odd name, but I cannot think of a better one – hopefully, however, ATPases will not be renamed 'water kinases'!), this new enzyme is believed to be vital for the subsequent dissimulation of the starch, perhaps because the addition of phosphate groups assists starch degradation. The second new protein relates to starch degradation. Niittylä et al. (2004) have very recently discovered a previously unknown maltose transporter protein in chloroplast envelopes. Combined with preceding evidence that a chloroplast β-amylase is quite important for starch degradation (Scheidig et al. 2002), it now seems likely that chloroplasts at night are initially brewery-like in that most of the starch hydrolysis produces maltose. This maltose is then transported into the cytosol and used there to manufacture sucrose. The previous popular pathway of starch hydrolysis to glucose and transport of this sugar (Servaites and Geiger 2002) might be a minor one. However, one question remains: does the enzyme maltase exist in the cytosol of leaf cells?

4.2 Sucrose and Other Soluble Sugars

It is sometimes good to stand back, look at photosynthesis and wonder. When Salerno and Curatti (2003) did this, they found great difficulty in suggesting why most plants have adopted sucrose (a disaccharide of glucose and fructose) to be the major form of translocated photosynthate when many other soluble sugars may have been similarly suitable. Admittedly, some plants have chosen instead to translocate sugar alcohols such as sorbitol; the final enzyme in the biosynthesis of sorbitol in apple leaves has now been characterized (Zhou et al. 2003). A few other plants add one or two galactose or fructose units onto the sucrose before translocating it (see our 2000 review), but these are isolated exceptions. Sucrose easily predominates; thus it is little wonder that its biosynthesis and the regulation of this biosynthesis receive much attention.

One regulatory step receiving renewed attention is the interconversion between fructose-6-P and fructose-1,6-P_2 which involves three enzymes [viz. FBPase, ATP-linked phosphofructokinase (PFK) and pyrophosphate (PPi)-linked PFK] and the odd regulatory sugar-P, fructose-2,6-P_2. This sugar-P is both synthesised and degraded by a single, appropriately controlled enzyme (Markham and Kruger 2002). A clear description of the interplay between these enzymes is still emerging (Nielsen and Stitt 2001; Trevanion 2002). Meantime, a new role for the PPi-linked PFK has been proposed: Costa dos Santos et al. (2003) have evidence that, in maize roots, its reversible reaction operates in the direction of PPi synthesis to supply PPi to the tonoplast H^+-pyrophosphatase that pumps protons into the vacuole, thereby generating a pH in the vacuole 2 units lower than that in the cytosol. An extension of this proposal, involving the inclusion of the old classical ATP-linked PFK to supply the necessary fructose-1,6-P_2, is entertained in Fig. 3. If this system operated in leaves (and there is no reason why it should not), the acidification of the vacuole might therein promote hydrolysis of sucrose by acid invertase, releasing hexoses that exit and interact with cytosolic hexokinase in the newly emerging signal transduc-

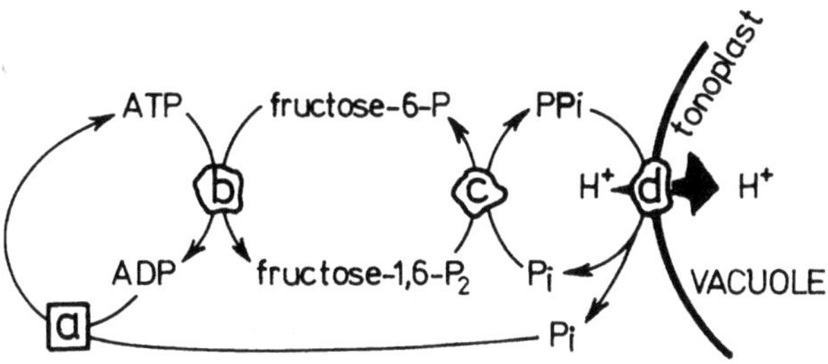

Fig. 3. Possible utilisation of PFKs for vacuolar acidification. Photosynthetically generated ATP (*a*), delivered either directly or indirectly (via a sugar-P shuttle system) to the cytosol, could be used by ATP-linked PFK (*b*) to supply fructose-1,6-P_2 to the PPi-linked PFK (*c*) which, by catalysing its reverse reaction, generates the PPi used by the tonoplast H^+-pyrophosphatase (*d*). Section involving *c* and *d* is based on Costa dos Santos et al. (2003)

tion pathways (see below). The central steps of sucrose synthesis would then be intertwined with the signal transduction system that influences the transcription of genes, including those of the Calvin cycle enzymes.

The remainder of the sucrose biosynthetic pathway involves the conversion of two molecules of fructose-6-P to one of sucrose. A cytosolic P-glucomutase normally operative in this pathway has been tenuously considered to be beneficial (Lytovchenko et al. 2002), but not necessarily essential since its chloroplast counterpart may be able to indirectly substitute for it in an emergency (Lytovchenko and Fernie 2003). Considerable attention is still given to the enzyme two steps further along the pathway, viz. sucrose-P synthase. The phosphorylation/dephosphorylation-based regulation of this enzyme, like so many others, now seems to be modulated by light (Pagnussat et al. 2002). It is still popular to genetically engineer plants to boost their sucrose-P synthase content. Recently, *Arabidopsis* became more freezing tolerant (Strand et al. 2003), old leaves of tobacco photosynthesised faster (Baxter et al. 2003) and tomato plants experienced a 58% increase in photosynthetic capacity (Lunn et al. 2003) when this was done.

Another popular topic relating to sucrose is the ability of this sugar, and other soluble sugars, to initiate signal transduction pathways (which terminate in the regulation of gene expression) when they accumulate in cells. Both sucrose and glucose, and possibly other sugars, can initiate these pathways, of which there are probably several (Xiao et al. 2000; Sinha et al. 2002; Tiessen et al. 2003). Most attention is given to the ability of the ubiquitous enzyme hexokinase to 'sense' the glucose concentration and, when this is substantial, to initiate a signalling pathway into the nucleus where the transcription of genes for Calvin cycle enzymes is reduced (see Fig. 2 in our 2000 review), while the transcription of genes for other proteins such as sucrose synthase (Ciereszko and Kleczkowski 2002), and possibly cytosolic glucose-6-P dehydrogenase (Hauschild and von Schaewen 2003), and a protein involved in ADPG-PPase activation (Tiessen et al. 2003) is increased. There has been some debate about how hexokinase does this, but an ingenious experiment by Moore et al. (2003) in which mutant hexokinases were introduced into mutant *Arabidopsis* plants demonstrated that the glucose-sensing function of the hexokinase was independent of this protein's catalytic ability (see Frommer et al. 2003; Harrington and Bush 2003). Clearly, the renewed attention being given to this famous old enzyme is worthwhile, and the discovery of a new form of hexokinase in the stroma of moss chloroplasts (Olsson et al. 2003) can only enliven the attention it receives in the future.

Before leaving soluble sugars, we should note that the mysterious trehalose (a disaccharide of two glucose units linked C1 to C1, not to be confused with the Weihenstephan brewery's maltose which is a disaccharide of two glucose units linked C1 to C4) is still arousing curiosity. It exists in fungi and some invertebrate animals, but no more than trace amounts have been detected in any plant except the primitive, desiccation-tolerant lycophyte *Selaginella lepidophylla* (resurrection plant). Nevertheless, genes encoding the two enzymes required for its synthesis from UDP-glucose and glucose-6-P, via trehalose-6-P (Penna 2003), appear to be widespread in plants, and investigations have demonstrated that the first, detected in *Arabidopsis* (Vogel et al. 2001), is essential for the development of embryos (Eastmond et al. 2002; Schluepmann et al. 2003). Presumably, trehalose is hydrolysed by a trehalase (Wingler 2002) as fast as it is formed in all plants except *S. lepidophylla*. In this plant, and other organisms where it accumulates, it confers tolerance to abiotic stresses such as drought; thus Garg et al. (2002) decided to introduce the two trehalose biosynthetic genes from a bacterium into rice and look for a similar situation. Indeed, their plants had elevated levels of trehalose, were more resistant to drought, salinity and cold, and even displayed an elevated capacity for photosynthesis. Meantime, trehalose still remains mysterious, especially since electronic database searches give no reports of it in any moss or alga except *Euglena gracilis* (Porchia et al. 1999) (but *E. gracilis* is equally mysterious, being considered by systematists to be more closely related to certain protozoa than to algae). Such database searches also give no reports of it in seeds, yet one might expect maturing seeds, destined to survive long periods of desiccation, to benefit by synthesising some.

4.3 Isoprene

A somewhat overlooked extension of photosynthetic carbon metabolism, alluded to in our 1996 review, is the manufacture of isoprene (2-methyl-1,3-butadiene) by forest plants, including ferns and many trees. This gas, which has value as a cellular thermoprotective agent (Brüggemann and Schnitzler 2002), possibly in association with photorespiration (Peñuelas and Llusià 2002), is easily emitted into the atmosphere. It is synthesised from the same 5-carbon precursor molecule as that used to build isoprenoids such as carotenoids and steroids, and most of this precursor is, in turn, synthesised in the chloroplast (Rosenstiel et al. 2002) in eight steps (Fellermeier et al. 2003) that begin with two molecules of GAP supplied by the Calvin cycle (one of the GAPs is first converted to a pyruvate). Sharkey (1996) earlier reported that some trees direct up to 15% of their assimilated CO_2 into isoprene, and since most of the Earth's photosynthesis is performed by forests, isoprene should clearly be included with starch and sucrose as an important end-product of the Calvin cycle.

5 Pathways and Processes That Assist the Calvin Cycle

5.1 Uptake of Inorganic Carbon

Terrestrial plants take up inorganic carbon (Ci) as CO_2, while aquatic plants, algae and cyanobacteria have the additional option of taking up bicarbonate (HCO_3^-). If HCO_3^- is selected it must, sooner or later, be converted to CO_2 because this is the form of Ci used by the Calvin cycle's Rubisco. The interconversion between CO_2 and HCO_3^- involves the two reversible reactions: $CO_2 + H_2O \rightarrow H_2CO_3 \rightarrow H^+ + HCO_3^-$, the first of which is slow, but is accelerated by the enzyme carbonic anhydrase (CA) if present. Algae and cyanobacteria, both in laboratories and in the ocean, variously use intracellular and extracellular (periplasmic) forms of this enzyme, plus plasmalemma-located, ATP-powered (directly or indirectly) transporter proteins for either CO_2 or HCO_3^- to not only deliver Ci as CO_2 to Rubisco, but also to actively pump the Ci into the cell so that the concentration of CO_2 in the vicinity of Rubisco is much higher than that in the surrounding water (or, for that matter, much higher than that in the air above) (Ghoshal et al. 2002; Tortell and Morel 2002; Rost et al. 2003). The concepts and research on this topic, last covered in our 2000 review, have been more recently massively reviewed (see Price and Badger 2002).

The ATP that powers Ci uptake is ultimately derived from photosynthetic electron transport [although in the odd case of the eustigmatophyte alga *Nannochloropsis gaditana* mitochondrial respiration also contributes (Huertas et al. 2002)]. In an intriguing report, Tchernov et al. (2003) partially reversed the functions of ATP and Ci uptake by providing evidence that many microalgae may take up much more Ci than they need, such that excess Ci leaks back out, but the cell still benefits because the associated ATP consumption represents yet another avenue for the dissipation of excess harvested light energy. Meanwhile, from other studies there are indications that extracellular structures, as well as extracellular CA, might assist Ci uptake. There is tenuous (Buitenhuis et al. 1999) but not universal (Herfort et al. 2002) support for an early proposal that the manufacture of $CaCO_3$ coccoliths by the widespread marine coccolithophorid microalgae promotes the conversion of seawater HCO_3^- to CO_2 for Rubisco by providing the necessary H^+ (Fig. 4). Now, a role for the elegant frustule of diatoms is also under consideration, because Milligan and Morel (2002) have evidence that the frustule's biosilica appropriately buffers the pH of the microenvironment of the extracellular CA.

Moving up to plants, it seems that those that have moved to aquatic environments, such as the fresh-water *Egeria densa* (Lara et al. 2002) and the marine seagrass *Ruppia cirrhosa*

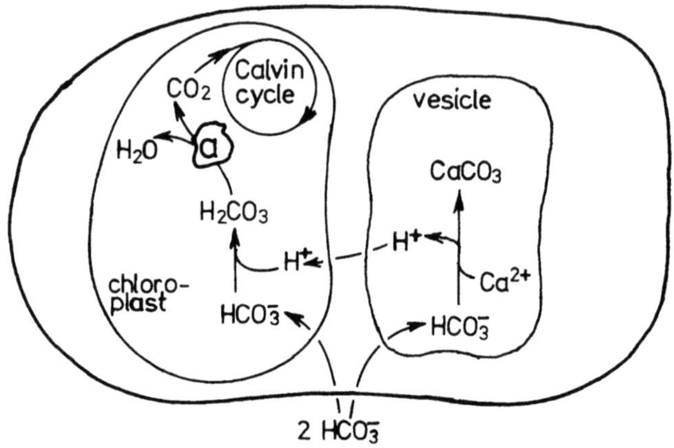

Fig. 4. Dual use of HCO_3^- by coccolithophorid microalgae such as *Emiliania huxleyi*. Coccoliths are synthesised from Ca^{2+} and HCO_3^- intracellularly in Golgi vesicles (before being transported to the cell surface). The H^+ released conceivably joins HCO_3^- in chloroplasts, forming H_2CO_3, the substrate of carbonic anhydrase (*a*). (Adapted from Buitenhuis et al. 1999)

(Hellblom and Axelsson 2003), have copied algae to some extent in that they can utilise HCO_3^- by acidifying the solution adjacent to the leaf surface and (in the case of the seagrass) producing an extracellular CA, thereby enhancing the conversion of HCO_3^- to CO_2, which then simply diffuses into the cells. In fact, the seagrass mechanism is little different to that used by some macroalgae that succeed with simple CO_2 diffusion as the final step (Snoeijs et al. 2002).

Most plants, however, are terrestrial; thus CO_2 is their only source of Ci. Most are 'C_3 plants', and a role for the CA that their leaves contain is, as ever, temptingly enigmatic. It may be important (Bernacchi et al. 2002) because, in CO_2's journey to the chloroplast, it must traverse the cytosol where at least some of it will be converted to HCO_3^-. However, we are currently distracted by a new group of proteins: the odd, membrane-located transporter proteins that facilitate the movement of water across membranes. Designated 'aquaporins', they have now been found to facilitate the entry of CO_2 into the photosynthetic cells of bean and tobacco (Terashima and Ono 2002; Uehlein et al. 2003), thereby assisting its journey to Rubisco. One now wonders whether chloroplast-envelope membranes also possess aquaporins.

5.2 Concentrating CO_2 Near Terrestrial Rubisco: C_4 Photosynthesis

The universality of the Calvin cycle was briefly challenged 22 years after its discovery when Hatch and Slack (1966) reported that sugar-cane leaves appeared to utilise an alternative cycle centred on the carboxylation of P-enolpyruvate (PEP) rather than RuBP. However, it soon became apparent that these leaves do contain the Calvin cycle in tough, photosynthetic bundle-sheath cells, and that they need it to grow (see our first, 1974 review). The high capacity for PEP carboxylation, located in cooperating adjacent mesophyll cells that are distinctive in lacking the Calvin cycle, was soon recognised to be the first component in a CO_2-collection-and-concentration system appropriately named 'the C_4 pathway of photosynthesis'.

It is sobering to be reminded that the sum of the reactions that constitute the C_4 pathway is nothing more than the hydrolysis of two ATP (Miyao 2003). The consequence, however, is the substantial elevation of the concentration of CO_2 in the vicinity of Rubisco (von Caemmerer and Furbank 2003), and this profoundly alters the plant's physiology: photorespiration is deterred (Kiirats et al. 2002), Rubisco works faster and water use efficiency is heightened. The latter allows C_4 plants to succeed well in relatively warm, dry habitats. By contrast, they do poorly in cool habitats, at least in part because their content of Rubisco, which is less than that of C_3 plants, is insufficient when it is slowed by low temperature (Kubien et al. 2003). The tendency for the C_4 plant's PEP carboxylase to lose activity in cooled leaves (Chinthapalli et al. 2003) may also contribute. Incidentally, further studies on the proteins that mediate the activation of this enzyme by phosphorylating it at sunrise (Bakrim et al. 2001; Echevarría et al. 2001; Saze et al. 2001) and dephosphorylating it at sunset (Dong et al. 2001) have been reported.

Exceptions to the above include a species of *Miscanthus* that maintains a higher content of Rubisco when cool (Naidu et al. 2003), and the high-altitude grass *Muhlenbergia richardsonis* that grows flat in order to take advantage of the much higher temperature of the thin boundary layer of air at ground level when it is calm in such localities (Sage and Sage 2002). C_4 plants also do poorly in shade because much of the photochemical activity is restricted to the bundle-sheath cells surrounding veins. Ogle (2003) has calculated that veins would need to be uncommonly close together before C_4 plants could venture into the shade.

Strange C_4 plants in the family Chenopodiaceae are being discovered in Asia. Following the surprise description of single-celled C_4 photosynthesis in *Borszczowia aralocaspica*, in which Calvin-cycle-containing chloroplasts are at one end of an elongated cell while Calvin-cycle-lacking, PEP-regenerating chloroplasts are at the opposite end (see our last review), there is now *Bienertia cycloptera*, also with single-celled C_4 photosynthesis, but in this plant the Calvin-cycle chloroplasts are embedded in a 'bubble' of cytoplasm surrounded by a large central vacuole, but in contact (via cytoplasmic strands) with a layer of peripheral cytoplasm in which the PEP-regenerating chloroplasts are located (Voznesenskaya et al. 2002). Finally, there are species of *Salsola* that are C_4 plants in which C_4 photosynthesis even occurs in green cotyledons following seed germination (Voznesenskaya et al. 2003).

The discovery of plants with single-celled C_4 photosynthesis has created excitement because such plants reflect the vision of genetic engineers planning to improve C_3 crop plants by transforming their single type of photosynthetic cell with one or more of the genes that encode enzymes of the C_4 pathway. However, such excitement is probably premature for two reasons, as pointed out by von Caemmerer (2003): firstly, these two odd plants have quite low photosynthetic rates, even lower than C_3 crop plants, and secondly, C_3-plant cells tend to have chloroplasts appressed to the plasma membrane facing intercellular air spaces, thereby negating any possibility of locally concentrating CO_2. There is no impressive arrangement of two types of chloroplasts as in the two chenopods. Reports of C_3 plants transformed with C_4 enzymes continue to appear, but no clear-cut benefit has yet emerged.

A major reorientation of our perception of C_4 photosynthesis is warranted following the recent demonstration by Hibberd and Quick (2002) that the sheath of green cells surrounding vascular strands in the petioles of typical C_3 plants such as tobacco and celery (the latter obvious to all culinary chefs slicing celery stalks into fresh salads) performs photosynthesis with strong C_4 characteristics (Fig. 5). Remarkably, these C_3 bundle-

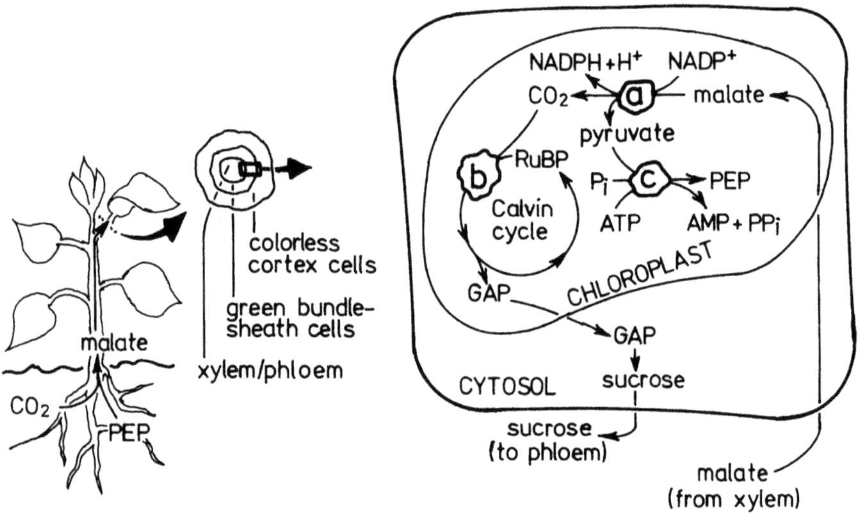

Fig. 5. CO_2 for photosynthesis in the petioles of C_3 plants is obtained by decarboxylation of malate delivered from the xylem, and previously synthesised in lower plant parts (e.g. roots) from respiratory CO_2. The biochemical and transport processes that balance the PEP produced in the bundle-sheath chloroplast and the PEP consumed in the plant root are not shown. *a* NADP-linked malic enzyme; *b* Rubisco; *c* pyruvate Pi dikinase

sheath cells contain high activities of malate-decarboxylating enzymes that release CO_2 (for Rubisco) from malate, and of the 'C_4 enzyme' pyruvate Pi dikinase that generates PEP from pyruvate (the other product of decarboxylation), and which is probably regulated by the odd phosphorylation system known from C_4 plants (Chastain et al. 2002). Quite possibly, these C_3 bundle-sheath cells are the evolutionary ancestors of the bundle-sheath cells of genuine C_4 plants (Raven 2002). It is considered that the C_3-plant petiole cells retrieve respiratory CO_2 evolved in the rhizosphere and lower parts of the plant (and, somewhat ironically, this CO_2 must therefore have already met Rubisco during conventional C_3 photosynthesis by a leaf of the same plant earlier in its life!). The system envisaged is one whereby three root enzymes generate a malate from the respired CO_2 and a PEP, and this malate is then transported through the xylem to the petiole. Differences from traditional photosynthesis are: (1) the task of mesophyll cells is performed by a distant part of the plant, (2) PEP regeneration is in the same chloroplast as the Calvin cycle, and (3) no boost to the concentration of CO_2 near Rubisco has been mentioned. Nevertheless, it is not inconceivable that, in the future, petioles will be looked upon as 'C_4 organs within C_3 plants'.

5.3 Night-Time Preparations for the Calvin Cycle: Crassulacean Acid Metabolism (CAM)

CAM plants, as is well known, achieve the intuitively impossible task of organising the daytime delivery of CO_2 to the Calvin cycle behind closed stomata. Plants regularly subjected to severe water stress such as desert cacti, rainforest epiphytes (Pierce et al. 2002) and halophytes (Broetto et al. 2002) do this. Their sleuth is to collect CO_2 during the preceding night through open stomata; this CO_2 is joined to PEP (catalysed by PEP carboxylase), the oxaloacetate produced is reduced to malate, and this malate is then transported into the vacuole and accumulated there. Next day the malate is moved out into the cytosol and decarboxylated to supply the Calvin cycle with CO_2 while, simultaneously, the stomata are closed to conserve water. Later in the day they open their stomata a little and complete their day's photosynthesis by directly fixing atmospheric CO_2, so that, overall, typical CAM plants collect three-quarters of their CO_2 at night, and the remaining quarter during late afternoons (Winter and Holtum 2002) when Rubisco is maximally activated (Griffiths et al. 2002). Both carboxylases, i.e. Rubisco and PEP carboxylase, are important components in the molecular machinery that establishes the endogenous circadian rhythm which, in turn, sets in motion the day/night phases of CAM (Wyka and Lüttge 2003).

Active membrane-located transporter proteins are critical for CAM, and are being anxiously sought. The nocturnal entry of malate into vacuoles could conceivably be mediated by the tonoplast dicarboxylate transporter recently discovered by Emmerlich et al. (2003), but these authors made their discovery with the non-CAM plant *Arabidopsis*, and noted that the transporter was eerily similar to a human Na^+/dicarboxylate cotransporter. A different and possibly more serious candidate transporter protein is the voltage-gated malate-ion channel found in tonoplasts of the CAM plant *Kalanchoë daigremontiana* by Hafke et al. (2003). It was recorded to be present at a density of one channel per 5 μm^2, sufficient to accommodate observed rates of vacuolar malate uptake at night. Another tonoplast transporter recently proposed (McRae et al. 2002) is a sucrose uniporter that permits the diurnal uptake of this end-product of photosynthesis into vacuoles where it is hydrolysed into glucose and fructose; CAM plants such as pineapple prefer to accumulate these soluble sugars rather than starch in chloroplasts.

The conquest of stressful habitats by CAM plants does not mean that these plants are immune from stress. Their CAM needs to be accompanied by other biophysical and biochemical mechanisms that deal with potential oxidative stress and photoinhibition (Broetto et al. 2002; Lu et al. 2003). Nevertheless, the success of CAM is unquestionable, and has fascinated plant physiologists to such an extent that four reviews have recently appeared, dealing with the 407 CAM plants investigated to date (Sayed 2001), the consequences of the high CO_2 concentration and often-overlooked high O_2 concentration that develop in leaves during the day (Lüttge 2002), a Congress (Holtum 2002), and a brief history of CAM (that includes an unfortunate choice of words incorrectly implying that CAM occurs in nutshells!) (Black and Osmond 2003).

5.4 Mitochondrial Metabolism

The other famous cycle of carbon metabolism, viz. the respiratory Krebs cycle, must operate to some extent in photosynthetic cells because there are beneficial interactions with the Calvin cycle (Raghavendra and Padmasree 2003) and because the Krebs cycle is the source of some of the carbon skeletons for amino acid biosynthesis. In fact, Peisker and Apel (2001) estimate that 71% of mitochondrial respiration is sustained by C_3 plant leaves in the light. Loreto et al. (2001) imply that an even greater percentage may be sustained in C_4 plant leaves, but (as expected) most of the CO_2 released by the Krebs cycle is recaptured by PEP carboxylase. Even in C_3 plants a good portion of this CO_2 is caught by Rubisco before escaping from the leaf (Pinelli and Loreto 2003). Meanwhile, examination of a mutant wild tomato (*Lycopersicon pennellii*) depleted in aconitase (the second enzyme of the Krebs cycle) now indicates that these Krebs cycle activities may be excessive. The mutant tomato exhibited a reduced flux through the Krebs cycle, but nevertheless photosynthesised faster and yielded more fruit (Carrari et al. 2003).

5.5 Photorespiration

The Calvin cycle not only faced a challenge 22 years after its discovery (see Sect. 5.2), but also suffered a humbling experience 5 years later when it was revealed (see Ogren 2003) that its CO_2-fixing enzyme Rubisco was not only slow, but also unfaithful: on about one occasion in three it accepted O_2, rather than CO_2, as substrate, thereby producing glycolate-P and initiating photorespiration. Schloss (2002) declares that this discovery 'stunned the biochemical community', in part because an oxygenase-enzyme devoid of any organic cofactor was unheard of, but Benson (2002), on the other hand, states that he 'failed to get excited' by the discovery because he knew how labile RuBP is. Recently, Lilley et al. (2003) have brought Rubisco's oxygenase activity to light by showing that, in the presence of Mn^{2+} instead of Mg^{2+}, it displays far-red chemiluminescence, quite a feat for an enzyme classically considered to be one of the catalysts of 'the dark reactions' of photosynthesis.

The glycolate-P produced by Rubisco is hydrolysed to glycolate and Pi within the chloroplast by a specific phosphatase which finally has been purified (Mamedov et al. 2001). The glycolate is exported to the cytosol, but the Pi is not, thereby making the chloroplast independent of cytosolic Pi for the photorespiratory component of its carbon metabolism (Holbrook and Keys 2003) [this is in contrast to its delicate dependency on cytosolic Pi when GAP is being generated (Walker 2003)]. The exported glycolate is processed by reactions in peroxisomes (extensively reviewed by Igamberdiev and Lea 2002) and mitochondria, such that one molecule each of CO_2 and glycerate are derived from two molecules of glycolate. The CO_2 represents the photorespiratory CO_2 loss, which is difficult to measure and may only be about one-quarter of the initially fixed CO_2 in C_3 plants (Yoshimura et al. 2001), rather than close to one-half as generally suspected.

Much attention is still being given to the roles of the mitochondrion in photorespiration. It is home to the multi-enzyme complex (reviewed by Bauwe and Kolukisaoglu 2003) that converts two glycine (derived from the two glycolate) to one serine in reactions releasing one CO_2, one NH_3 and reducing one NAD^+. The resultant NADH is possibly reoxidised by the plant-specific mitochondrial NADH dehydrogenase located on the matrix side of the inner membrane; unlike other mitochondrial membrane proteins, this one is encoded by a gene whose expression is somehow (but appropriately) light activated (see Møller 2002). Reducing equivalents might then be funnelled to O_2 by the alternative oxidase (also specific to plant mitochondria) that is (also appropriately) activated by the photorespiratory intermediates glyoxylate and hydroxypyruvate (Pastore et al. 2001), and would best serve the purpose of linking photorespiration to the dissipation of excess harvested light energy (Millenaar and Lambers 2003) when leaves are stressed by high light and low water supply, as is common for savannah trees (Franco and Lüttge 2002). Such energy dissipation would occur because this electron transport pathway is 'energy wasteful', i.e. it is not linked to ATP synthesis (Møller 2002). Finally, a remarkable observation is that a *variant* of one of the enzymes in the multi-enzyme complex mentioned above is used by plants for a totally unrelated task: that of catalysing the synthesis of the signalling molecule nitric oxide (NO) which participates in defence against pathogens (Chandok et al. 2003), and which has skyrocketed into significance in mammalian biochemistry during the past decade.

6 A Few Departing Comments About Chloroplasts, the Calvin Cycle's Organelle

Somehow our excitement over chloroplasts in recent years has been less than in days gone by, such as 12 years post-Calvin cycle when the hunt for the first photosynthetically competent isolated chloroplast was on, and the word 'envelope' was adopted as the name for the organelle's pair of surrounding membranes (see Walker 2003). Much has been learnt since that time, including the often-forgotten statistic that chloroplasts occupy a greater volume than the cytosol in a typical leaf cell (Winter et al. 1993), the fact that many biosynthetic pathways (including those for fatty acids, some isoprenoids, and the shikimate pathway) reside beside the Calvin cycle in the chloroplast stroma (see Fig. 3 in our 1998 review), and the existence on the envelope of critical transporter proteins that interconnect the chloroplast's metabolism with the rest of the cell; most recently a transporter for xylulose-5-P has been revealed (Eicks et al. 2002). Now we find ourselves concluding by noting that there is an unusual algal chloroplast that still possesses a cyanobacterial-like peptidoglycan wall (Raven 2003), and that there are regular chloroplasts with odd functions or in unexpected places. Chloroplasts in the outer layer of stems and twigs of woody plants photosynthesise (Schmidt et al. 2000), but the benefit may not so much be sugar synthesis, but rather O_2 supply to the respiration of underlying non-photosynthetic tissues (Pfanz et al. 2002). Perhaps the photosynthesis by these stems and twigs is C_4-like (Hibberd and Quick 2002) and uses CO_2 delivered from root respiration (Fig. 5), in which case (ironically) root respiration would be indirectly helping to supply O_2 for shoot respiration. Finally, another case of chloroplasts being beneficial by photosynthetically generating O_2 for respiration is the developing embryo of the legume seed (Rolletschek et al. 2003).

The most bizarre chloroplasts are algal chloroplasts absconded by marine molluscs. A long-term (months) endosymbiotic association is established between the intracellular photosynthesising chloroplasts (which colour the molluscs green) and the mollusc (Green et al. 2000), with the latter obtaining more than half of its 'dietary' carbohydrate from the former (Raven et al. 2001). More extreme is the recently discovered, non-symbiotic acquisition of diatom chloroplasts by the common benthic foraminifer *Nonionella stella*, in the dark depths of the ocean, probably for the purpose of adding the chloroplast's nitrate reductase to its repertoire of nitrogen-assimilation enzymes (Grzymski et al. 2002).

It may seem strange to finalise these reviews with chloroplasts caught in continuous darkness, but it is symbolic of the bright/dark dichotomy with which we conclude. How absolutely wonderful it must have been, 50 years ago, for the Calvin cycle discoverers to have elucidated the biochemical outline of Earth's CO_2-to-sugars support system. Is it not somewhat naïve, just 50 years later, to connive to twist it to do better? The potential is not great. Walker (1995) has questioned the wisdom of such plans (see our 1998 review), and his wisdom has not been contested. Can we not be pleased with the Calvin cycle as it is, and be content with simply discovering more about its place in the chloroplast, the cell, the photosynthetic organism and the cycles of life on Earth?

Acknowledgement. We thank Vivienne Sofia Kelly (born in Münster) for her exceptionally fastidious typing of the list of 173 references.

References

Andralojc JP, Keys AJ, Kossmann J, Parry MAJ (2002) Elucidating the biosynthesis of 2-carboxyarabinitol 1-phosphate through reduced expression of chloroplastic fructose 1,6-bisphosphate phosphatase and radiotracer studies with $^{14}CO_2$. Proc Natl Acad Sci USA 99:4742–4747

Aoki N, Hirose T, Scofield GN, Whitfeld PR, Furbank RT (2003) The sucrose transporter gene family in rice. Plant Cell Physiol 44:223–232

Ashida H, Saito Y, Kojima C, Kobayashi K, Ogasawara N, Yokota A (2003) A functional link between RuBisCO-like protein of *Bacillus* and photosynthetic RuBisCO. Science 302:286–290

Bakrim N, Brulfert J, Vidal J, Chollet R (2001) Phospho*enol*pyruvate carboxylase kinase is controlled by a similar signaling cascade in CAM and C_4 plants. Biochem Biophys Res Commun 286:1158–1162

Balmer Y, Koller A, del Val G, Manieri W, Schürmann P, Buchanan BB (2003) Proteomics gives insight into the regulatory function of chloroplast thioredoxins. Proc Natl Acad Sci USA 100:370–375

Baroja-Fernández E, Muñoz FJ, Akazawa T, Pozueta-Romero J (2001) Reappraisal of the currently prevailing model of starch biosynthesis in photosynthetic tissues: a proposal involving the cytosolic production of ADP-glucose by sucrose synthase and occurrence of cyclic turnover of starch in the chloroplast. Plant Cell Physiol 42:1311–1320

Bassham JA (2003) Mapping the carbon reduction cycle: a personal retrospective. Photosyn Res 76:35–52

Bassham JA, Benson AA, Kay LD, Harris AZ, Wilson AT, Calvin M (1954) The path of carbon in photosynthesis. 21. The cyclic regeneration of carbon dioxide acceptor. J Am Chem Soc 76:1760–1770

Bauwe H, Kolukisaoglu Ü (2003) Genetic manipulation of glycine decarboxylation. J Exp Bot 54:1523–1535

Baxter CJ, Foyer CH, Turner J, Rolfe SA, Quick WP (2003) Elevated sucrose-phosphate synthase activity in transgenic tobacco sustains photosynthesis in older leaves and alters development. J Exp Bot 54:1813–1820

Benson AA (2002) Following the path of carbon in photosynthesis: a personal story. Photosyn Res 73:29–49

Bernacchi CJ, Portis AR, Nakano H, von Caemmerer S, Long SP (2002) Temperature response of mesophyll conductance. Implications for the determination of Rubisco enzyme kinetics and for limitations to photosynthesis in vivo. Plant Physiol 130:1992–1998

Black CC, Osmond CB (2003) Crassulacean acid metabolism photosynthesis: 'working the night shift'. Photosyn Res 76:329–341

Broetto F, Lüttge U, Ratajczak R (2002) Influence of light intensity and salt-treatment on mode of photosynthesis and enzymes of the antioxidative response system of *Mesembryanthemum crystallinum*. Funct Plant Biol 29:13–23

Brüggemann N, Schnitzler J-P (2002) Comparison of isoprene emission, intercellular isoprene concentration and photosynthetic performance in water-limited oak (*Quercus pubescens* Willd. and *Quercus robur* L.) saplings. Plant Biol 4:456–463

Buitenhuis ET, de Baar HJW, Veldhuis MJW (1999) Photosynthesis and calcification by *Emiliania huxleyi* (Prymnesiophyceae) as a function of inorganic carbon species. J Phycol 35:949–959

Carrari F, Nunes-Nesi A, Gibon Y, Lytovchenko A, Loureiro ME, Fernie AR (2003) Reduced expression of aconitase results in an enhanced rate of photosynthesis and marked shifts in carbon partitioning in illuminated leaves of wild species tomato. Plant Physiol 133:1322–1335

Chandok MR, Ytterberg AJ, van Wijk KJ, Klessig DF (2003) The pathogen-inducible nitric oxide synthase (iNOS) in plants is a variant of the P protein of the glycine decarboxylase complex. Cell 113:469–482

Chastain CJ, Fries JP, Vogel JA, Randklev CL, Vossen AP, Dittmer SK, Watkins EE, Fiedler LJ, Wacker SA, Meinhover KC, Sarath G, Chollet R (2002) Pyruvate, orthophosphate dikinase in leaves and chloroplasts of C_3 plants undergoes light-/dark-induced reversible phosphorylation. Plant Physiol 128:1368–1378

Chinthapalli B, Murmu J, Raghavendra AS (2003) Dramatic difference in the responses of phospho*enol*pyruvate carboxylase to temperature in leaves of C_3 and C_4 plants. J Exp Bot 54:707–714

Chueca A, Sahrawy M, Pagano EA, López Gorge J (2002) Chloroplast fructose-1,6-bisphosphatase: structure and function. Photosyn Res 74:235–249

Ciereszko I, Kleczkowski LA (2002) Glucose and mannose regulate the expression of a major sucrose synthase gene in *Arabidopsis* via hexokinase-dependent mechanisms. Plant Physiol Biochem 40:907–911

Collin V, Issakidis-Bourguet E, Marchand C, Hirasawa M, Lancelin J-M, Knaff DB, Miginiac-Maslow M (2003) The *Arabidopsis* plastidial thioredoxins. New functions and new insights into specificity. J Biol Chem 278:23747–23752

Costa dos Santos A, da-Silva WS, de Meis L, Galina A (2003) Proton transport in maize tonoplasts supported by fructose-1,6-bisphosphate cleavage. Pyrophosphate-dependent phosphofructokinase as a pyrophosphate-regenerating system. Plant Physiol 133:885–892

Crafts-Brandner SJ, Salvucci ME (2002) Sensitivity of photosynthesis in a C_4 plant, maize, to heat stress. Plant Physiol 129:1773–1780

Crevillén P, Ballicora MA, Mérida A, Preiss J, Romero JM (2003) The different large subunit isoforms of *Arabidopsis thaliana* ADP-glucose pyrophosphorylase confer distinct kinetic and regulatory properties to the heterotetrameric enzyme. J Biol Chem 278:28508–28515

Dong L, Ermolova NV, Chollet R (2001) Partial purification and biochemical characterization of a heteromeric protein phosphatase 2A holoenzyme from maize (*Zea mays* L.) leaves that dephosphorylates C_4 phospho*enol*pyruvate carboxylase. Planta 213:379–389

Du Y-C, Peddi SR, Spreitzer RJ (2003) Assessment of structural and functional divergence far from the large subunit active site of ribulose-1,5-bisphosphate carboxylase/oxygenase. J Biol Chem 278:49401–49405

Eastmond PJ, van Dijken AJH, Spielman M, Kerr A, Tissier AF, Dickinson HG, Jones JDG, Smeekens SC, Graham IA (2002) Trehalose-6-phosphate synthase 1, which catalyses the first step in trehalose synthesis, is essential for *Arabidopsis* embryo maturation. Plant J 29:225–235

Echevarría C, Garcia-Mauriño S, Alvarez R, Soler A, Vidal J (2001) Salt stress increases the Ca^{2+}-independent phospho*enol*pyruvate carboxylase kinase activity in *Sorghum* leaves. Planta 214:283–287

Eicks M, Maurino V, Knappe S, Flügge U-I, Fischer K (2002) The plastidic pentose phosphate translocator represents a link between the cytosolic and the plastidic pentose phosphate pathways in plants. Plant Physiol 128:512–522

Emmerlich V, Linka N, Reinhold T, Hurth MA, Traub M, Martinoia E, Neuhaus HE (2003) The plant homolog to the human sodium/dicarboxylic cotransporter is the vacuolar malate carrier. Proc Natl Acad Sci USA 100:11122–11126

Entus R, Poling M, Herrmann KM (2002) Redox regulation of *Arabidopsis* 3-Deoxy-D-*arabino*-heptulosonate 7-phosphate synthase. Plant Physiol 129:1866–1871

Fellermeier M, Sagner S, Spiteller P, Spiteller M, Zenk MH (2003) Early steps of deoxyxylulose phosphate pathway in chromoplasts of higher plants. Phytochemistry 64:199–207

Franco AC, Lüttge U (2002) Midday depression in savanna trees: coordinated adjustments in photochemical efficiency, photorespiration, CO_2 assimilation and water use efficiency. Oecologia 131:356–365

Frommer WB, Schulze WX, Lalonde S (2003) Hexokinase, jack-of-all-trades. Science 300:261–263

Garg AK, Kim J-K, Owens TG, Ranwala AP, Choi YD, Kochian LV, Wu RJ (2002) Trehalose accumulation in rice plants confers high tolerance levels to different abiotic stresses. Proc Natl Acad Sci USA 99:15898–15903

Gerhardt S, Echt S, Busch M, Freigang J, Auerbach G, Bader G, Martin WF, Bacher A, Huber R, Fischer M (2003) Structure and properties of an engineered transketolase from maize. Plant Physiol 132:1941–1949

Ghoshal D, Husic HD, Goyal A (2002) Dissolved inorganic carbon concentration mechanism in *Chlamydomonas moewusii*. Plant Physiol Biochem 40:299–305

Giegé P, Heazlewood JL, Roessner-Tunali U, Millar AH, Fernie AR, Leaver CJ, Sweetlove LJ (2003) Enzymes of glycolysis are functionally associated with the mitochondrion in *Arabidopsis* cells. Plant Cell 15:2140–2151

Gómez-Casati DF, Cortassa S, Aon MA, Iglesias AA (2003) Ultrasensitive behavior in the synthesis of storage polysaccharides in cyanobacteria. Planta 216:969–975
Graciet E, Lebreton S, Camadro J-M, Gontero B (2002) Thermodynamic analysis of the emergence of new regulatory properties in a phosphoribulokinase-glyceraldehyde 3-phosphate dehydrogenase complex. J Biol Chem 277:12697–12702
Green BJ, Li W-Y, Manhart JR, Fox TC, Summer EJ, Kennedy RA, Pierce SK, Rumpho ME (2000) Mollusc–algal chloroplast endosymbiosis. Photosynthesis, thylakoid protein maintenance, and chloroplast gene expression continue for many months in the absence of the algal nucleus. Plant Physiol 124:331–342
Griffiths H, Helliker B, Roberts A, Haslam RP, Girnus J, Robe WE, Borland AM, Maxwell K (2002) Regulation of Rubisco activity in Crassulacean acid metabolism plants: better late than never. Funct Plant Biol 29:689–696
Grzymski J, Schofield OM, Falkowski PG, Bernhard JM (2002) The function of plastids in the deep-sea benthic foraminifer, *Nonionella stella*. Limnol Oceanogr 47:1569–1580
Hafke JB, Hafke Y, Smith JAC, Lüttge U, Thiel G (2003) Vacuolar malate uptake is mediated by an anion-selective inward rectifier. Plant J 35:116–128
Harrington GN, Bush DR (2003) The bifunctional role of hexokinase in metabolism and glucose signaling. Plant Cell 15:2493–2496
Hatch MD, Slack CR (1966) Photosynthesis by sugar-cane leaves. A new carboxylation reaction and the pathway of sugar formation. Biochem J 101:103–111
Hauschild R, von Schaewen A (2003) Differential regulation of glucose-6-phosphate dehydrogenase isoenzyme activities in potato. Plant Physiol 133:47–62
Hellblom F, Axelsson L (2003) External HCO_3^- dehydration maintained by acid zones in the plasma membrane is an important component of the photosynthetic carbon uptake in *Ruppia cirrhosa*. Photosyn Res 77:173–181
Hendriks JHM, Kolbe A, Gibon Y, Stitt M, Geigenberger P (2003) ADP-glucose pyrophosphorylase is activated by posttranslational redox-modification in response to light and to sugars in leaves of *Arabidopsis* and other plant species. Plant Physiol 133:838–849
Herfort L, Thake B, Roberts J (2002) Acquisition and use of bicarbonate by *Emiliania huxleyi*. New Phytol 156:427–436
Herter S, Fuchs G, Bacher A, Eisenreich W (2002) A bicyclic autotrophic CO_2 fixation pathway in *Chloroflexus aurantiacus*. J Biol Chem 277:20277–20283
Hibberd JM, Quick WP (2002) Characteristics of C_4 photosynthesis in stems and petioles of C_3 flowering plants. Nature 415:451–454
Holbrook GP, Keys AJ (2003) Evidence for recycling of inorganic phosphate by wheat chloroplasts during photosynthesis at air levels of CO_2 and O_2. J Plant Physiol 160:1351–1360
Holtum JAM (2002) Crassulacean acid metabolism: plasticity in expression, complexity of control. Funct Plant Biol 29:657–661
Huertas IE, Colman B, Espie GS (2002) Mitochondrial-driven bicarbonate transport supports photosynthesis in a marine microalga. Plant Physiol 130:284–291
Hügler M, Krieger RS, Jahn M, Fuchs G (2003) Characterization of acetyl-CoA/propionyl-CoA carboxylase in *Metallosphaera sedula*. Carboxylating enzyme in the 3-hydroxypropionate cycle for autotrophic carbon fixation. Eur J Biochem 270:736–744
Igamberdiev AU, Lea PJ (2002) The role of peroxisomes in the integration of metabolism and evolutionary diversity of photosynthetic organisms. Phytochemistry 60:651–674
Ishijima S, Uchibori A, Takagi H, Maki R, Ohnishi M (2003) Light-induced increase in free Mg^{2+} concentration in spinach chloroplasts: measurement of free Mg^{2+} by using a fluorescent probe and necessity of stromal alkalinization. Arch Biochem Biophys 412:126–132
Ito H, Iwabuchi M, Ogawa K (2003) The sugar-metabolic enzymes aldolase and triose-phosphate isomerase are targets of glutathionylation in *Arabidopsis thaliana*: detection using biotinylated glutathione. Plant Cell Physiol 44:655–660

Jiang H, Xu D-Q (2001) The cause of the difference in leaf net photosynthetic rate between two soybean cultivars. Photosynthetica 39:453–459

Kanao T, Kawamura M, Fukui T, Atomi H, Imanaka T (2002) Characterization of isocitrate dehydrogenase from the green sulfur bacterium *Chlorobium limicola*. A carbon dioxide-fixing enzyme in the reductive tricarboxylic acid cycle. Eur J Biochem 269:1926–1931

Kiirats O, Lea PJ, Franceschi VR, Edwards GE (2002) Bundle sheath diffusive resistance to CO_2 and effectiveness of C_4 photosynthesis and refixation of photorespired CO_2 in a C_4 cycle mutant and wild-type *Amaranthus edulis*. Plant Physiol 130:964–976

Kubien DS, von Caemmerer S, Furbank RT, Sage RF (2003) C_4 photosynthesis at low temperature. A study using transgenic plants with reduced amounts of Rubisco. Plant Physiol 132:1577–1585

Kühn C (2003) A comparison of the sucrose transporter systems of different plant species. Plant Biol 5:215–232

Laby RJ, Kim D, Gibson SI (2001) The *ram1* mutant of *Arabidopsis* exhibits severely decreased β-amylase activity. Plant Physiol 127:1798–1807

Lara MV, Casati P, Andreo CS (2002) CO_2-concentrating mechanisms in *Egeria densa*, a submersed aquatic plant. Physiol Plant 115:487–495

Law RD, Crafts-Brandner SJ, Salvucci ME (2001) Heat stress induces the synthesis of a new form of ribulose-1,5-bisphosphate carboxylase/oxygenase activase in cotton leaves. Planta 214:117–125

Lebreton S, Graciet E, Gontero B (2003) Modulation, via protein–protein interactions, of glyceraldehyde-3-phosphate dehydrogenase activity through redox phosphoribulokinase regulation. J Biol Chem 278:12078–12084

Lemaire SD, Collin V, Keryer E, Quesada A, Miginiac-Maslow M (2003) Characterization of thioredoxin *y*, a new type of thioredoxin identified in the genome of *Chlamydomonas reinhardtii*. FEBS Lett 543:87–92

Lilley RMcC, Wang XQ, Krausz E, Andrews TJ (2003) Complete spectra of the far-red chemiluminescence of the oxygenase reaction of Mn^{2+}-activated ribulose-bisphosphate carboxylase/oxygenase establish excited Mn^{2+} as the source. J Biol Chem 278:16488–16493

Loreto F, Velikova V, di Marco G (2001) Respiration in the light measured by $^{12}CO_2$ emission in $^{13}CO_2$ atmosphere in maize leaves. Aust J Plant Physiol 28:1103–1108

Lu C, Qiu N, Lu Q, Wang B, Kuang T (2003) PSII photochemistry, thermal energy dissipation, and the xanthophyll cycle in *Kalanchoë daigremontiana* exposed to a combination of water stress and high light. Physiol Plant 118:173–182

Lunn JE, Gillespie VJ, Furbank RT (2003) Expression of a cyanobacterial sucrose-phosphate synthase from *Synechocystis* sp. PCC 6803 in transgenic plants. J Exp Bot 54:223–237

Lüttge U (2002) CO_2-concentrating: consequences in Crassulacean acid metabolism. J Exp Bot 53:2131–2142

Lytovchenko A, Fernie AR (2003) Photosynthetic metabolism is severely impaired on the parallel reduction of plastidial and cytosolic isoforms of phosphoglucomutase. Plant Physiol Biochem 41:193–200

Lytovchenko A, Sweetlove L, Pauly M, Fernie AR (2002) The influence of cytosolic phosphoglucomutase on photosynthetic carbohydrate metabolism. Planta 215:1013–1021

Mamedov TG, Suzuki K, Miura K, Kucho K-i, Fukuzawa H (2001) Characteristics and sequence of phosphoglycolate phosphatase from a eukaryotic green alga *Chlamydomonas reinhardtii*. J Biol Chem 276:45573–45579

Markham JE, Kruger NJ (2002) Kinetic properties of bifunctional 6-phosphofructo-2-kinase/fructose-2,6-bisphosphatase from spinach leaves. Eur J Biochem 269:1267–1277

Martinsuo P, Pursiheimo S, Aro E-M, Rintamäki E (2003) Dithiol oxidant and disulfide reductant dynamically regulate the phosphorylation of light-harvesting complex II proteins in thylakoid membranes. Plant Physiol 133:37–46

McRae SR, Christopher JT, Smith JAC, Holtum JAM (2002) Sucrose transport across the vacuolar membrane of *Ananas comosus*. Funct Plant Biol 29:717–724

Millenaar FF, Lambers H (2003) The alternative oxidase: in vivo regulation and function. Plant Biol 5:2–15

Milligan AJ, Morel FMM (2002) A proton buffering role for silica in diatoms. Science 297:1848–1850

Miyao M (2003) Molecular evolution and genetic engineering of C_4 photosynthetic enzymes. J Exp Bot 54:179–189

Møller IM (2002) A new dawn for plant mitochondrial NAD(P)H dehydrogenases. Trends Plant Sci 7:235–237

Moore B, Zhou L, Rolland F, Hall Q, Cheng W-H, Liu Y-X, Hwang I, Jones T, Sheen J (2003) Role of the *Arabidopsis* glucose sensor HXK1 in nutrient, light, and hormonal signaling. Science 300:332–336

Moroney JV, Bartlett SG, Samuelsson G (2001) Carbonic anhydrases in plants and algae. Plant Cell Environ 24:141–153

Murchie EH, Yang J, Hubbart S, Horton P, Peng S (2002) Are there associations between grain-filling rate and photosynthesis in the flag leaves of field-grown rice? J Exp Bot 53:2217–2224

Naidu SL, Moose SP, AL-Shoaibi AK, Raines CA, Long SP (2003) Cold tolerance of C_4 photosynthesis in *Miscanthus* × *giganteus*: adaptation in amounts and sequence of C_4 photosynthetic enzymes. Plant Physiol 132:1688–1697

Nielsen TH, Stitt M (2001) Tobacco transformants with strongly decreased expression of pyrophosphate:fructose-6-phosphate 1-phosphotransferase expression in the base of their young growing leaves contain much higher levels of fructose-2,6-bisphosphate but no major changes in fluxes. Planta 214:106–116

Niittylä T, Messerli G, Trevisan M, Chen J, Smith AM, Zeeman SC (2004) A previously unknown maltose transporter essential for starch degradation in leaves. Science 303:87–89

Ocón E, Hermoso R, Chueca A, Lázaro JJ (2001) Cross-linking between chloroplast fructose-1,6-bisphosphatase and thioredoxin *f*. Physiol Plant 113:452–460

Ogle K (2003) Implications of interveinal distance for quantum yield in C_4 grasses: a modeling and meta-analysis. Oecologia 136:532–542

Ogren WL (2003) Affixing the O to Rubisco: discovering the source of photorespiratory glycolate and its regulation. Photosyn Res 76:53–63

Okano Y, Mizohata E, Xie Y, Matsumura H, Sugawara H, Inoue T, Yokota A, Kai Y (2002) X-ray structure of *Galdieria* Rubisco complexed with one sulfate ion per active site. FEBS Lett 527:33–36

Olsson T, Thelander M, Ronne H (2003) A novel type of chloroplast stromal hexokinase is the major glucose-phosphorylating enzyme in the moss *Physcomitrella patens*. J Biol Chem 278:44439–44447

Pagnussat GC, Fiol DF, Salerno GL (2002) A CDPK type protein kinase is involved in rice SPS light modulation. Physiol Plant 115:183–189

Parry MAJ, Andralojc PJ, Mitchell RAC, Madgwick PJ, Keys AJ (2003) Manipulation of Rubisco: the amount, activity, function and regulation. J Exp Bot 54:1321–1333

Pastore D, Trono D, Laus MN, di Fonzo N, Passarella S (2001) Alternative oxidase in durum wheat mitochondria. Activation by pyruvate, hydroxypyruvate and glyoxylate and physiological role. Plant Cell Physiol 42:1373–1382

Peisker M, Apel H (2001) Inhibition by light of CO_2 evolution from dark respiration: comparison of two gas exchange methods. Photosyn Res 70:291–298

Penna S (2003) Building stress tolerance through over-producing trehalose in transgenic plants. Trends Plant Sci 8:355–357

Peñuelas J, Llusià J (2002) Linking photorespiration, monoterpenes and thermotolerance in *Quercus*. New Phytol 155:227–237

Pfanz H, Aschan G, Langenfeld-Heyser R, Wittmann C, Loose M (2002) Ecology and ecophysiology of tree stems: corticular and wood photosynthesis. Naturwissenschaften 89:147–162

Pierce S, Winter K, Griffiths H (2002) The role of CAM in high rainfall cloud forests: an in situ comparison of photosynthetic pathways in Bromeliaceae. Plant Cell Environ 25:1181–1189

Pinelli P, Loreto F (2003) $^{12}CO_2$ emission from different metabolic pathways measured in illuminated and darkened C_3 and C_4 leaves at low, atmospheric and elevated CO_2 concentration. J Exp Bot 54:1761–1769

Porchia AC, Fiol DF, Salerno GL (1999) Differential synthesis of sucrose and trehalose in *Euglena gracilis* cells during growth and salt stress. Plant Sci 149:43–49

Portis AR (2003) Rubisco activase – Rubisco's catalytic chaperone. Photosyn Res 75:11–27

Price GD, Badger MR (2002) Advances in understanding how aquatic photosynthetic organisms utilize sources of dissolved inorganic carbon for CO_2 fixation. Funct Plant Biol 29:117–121

Qi J, Isupov MN, Littlechild JA, Anderson LE (2001) Chloroplast glyceraldehyde-3-phosphate dehydrogenase contains a single disulfide bond located in the C-terminal extension to the B subunit. J Biol Chem 276:35247–35252

Raghavendra AS, Padmasree K (2003) Beneficial interactions of mitochondrial metabolism with photosynthetic carbon assimilation. Trends Plant Sci 8:546–553

Raines CA (2003) The Calvin cycle revisited. Photosyn Res 75:1–10

Raven JA (2002) Evolutionary options. Nature 415:375–377

Raven JA (2003) Carboxysomes and peptidoglycan walls of cyanelles: possible physiological functions. Eur J Phycol 38:47–53

Raven JA, Girard-Bascou J (2001) Algal model systems and the elucidation of photosynthetic metabolism. J Phycol 37:943–950

Raven JA, Walker DI, Jensen KR, Handley LL, Scrimgeour CM, McInroy SG (2001) What fraction of the organic carbon in sacoglossans is obtained from photosynthesis by kleptoplastids? An investigation using the natural abundance of stable carbon isotopes. Mar Biol 138:537–545

Regierer B, Fernie AR, Springer F, Perez-Melis A, Leisse A, Koehl K, Willmitzer L, Geigenberger P, Kossmann J (2002) Starch content and yield increase as a result of altering adenylate pools in transgenic plants. Nature Biotech 20:1256–1260

Ritte G, Lloyd JR, Eckermann N, Rottmann A, Kossmann J, Steup M (2002) The starch-related R1 protein is an α-glucan, water dikinase. Proc Natl Acad Sci USA 99:7166–7171

Robinson JJ, Scott KM, Swanson ST, O'Leary MH, Horken K, Tabita FR, Cavanaugh CM (2003) Kinetic isotope effect and characterization of form II RubisCO from the chemoautotrophic endosymbionts of the hydrothermal vent tubeworm *Riftia pachyptila*. Limnol Oceanogr 48:48–54

Rolletschek H, Weber H, Borisjuk L (2003) Energy status and its control on embryogenesis of legumes. Embryo photosynthesis contributes to oxygen supply and is coupled to biosynthetic fluxes. Plant Physiol 132:1196–1206

Rosenstiel TN, Fisher AJ, Fall R, Monson RK (2002) Differential accumulation of dimethylallyl diphosphate in leaves and needles of isoprene- and methylbutenol-emitting and nonemitting species. Plant Physiol 129:1276–1284

Rost B, Riebesell U, Burkhardt S, Sültemeyer D (2003) Carbon acquisition of bloom-forming marine phytoplankton. Limnol Oceanogr 48:55–67

Sage RF (2002) Variation in the k_{cat} of Rubisco in C_3 and C_4 plants and some implications for photosynthetic performance at high and low temperature. J Exp Bot 53:609–620

Sage RF, Sage TL (2002) Microsite characteristics of *Muhlenbergia richardsonis* (Trin.) Rydb., an alpine C_4 grass from the White Mountains, California. Oecologia 132:501–508

Sainis JK, Dani DN, Dey GK (2003) Involvement of thylakoid membranes in supramolecular organization of Calvin cycle enzymes in *Anacystis nidulans*. J Plant Physiol 160:23–32

Salamone PR, Kavakli IH, Slattery CJ, Okita TW (2002) Directed molecular evolution of ADP-glucose pyrophosphorylase. Proc Natl Acad Sci USA 99:1070–1075

Salerno GL, Curatti L (2003) Origin of sucrose metabolism in higher plants: when, how and why? Trends Plant Sci 8:63–69

Sánchez R, Cejudo FJ (2003) Identification and expression analysis of a gene encoding a bacterial-type phospho*enol*pyruvate carboxylase from *Arabidopsis* and rice. Plant Physiol 132:949–957

Sayed OH (2001) Crassulacean acid metabolism 1975–2000, a check list. Photosynthetica 39:339–352

Saze H, Ueno Y, Hisabori T, Hayashi H, Izui K (2001) Thioredoxin-mediated reductive activation of a protein kinase for the regulatory phosphorylation of C_4-form phosphoenolpyruvate carboxylase from maize. Plant Cell Physiol 42:1295–1302

Scheibe R, Wedel N, Vetter S, Emmerlich V, Sauermann S-M (2002) Co-existence of two regulatory NADP-glyceraldehyde 3-P dehydrogenase complexes in higher plant chloroplasts. Eur J Biochem 269:5617–5624

Scheidig A, Fröhlich A, Schulze S, Lloyd JR, Kossmann J (2002) Downregulation of a chloroplast-targeted β-amylase leads to a starch-excess phenotype in leaves. Plant J 30:581–591

Schloss JV (2002) Oxygen toxicity from plants to people. Planta 216:38–43

Schluepmann H, Pellny T, van Dijken A, Smeekens S, Paul M (2003) Trehalose 6-phosphate is indispensable for carbohydrate utilization and growth in *Arabidopsis thaliana*. Proc Natl Acad Sci USA 100:6849–6854

Schmidt J, Batic F, Pfanz H (2000) Photosynthetic performance of leaves and twigs of evergreen holly (*Ilex aquifolium* L.). Phyton 40:179–190

Schneider A, Häusler RE, Kolukisaoglu Ü, Kunze R, van der Graaff E, Schwacke R, Catoni E, Desimone M, Flügge U-I (2002) An *Arabidopsis thaliana* knock-out mutant of the chloroplast triose phosphate/phosphate translocator is severely compromised only when starch synthesis, but not starch mobilisation is abolished. Plant J 32:685–699

Servaites JC, Geiger DR (2002) Kinetic characteristics of chloroplast glucose transport. J Exp Bot 53:1581–1591

Sharkey TD (1996) Isoprene synthesis by plants and animals. Endeavour 20:74–78

Sinha AK, Hofmann MG, Römer U, Köckenberger W, Elling L, Roitsch T (2002) Metabolizable and non-metabolizable sugars activate different signal transduction pathways in tomato. Plant Physiol 128:1480–1489

Smidansky ED, Clancy M, Meyer FD, Lanning SP, Blake NK, Talbert LE, Giroux MJ (2002) Enhanced ADP-glucose pyrophosphorylase activity in wheat endosperm increases seed yield. Proc Natl Acad Sci USA 99:1724–1729

Smidansky ED, Martin JM, Hannah LC, Fischer AM, Giroux MJ (2003) Seed yield and plant biomass increases in rice are conferred by deregulation of endosperm ADP-glucose pyrophosphorylase. Planta 216:656–664

Snoeijs P, Klenell M, Choo KS, Comhaire I, Ray S, Pedersén M (2002) Strategies for carbon acquisition in the red marine macroalga *Coccotylus truncatus* from the Baltic Sea. Mar Biol 140:435–444

Sparla F, Pupillo P, Trost P (2002) The C-terminal extension of glyceraldehyde-3-phosphate dehydrogenase subunit B acts as an autoinhibitory domain regulated by thioredoxins and nicotinamide adenine dinucleotide. J Biol Chem 277:44946–44952

Spreitzer RJ, Salvucci ME (2002) Rubisco: structure, regulatory interactions, and possibilities for a better enzyme. Annu Rev Plant Biol 53:449–475

Strand Å, Foyer CH, Gustafsson P, Gardestrom P, Hurry V (2003) Altering flux through the sucrose biosynthesis pathway in transgenic *Arabidopsis thaliana* modifies photosynthetic acclimation at low temperatures and the development of freezing tolerance. Plant Cell Environ 26:523–535

Sudo E, Makino A, Mae T (2003) Differences between rice and wheat in ribulose-1,5-bisphosphate regeneration capacity per unit of leaf-N content. Plant Cell Environ 26:255–263

Tchernov D, Silverman J, Luz B, Reinhold L, Kaplan A (2003) Massive light-dependent cycling of inorganic carbon between oxygenic photosynthetic microorganisms and their surroundings. Photosyn Res 77:95–103

Terashima I, Ono K (2002) Effects of $HgCl_2$ on CO_2 dependence of leaf photosynthesis: evidence indicating involvement of aquaporins in CO_2 diffusion across the plasma membrane. Plant Cell Physiol 43:70–78

Tiessen A, Hendriks JHM, Stitt M, Branscheid A, Gibon Y, Farré EM, Geigenberger P (2002) Starch synthesis in potato tubers is regulated by post-translational redox modification of ADP-glucose pyrophosphorylase: a novel regulatory mechanism linking starch synthesis to the sucrose supply. Plant Cell 14:2191–2213

Tiessen A, Prescha K, Branscheid A, Palacios N, McKibbin R, Halford NG, Geigenberger P (2003) Evidence that SNF1-related kinase and hexokinase are involved in separate sugar-signalling pathways modulating post-translational redox activation of ADP-glucose pyrophosphorylase in potato tubers. Plant J 35:490–500

Tortell PD, Morel FMM (2002) Sources of inorganic carbon for phytoplankton in the eastern subtropical and equatorial Pacific Ocean. Limnol Oceanogr 47:1012–1022

Trevanion SJ (2002) Regulation of sucrose and starch synthesis in wheat (*Triticum aestivum* L.) leaves: role of fructose 2,6-bisphosphate. Planta 215:653–665

Uehlein N, Lovisolo C, Siefritz F, Kaldenhoff R (2003) The tobacco aquaporin NtAQP1 is a membrane CO_2 pore with physiological functions. Nature 425:734–737

Van Heerden PDR, Krüger GHJ, Loveland JE, Parry MAJ, Foyer CH (2003) Dark chilling imposes metabolic restrictions on photosynthesis in soybean. Plant Cell Environ 26:323–337

Vogel G, Fiehn O, Jean-Richard-dit-Bressel L, Boller T, Wiemken A, Aeschbacher RA, Wingler A (2001) Trehalose metabolism in *Arabidopsis*: occurrence of trehalose and molecular cloning and characterization of trehalose-6-phosphate synthase homologues. J Exp Bot 52:1817–1826

von Caemmerer S (2003) C_4 photosynthesis in a single C_3 cell is theoretically inefficient but may ameliorate internal CO_2 diffusion limitations of C_3 leaves. Plant Cell Environ 26:1191–1197

von Caemmerer S, Furbank RT (2003) The C_4 pathway: an efficient CO_2 pump. Photosyn Res 77:191–207

Voznesenskaya EV, Franceschi VR, Kiirats O, Artyusheva EG, Freitag H, Edwards GE (2002) Proof of C_4 photosynthesis without Kranz anatomy in *Bienertia cycloptera* (Chenopodiaceae). Plant J 31:649–662

Voznesenskaya EV, Franceschi VR, Artyusheva EG, Black CC, Pyankov VI, Edwards GE (2003) Development of the C_4 photosynthetic apparatus in cotyledons and leaves of *Salsola richteri* (Chenopodiaceae). Int J Plant Sci 164:471–487

Walker DA (1995) Manipulating photosynthetic metabolism to improve crops: an inversion of ends and means. J Exp Bot 46:1253–1259

Walker DA (2003) Chloroplasts in envelopes: CO_2 fixation by fully functional intact chloroplasts. Photosyn Res 76:319–327

Wang Y-L, Zhou J-H, Wang Y-F, Bao J-S, Chen H-B (2001) Properties of hybrid enzymes between *Synechococcus* large subunits and higher plant small subunits of ribulose-1,5-bisphosphate carboxylase/oxygenase in *Escherichia coli*. Arch Biochem Biophys 396:35–42

Whitney SM, Andrews TJ (2001) Plastome-encoded bacterial ribulose-1,5-bisphosphate carboxylase/oxygenase (RubisCO) supports photosynthesis and growth in tobacco. Proc Natl Acad Sci USA 98:14738–14743

Whitney SM, Andrews TJ (2003) Photosynthesis and growth of tobacco with a substituted bacterial Rubisco mirror the properties of the introduced enzyme. Plant Physiol 133:287–294

Wingler A (2002) The function of trehalose biosynthesis in plants. Phytochemistry 60:437–440

Winter K, Holtum JAM (2002) How closely do the $\delta^{13}C$ values of Crassulacean acid metabolism plants reflect the proportion of CO_2 fixed during day and night? Plant Physiol 129:1843–1851

Winter H, Robinson DG, Heldt HW (1993) Subcellular volumes and metabolite concentrations in barley leaves. Planta 191:180–190
Wyka TP, Lüttge UE (2003) Contribution of C_3 carboxylation to the circadian rhythm of carbon dioxide uptake in a Crassulacean acid metabolism plant *Kalanchoë daigremontiana*. J Exp Bot 54:1471–1479
Xiao W, Sheen J, Jang J-C (2000) The role of hexokinase in plant sugar signal transduction and growth and development. Plant Mol Biol 44:451–461
Yoshimura Y, Kubota F, Hirao K (2001) Estimation of photorespiration rate by simultaneous measurements of CO_2, gas exchange rate, and chlorophyll fluorescence quenching in the C_3 plant *Vigna radiata* (L.) Wilczek and the C_4 plant *Amaranthus mongostanus* L. Photosynthetica 39:377–382
Zeeman SC, Tiessen A, Pilling E, Kato KL, Donald AM, Smith AM (2002) Starch synthesis in *Arabidopsis*. Granule synthesis, composition, and structure. Plant Physiol 129:516–529
Zhang N, Kallis RP, Ewy RG, Portis AR (2002) Light modulation of Rubisco in *Arabidopsis* requires a capacity for redox regulation of the larger Rubisco activase isoform. Proc Natl Acad Sci USA 99:3330–3334
Zhou R, Cheng L, Wayne R (2003) Purification and characterization of sorbitol-6-phosphate phosphatase from apple leaves. Plant Sci 165:227–232

Grahame J. Kelly
School of Life Sciences
Queensland University of Technology
Brisbane, Queensland 4000, Australia

Physiology of Ectomycorrhiza (ECM)

Mika Tarkka, Uwe Nehls, and Rüdiger Hampp

Since our last progress report, investigations into ectomycorrhizal development and functioning have focused more and more on partner-specific regulation of gene expression (for recent reviews see, e.g., Tagu et al. 2000; Wiemken and Boller 2002). In addition, significant progress has been achieved with regard to both the physiology of ectomycorrhizal fungi (for reviews see Nehls et al. 2001; Javelle et al. 2004) and plant–fungus signalling (for a review see Martin et al. 2001). With regard to carbon allocation toward the fungal partner, sources other than host photosynthesis have become increasingly obvious.

1 Development of Ectomycorrhiza (ECM)

The developmental regulation of ECM formation has been investigated from the sensing of the host to the changes in hyphal growth pattern. Key regulating substances and putative fungal ECM-related signalling genes have been recently characterized.

1.1 Signalling Compounds and Genes

The substances leading the growth of fungal hyphae towards plant roots are largely unknown. A specifically affecting chemotropic substance, the flavonoid rutin, has been isolated from eucalypt root exudates (Lagrange et al. 2001). Rutin induces the growth of certain *Pisolithus tinctorius* strains at concentrations as low as 1 pM in Petri dish cultures.

The production of root hairs is mostly halted during the formation of ECM. Hypaphorine, an indolic compound from *Pisolithus tinctorius*, is an auxin analogue that suppresses root hair elongation (Ditengou et al. 2000). Hyphaphorine effect on root hairs can be restored by indole-3-acetic acid, indicating that they bind to the same targets. This was proved by showing

Progress in Botany, Vol. 66

that horseradish peroxidase C serves as a common target protein for both hypaphorine and indole-3-acetic acid (Kawano et al. 2001).

The early changes in fungal gene expression, due to non-characterized plant-derived substances, have been studied in *Laccaria bicolor* hyphae growing in suspension culture around red pine roots. Kim et al. (1999a) isolated several differentially expressed genes from *L. bicolor*. One of the isolated genes, *Lbaut7*, whose homologue in budding yeast functions as an adapter between microtubules and autophagosomes, was found to be upregulated in 6 h of interaction (Kim et al. 1999b). The results indicate that the autophagocytosis could be activated during the early phase of ECM formation, possibly to facilitate nutrient transfer processes. For efficient transport of substrates, fungal hyphae use peristaltic movements of vacuoles. The movement of material is probably regulated by dolipore septa, since there often appears a change in the rate of transfer between two fungal cells (Ashford et al. 1999). Microtubules regulate vacuole motility in the hyphae of *P. tinctorius* (Hyde et al. 1999), and LbAut7 may function as an adapter between the vacuoles and microtubule cytoskeleton.

The root and fungal morphogenesis that leads to the development of ECM is backed up by changes in signalling pathways. It has been discovered that the small GTPases from the ras superfamily are important regulators of hyphal morphology and vesicle transport in the fungal hyphae. As an upstream regulator, a *ras* homologue was isolated from *L. bicolor* by Sundaram et al. (2001). The expression of the *Lbras* gene is upregulated after 48 h incubation in suspension culture both around pine roots and in developed ectomycorrhizal tissues. The functionality of the LbRas protein was verified in budding yeast and fibroblast cell cultures. On the contrary, the *ras* homologues of *Suillus bovinus*, *Sbras1* and *Sbras2* are constitutively expressed in vegetative and symbiotic hyphae (Raudaskoski et al. 2001). LbRas was localized to the fungal plasma membrane in symbiotic tissues (Sundaram et al. 2001).

The morphology of hyphae in the ectomycorrhizal mantle and the Hartig net is completely altered from that of pure culture hyphae. The symbiotic hyphae often swell and branch and grow in a finger-like form. These hyphae have a reorganized cytoskeleton. The reorganization of hyphal microtubules and microfilaments does not involve differential regulation of tubulins or actins (Tarkka et al. 2000). To study the regulation of the hyphal morphogenesis, a *Cdc42* homologue was isolated from *Suillus bovinus* (Gorfer et al. 2001). Cdc42 is a small ras-superfamily GTP-binding protein that, among its other functions, controls the organization of actin cytoskeleton in eukaryotes. SbCdc42 was co-localized with the actin cytoskeleton on the plasma membrane of the growing tips and swelling regions of the fungal hyphae, indicating a role for this signalling protein in fungal morphogenesis. The small GTP-binding proteins also regulate vesicle and endosome dynamics in hyphal cells. The vacuolar tubular system of *P. tinctorius* forms more rapidly when an activator of the small GTPases is applied to the culture medium (Hyde et al. 2002).

The differentiation of fruit bodies involves a complex morphogenetic program. Usually, at the maturation of fruit bodies, fungal cells show altered, swollen morphology. Therefore, similar changes in fungal gene expression pattern could be expected during either fruit body or ECM formation. Nehls et al. (1999a) identified two genes that are transcribed similarly in ECM and fruit bodies. One of them appeared to be 50-fold downregulated, the other one 30-fold upregulated during both developmental processes. The second, upregulated gene showed vague homology to plant extensins, indicating that it may encode a fungal cell wall protein.

1.2 Hormones

Dichotomous branching of pine short roots is under the control of the plant hormones auxin and ethylene. In some pine species, such as *Pinus pinaster*, short root branching is a spontaneous developmental process, while in Scots pine it takes place during the establishment of ECM. Kaska et al. (1999) used split root cultures for a detailed analysis of the role of different plant hormones in the tip bifurcation process. Auxin transport inhibitors (ATI) and ethylene releasing compounds were shown to activate the tip branching. The spontaneous and ATI-related tip branching could both be suppressed by an ethylene synthesis inhibitor, indicating the necessary role for ethylene in this process. High nutrient levels were also able to inhibit root bifurcation.

Mycorrhizal infection causes changes in polarity of the cells in the pine root cortex in that the size of the cortical cells is reduced radially and increased axially (Laurans et al. 2001). The root cortical cells change more markedly during the interaction with an auxin overproducer strain of *H. cylindrosporum*. A genetic screen for auxin-relatedly expressed genes in *P. pinaster* has yielded several other genes, such as a peroxidase gene, the expression of which is upregulated (Charvet-Candela et al. 2002a). The transcription factor homologue *Pp-iaa88* is upregulated as soon as the fungal hyphae reach the pine root surface, and its expression increases during ECM establishment (Charvet-Candela et al. 2002b). The predicted amino acid sequence of *Pp-C61*, on the other hand, shows no homology to known proteins. *Pp-C61* is upregulated by auxin, ethylene, abscisic acid and in the ECM, indicating a complex form of transcriptional control (Reddy et al. 2003).

For many plant species, the induction of somatic embryos and their subsequent rooting has proved to be difficult. Ectomycorrhizal fungi have been investigated as inductors for embryo germination and root induction. *Pisolithus tinctorius* enhances Scots pine somatic embryo germination in vitro (Niemi and Haggman 2002). This induction succeeds only in the absence of direct contact between the organisms. *Paxillus involutus* and *P. tinctorius* were shown to accelerate root induction and subsequent root growth of Scots pine somatic embryos (Niemi et al. 2002b). Certain polyamines showed synergistic effects with the fungi in this respect. The induced rooting of fascicular shoots by the same fungi could also not be explained as a sheer auxin effect, since in vitro production rates of indole acetic acid (IAA) by the two fungi did not correlate with rooting responses (Niemi et al. 2000). The polyamine ratios of *P. involutus* differ during the ageing of mycelium, and there exists a large release of polyamines from mycelia in an age-dependent manner (Fornale et al. 1999). The production rate of the in vitro rooting-effective polyamine, putrescine, was also shown to be age-dependent, indicating a need for optimization of growth conditions for optimal rooting conditions (Niemi et al. 2002a). Polyamine contents of mycorrhizas were unfortunately not addressed.

1.3 Fungal Cell Wall Composition

The composition of cell wall polysaccharides of ectomycorrhizal fungi is poorly known. The neutral exopolysaccharides of *Thelephora terrestris* in pure culture were shown to consist of 1–6-linked alpha-d-mannose-P units with a variety of side chains, beta (1–3) and beta-1–6)-glucans, and a polysaccharide composed from mannose, galactose, fucose and xylose (Osaku et al. 2002).

From the hyphal cell walls, the most investigated group of proteins have been hydrophobins. These small cysteine-rich proteins cover the hyphal cells and form rodlets on aerial hyphae. Hydrophobins are involved in the adhesion and aggregation of hyphae, and have been shown to be virulence factors in some plant pathogenic fungi. In *P. tinctorius*, the HydPt-1 hydrophobin localizes on the hyphal surfaces in vegetative and symbiotic hyphae (Tagu et al. 2001), indicating a similar function for hydrophobins in *P. tinctorius* as in other studied fungi. One of the *P. tinctorius* hydrophobin genes, *hydPt-3*, is upregulated during the formation of ECM (Duplessis et al. 2001). The expression of *hydPt-3* is not modulated by carbon or nitrogen, indicating that other, possibly growth pattern-related factors are involved in its regulation. A hydrophobin protein and the corresponding gene have also been isolated from the fungus *Tricholoma terreum*. The transcription of the *T. terreum hyd1* gene is stronger in Scots pine ECMs than in the hyphae on the non-host Norway spruce, and the *T. terreum* hydrophobin protein was localized in aerial, mantle and Hartig net hyphae (Mankel et al. 2002).

During the formation of *Pisolithus tinctorius* ECM, several polypeptides in the apparent MW range of 30 kDa are upregulated. The cDNAs corresponding to six of these proteins have been isolated by Laurent et al. (1999). Interestingly, the predicted PtSRAP amino acid sequences include an RGD motif, common to cell adhesion proteins in animal cells. Since the expression of the PtSRAP genes is drastically upregulated during mantle formation, they may be involved in hyphal aggregation.

1.4 Oxidases and Defence

It has been suggested that a short induction of plant defence exists at the stage of hyphal penetration into the root. Suspension culture studies have shown that Norway spruce cells are able to respond to fungal elicitors from ECM fungi. Concomitantly a short pulse of hydrogen peroxide is produced and followed by apparent changes in protein phosphorylation patterns. These responses, which are similar to those taking place during the early response against plant pathogens, have been shown to be backed up by the activation of G-protein signalling (Hebe et al. 1999).

The value of the suspension culture studies has to be tested with living organisms. As an informative approach, the expression of a putative defensin homologue from Norway spruce was analysed. The defensin protein is accumulated as soon as the spruce root pathogen *Heterobasidion annosum* infects the root system, but not during the infection with *L. bicolor*, indicating a differential response already at the early stage of infection (Fossdal et al. 2003). To further analyse the expression of the defensin gene, reporter studies would be needed.

The oxidative cross-linking of plant cell wall through secreted peroxidases is an important way of regulating cell wall rigidity. The regulation of the peroxidase family in pine has received some attention. Short roots are the predominant root type for ECM formation in pine, and the proteome of short roots differs from the other root types. Nine short-root-specific proteins have been identified as peroxidases, and a gene encoding one of them, *Psyp1*, has been isolated (Tarkka et al. 2001). *Psyp1* expression is downregulated during mycorrhization. This could suggest a change in the cross-linking status of the plant cell wall, facilitating the growth of fungal hyphae inside the root. Interestingly, during the genetic screen of auxin-treated *Pinus pinaster* seedlings, a peroxidase clone *PpPrx75* was isolated

(Charvet-Candela et al. 2002b). Its expression rate during the short root or ECM formation was not assessed, but it was shown to be upregulated by auxins, ethylene, abscisic acid and the plant flavonone quercetin.

During the exposure to environmental stresses, plants and fungal hyphae produce reactive oxygen species which lead to changes in gene expression. For example, the overall peroxidase activity of Scots pine fine roots is higher in polluted than in non-polluted soils (Markkola et al. 2002). *P. involutus* hyphae are able to take up cadmium and bind the heavy metal on cell walls, or transport it into the cytoplasm or vacuole (Blaudez et al. 2000). Cadmium stress has an inducing effect on superoxide dismutase (SOD) activity and gene expression in *P. involutus* (Jacob et al. 2001; Ott et al. 2002). The antioxidative system of *P. involutus* was investigated under cadmium stress and glutathione was shown to be the potential oxidative stress-responsive antioxidant. In contrast to glutathione, no ascorbate was detected in the stressed or unstressed hyphae. The hydrogen peroxide scavengers catalase and glutathione-dependent peroxidase did not significantly change their activities during Cd application (Ott et al. 2002).

2 Ectomycorrhizal Transcriptome

Differential gene expression during ectomycorrhizal development and function has been investigated intensely over the past decade (Kim et al. 1999a; Nehls et al. 1999a; Voiblet et al. 2001; Polidori et al. 2002), revealing a first glance of events physiologically important for the development and function of this symbiosis (for a review see Tagu et al. 2000). These investigations have enabled the formulation of different working hypotheses that could explain structural and functional adaptations observed during plant–fungal interactions (Tagu et al. 1998; Martin et al. 2001; Nehls et al. 2001; Javelle et al. 2004; Nehls 2004). Nevertheless, all hypotheses are based on the rather small number of genes that have been investigated. To overcome this limitation, a number of expressed sequence tag (EST) projects have been initiated using a small set of ectomycorrhizal plant (*Betula*, A. Tunlid, pers. comm.; *Populus*, Sterky et al. 1998; Kohler et al. 2003) and fungal (*Paxillus*: Le Quere et al. 2002; A. Tunlid, pers. comm.; *Laccaria*: Podila et al. 2002; Peter et al. 2003; *Amanita*: U. Nehls et al., unpubl.; *Hebeloma*: Wipf et al. 2002, C. Plassard, pers. comm.; *Tuber*: Lacourt et al. 2002) systems. Typical for EST projects of filamentous fungi is that a very large number of ESTs (50–65%) encode sequences representing unknown genes (Skinner et al. 2001), compared to only 20–25 % in the case of plant ESTs (Ronning et al. 2003).

These EST sets are meanwhile used to obtain a more global overview of the modulation of plant and fungal gene expression during ectomycorrhizal development. The first ECM micro-array analysis was conducted by hybridizing 850 cDNAs from 4-day-old mycorrhizas of *Eucalyptus globulus* with *Pisolithus tinctorius* with the host, fungus and mycorrhizal cDNA pools (Voiblet et al. 2001). Significantly different expression levels were detected for 17% of the genes investigated, but none of them showed an ECM-specific expression pattern.

In addition to large EST collections, genome projects have been initiated at the fungal site for *Laccaria bicolor* (Lammers et al. 2003, finished 2004) and at the plant site for *Populus trichocarpa* (News in brief 2002, finished in 2003). In a few years from now, this genome information will enable the generation of genome-wide micro-arrays, and a detailed picture of plant and fungal gene expression during ectomycorrhizal development and function will thus become available.

3 Carbohydrate and Nitrogen Relationships in Ectomycorrhiza

3.1 Host Photosynthesis and Carbon Cycling

The basis for mycorrhizal functioning is the supply of sugar by the host plant. This is mainly assayed by experiments where the host plant is fed with labeled (^{14}C) CO_2, and the transfer of label to fungal hyphae is monitored.

Wu et al. (2002) determined ^{14}C transfer from *Pinus densiflora* seedlings to extraradical mycelia, generating root autoradiographs with an imaging plate. In a time course experiment they detected label in shoots, roots, mycorrhizas and extraradical hyphae within 1 day. After 3 days a maximum in radiation was determined in underground parts. Mycorrhizal (myc) roots accumulated 2.6 times more label than non-myc controls, mycorrhizas 5.2 times more than non-myc fine roots. Thirteen days after the pulse treatment extraradical mycelia contained 24% of the total label, which proves the high sink strength of these structures for photoassimilates.

A digital autoradiographic technique that allows for a simultaneous visualization and quantification of spatial and temporal changes was used by Leake et al. (2001) to study carbon allocation patterns in ectomycorrhizal mycelia. In order to analyze the effect of litter patches (nutrient resource quality) on carbon allocation, seedlings of *Pinus sylvestris* grown in microcosm on non-sterile peat and in the absence or presence of *Paxillus involutus* and *Suillus bovinus* were pulse labeled with (^{14}C) CO_2. Litter patches stimulated growth of ectomycorrhizal mycelium which was obviously supported by carbon allocation; up to 60% of carbon transferred from the host to the external mycelium was delivered to mycelium in patches. As mycelia of most recently colonized patches received the largest amount, these appear to be the most active sinks. Leake et al. (2001) also give evidence that the interaction of different fungi can determine sink strength. The presence of saprotrophic fungi such as the wood decomposer *Phanerochaete velutina* reduced the allocation of carbon to the mycorrhizal fungus. From these observations it can be concluded that mycorrhizal mycelia consume carbon according to their nutrient foraging activity, and that the close spatial vicinity of wood-decomposing fungi could open access to an alternative carbon source for the ectomycorrhizal fungus. If such an interaction does not interfere with nutrient transfer to the host plant, such alternative sources of carbon could reduce the load on photoassimilate supply by the host plant.

The importance of photoassimilate supply to the fungus for nutrient acquisition by the fungus was also confirmed in double-labeling experiments using seedlings of *Populus tremula* × *Populus alba*/+/– *Laccaria* and of *Pinus sylvestris* +/– *Suillus bovinus* (Bücking and Heyser 2001, 2003). X-ray microanalysis and microautoradiographs after pulse-labeling with (^{33}P)inorganic phosphate (Pi) and $^{14}CO_2$ of plants kept in darkness or under illumination showed that shading (less carbon allocation to the fungus) caused decreased Pi absorption by the mycobiont and reduced transfer to the host, while increasing the number of polyphosphate granules in fungal hyphae. Thus, polyphosphate formation in fungal mycelia could be a response to limited carbon supply.

The determination of carbon balances is another approach to quantify carbon allocation to the root system. For stands of *Pseudotsuga menziessii* var. glauca, McDowell et al. (2001) reported that carbon allocation to fine roots and mycorrhizal production made up 47–59% of the total belowground carbon allocation, depending on nutrient availability.

The extent to which the host has to pay for nutrient delivery by the fungus depends on the fungal partner. Bidartondo et al. (2001) used combinations of *Pinus muricata*/*Paxillus involutus*/*Suillus pungens* + *Rhizopogon* species in microcosms to investigate the availability of nutrients [addition of ammonium/apatite (Pi, Ca supply)] on carbon allocation.

P. involutus caused the lowest host biomass production (highest cost), obviously because high rates of N transfer coincided with high rates of carbon consumption. The high sink strength of the fungus correlated with a more rapid and intense colonization of the substrate at low mycelial respiration; nutrient (N) addition resulted in increased respiration and mycelial biomass. The authors conclude that ammonium uptake causes a high carbon cost, and that this could be the reason for the observed, reduced sporocarp formation and mycelial growth of ECM-forming fungi commonly found after high levels of nitrogen addition. It is, however, also reasonable to assume that increased N supply to the host plant results in a decreased allocation of photoassimilates (sucrose) to the root, as shoot growth is favoured under these conditions (Hampp et al. 1995). This way the fungus would suffer from a decreased carbon supply, with the same consequences as reported by Bidartondo et al. (2001).

3.2 Alternative Carbon Sources

Ectomycorrhizal fungi are also able to make use of organic carbon in the soil. This can even lead to soil carbon depletion. Introduction of *Pinus radiata* and their accompanying ectomycorrhizal fungi into grasslands (Ecuador) caused a significant loss of soil C, together with a large reduction in the number of fungal species [from about 100 to 3 (Chapela et al. 2001)]. This illustrates the dynamic role of ectomycorrhizal fungi in soil C processing and is also in accordance with the assumed saprotrophic capabilities of ectomycorrhizal fungi (Wiemken and Boller 2002).

A comparison of carbon turnover by different types of mycorrhiza was made by Cornelissen et al. (2001). They analyzed the relative growth rate (RGR) of host seedlings, foliar nutrient concentration and leaf litter decomposability. All three parameters were poor with ericoid mycorrhiza, but high with arbuscular mycorrhiza. Ectomycorrhizas caused intermediate RGR, higher foliar contents of N and P, and were intermediate to poor with regard to leaf litter decomposition. The authors conclude that, for subsets of the temperate flora, ericoid and ECMs are linked with low, arbuscular mycorrhizas with high ecosystem carbon turnover.

In order to investigate the role of mycorrhization in the uptake of organic acids/amino acids from the soil, Blaudez et al. (2001) compared the uptake of label from externally fed ^{14}C glutamate and malate into roots, fungus and mycorrhiza using birch seedlings and *Paxillus involutus*. Mycorrhization increased uptake of label significantly, especially in 6- to 15-day-old mycorrhized roots. In both free-living mycelia and ECMs, label was mainly found in glutamine, while in non-mycorrhizal roots it occurred predominantly in citrulline and insoluble compounds; mycorrhiza formation thus alters the use of exogenously supplied carbon.

However, there are also other modes of carbon transfer. Using *Suillus variegatus*/*Pinus sylvestris* seedlings together with the wood-decomposing fungus *Hypholoma fasciculare* in microcosm studies, Lindahl et al. (2001) showed that *S. variegatus* mycelium extending from pine seedlings and saprotrophic mycelium growing out into the soil from wood blocks of birch were able to capture ^{32}P from each other, depending on support-related intensity of growth (vigor of seedlings, size of wood blocks). ECM-forming fungi can also exploit dead nematodes. In microcosms containing *Betula pendula* +/– mycorrhiza (*Paxillus involutus*), +/– dead nematodes with known P and N contents, mycorrhized plants produced greater yields and had larger N and P content in the presence of nematode necromass (Perez-Morena and Read 2001a). Obviously, the symbiotic system exploited N and P more effectively, in addition to a high transfer rate to the plant. ECMs are even able to transfer nutrients from living animals. *Laccaria bicolor* has been shown to attack the bodies of living as well as dead collembola, and to facilitate transfer of nitrogen from the animals to its host plant (Klironomos and Hart 2001). Thus even more complex nutrient sources may be accessible to some mycorrhizal associations. These findings implicate the potential of ECM-forming fungi in nutrient recycling of the soil mesofauna.

As plants are in part interconnected by hyphae of mycorrhiza-forming fungi, carbon transfer can also be from fungus to plant. This is obviously the case in heterotrophic plants. Some derive most or all of their carbon from ectomycorrhizal fungi, i.e. they are nursed through ectomycorrhizal connections. This is possibly a role for subdominant or rare ECM fungal species. By means of ^{14}C labeling heterotrophic *Corallorhiza trifida* was shown to be fed by a shared fungal network connected to *Betula pendula* or *Salix repens* but not to *Pinus sylvestris* (McKendrick et al. 2000). Similar highly specific nurse functions were found for *Rhizopogon* species fungi and mycoheterotrophic plants such as *Sarcodes sanguinea* and *Pterospora andromedea* (Bruns and Read 2000), as well as for *Monotropa uniflora* and ECM fungi of the Russulaceae (Young et al. 2002).

3.3 Effects of Elevated CO_2

Net primary production of trees on sites with moderate soil fertility can be significantly increased by elevated CO_2 over a given period of time [e.g. Woodward 2002: Free Air CO_2 Enrichment technique (FACE) experiment]. Under such conditions, growth could still be improved by nitrogen supply; however, this resulted in little change in the shoot-to-root ratio (Niklaus et al. 2001). In general, responses of plants depend on the length of exposure: many short-term experiments (days, weeks, months) show transient effects only, which disappear, or are less obvious, after long-term exposure (Woodward 2002). This variation in response has to be taken into account when investigating the effects of elevated CO_2 on mycorrhiza formation/functioning. At least in the short term there is a body of evidence that a larger supply of carbon to the root system benefits the symbiotic interaction with regard to solute/metabolite exchange. We found only a few recent publications on the effect of long-term exposures to elevated CO_2.

Individual trees of a 37-year-old *Picea abies* forest that had been fertilized with nitrogen for 15 years were exposed to 700 ppm in whole-tree chambers +/– fertilization for 3 years (Franssen et al. 2001). After this period, ECM community structure was compared with that at the beginning of the treatment. Morphotyping and molecular analyses showed significant effects on ECM community structure but not on species richness (total numbers of taxa found) in response to CO_2, i.e. there was mainly a change in the abundance of a few common species. In accordance with the adverse effects of fertilization on mycorrhiza development, responses were stronger in the unfertilized plot.

Biomass analysis of a poplar plantation exposed to 550 ppm (FACE technique, see above) showed an increased below-ground allocation of biomass (Lukac et al. 2003). Root biomass increased by 47–76%, and that of fine roots by 35–84%. The response of the mycorrhizal community was more varied. The rate of infection increased only in *Populus alba*, not in *Populus nigra* [higher infection only for arbuscular mycorrhiza (AM)] and *Populus* × *euramericana*.

Using model ecosystems, 2- to 4-year-old beech and spruce trees were exposed in open-top chambers for 4 years to ambient and elevated (ambient + 200 ppm: 566 ppm) CO_2 in nutrient-poor and -rich soil (Wiemken et al. 2001). Root biomass was not affected after the 4-years' exposure, but elevated CO_2 stimulated fine root production, especially in the nutrient-rich soil. The biomass of ECM-forming fungi (determined via ergosterol content) was much higher in nutrient-poor soil but was not changed by the CO_2 treatment.

In summary, the data indicate that it is mainly the nutrient supply that affects ectomycorrhization in the long run, while the effects of elevated CO_2 are much dependent on the host plant and the degree to which it allocates a surplus of photoassimilates to the root system.

3.4 Fungal Carbohydrate Metabolism

Ectomycorrhizal fungi produce a series of fungus-specific sugars and sugar alcohols, some of which are exuded. A ^{13}C analysis of *Cantharellus cibarius* exudates yielded mannitol and trehalose which are possibly the basis for reproduction of *Pseudomonas* bacteria inside fruit bodies (Rangel-Castro et al. 2002). Labeled storage products detected were mannitol, trehalose and arginine. According to labeling patterns, the mannitol pathway was connected to trehalose production.

As in drought resistance (Shi et al. 2002), fungus-specific carbohydrates play a role as compatible solutes in freezing tolerance (Tibbett et al. 2002). *Hebeloma* strains of arctic/temperate origin were grown at 22, 12, 6 and 2 °C and frozen to –5 °C after pre-conditioning at 2 or 22 °C. Temperate strains exhibited a clear increase in pool sizes of trehalose, mannitol and arabitol, especially after pre-conditioning.

Distinct enzymes of fungal primary metabolism were investigated by Agostini et al. (2001), Grotjohann et al. (2001) and Balasubramanian et al. (2002). Activities of the enzymes of the citric acid cycle were assayed in *Suillus bovinus* (Grotjohann et al. 2001) and kinetic constants are given for citrate synthase, aconitase, isocitrate dehydrogenase, succinate dehydrogenase, fumarase and malate dehydrogenase. Two enzymes, namely 2-oxoglutatrate dehydrogenase and succinyl CoA synthetase, could not be detected and the authors assumed that this could be due either to technical problems or, less probable, to an incomplete tricarboxylic acid cycle in this fungus. Hexokinase, a key enzyme of hexose activation, was cloned from white truffel (*Tuber borchii*) and enzyme kinetics determined by overexpression in *E. coli* (Agostini et al. 2001). From three distinct proteins, one (hxk-1) was characterized. Hxk-1 was not inhibited by trehalose 6-phosphate and glucose 6-phosphate. Malate synthase, an enzyme of the glyoxylate pathway, was investigated in mycorrhizas formed in vitro between *Laccaria bicolor* and *Pinus resinosa* (Balasubramanian et al. 2002). Expression of the respective gene was symbiosis-regulated, and dependent on the availability of carbon sources, especially glucose. As upregulation occurred in the preinfection stage, the proposed role is the utilization of 2-carbon-compounds formed from catabolic processes early in the interaction, which could facilitate the formation of the symbiotic hyphal structures.

3.5 Impact of Nitrogen Quality and Quantity on Ectomycorrhizal Community Structure

In non-polluted forest ecosystems, nitrogen is available mainly in the litter layer. Here nitrate (as a result of denitrification), ammonium, free amino acids, proteins and nucleic acids are found.

The ability of forest plants to utilize organic nitrogen sources directly (without the help of ectomycorrhizal fungi) has been often overlooked. Trees have a remarkable ability to take up amino acids (Nasholm et al. 1998; Persson and Nasholm 2001) and this capacity is not modulated by mycorrhiza formation (Persson and Nasholm 2001). Nevertheless, plants have a very limited capacity to utilize complex organic substances that are usually found in the organic layer (Jonasson and Shaver 1999; Perez-Moreno and

Read 2001b). Furthermore, the amount of free ammonium or amino acids in soil water is usually rather limited, due to the association of these nitrogen sources to humic substances (Yu et al. 2002). Degradation of litter by bacteria and fungi, as well as forest-floor animals, is thus necessary to meet the nitrogen demand of forest plants. In addition, the association with ectomycorrhizal fungi could overcome limitations of forest plants regarding nitrogen uptake due to genetic determination (Mari et al. 2003).

The ability of ectomycorrhizal fungi to utilize nitrogen sources differs remarkably. Fast-growing ectomycorrhizal fungi, with a limited capacity to utilize complex organic nitrogen sources, are often found after afforestation of nutrient-rich grasslands, while slow-growing fungi, with the ability to degrade complex organic sources, are associated with mature forest ecosystems (Eaton and Ayres 2002). *Amanita* species have a profound ability to utilize protein as a nitrogen source (Sawyer et al. 2003a,b). At least two proteases AmProt1 and AmProt2 are excreted by *A. muscaria* (Nehls et al. 2001). AmProt1 has a narrow pH-optimum around 3.0 and is released at pH values up to 5.4. AmProt2 reveals a broad pH-optimum between 3 and 6 and is only excreted at pH values between 5.4 and 6.3. Since forest litter layers are intensely colonized by biofilm-forming bacteria (see below), where the microenvironment is adapted to bacterial growth (e.g. pH 5–6), expression of a protease that is active at a less acidic pH would favor the mobilization of bacteria-derived proteins by ectomycorrhizal fungi.

Nitrogen availability in soil has a large impact on the ectomycorrhizal fungal community structure (Lilleskov et al. 2002a,b). Low nitrogen content in forest soils is often accompanied by a larger fungal biodiversity. Nitrogen fertilization of N-limited stands (at moderate levels) leads to a better fungal growth and productivity of the stand without dramatic changes in fungal biodiversity. In contrast, nitrogen excess in the soil dramatically reduces fungal biodiversity (Wollecke et al. 1999; Baum and Makeschin 2000) but also biomass production of the extraradical mycelium (Nilsson and Wallander 2003) responsible for nutrient uptake.

In summary, litter type and nitrogen availability together with the different abilities of ectomycorrhizal fungi to utilize nitrogen sources might explain the ectomycorrhizal fungal biodiversity of different forest ecosystems at a functional level (Conn and Dighton 2000).

3.6 Nitrogen Uptake and Assimilation by Ectomycorrhizal Fungi

One aspect of the capability of an organism to utilize different nitrogen sources is its uptake capacity for a given source. While several different transporters were cloned from ectomycorrhizal fungi, nutrient transporters from ectomycorrhizal plants have not yet been focused on.

Nitrate transporter genes have been identified so far from *Hebeloma cylindrosporum* (Jargeat et al. 2003) and *Amanita muscaria* (Nehls, unpubl.). In *H. cylindrosporum* the gene is expressed at a high level in the presence of nitrate and organic nitrogen, but repressed in the presence of ammonium (the favored nitrogen source of most fungi).

Two different nitrate reductase (NR) genes (catalyzing the first step of nitrate assimilation) were identified in *H. cylindrosporum* (Jargeat et al. 2000, 2003). Nitrate reductases have been isolated also from the ectomycorrhizal fungus *Tuber borchii* and its plant partner *Tilia platyphyllos* (Guescini et al. 2003). Comparative expression analysis of plant and fungal genes revealed that the transcript level of fungal NR in ECMs is eight times higher compared to free-living mycelia, whereas plant NR transcription was downregulated in symbiosis. This result indicates that the fungal partner plays a fundamental role in nitrate assimilation by ECMs.

As in ascomycetes, the genes necessary for nitrate utilization are clustered in the genome of *H. cylindrosporum* (Jargeat et al. 2003). Three ammonium transporter genes of *H. cylindrosporum* were isolated and functionally characterized in yeast (Javelle et al. 2001, 2003). They revealed K_m values smaller than 0.1 μM (HcAMT1 and HcAMT2) and 12 μM (HcAMT3). Under ammonium excess, the transcription of *AMT1* and *AMT2* is highly repressed while the transcript level of *AMT3* is only slightly affected. Since the K_m value of AMT3 is about 100-fold higher than that of the other ammonium transporter, AMT3 activity is favored at high ammonium concentrations. An ammonium transporter gene (*TbAMT1*), revealing a K_m value of 2 μM when heterologously expressed in yeast, was also isolated from the ascomycete *Tuber borchii* (Montanini et al. 2002). The expression of the *TbAMT1* gene is quite atypical for fungi. Independent of whether the hyphae are exposed to ammonium (good nitrogen source) or nitrate (bad nitrogen source), the expression level stays similar. Furthermore, gene expression is only slowly and slightly increased after nitrogen depletion.

Amino acids (as a result of protein degradation) are frequently found in forest soils and are thus of great importance for nitrogen nutrition (Wallenda and Read 1999). Fungal amino acid importer genes have been isolated to date from *A. muscaria* (Nehls et al. 1999b) and *H. cylindrosporum* (Wipf et al. 2002). As determined by heterologous expression in yeast, these genes encode high-affinity H^+/amino acid symporters with a broad amino acid spectrum. *A. muscaria* AmAAP1 has a higher affinity to basic and aromatic compared to acidic or neutral amino acids. These differences in affinity might reflect the fact that basic amino acids are present in soil in significantly lower concentrations (8–30 μM) than neutral (80–100 μM) ones (Scheller 1996).

In fungal hyphae, ammonium (as a result of nitrate assimilation, ammonium uptake or amino acid catabolism) can be incorporated into the key nitrogen donors glutamate by NADP-dependent glutamate dehydrogenase (*GDH*) and glutamine by glutamine synthetase (*GS*). At low ammonium concentrations, both *GDH* and *GS* genes are transcribed (Javelle et al. 2003), resulting in elevated ammonium assimilation capacities. At high ammonium concentrations, *GDH* expression is efficiently repressed. Under these conditions, only GS activity enables ammonium assimilation.

3.7 Regulation of Fungal Nitrogen Export by the Hyphal Nitrogen Status

The nitrogen-dependent expression profile of nitrogen importer genes of basidiomycotic ectomycorrhizal fungi (*A. muscaria*: Nehls et al. 1999b; *H. cylindrosporum*: Javelle et al. 2001, 2003; Wipf et al. 2002) resembles that of nitrogen importers of the well-investigated ascomycetes. Here, nitrogen-dependent gene repression is presumably regulated by the internal nitrogen status of cells, i.e. the intracellular ammonium concentration (TerSchure et al. 2000) and/or the activity of the GS (Sophianopoulou and Diallinas 1995). This type of regulation is also found in ectomycorrhizal fungi. In *H. cylindrosporum*, *AMT1* and *AMT2* expression is supposed to be controlled by the intracellular glutamine level, whereas *GDH* expression is controlled by the ammonium level (Javelle et al. 2003).

A regulation of nitrogen import systems by the internal nitrogen status of the hyphae could explain the efficient nitrogen uptake by extramatrical hyphae as well as the net export of nitrogen at the plant/fungus interface (Nehls 2004). Since the nitrogen content of forest soil is quite low and fractions of the nitrogen are transported to other parts of the growing fungal colony (e.g. mycorrhizas), *soil-growing hyphae* are presumably nitrogen-limited, resulting in a low endogenous nitrogen status and a strong expression of nitrogen importer genes. On the other hand, the *mycorrhizal zone of interaction* (mantle, Hartig net) is well supplied with nitrogen by the large mass of soil-growing hyphae, thus revealing a high nitrogen status and a strongly reduced nitrogen importer gene expression. This nitrogen-dependent repression of amino acid transporter gene expression in mycorrhizas (Nehls et al. 1999b; Wipf et al. 2002), together with posttranslational events (e.g. increased degradation of plasma membrane transport proteins) which are described for yeasts (Springael and Andre 1998), could thus result in a highly reduced fungal capacity for re-uptake of amino acids at the plant/fungus interface. As shown for a yeast mutant lacking arginine uptake activity, the reduced re-import capacity for this amino acid resulted in a net arginine loss of the cells (Grenson 1973). In combination with efflux mechanisms (e.g. nitrogen leakage), this would thus result in a net export of nitrogen.

4 Mycorrhiza and Drought/Stress Tolerance

4.1 Water Stress

The role of mycorrhizal fungi (AM and ECM) in improving plant–water relationships is still a matter of debate, although recent data have shown that mycorrhization has an influence on water channel proteins in host fine roots. Mycorrhization of fine roots of spruce seedlings with *Amanita muscaria* resulted in a decreased expression of fine-root-associated plant aquaporins (Marjanovic et al., unpubl.), suggesting that the fungal partner improves water supply which reduces the density of aquaporins per membrane area. This is in accordance with the determination of root hydraulic conductivity in other ectomycorrhizal systems. Ectomycorrhizas of *Ulmus americana* seedlings showed increased root water flow, and kinetics were studied with intact plants applying temperatures between 4 and 20 °C and in the presence of $HgCl_2$, an inhibitor of aquaporins (Mushin and Zwiazek 2002a). Hydraulic conductance and conductivity declined with decreasing temperature. Both parameters were higher in mycorrhizal seedlings at all temperatures, i.e. mycorrhization could be especially important to improve water uptake from cold soils. However, the advantage at low temperatures cannot be generalized. In studies where aspen and white spruce (with and without ECM-forming fungi) were exposed to low soil temperatures at night (4–8 versus 20 °C), mycorrhized plants showed a generally higher root hydraulic conductivity than non-mycorrhized controls, but no effect with regard to soil temperature (Mushin and Zwiazek 2002b). Shoot water potential was only increased in aspen, not in spruce.

Evidence for the beneficial effect of ECM on host water status comes also from field studies. *Alnus glutinosa*, grown in eutrophic peatland (The Netherlands), exhibited decreased numbers of mycorrhized root tips, and this was accompanied by increased desiccation sensitivity (Baar et al. 2002). However, drought resistance of a host plant depends not only on the degree of mycorrhization, but also on the type of ECM, i.e. the drought resistance of the fungal partner. In water-exclusion (duration 2–3 months) experiments performed under field conditions, mycorrhization of roots of five different ecotypes of beech was investigated in relation to plants exposed to natural precipitation (Shi et al. 2002). Decreased water availability did not affect the degree of mycorrhization, but did affect the species composition of the ECM-forming fungal community. ECM fungi, the presence of which was not altered, and those increasing in number, all showed species-specific accumulation patterns of compatible solutes such as fungus-specific sugars and sugar alcohols. Plants with more drought-tolerating mycorrhizas had less negative leaf pre-dawn water potentials. The data show that, within certain limits, an increase in drought causes a shift in plant/fungus communities.

In a comparable experiment, where *Pinus halepensis* was subjected to water stress by withholding irrigation for 4 months, followed by rehydration for 30 days, mycorrhization could not reduce drought effects on plant growth, but facilitated recovery after rehydration (Morte et al. 2001). Effects of drought on parameters such as stomatal conductance, leaf water potential and leaf turgor were less severe in mycorrhized plants, which again is indicative of improved water uptake from dry soils.

4.2 Salinity

Mycorrhization can also compensate to some degree for effects of soil salinity. Ectomycorrhizas formed between *Hebeloma crustuliniforme* and white spruce seedlings were exposed to 25 mM NaCl (Mushin and Zwiazek 2002b). Again, mycorrhization improved root hydraulic conductance, but also root branching, chlorophyll content and growth. Accordingly, salt concentrations were significantly lower and root hydraulic conductance several-fold higher in mycorrhized seedlings. Removal of the fungal sheath decreased hydraulic conductance by 50%. By both reduction of Na content and increasing P and N absorption ectomycorrhization should increase resistance of plants growing in salinized soil. Fungi by themselves can tolerate relatively high salt concentrations.

Chen et al. (2001) tested 18 *Pisolithus* isolates for their salt resistance to be used as inoculants for outplanting compatible, salt-resistant host trees at saline sites. Most *Pisolithus* isolates tolerated NaCl concentrations of 200 mM (50% growth inhibition), and an Na_2SO_4 concentration of 100 mM; i.e. they were broadly resistant to salinity.

5 Mycorrhiza/Bacteria Interactions

Soil bacteria can have profound effects on mycorrhiza establishment. During recent years, several novel bacteria exerting positive effects on mycorrhiza formation, mycorrhiza helper bacteria (MHB), have been characterized. The variety in the microbial populations around mycorrhizal roots of different fungal and plant partners and in different sections of soil profile has also been assessed.

5.1 Mycorrhiza Helper Bacteria (MHB) and Antagonists

The traditional forms of MHB, fluorescent pseudomonads and sporulating bacilli, still maintain their part in the recent literature. However, bacteria from such groups as *Rhodococcus* and *Burkholderia* have also appeared to show a positive effect on mycorrhization.

The establishment of mycorrhiza between *Acacia holosericea* and *Pisolithus alba* and root and shoot biomasses were all improved due to the inoculation of two pseudomonad strains (Founoune et al. 2002a). The bacteria showed a positive effect on hyphal growth, which was thought to underlie the MHB effect. In the next study, several fluorescent pseudomonad strains were collected from mycorrhizosphere soil, from the galls or from the roots of *A. holosericea* (Founoune et al. 2002b). Fourteen of the bacterial strains were shown to promote mycorrhiza development and, in addition, several of them also showed a positive effect on shoot biomass. When the bacteria around Australian *Acacia* species were studied, the MHB effect varied from 45–70% improvement of ECM formation with *P. alba* (Duponnois and Plenchette 2003). Interestingly, the studied MHB had a positive effect on the fungal growth of *Pisolithus*, but not on *Scleroderma* isolates. These bacteria also enhanced arbuscular mycorrhizal development, making them good candidates for nursery experiments.

The mechanisms behind MHB action are poorly understood. The investigations on *Pinus sylvestris–Lactarius rufus* mycorrhizas have given some insight into the several strategies of the MHB. Two mycorrhizosphere bacterial isolates from *Burkholderia* and one *Rhodococcus* isolate increased the formation of secondary mycorrhizal lateral roots in pine, while a *Paenibacillus* isolate stimulated infection of both first- and second-order lateral roots, indicating a differential hormonal effect by the bacterial strains on the root system (Poole et al. 2001). Further, only the *Burkholderia* isolates concentrated on the ectomycorrhizal roots and were capable of effective spread on Scots pine root systems. The hyphal-growth-promoting effect has been mostly measured by Petri dish tests or by following the growth of mycelium on plant roots and in soil in microcosm experiments. In a long-term survey, Brule et al. (2001) trapped fungal mycelium around Douglas fir seedlings. Their results showed that long-term survival of *Laccaria bicolor* isolate was improved by a fluorescent pseudomonad only in autoclaved soil. This indicates that in severe stress conditions, such as contaminated soils, the MHB may be important for the survival of fungal inocula, but under natural conditions the forest soil microbes may rapidly displace the introduced bacteria, and no effect can be expected.

The bacterial populations affect plant and fungal growth in several ways. *Bacillus* isolates from the mycorrhizas of Scots pine and *Suillus luteus* showed variability in their potential to serve as MHB. The bacterial strains were mostly able to promote hyphal growth, being the probable mechanism behind the MHB effect (Bending et al. 2002). A few antagonistic bacteria were also isolated from the groups of *Burkholderia* and *Serratia*. The bacterial community of *Quercus robur* mycorrhizas with *Tuber borchii* was studied by Sbrana et al. (2002). Out of the large collection of bacterial strains, 101 strains showed antifungal activity, among them all of the isolated actinomycete strains. The supposed MHB were shown to sporulate, and they improved growth of *T. borchii* mycelium by up to 78%. However, no mycorrhization experiments were conducted to secure the MHB identity of the spore-formers. In the only report on hyphal-growth-promoting actinomycetes (Becker et al. 1999), the investigated streptomycete strain induced the growth of ectomycorrhizal fungi *L. bicolor* and *Cenococcum geophilum* on Petri dish cultures, but inhibited the hyphal extension of plant pathogenic *Armillaria* species. The inhibitory mechanisms of bacterial strains may vary. The decrease in *Heterobasidion annosum* growth on iron-deficient medium by fluorescent pseudomonad strains 92 and BBc6 was caused by strong siderophore production by the bacteria (Gamalero et al. 2003).

The ECM fungus *L. bicolor* harbors intracellular bacteria. Bertaux et al. (2003) showed that these apparently belong to the genus *Paenibacillus*. In fungal suspension culture, the bacteria were variable in number, up to 50 per fungal cell, and occurred in only a few fungal cells. Unfortunately, the study has as yet provided no information concerning the viability of the colonized fungal cells; this has to be analysed in order to confirm that the intracellular bacteria are endosymbionts and not saprotrophs.

5.2 Bacterial Populations of the Rhizosphere

The microbial populations in the mycorrhizal root tips, in the mycorrhizosphere and 'bulk soil' differ qualitatively from each other. The utilization patterns of carbon and nitrogen sources have been studied in the mycorrhizosphere around Scots pine seedlings (Heinonsalo et al. 2001). The bacterial isolates from the mycorrhizosphere used a wider range of carbon sources than those from the mineral soil bacteria. As a novel group of prokaryotes, Archaea were identified from Scots pine mycorrhizospheres and mycorrhizas with *Suillus bovinus* and *Paxillus involutus* (Bomberg et al. 2003). The non-mycorrhizal short roots of pine did not yield any archaeal sequences, suggesting that their spread is at least partially dependent on hyphal growth.

Tuber borchii fruit bodies host a variety of bacteria. The studied bacterial isolates belong to fluorescent pseudomonads and sporulating bacilli with cellulolytic and chitinolytic activities (Gazzanelli et al. 1999; Barbieri et al. 2001; Citterio et al. 2001). The authors suggest that these enzyme activities could be partly responsible for ascus opening in the *T. borchii* fruit bodies, although evidence remains elusive. The vegetative hyphae of *T. borchii* also contain bacteria. rDNA analysis and fluorescent in-situ hybridization (FISH) showed a novel *Cytophaga–Flexibacter–Bacterioides* phylogroup growing on the hyphae (Barbieri et al. 2000). FISH has also been used for the analysis of beech mycorrhizosphere and mycorrhizal mantle surface (Mogge et al. 1999). A large group of proteobacteria were detected on mantle surfaces, the most common being alpha- and beta-proteobacteria. Two studied *Laccaria* species were shown to harbor differing proteobacterial communities, emphasizing the specificity that exists between the microbial partners.

An important part of bacterial community studies concentrates on the changes due to introduced organisms. This is becoming important as more and more genetically engineered bacteria with desired abilities, such as strong antagonism against pathogenic organisms or toxin degradation potential, exist. It is important to keep in mind that, for example, plant growth-promoting rhizobacteria (PGPR) should have a neutral or positive effect on other plant-beneficial microbial populations. According to recent data, the induction of plant growth may (Hoflich et al. 2001) or may not (Probanza et al. 2001) be connected with improved mycorrhiza development, and when loblolly pine seedlings were inoculated with PGPR, the degree of ectomycorrhization was actually found to decrease (Vonderwell et al. 2001). The authors of the last case suggest a higher metabolic cost for mycorrhizal maintenance in the presence of the rhizobacteria, although antibiosis or competition for nutrients might also give an explanation for the observed reduction in mycorrhization. The introduction of an antagonistic *Pseudomonad fluorescens* strain to the rhizosphere of silver birch resulted in induced plant growth (Bjorklof et al. 2003). The microbial community of the rhizosphere buffered the antagonistic effect of the novel bacterium, although slight changes in the rate of mycorrhization and in the composition of indigenous bacterial communities were evident.

The detoxification of contaminated soils by microbes may facilitate the growth of plants and fungal hyphae. This phenomenon has been shown in m-toluate-containing soils inoculated with a toluate-degrading pseudomonad strain (Sarand et al. 1999a,b). Inversely, the external hyphae of mycorrhizal fungi may also affect bacterial communities in toxic soils. Petrolium-contaminated soils were shown by Heinonsalo et al. (2000) to be most effectively detoxified around the soil colonized by mycorrhizal pine roots, areas that also supported the highest numbers of bacteria. The mycorrhizal fungi themselves may also be used for 'biodegradation' of certain substances, such as halogenated biphenyls (Green et al. 1999).

6 New Techniques: Genetic Manipulation of Plants and Fungi

In order to verify the importance of the differentially regulated genes, transformation systems for trees and fungi would be needed. *Agrobacterium*-mediated gene transfer is an established method for birch and poplar, but is still extremely time-consuming in conifer species. Until recently, only ballistic or protoplast-based transformation has been possible for the ECM fungi. When these techniques are used, the copy number of the transgene is often high. *Agrobacterium* has been recently applied for the transformation of symbiotic fungi (Pardo et al. 2002). Phleomycin resistance could be transferred to *S. bovinus*, *H. cylindrosporum* and *P. involutus*. Integration of foreign DNA has been verified at first in *S. bovinus* (Hanif et al. 2002) and most recently in *H. cylindrosporum* (Combier et al. 2003). In transformation experiments, single integration of foreign DNA is desired, since side effects, such as multicopy suppression, must be avoided. In the *Agrobacterium* experiments the foreign DNA was detected once in the genomes of these two fungi, suggesting that this form of gene transfer has a great potential in ECM fungi. The integrative plasmid used by Combier et al. (2003) contained sequences that could later be used for PCR-based screening of non-mycorrhizal mutants, opening new vistas for ECM research.

Retrotransposons (RTNs) have proved to be good genetic markers and tools for mutagenesis. The first RTNs of ECM fungi have been characterized in *Tricholoma matsutake* (Murata and Yamada 2000). The reverse transcriptase domain of the RTN marY1, which belongs to the gypsy group of RTNs, was shown to exist in the worldwide population of *T. matsutake* and *T. magnivelare* species. *MarY1* was successfully expressed in budding yeast, indicating that it can be modified in this heterologous host (Murata and Miyazaki 2001).

For proper transgene expression, suitable promoter and terminator sequences may not always be sufficient. This was the case for the *HydPt-1* gene of *P. tinctorius*, which was introduced into *Hebeloma cylindrosporum* to test whether its expression leads to a cell wall phenotype in *Hebeloma cylindrosporum*. Introns were required for the expression of the hydrophobin gene in *H. cylindrosporum*. Even then no HydPt-1 protein accumulation could be detected, indicating that transgene expression is tightly regulated in homobasidiomycetes (Tagu et al. 2002).

7 Conclusions and Perspectives

Mycorrhizal research has primarily two points of focus, ecology and molecular genetics. The latter approach is just switching from pure collection of data (differential expression of genes in different stages of mycorrhiza development) towards the annotation of the expressed sequence tags (ESTs) obtained. This step is still in its infancy, but will become increasingly available with the completion of sequencing projects with some selected model fungi (e.g. *Laccaria bicolor*). Another important step towards the understanding of mycorrhiza function is genetic manipulation of the mycorrhizal partners. With regard to the host, poplar has developed into a model plant with relatively easy access to specific mutants by *Agrobacterium*-based transformation. The situation is more difficult as far as the fungal partner is concerned. Up to now, only very few fungal species have been transformed, *Hebeloma* being the most important ECM-forming fungus.

References

Agostini D, Polidori E, Palma F, Ceccaroli P, Saltarelli R, Tonelli D, Stocchi V (2001) Cloning, expression and characterization of the *hxk-1* gene from the white truffle *Tuber borchii* Vittad.: a first step toward understanding sugar metabolism. Fungal Genet Biol 33:15–23

Ashford AE, Vesk PA, Orlovich DA, Markovina AL, Allaway WG (1999) Dispersed polyphosphate in fungal vacuoles in *Eucalyptus pilularis/Pisolithus tinctorius* ectomycorrhizas. Fungal Genet Biol 28:21–33

Baar J, Bastiaans T, van de Coevering MA, Roelofs JGM (2002) Ectomycorrhizal root development in wet alder carr forests in response to desiccation and eutropication. Mycorrhiza 12:147–151

Balasubramanian S, Kim SJ, Podila GK (2002) Differential expression of a malate synthase gene during the preinfection stage of symbiosis in the ectomycorrhizal fungus *Laccaria bicolor*. New Phytol 154:517–527

Barbieri E, Potenza L, Rossi I, Sisti D, Giomaro G, Rossetti S, Beimfohr C, Stocchi V (2000) Phylogenetic characterization and in situ detection of a *Cytophaga–Flexibacter–Bacteroides* phylogroup bacterium in *Tuber borchii* Vittad, ectomycorrhizal mycelium. Appl Environ Microbiol 66:5035–5042

Barbieri E, Potenza L, Stocchi V (2001) Molecular characterization of cellulosolytic-chitinolytic bacteria associated with fruitbodies of the ectomycorrhizal fungus *Tuber borchii* Vittad. Symbiosis 30:123–139

Baum C, Makeschin F (2000) Effects of nitrogen and phosphorus fertilization on mycorrhizal formation of two poplar clones (*Populus trichocarpa* and *P.-tremula* × *tremuloides*). J Plant Nutr Soil Sci 163:491–497

Becker DM, Bagley ST, Podila GK (1999) Effects of mycorrhizal-associated streptomycetes on growth of *Laccaria bicolor*, *Cenococcum geophilum*, and *Armillaria* species and on gene expression in *Laccaria bicolor*. Mycologia 91:718–723

Bending GD, Poole EJ, Whipps JM, Read DJ (2002) Characterisation of bacteria from *Pinus sylvestris–Suillus luteus* mycorrhizas and their effects on root–fungus interactions and plant growth. FEMS Microbiol Ecol 39:219–227

Bertaux J, Schmid M, Chemidlin Prevost-Boure N, Churin JL, Hartmann A, Garbaye J, Frey-Klett P (2003) In situ identification of intracellular bacteria related to *Paenibacillus* spp. in the mycelium of the ectomycorrhizal fungus *Laccaria bicolor* S238 N. Appl Environ Microbiol 69:4243–4248

Bidartondo MI, Ek H, Wallander H, Söderström B (2001) Do nutrient additions alter carbon sink strength of ectomycorrhizal fungi? New Phytol 151:543–550

Bjorklof K, Sen R, Jorgensen KS (2003) Maintenance and impacts of an inoculated mer/luc-tagged *Pseudomonas fluorescens* on microbial communities in birch rhizospheres developed on humus and peat. Microbial Ecol 45:39–52

Blaudez D, Botton B, Chalot M (2000) Cadmium uptake and subcellular compartmentation in the ectomycorrhizal fungus *Paxillus involutus*. Microbiology 146:1109–1117

Blaudez D, Botton B, Dizengremel P, Chalot M (2001) The fate of [C-14]glutamate and [C-14]malate in birch roots is strongly modified under inoculation with *Paxillus involutus*. Plant Cell Environ 24:449–457

Bomberg M, Jurgens G, Saano A, Sen R, Timonen S (2003) Nested PCR detection of Archaea in defined compartments of pine mycorrhizospheres developed in boreal forest humus microcosms. FEMS Microbiol Ecol 43:163–171

Brule C, Frey-Klett P. Pierrat JC, Gerard F, Lomoine MC, Rousselet JL, Sommer J, Garbaye J (2001) Survival in the soil of the ectomycorrhizal fungus *Laccaria bicolor* and effect of a mycorrhiza helper *Pseudomonas fluorescens*. Soil Biol Biochem 33:1683–1694

Bruns TD, Read DJ (2000) In vitro germination of nonphotosynthetic, myco-heterotrophic plants stimulated by fungi isolated from the adult plants. New Phytol 148:335–342

Bücking H, Heyser W (2001) Microautoradiographic localization of phosphate and carbohydrates in mycorrhizal roots of *Populus tremula* × *Populus alba* and the implications for transfer processes in ectomycorrhizal associations. Tree Physiol 21:101–107

Bücking H, Heyser W (2003) Uptake and transfer of nutrients in ectomycorrhizal associations: interactions between photosynthesis and phosphate nutrition. Mycorrhiza 13:59–68

Chapela IH, Osher LJ, Horton TR, Henn MR (2001) Ectomycorrhizal fungi introduced with exotic pine plantations induce soil carbon depletion. Soil Biol Biochem 33:1733–1740

Charvet-Candela V, Hitchin S, Reddy MS, Cournoyer B, Marmeisse R, Gay G (2002a) Characterization of a *Pinus pinaster* cDNA encoding an auxin up-regulated putative peroxidase in roots. Tree Physiol 22:231–238

Charvet-Candela V, Hitchin S, Ernst D, Sandermann H, Marmeisse R, Gay G (2002b) Characterization of an Aux/IAA cDNA upregulated in *Pinus pinaster* roots in response to colonization by the ectomycorrhizal fungus *Hebeloma cylindrosporum*. New Phytol 154:769–777

Chen DM, Ellul S, Herdman K, Cairney JWG (2001) Influence of salinity on biomass production by Australian *Pisolithus* spp. isolates. Mycorrhiza 11:231–236

Citterio B, Malatesta M, Battistelli S, Marcheggiani F, Baffone W, Saltarelli R, Stocchi V, Gazzanelli G (2001) Possible involvement of *Pseudomonas fluorescens* and *Bacillaceae* in structural modifications of *Tuber borchii* fruit bodies. Can J Microbiol 47:264–268

Combier J-P, Melayah D, Raffier C, Gay G, Marmeisse R (2003) *Agrobacterium tumefaciens*-mediated transformation as a tool for insertional mutagenesis in the symbiotic ectomycorrhizal fungus *Hebeloma cylindrosporum*. FEMS Microbiol Lett 220:141–148

Conn C, Dighton J (2000) Litter quality influences on decomposition, ectomycorrhizal community structure and mycorrhizal root surface acid phosphatase activity. Soil Biol Biochem 32:489–496

Cornelissen JHC, Aerts R, Cerabolini B, Werger MJA, van der Heijden MGA (2001) Carbon cycling traits of plant species are linked with mycorrhizal strategy. Oekologia 129:611–619

Ditengou FA, Beguiristain T, Lapeyrie F (2000) Root hair elongation is inhibited by hypaphorine, the indole alkaloid from the ectomycorrhizal fungus *Pisolithus tinctorius*, and restored by indole-3-acetic acid. Planta 211:722–728

Duplessis S, Sorin C, Voiblet C, Palin B, Martin F, Tagu D (2001) Cloning and expression analysis of a new hydrophobin cDNA from the ectomycorrhizal basidiomycete *Pisolithus*. Curr Genet 39:335–339

Duponnois R, Plenchette C (2003) A mycorrhiza helper bacterium enhances ectomycorrhizal and endomycorrhizal symbiosis of Australian *Acacia* species. Mycorrhiza 13:85–91

Eaton GK, Ayres MP (2002) Plasticity and constraint in growth and protein mineralization of ectomycorrhizal fungi under simulated nitrogen deposition. Mycologia 94:921–932

Fornale S, Sarjala T, Bagni N (1999) Endogenous polyamine content and metabolism in the ectomycorrhizal fungus *Paxillus involutus*. New Phytol 143:581–587

Fossdal CG, Nagy NE, Sharma P, Lonneborg A (2003) The putative gymnosperm plant defensin polypeptide (SPI1) accumulates after seed germination, is not readily released, and the SPI1 levels are reduced in *Pythium dimorphum*-infected spruce roots. Plant Mol Biol 52:291–302

Founoune H, Duponnois R, Ba AM, Sall S, Branget I, Lorquin J, Neyra M, Chotte JL (2002a) Mycorrhiza helper bacteria stimulated ectomycorrhizal symbiosis of *Acacia holosericea* with *Pisolithus alba*. New Phytol 153:81–89

Founoune H, Duponnois R, Meyer JM, Thioulouse J, Masse D, Chotte JL, Neyra M (2002b) Interactions between ectomycorrhizal symbiosis and fluorescent pseudomonads on *Acacia holosericea*: isolation of mycorrhiza helper bacteria (MHB) from a Soudano-Sahelian soil. FEMS Microbiol Ecol 41:37–46

Fransson PMA, Taylor AFS, Finlay RD (2001) Elevated atmospheric CO_2 alters root symbiont community structure in forest trees. New Phytol 152:431–442

Gamalero E, Fracchia L, Cavaletto M, Garbaye J, Frey-Klett P, Varese GC, Martinotti MG (2003) Characterization of functional traits of two fluorescent pseudomonads isolated from basidiomes of ectomycorrhizal fungi. Soil Biol Biochem 35:55–65

Gazzanelli G, Malatesta M, Pianetti A, Baffone W, Stocchi V, Citterio B (1999) Bacteria associated to fruitbodies of the ectomycorrhizal fungus *Tuber borchii* Vittad. Symbiosis 26:211–222

Gorfer M, Tarkka MT, Hanif M, Pardo AG, Laitiainen E, Raudaskoski M (2001) Characterization of small GTPases Cdc42 and Rac and the relationship between Cdc42 and actin cytoskeleton in vegetative and ectomycorrhizal hyphae of *Suillus bovinus*. Mol Plant Microbe Interact 14:135–144

Green NA, Meharg AA, Till C, Troke J, Nicholson JK (1999) Degradation of 4-fluorobiphenyl by mycorrhizal fungi as determined by F-19 nuclear magnetic resonance spectroscopy and C-14 radiolabelling analysis. Appl Environ Microbiol 65:4021–4027

Grenson M (1973) Specificity and regulation of the uptake and retention of amino acids and pyrimidines in yeast. In: Vanek Z, Hostalek Z, Culdin J (eds) Genetics of industrial microorganisms. Academica, Prague, pp 179–193

Grotjohann N, Huang Y, Kowallik W (2001) Tricarboxylic acid cycle enzymes of the ectomycorrhizal basidiomycete, *Suillus bovinus*. Z Naturforsch C 56:334–342

Guescini M, Pierleoni R, Palma F, Zeppa S, Vallorani L, Potenza L, Sacconi C, Giomaro G, Stocchi V (2003) Characterization of the *Tuber borchii* nitrate reductase gene and its role in ectomycorrhizae. Mol Genet Genom 269:807–816

Hampp R, Schaeffer C, Wallenda T, Stülten C, Johann P, Einig W (1995) Changes in carbon partitioning/allocation due to ectomycorrhiza formation: biochemical evidence. Can J Bot 73:S548–S556

Hanif M, Pardo AG, Gorfer M, Raudaskoski M (2002) T-DNA transfer and integration in the ectomycorrhizal fungus *Suillus bovinus* using hygromycin B as a selectable marker. Curr Genet 41:183–188

Hebe G, Hager A, Salzer P (1999) Initial signalling processes induced by elicitors of ectomycorrhiza-forming fungi in spruce cells can also be triggered by G-protein-activating mastoparan and protein phosphatase-inhibiting cantharidin. Planta 207:418–425

Heinonsalo J, Jorgensen KS, Haahtela K, Sen R (2000) Effects of *Pinus sylvestris* root growth and mycorrhizosphere development on bacterial carbon source utilization and hydrocarbon oxidation in forest and petroleum-contaminated soils. Can J Microbiol 46:451–464

Heinonsalo J, Jorgensen KS, Sen R (2001) Microcosm-based analyses of Scots pine seedling growth, ectomycorrhizal fungal community structure and bacterial carbon utilization profiles in boreal forest humus and underlying illuvial mineral horizons. FEMS Microbiol Ecol 36:73–84

Hoflich G, Munzenberger B, Busse J (2001) Importance of inoculated rhizosphere bacteria and ectomycorrhizal fungi on growth of pine seedlings in different soils. Forstwiss Centralbl 120:68–79

Hyde GJ, Davies D, Perasso L, Cole L, Ashford AE (1999) Microtubules, but not actin microfilaments, regulate vacuole motility and morphology in hyphae of *Pisolithus tinctorius*. Cell Motil Cytoskeleton 42:114–124

Hyde GJ, Davies D, Cole L, Ashford AE (2002) Regulators of GTP-binding proteins cause morphological changes in the vacuole system of the filamentous fungus *Pisolithus tinctorius*. Cell Motil Cytoskeleton 51:133–146

Jacob C, Courbot M, Brun A, Steinman HM, Jacquot JP, Botton B, Chalot M (2001) Molecular cloning, characterization and regulation by cadmium of a superoxide dismutase from the ectomycorrhizal fungus *Paxillus involutus*. Eur J Biochem 268:3223–3232

Jargeat P, Gay G, Debaud J.C, Marmeisse R (2000) Transcription of a nitrate reductase gene isolated from the symbiotic basidiomycete fungus *Hebeloma cylindrosporum* does not require induction by nitrate. Mol Gen Genet 263:948–956

Jargeat P, Rekangalt D, Verner MC, Gay G, Debaud JC, Marmeisse R, Fraissinet-Tachet L (2003) Characterisation and expression analysis of a nitrate transporter and nitrite reductase genes, two members of a gene cluster for nitrate assimilation from the symbiotic basidiomycete *Hebeloma cylindrosporum*. Curr Genet 43:199–205

Javelle A, Rodriguez-Pastrana BR, Jacob C, Botton B, Brun A, Andre B, Marini AM, Chalot M (2001) Molecular characterization of two ammonium transporters from the ectomycorrhizal fungus *Hebeloma cylindrosporum*. FEBS Lett 505:393–398

Javelle A, Morel M, Rodriguez-Pastrana BR, Botton B, Andre B, Marini AM, Brun A, Chalot M (2003) Molecular characterization, function and regulation of ammonium transporters (Amt) and ammonium-metabolizing enzymes (GS, NADP-GDH) in the ectomycorrhizal fungus *Hebeloma cylindrosporum*. Mol Microbiol 47:411–430

Javelle A, Chalot M, Brun A, Botton B (2004) Nitrogen transport and metabolism in mycorrhizal fungi and mycorrhizas. In: Varma A, Abbott L, Werner D, Hampp R (eds) Plant surface microbiology. Springer, Berlin Heidelberg New York, pp 393–430

Jonasson S, Shaver GR (1999). Within-stand nutrient cycling in arctic and boreal wetlands. Ecology 80:2139–2150

Kaska DD, Myllyla R, Cooper JB (1999) Auxin transport inhibitors act through ethylene to regulate dichotomous branching of lateral root meristems in pine. New Phytol 142:49–58

Kawano T, Kawano N, Hosoya H, Lapeyrie F (2001) Fungal auxin antagonist hypaphorine competitively inhibits indole-3-acetic acid-dependent superoxide generation by horseradish peroxidase. Biochem Biophys Res Commun 288:546–551

Kim SJ, Hiremath ST, Podila GK (1999a) Cloning and identification of symbiosis-regulated genes from the ectomycorrhizal *Laccaria bicolor*. Mycol Res 2:168–172
Kim SJ, Bernreuther D, Thumm M, Podila GK (1999b) *LB-AUT7*, a novel symbiosis-regulated gene from an ectomycorrhizal fungus, *Laccaria bicolor*, is functionally related to vesicular transport and autophagocytosis. J Bacteriol 181:1963–1967
Klironomos JN, Hart MM (2001) Animal nitrogen swap for plant carbon. Nature 410:651–652
Kohler A, Delaruelle C, Martin D, Encelot N, Martin F (2003) The poplar root transcriptome: analysis of 7000 expressed sequence tags. FEBS Lett 542:37–41
Lacourt I, Duplessis S, Abba S, Bonfante P, Martin F (2002) Isolation and characterization of differentially expressed genes in the mycelium and fruit body of *Tuber borchii*. Appl Environ Microbiol 68:4574–4582
Lagrange H, Jay-Allgmand C, Lapeyrie F (2001) Rutin, the phenolglycoside from eucalyptus root exudates, stimulates *Pisolithus* hyphal growth at picomolar concentration. New Phytol 149:349–355
Lammers P, Tuskan GA, DiFazio SP, Podila GK, Martin F (2004) Mycorrhizal symbionts of *Populus* to be sequenced by the United States Department of Energy's Joint Genome Institute. Mycorrhiza 14:63–64
Laurans F, Pepin R, Gay G (2001) Fungal auxin overproduction affects the anatomy of *Hebeloma cylindrosporum–Pinus pinaster* ectomycorrhizas. Tree Physiol 21:533–540
Laurent P, Voiblet C, Tagu D, de Carvalho D, Nehls U, De Bellis R, Balestrini R, Bauw G, Bonfante P, Martin F (1999) A novel class of ectomycorrhiza-regulated cell wall polypeptides in *Pisolithus tinctorius*. Mol Plant Microbe Interact 12:862–871
Le Quere A, Johansson T, Tunlid A (2002) Size and complexity of the nuclear genome of the ectomycorrhizal fungus *Paxillus involutus*. Fungal Genet Biol 36:234–241
Leake JR, Donnelly DP, Saunders EM, Boddy L, Read DJ (2001) Rates and quantities of carbon flux to ectomycorrhizal mycelium following C-14 pulse labeling of *Pinus sylvestris* seedlings: effects of litter patches and interactions with a wood-decomposer fungus. Tree Physiol 21:71–82
Lilleskov EA, Fahey TJ, Horton TR, Lovett GM (2002a) Belowground ectomycorrhizal fungal community change over a nitrogen deposition gradient in Alaska. Ecology 83:104–115
Lilleskov EA, Hobbie EA, Fahey TJ (2002b) Ectomycorrhizal fungal taxa differing in response to nitrogen deposition also differ in pure culture organic nitrogen use and natural abundance of nitrogen isotopes. New Phytol 154:219–231
Lindahl B, Stenlid J, Finlay R (2001) Effects of resource availability on mycelial interactions and P-32 transfer between a saptropic and ectomycorrhizal fungus in soil microcosms. FEMS Microbiol Ecol 38:43–52
Lukac M, Calfapietra C, Godbold DL (2003) Production, turnover and mycorrhizal colonization of root systems of three *Populus* species grown under CO_2 (POPFACE). Global Change Biol 9:838–848
Mankel A, Krause K, Kothe E (2002) Identification of a hydrophobin gene that is developmentally regulated in the ectomycorrhizal fungus *Tricholoma terreum*. Appl Environ Microbiol 68:1408–1413
Mari S, Jonsson A, Finlay R, Ericsson T, Kahr M, Eriksson G (2003) Genetic variation in nitrogen uptake and growth in mycorrhizal and nonmycorrhizal *Picea abies* (L.) Karst. seedlings. For Sci 49:258–267
Markkola AM, Tarvainen O, Ahonen-Jonnarth U, Strommer R (2002) Urban polluted forest soils induce elevated root peroxidase activity in Scots pine (*Pinus sylvestris* L.) seedlings. Environ Pollut 116:273–278
Martin F, Duplessis S, Ditengou F, Lagrange H, Voiblet C, Lapeyrie F (2001) Developmental cross talking in the ectomycorrhizal symbiosis: signals and communication genes. New Phytol 151:145–154
McDowell NG, Balster NJ, Marshall JD (2001) Belowground carbon allocation of Rocky Mountain Douglas-fir. Can J Forest Res 31:1425–1436

McKendrick SL, Leake JR, Taylor DL, Read DJ (2000) Symbiotic germination and development of myco-heterotrophic plants in nature: transfer of carbon from ectomycorrhizal *Salix repens* and *Betula pendula* to the orchid *Corallorhiza trifida* through shared hyphal connections. New Phytol 145:539–548
Mogge B, Loferer C, Agerer R, Hutzler P, Hartmann A (1999) Bacterial community structure and colonization patterns of *Fagus sylvatica* L-ectomycorrhizospheres as determined by fluorescence in situ hybridization and confocal laser scanning microscopy. Mycorrhiza 9:271–278
Montanini B, Moretto N, Soragni E, Percudani R, Ottonello S (2002) A high-affinity ammonium transporter from the mycorrhizal ascomycete *Tuber borchii*. Fungal Genet Biol 36:22–34
Morte A, Diaz G, Rodriguez P, Alarcon JJ, Sanchez-Blanco MJ (2001) Growth and water relations in mycorrhizal and nonmycorrhizal *Pinus halepensis* plants in response to drought. Biol Plant 44:263–267
Murata H, Miyazaki Y (2001) Expression of *marY1*, a gypsy-type LTR-retroelement from the ectomycorrhizal homobasidiomycete *Tricholoma matsutake*, in the budding yeast *Saccharomyces cerevisiae*. Biosci Biotechnol Biochem 65:993–995
Murata H, Yamada A (2000) *marY1*, a member of the gypsy group of long terminal repeat retroelements from the ectomycorrhizal basidiomycete *Tricholoma matsutake*. Appl Environ Microbiol 66:3642–3645
Mushin TM, Zwiazek JJ (2002a) Ectomycorrhizas increase apoplastic water transport and root hydraulic conductivity in *Ulmus americana* seedlings. New Phytol 152:153–158
Mushin TM, Zwiazek JJ (2002b) Colonization with *Hebeloma crustuliniforme* increases water conductance and limits shoot sodium uptake in white spruce (*Picea glauca*) seedlings. Plant Soil 238:217–225
Nasholm T, Ekblad A, Nordin A, Giesler R, Hogberg M, Bogberg P (1998) Boreal forest plants take up organic nitrogen. Nature 392:914–916
Nehls U (2003) Ectomycorrhizal development and function – transcriptome analysis. New Phytol 159:5–7
Nehls U (2004) Carbohydrates and nitrogen: nutrients and signals in ectomycorrhizas. In: Varma A, Abbott L, Werner D, Hampp R (eds) Plant surface microbiology. Springer, Berlin Heidelberg New York, pp 373–392
Nehls U, Mikolajewski S, Ecke M, Hampp R (1999a) Identification and expression analysis of two fungal cDNAs regulated by ectomycorrhiza and fruit body formation. New Phytol 144:195–202
Nehls U, Kleber R, Wiese J, Hampp R (1999b) Isolation and characterization of a general amino acid permease from the ectomycorrhizal fungus *Amanita muscaria*. New Phytol 144:343–349
Nehls U, Bock A, Einig W, Hampp R (2001) Excretion of two proteases by the ectomycorrhizal fungus *Amanita muscaria*. Plant Cell Environ 24:741–747
News in brief (2002) Geneticists branch out to sequence poplar genome. Nature 415:725
Niemi K, Haggman H (2002) *Pisolithus tinctorius* promotes germination and forms mycorrhizal structures in Scots pine somatic embryos in vitro. Mycorrhiza 12:263–267
Niemi K, Salonen M, Ernstsen A, Heinonen-Tanski H, Haggman K (2000) Application of ectomycorrhizal fungi in rooting of Scots pine fascicular shoots. Can J Forest Sci 30:1221–1230
Niemi K, Haggman H, Sarjala T (2002a) Effects of exogenous diamines on the interaction between ectomycorrhizal fungi and adventitious root formation in Scots pine in vitro. Tree Physiol 22:373–381
Niemi K, Vuorinen T, Ernstsen A, Haggman H (2002b) Ectomycorrhizal fungi and exogenous auxins influence root and mycorrhiza formation of Scots pine hypocotyl cuttings in vitro. Tree Physiol 22:1231–1239

Niklaus PA, Wohlfender M, Siegwolf R, Körner C (2001) Effects of six years of atmospheric CO_2 enrichment on plant, soil and soil microbial C of a calcareous grassland. Plant Soil 233:189–2002

Nilsson LO, Wallander H (2003) Production of external mycelium by ectomycorrhizal fungi in a Norway spruce forest was reduced in response to nitrogen fertilization. New Phytol 158:409–416

Osaku CA, Sassaki GL, Zancan GT, Iacomini M (2002) Studies on neutral exopolysaccharides produced by the ectomycorrhizal fungus *Thelephora terrestris*. FEMS Microbiol Lett 216:145–149

Ott T, Fritz E, Polle A, Schutzendubel A (2002) Characterisation of antioxidative systems in the ectomycorrhiza-building basidiomycete *Paxillus involutus* (Bartsch) Fr. and its reaction to cadmium. FEMS Microbiol Ecol 42:359–366

Pardo AG, Hanif M, Raudaskoski M, Gorfer M (2002) Genetic transformation of ectomycorrhizal fungi mediated by *Agrobacterium tumefaciens*. Mycol Res 106:132–137

Perez-Moreno J, Read DJ (2001a) Nutrient transfer from soil nematodes to plants: a direct pathway provided by the mycorrhizal mycelial network. Plant Cell Environ 24:1219–1226

Perez-Moreno J, Read DJ (2001b) Exploitation of pollen by mycorrhizal mycelial systems with special reference to nutrient recycling in boreal forests. Proc R Soc Lond B Biol Sci 268:1329–1335

Persson J, Nasholm T (2001) Amino acid uptake: a widespread ability among boreal forest plants. Ecol Lett 4:434–438

Peter M, Courty PE, Kohler A, Christine Delaruelle C, David Martin D, Tagu D, Frey-Klett P, Duplessis S, Chalot M, Podila G, Martin F (2003) Analysis of expressed sequence tags from the ectomycorrhizal basidiomycetes *Laccaria bicolor* and *Pisolithus microcarpus*. New Phytol 159:117–129

Podila GK, Zheng J, Balasubramanian S, Sundaram S, Hiremath S, Brand JH, Hymes MJ (2002) Fungal gene expression in early symbiotic interactions between *Laccaria bicolor* and red pine. Plant Soil 244:117–128

Polidori E, Agostini D, Zeppa S, Potenza L, Palma F, Sisti D, Stocchi V (2002) Identification of differentially expressed cDNA clones in *Tilia platyphyllos–Tuber borchii* ectomycorrhizae using a differential screening approach. Mol Genet Genom 266:858–864

Poole EJ, Bending GD, Whipps JM, Read DJ (2001) Bacteria associated with *Pinus sylvestris–Lactarius rufus* ectomycorrhizas and their effects on mycorrhiza formation in vitro. New Phytol 151:743–751

Probanza A, Mateos JL, Garcia JAL, Ramos B, deFelipe MR, Manero FJG (2001) Effects of inoculation with PGPR *Bacillus* and *Pisolithus tinctorius* on *Pinus pinea* L. growth, bacterial rhizosphere colonization, and mycorrhizal infection. Microbial Ecol 41:140–148

Rangel-Castro JI, Danell E, Pfeffer PE (2002) A C-13-NMR study of exudation and storage of carbohydrates and amino acids in the ectomycorrhizal edible mushroom *Cantharellus cibarius*. Mycologia 94:190–199

Raudaskoski M, Pardo A, Tarkka M, Gorfer M, Hanif M, Laitiainen E (2001) Small GTPases, cytoskeleton and signal transduction in filamentous homobasidiomycetes. NATO Sci Ser I Life Behav Sci 328:123–136

Reddy SM, Pandey AK, Melayah D, Marmeisse R, Gay G (2003) The auxin responsive gene *Pp-C61* is up-regulated in *Pinus pinaster* roots following inoculation with ectomycorrhizal fungi. Plant Cell Environ 26:681–691

Ronning CM, Stegalkina SS, Ascenzi RA, Bougri O, Hart AL, Utterbach TR, Vanaken SE, Riedmüller SB, White JA, Cho J, Pertea GM, Lee Y, Karamycheva S, Sultana R, Tsai J, Quackenbush J, Griffiths HM, Restrepo S, Smart CD, Fry WE, van der Hoeven R, Tanksley S, Zhang P, Jin H, Yamamoto ML, Baker BJ, Buell RC (2003) Comparative analyses of potato expressed sequence tag libraries. Plant Physiol 131:419–429

Sarand I, Haario H, Jorgensen KS, Romantschuk M (1999a) Effect of inoculation of a *TOL* plasmid containing mycorrhizosphere bacterium on development of Scots pine seed-

lings, their mycorrhizosphere and the microbial flora in *m*-toluate-amended soil. FEMS Microbiol Ecol 31:127–141
Sarand I, Timonen S, Koivula T, Peltola R, Haahtela K, Sen R, Romantschuk M (1999b) Tolerance and biodegradation of *m*-toluate by Scots pine, a mycorrhizal fungus and fluorescent pseudomonads individually and under associative conditions. J Appl Microbiol 86:817–826
Sawyer NA, Chambers SM, Cairney JW (2003a) Utilisation of inorganic and organic nitrogen sources by *Amanita* species native to temperate eastern Australia. Mycol Res 107:413–420
Sawyer NA, Chambers SM, Cairney JW (2003b) Variation in nitrogen source utilisation by nine *Amanita muscaria* genotypes from Australian *Pinus radiata* plantations. Mycorrhiza 13:217–221
Sbrana C, Agnolucci M, Bedini S, Lepera A, Toffanin A, Giovannetti M, Nuti MP (2002) Diversity of culturable bacterial populations associated to *Tuber borchii* ectomycorrhizas and their activity on *T. borchii* mycelial growth. FEMS Microbiol Lett 211:195–201
Scheller E (1996) Aminosäuregehalte von Ap- und Ah-Horizonten verschiedener Böden und deren Huminsäuren- und Fulvosäuren-Fraktion. Mitt Dtsch Bodenkundl Ges 81:201–204
Shi L, Guttenberger M, Kottke I, Hampp R (2002) The effect of drought on mycorrhizas of beech (*Fagus sylvatica* L.): changes in community structure, and the content of carbohydrates and nitrogen storage bodies of the fungi. Mycorrhiza 12:303–311
Skinner W, Keon J, Hargreaves J (2001) Gene information for fungal plant pathogens from expressed sequences. Curr Opin Microbiol 4:381–386
Sophianopoulou V, Diallinas G (1995) Amino acid transporters of lower eukaryotes: regulation, structure and topogenesis. FEMS Microbiol Rev 16:53–75
Springael JY, Andre B (1998) Nitrogen-regulated ubiquitination of the Gapl permease of *Saccharomyces cerevisiae*. Mol Biol Cell 9:1253–1263
Sterky F, Regan S, Karlsson J, Hertzberg M, Rohde A, Holmberg A, Amini B, Bhalerao R, Larsson M, Villarroel R, Van Montagu M, Sandberg G, Olsson O, Teeri TT, Boerjan W, Gustafsson P, Uhlen M, Sundberg B, Lundeberg J (1998) Gene discovery in the wood-forming tissues of poplar: analysis of 5,692 expressed sequence tags. Proc Natl Acad Sci USA 95:13330–13335
Sundaram S, Kim SJ, Suzuki H, Mcquattie CJ, Hiremath ST, Podila GK (2001) Isolation and characterization of a symbiosis-regulated *ras* from the ectomycorrhizal fungus *Laccaria bicolor*. Mol Plant Microbe Interact 14:618–628
Tagu D, Kottke I, Martin F (1998) Hydrophobins in ectomycorrhizal symbiosis: hypothesis. Symbiosis 25:5–18
Tagu D, Lapeyrie F, Ditengou F, Lagrange H, Laurent P, Missoum N, Nehls U, Martin F (2000) Molecular aspects of ectomycorrhiza development. In: Podila G, Douds DD Jr (eds) Current advances in mycorrhizal research. American Phytopathological Society, Saint Paul, Minnesota, pp 69–90
Tagu D, de Bellis R, Balestrini R, de Vries OMH, Piccoli G, Stocchi V, Bonfante P, Martin F (2001) Immunolocalization of hydrophobin HYDPt-1 from the ectomycorrhizal basidiomycete *Pisolithus tinctorius* during colonization of *Eucalyptus globulus* roots. New Phytol 149:127–135
Tagu D, Marmeisse R, Baillet Y, Riviere S, Palin B, Bernardini F, Mereau A, Gay G, Balestrini R, Bonfante P, Martin F (2002) Hydrophobins in ectomycorrhizas: heterologous transcription of the *Pisolithus HydPt-1* gene in yeast and *Hebeloma cylindrosporum*. Eur J Histochem 46:23–29
Tarkka MT, Vasara R, Gorfer M, Raudaskoski M (2000) Molecular characterization of actin genes from homobasidiomycetes: two different actin genes from *Schizophyllum commune* and *Suillus bovinus*. Gene 251:27–35
Tarkka MT, Nyman TA, Kalkkinen N, Raudaskoski M (2001) Scots pine expresses short-root-specific peroxidases during development. Eur J Biochem 268:86–92

TerSchure EG, van Rielb NAB, Verrips CT (2000) The role of ammonia metabolism in nitrogen catabolite repression in *Saccharomyces cerevisiae*. FEMS Microbiol Rev 24:67–83

Tibbett M, Sanders FE, Cairney JWG (2002) Low-temperature-induced changes in trehalose, mannitol and arabitol associated with enhanced tolerance to freezing in ectomycorrhizal basidiomycetes (*Hebeloma* spp.). Mycorrhiza 12:249–255

Voiblet C, Duplessis S, Encelot N, Martin F (2001) Identification of symbiosis-regulated genes in *Eucalyptus globulus–Pisolithus tinctorius* ectomycorrhiza by differential hybridization of arrayed cDNAs. Plant J 25:181–191

Vonderwell JD, Enebak SA, Samuelson LJ (2001) Influence of two plant growth-promoting rhizobacteria on loblolly pine root respiration and IAA activity. For Sci 47:197–202

Wallenda T, Read DJ (1999) Kinetics of amino acid uptake by ectomycorrhizal roots. Plant Cell Environ 22:179–187

Wiemken V, Boller T (2002) Ectomycorrhiza: gene expression, metabolism and the wood-wide web. Curr Opin Plant Biol 5:1–7

Wiemken V, Laczko E, Ineichen K, Boller T (2001) Effects of elevated carbon dioxide and nitrogen fertilization on mycorrhizal fine roots and the soil microbial community in beech–spruce ecosystems on siliceous and calcareous soil. Microbial Ecol 42:126–135

Wipf D, Benjdia M, Tegeder M, Frommer WB (2002) Characterization of a general amino acid permease from *Hebeloma cylindrosporum*. FEBS Lett 528:119–124

Wipf D, Benjdia M, Rikirsch E, Zimmermann S, Tegeder M, Frommer WB (2003) An expression cDNA library for suppression cloning in yeast mutants, complementation of a yeast *his4* mutant, and EST analysis from the symbiotic basidiomycete *Hebeloma cylindrosporum*. Genome 46:177–181

Wollecke J, Munzenberger B, Huttl RF (1999) Some effects of N on ectomycorrhizal diversity of Scots pine (*Pinus sylvestris* L.) in northeastern Germany. Water Air Soil Pollut 116:135–140

Woodward FI (2002) Potential impacts of global CO_2 concentrations on plants. Curr Opin Plant Biol 5:207–211

Wu BY, Nara K, Hogetsu T (2002) Spatiotemporal transfer of carbon-14-labelled photosynthate from ectomycorrhizal *Pinus densiflora* seedlings to extraradical mycelia. Mycorrhiza 12:83–88

Young BW, Massicotte HB, Tackaberry LE, Baldwin QF, Egger KN (2002) *Monotropa uniflora*: morphological and molecular assessment of mycorrhizae retrieved from sites in the sub-boreal spruce biogeoclimatic zone in central British Columbia. Mycorrhiza 12:75–82

Yu Z, Zhang Q, Kraus TEC, Dahlgren RA, Anastasio C, Zasoski RJ (2002) Contribution of amino compounds to dissolved organic nitrogen in forest soils. Biogeochemistry 61:173–198

Mika Tarkka
Uwe Nehls
Rüdiger Hampp
Botanical Institute
Physiological Ecology of Plants
University of Tübingen
Auf der Morgenstelle 1
72076 Tübingen, Germany
e-mail: ruediger.hampp@uni-tuebingen.de

Network Dynamics in Plant Biology: Current Progress in Historical Perspective

Marc-Thorsten Hütt and Ulrich Lüttge

1 The Scene of Current Fashions: Systems Biology and Networks

In science there appear to be cyclic evolutions of the dominance of one of two contrasts, i.e.,

- generalists or specialists;
- the quest for understanding interactive functions or discovering and describing structures;
- integrating of the whole or dissecting of the parts.

The second of the two always tends to take over and dominate when there are methodological breakthroughs. In biology, we may illustrate this historically (see Sect. 4) by recalling the advents of enzymology/biochemistry and electron microscopy. We are currently confronted with it in ecology and by molecular cell biology.

With improved and miniaturized instrumentation for fieldwork and with large-scale and global remote sensing and data logging systems, we accelerate the acquisition of ecologically relevant data. As another example, envisage the consequences of fumigating a giant tree in a tropical rain forest and the harvest of thousands and thousands of insects coming down on the collecting mats. Our expectations of the number of insect species – all to be described and catalogued – on our planet are revolutionized (Erwin 1982, 1995). Moreover, our idea of the number of species on earth (May 1988) and our understanding of biodiversity (Linsenmair 1995; Kratochwil 1998) also need revolutionizing of our thinking (as in the references cited).

In molecular cell biology, we are moving with an accelerating pace , in a so-called post-genomic era barely 10 years after the vigorous onset of genome sequencing, from genomics to transcriptomics, proteomics, metabolomics, and what else? (A new term recently coined at a conference was 'channelomics'.) This is hastened by the development of large-scale screening approaches, such as microarray techniques, imaging, micro- and nanotechnologies etc. (Girke et al. 2003). The speed of acquisition and

Progress in Botany, Vol. 66

piling up of mountains of data – notably for which we can perform 'mining' in the web – is becoming bewildering. This bears the danger that data accumulation remains unreflected, as noted in a critical evaluation by I. Stengers: "...chose qui caractérise nos sciences depuis quelques décennies: l'inventivité technique va plus vite que la production des questions..." ("...characteristic of our sciences for several decades is that the invention of techniques is faster than the production of questions..."; Isabelle Stengers after de Perlinghi 2003). In fact, this is not all that new. On 25 July 1831 in a letter to Johann Wolfgang von Goethe, Christian Gottfried Nees von Esenbeck complained that "...unsere Wissenschaftsmänner vom Fach haben heut zu Tage sehr wenig Muße und wollen nur Resultate..." ("...these days our expert scientists have very little leisure and only want results...") (Kanz 2003). In those days, as well as now, it was and is certainly wrong to be satisfied with that.

On the time scale of years and decades one observes an alternation between the two contrasting attitudes of science pointed out above, with the first attitude often coming up when the diversity of experimental findings calls for unifying thoughts. Currently, in biology, this seems to be the case. We thus need approaches that can help to arrange our knowledge of details in a common conceptual framework. Two new concepts that have emerged over the last few years are 'systems biology' and 'networks'.

Systems biology is emerging from genome sequencing. It aims to collect comprehensive data on cells ('cellome') and organisms (Raikhel and Coruzzi 2003) with 'accurate measurement of thousands of molecules from complex biological samples' (Sweetlove et al. 2003). This very much sounds like and perhaps really is first of all a call for ever-more efficient machines and scientific data-collecting factories. However, the conceptual idea behind it is comprehensive modelling, i.e., 'the complete mathematical description of the metabolism of a model plant species' (Sweetlove et al. 2003; see also Raikhel and Coruzzi 2003). This is what we call *maximum model* (Hütt and Lüttge 2002). It aims to provide a perfectionist 'photographic' image of organisms. Its virtue – when based on comprehensive sets of data – is that it will allow fine-tuned simulations which can be regarded as a realistic alternative to actual experiments. It has several restraints though. First, notwithstanding the fast and considerable advances in technical developments noted above, it still appears utopian to expect to obtain complete data sets for organisms, i.e., not only cells but different cell types, tissues and organs. It is estimated, for example, that plants contain at least 200,000 different metabolites, posing immense technical analytical problems (Weckwerth 2003). Second, one needs to restrict oneself to so-called models (not to be confused with mathematical simulation models) that fulfil certain requirements. For higher plants, this model is *Arabidopsis thaliana*, with a small enough genome, a short annual life cycle, a small plant size easy to handle in cultures in vitro, high seed production, and readily offering itself for mutagenesis and genetic transformation. Notwithstanding the great progress obtained by restriction to such model organisms, are we not at the same time limiting our scope and view increasingly by concentrating on such models? The – often very much searched for – simplicity and typicality mask an understanding of diversity. For example, the esteemed international journal *Plant Physiology* dedicated a whole issue (December 2001) to a wide scope of exciting and important observations we would not make and insights we would not gain by restricting ourselves to *Arabidopsis*. Thus, while the '*in silico plant*' (Minorsky 2003) still remains idealistic, and even if utopia were reached eventually the outlook will be limited, there still remains much appeal of reductionism *sensu stricto*, i.e., with reduction of the degrees of freedom in approaches of mathematical *minimal models*. These are no less demanding on creative – rather than only management oriented – bioinformatics and requiring tools of nonlinear dynamics (Hütt and Lüttge 2002) than the maximum models.

The other new concept is networks. It emerged from a theoretical analysis of small-world scenarios, which appeared readily applicable to a multidisciplinary array of human activities in social sciences, economy, technology and engineering, physics and chemistry, and in the life sciences including plant biology (Watts and Strogatz 1998; Watts 1999; Strogatz 2001). While only a couple of years ago in plant biology thinking in the frame of networks appeared to be quite new and original, it has very rapidly become a fashion and it is hardly possible to read a paper in the literature on regulation, signaling and the like without stumbling over the term 'network'.

'Network' is also used in connection with systems biology. However, rather than dissecting the parts, one of the major accomplishments of network dynamics is that of gaining an overview by the global delineation of key nodes of systems as well as by collective performance where systems become nodes of super-networks and, thus, help to integrate the whole.

2 The Concept of Modules and Super-Networks

Analysis focusing on the collective behavior of networks consisting of dynamic units could complement efficiently current research in network topology and network reconstruction, where the aim is to obtain the full architecture of a particular network from experimental data and to find global observables for comparing different networks. From our point of view, both the theoretical framework of network dynamics and the acquisition of appropriate data are now at the point where this new view can fruitfully be adapted to the case of biology. The approach is invariant with respect to spatial and temporal scales. It can thus be applied as a unifying concept at all levels.

The theory of networks is established in mathematics in the field of 'graph theory'. Many applications of the mathematical concepts originate from studies of network architecture provided by (linear) systems theories in the early 1980s. In biology, observation of networks is familiar from food webs in ecology, species relationships in dynamics of populations, reaction networks in metabolism with pools and flows of metabolites (Jeong et al. 2000), the theory of neural networks developed in the 1960s and 1970s (see, e.g., Amit 1989), and genetic networks in relation to speciation and evolution (Bornholdt and Sneppen 1998; Bornholdt 2001).

A rather new field, however, is the study of *network dynamics*, where an ensemble of interacting (nonlinear) dynamic elements with a given architecture evolves, in time producing some 'spatio'-temporal pattern. This has to be distinguished from what is usually called 'dynamic networks', which consist of static elements with a dynamic (i.e., varying) architecture. In the latter case, the underlying theoretical concept is the theory of phase transitions with emphasis on percolation theory. In the former case, one has networks of dynamic elements and attempts to understand how network *structure* influences network *dynamics*. Some of the key ideas of this approach can be traced to the seminal work by Kauffman (1969) on the dynamics of networks consisting of binary elements (Boolean networks).

If we take the individual details we know as modules (Hartwell et al. 1999) or elements of networks – and notably, they may be 'sub-networks' already in themselves – we may knit them into networks, those networks into 'super-networks', and, thus, by iteration we may conceptually arrive at a complex hierarchical organization involving several (or even many) different scales in space and time. In this way, we obtain manageable conceptual entities at each level, putting details into context. As we will discuss in Section 3, the distinction between modular networks and hierarchical networks, which has emerged over the last few years, is the result of an intense and fruitful interplay between theory and experimental observation. Learning to translate mountains of, e.g., cDNA microarray data into the language of graph theory, apart from revolutionizing our view of regulation, also refined the categories of graph theory. In many ways such a description, when added to the methodological repertoire of bioinformatics, may help improve the interpretation of huge amounts of data and thus help the field advance from administrative management of data mountains to conceptual command.

3 The Theoretical Framework of Network Dynamics

One of the most important developments in theoretical biology over the past few years has certainly been the application of innovations in graph theory to the analysis of biological processes. In 1998, Watts and Strogatz proposed the construction of a graph with short average path length and – at the same time – a high clustering coefficient (Watts and Strogatz 1998). They triggered a wave of research activity, and it turned out that such 'small-world' graphs represent an adequate model for many technical, natural and biological systems (Fell and Wagner 2000; Newman 2001; Strogatz 2001). The construction of graphs with a scale-free degree distribution, as introduced by Barabási and Albert (1999), has been similarly successful for the modelling of real systems (Jeong et al. 2000; Wagner and Fell 2001). It has been shown that especially biological networks, such as metabolic, protein and gene networks, are based on scale-free topologies (Jeong et al. 2001; Farkas et al. 2003; Barabási and Oltvai 2004).

In this section, we will briefly summarize some concepts from graph theory and then move on to outlining the tools for implementing dynamics on a given graph. In Section 6, we apply this theoretical framework to a case study of simple forms of network dynamics. The guiding questions then will be to what extent and under what conditions are the observed dynamics a reflection of internal topological properties of the underlying graph.

Networks consist of *elements* (nodes, vertices) and *interactions* (connections, links) between them. The four classes of graphs most important for biological application are (1) regular, (2) random, (3) small-world and (4) scale-free graphs. We will briefly and qualitatively summarize prescriptions for constructing such graphs, as well as their most important properties. For more details, as well as a thorough mathematical discussion, see the excellent reviews by Albert and Barabási (2002) and Newman (2003), as well as the large outline of applications and open theoretical questions by Strogatz (2001). The recent, popular non-technical book by Barabási (2002) gives a very readable account of the appearances of graphs in everyday life.

In the formulation of biological models, *regular graphs* are used, for example, as approximations of spatially organized elements or as extreme cases, when network architecture is assumed to be of minor importance for the behavior of the system. For instance, a fully connected graph underlies the remarkable investigation of spontaneous synchronization of coupled phase oscillators by Kuramoto (1984). A system of elements with coupling to spatial neighbors (e.g., on a two-dimensional spatial lattice with the nodes being placed at all lattice sites) is an adequate starting point for investigating spatiotemporal pattern formation. It is clear, however, that often a network is not related to real (physical) space. For example, the elements could be biochemically active substances in a solution, with the network being defined by the capacity of each of the substances to interact with (e.g. activate or inhibit formation of) itself and the other substances. Such networks are called *relational networks* (as opposed to spatial networks). In these cases, the question of network architecture often becomes a key issue when one wants to understand the dynamic properties of a network.

Random graphs have been investigated by Erdös and Rényi (1960). These studies constituted a huge step forward in graph theory. In the Erdös-Rényi (ER) model, one has a system of N nodes. Each possible link has a probability p of being present in the graph. For this model of a random graph a large variety of mathematical properties can be obtained analytically. So, what quantities are useful for characterizing topological features of a graph?

On the level of individual nodes the most important observable is the *degree k* of a node, i.e., the number of nodes ('neighbors') linked to it. Averaging over all nodes one obtains the *average degree* of a graph. Often, particularly for scale-free graphs, instead of averaging it is more instructive to study the degree distribution $P(k)$, where $P(k)$ is the number of times the degree k is found in the graph. A different property of graph architecture is addressed with the so-called *clustering coefficient* that was introduced by Watts and Strogatz (1998). Their aim was to provide an efficient quantita-

Fig. 1. Typical examples of graphs for four of the classes discussed in the main text, together with their degree distributions. The classes are **a** random graphs, small-world graphs obtained by **b** rewiring and **c** link-adding, and **d** scale-free graphs. The examples show graphs with 100 nodes, while the degree distributions have been computed for graphs consisting of 1,000 nodes. For the random graph (**a**) the distribution $P(k)$ shows a pronounced peak, which approximates a Poisson distribution. Here, the connection probability has been set to p=0.05. The two types of small-world graphs are difficult to distinguish on the level of the graph itself. The degree distributions, however, show clear differences related to the respective prescriptions for generating these graphs. For the rewiring procedure (**b**) the degree distribution consists of isolated lines for low and intermediate rewiring probabilities. *Inset* shows an extreme case of very low rewiring probability, namely, $p=10^{-3}$. The signature of the regular backbone, a ring of nodes with links from each node to its eight nearest neighbors, from which the rewiring process is started, is clearly visible as an isolated line at k=8. Due to rewiring, two small lines appear at k=7 and k=9. *Main part of figure* gives the degree distribution for $p=10^{-2}$. There, lines with a larger deviation from the k=8 line are found. At the same value of p the other type of small-world graph (**c**) (obtained by the link-adding procedure) displays a degree distribution much more similar to the case of a random (ER) graph. The scale-free graph (**d**) has a completely different degree distribution, which can be described by a power law. *Inset* shows the same distribution in a double-logarithmic representation (a log-log plot). Already for 1,000 nodes the characteristic linear behavior in the log-log plot, which is the signature of a power-law distribution, is obvious

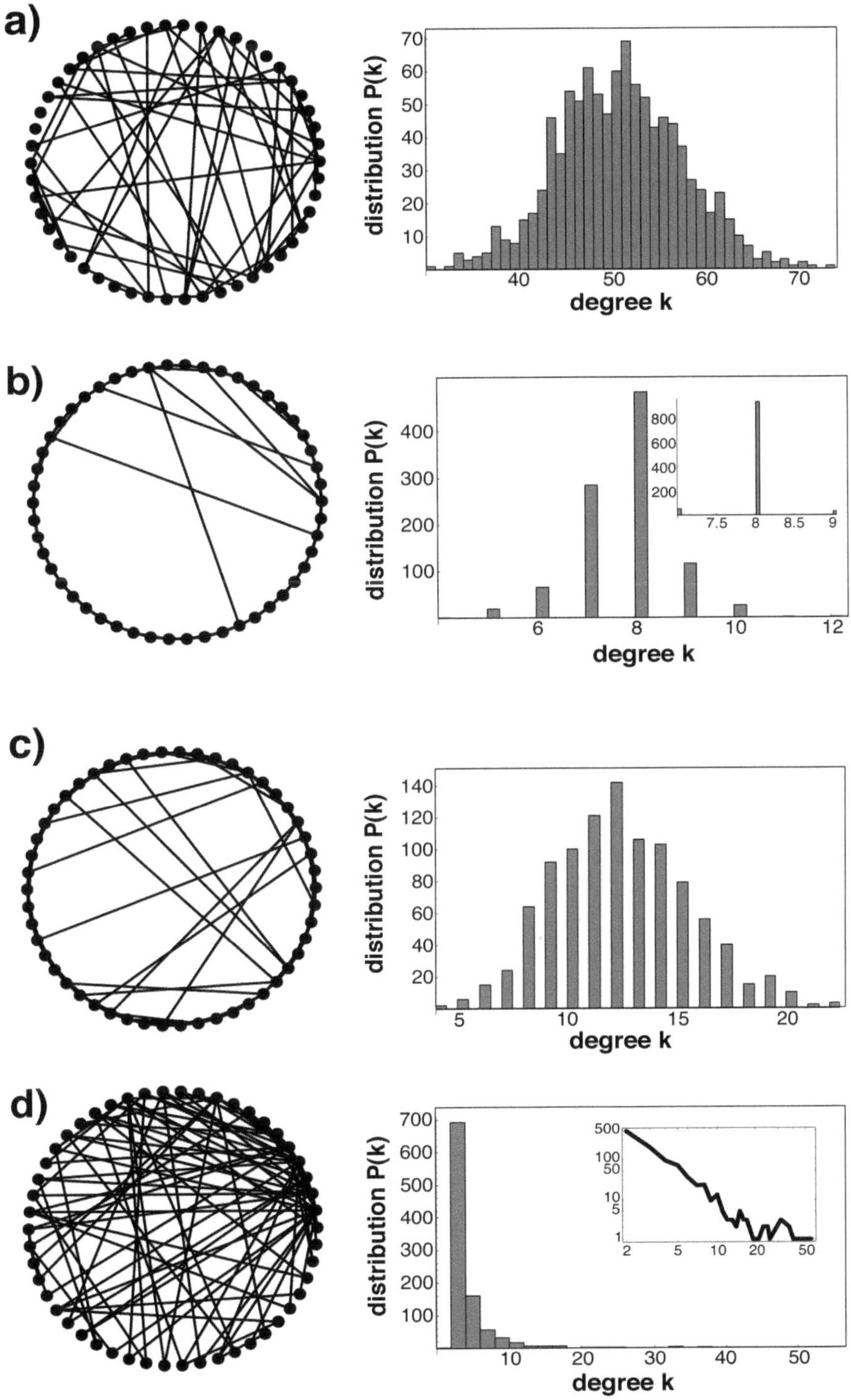
a)
distribution P(k)
70
60
50
40
30
20
10
40
50
60
70
degree k
b)
distribution P(k)
400
300
200
100
6
8
10
12
degree k
800
600
400
200
7.5
8
8.5
9
c)
distribution P(k)
140
120
100
80
60
40
20
5
10
15
20
degree k
d)
distribution P(k)
700
600
500
400
300
200
100
10
20
30
40
50
degree k
500
100
50
10
5
1
2
5
10
20
50

tive distinction between small-world graphs and other random graphs. As the degree introduced above, the clustering coefficient is an observable at the level of an individual node. Let us consider a node with degree k. The clustering coefficient C of this particular node is determined by the number of links the k neighbors have between them. When this number, say m, is normalized by the number of possible links between these k nodes, namely $k\,(k-1)/2$, one obtains C. Again, one can characterize the whole graph both by the average clustering coefficient and by the distribution $P(C)$ of clustering coefficients in this graph.

One might expect that in a random graph given by the ER model a wide range of degrees appears, each with a similar frequency. Counter-intuitively, for ER graphs the function $P(k)$ follows a Poisson distribution with a pronounced peak at some average degree k. Figure 1a gives an example of such a degree distribution for a graph consisting of 1,000 nodes. Furthermore, the average clustering coefficient is very low. This is the most striking difference to real (technical or natural) graphs, which usually, in addition to a certain amount of randomness, display a high clustering coefficient.

A remarkable topological property of random graphs that was discovered by Erdös and Rényi (1960) concerns the *connectedness* of a graph. A graph G is connected if any node can be reached from any other node of G by following the links present in G. Clearly, at very small connection probability p the random graph will not be connected. Conversely, close to $p=1$ connectedness is certain. However, how does connectedness emerge when p is increased from zero to one? Figure 2 shows the corresponding curve giving the percentage of connected graphs as a function of p, together with examples of disconnected graphs at low p with the largest connected component highlighted in grey.

Watts and Strogatz (1998) introduced a technique called *random rewiring*, which allowed them to define and investigate *small-world networks*. One starts from a regular network structure (e.g., a chain of nodes, where each node is connected to its k_0 nearest neighbors) and randomly rewires links in this graph, i.e., each link present in the regular 'backbone' has a probability p (the rewiring probability) of being disconnected from one of its nodes and randomly attached to another node. Watts and Strogatz (1998) named these structures 'small-world networks' because, typically, the distance (i.e., the number of vertices on the path) between two elements is rather small, in spite of the high clustering coefficient resulting from the regular backbone. This phenomenon is well known from social networks [cf. e.g., J. Guare's acclaimed theatre play *Six Degrees of Separation* (Guare 1990) and the seminal study of social networks by Milgram (1967)]. Watts (1999) argues that many social and presumably also many biological and biochemical networks are of this small-world type.

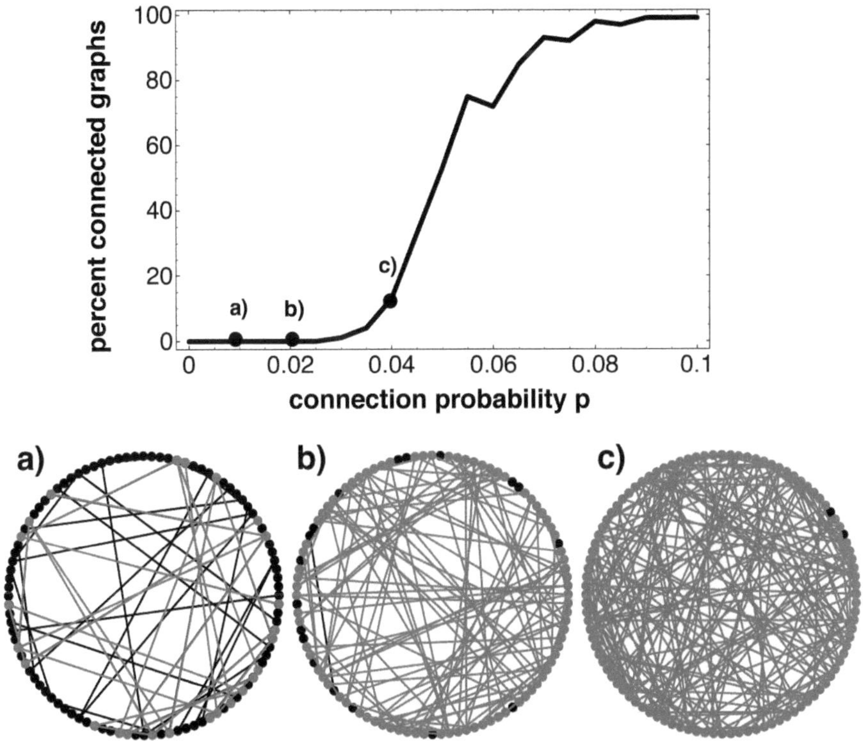

Fig. 2. Summary of the classical result of the Erdös-Rényi (ER) random graph, namely that connectedness emerges spontaneously when the connection probability p is increased and a critical value of p is surpassed. Percentage of connected graphs as a function of the connection probability p for the ER model is shown in *upper half of figure*, while the *lower part* gives three examples of disconnected graphs, which appear at small p. The *curve* has been obtained by generating 100 realizations of a graph consisting of 300 nodes for each value of p and then evaluating connectedness. Note that even for this comparatively small system size the phase transition-like behavior of the curve is clearly seen. The *three graphs* given in the *lower part of the figure* have been obtained for p=0.01 (**a**), 0.02 (**b**) and 0.04 (**c**), respectively. Each consists of 300 nodes. In all cases the largest connected component is shown in *gray*. One sees that the size of this component increases rapidly with p

A simple example of this phenomenon is seen in Fig. 3. There a typical small-world network is obtained by systematically (not randomly) 'rewiring' the fully connected individual clusters seen in Fig. 3a by breaking up one link and transferring it to a node of a neighboring cluster (Fig. 3b). An idea of the topological implications of this rewiring process is obtained when one considers how symmetry between the elements of a cluster is broken in the course of rewiring: Before, all elements were identical in terms of connectivity and arrangement. After the rewiring, four types of vertices exist in each cluster, easily distinguished by the number of connections.

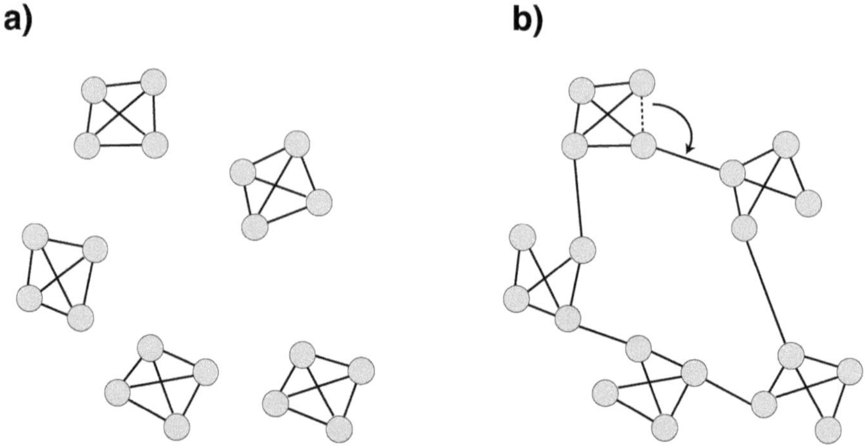

Fig. 3. Generation of hierarchical levels by interlinking formerly independent units. Each of the sub-networks in **a** has a specific dynamic behavior resulting from the nodes and interactions between them. Upon an external signal or as a result of spontaneous self-organization theses sub-networks can form a network on a larger scale (**b**). As a result of this 'rewiring' some collective behavior emerges. The resulting network has two length scales. (Adapted from Watts 1999)

With a few subsidiary conditions (cf. Watts 1999) this structure is called a *connected caveman network.*

An alternative method for obtaining small-world graphs is given by the *link-adding procedure*, which is described, e.g., in Strogatz (2001). There, the links of the regular backbone are not rewired, but links are randomly added to the graph. The equivalence of this procedure with respect to graph properties is discussed in Albert and Barabási (2002).

Currently, we see more and more networks at the genetic, protein or metabolic level becoming publicly available (see Sect. 5 for examples). As pointed out in Section 1, this is one of the striking signatures of current postgenomic research. We will come back to this aspect in Section 5. It is one of the main results of the last few years of intense studies of biological networks, particularly at the level of genes, proteins and metabolites, that a *scale-free graph* more accurately represents the topology of such networks (Jeong et al. 2000; Wagner and Fell 2001). The defining property of a scale-free graph is a degree distribution obeying a power law, $P(k) \propto k^{-\gamma}$, with the power law exponent γ. A prescription for generating scale-free graphs with $\gamma=3$ has been given by Barabási and Albert (1999). One starts with a small random graph consisting of N_0 nodes and iteratively adds nodes to this structure, where each added node is randomly linked to m other nodes in the graph. The probability of such a link to be formed, however, depends on the degree of the existing node: the higher the degree of a node in the graph the higher is the probability of a link to the newly added node. This rule is called *preferential attachment.*

Figure 1 summarizes these introductory comments on the different types of graphs by showing typical examples of such graphs, together with the corresponding degree distributions.

Maslow and Sneppen (2002) have been able to show the so-called disassortative property of biological networks: the 'hubs' in protein networks, i.e., the highly connected nodes, have a higher probability of linking to nodes with very low degree. The investigations by the Barabási group show that the class of scale-free graphs has to be subdivided according to the exponent γ, which characterizes the degree distribution (see Barabási and Oltvai 2004 for references). While the prescription based on preferential attachment leads to $\gamma = 3$, most biological networks yield $2<\gamma<3$. In these networks, one has a pronounced 'network-within-a-network' structure with hubs being present at all scales. For these cases, the term *hierarchical network* has been introduced (Ravasz et al. 2002), as already mentioned in Section 2.

Obviously, right now graph theory constitutes a promising framework for intense, biologically motivated theoretical studies as well as for experimental investigation guided by theoretical principles: model systems are established, data sets for a variety of biological networks are accessible (see, e.g., Giot et al. 2003; Li et al. 2004; Tong et al. 2004), and first ordering principles in the architecture of biological systems – especially via the notion of scale-free graphs – appear. One aspect, which is often neglected in current research efforts, is the investigation of *network dynamics.*

There, the nodes of a graph represent dynamic elements while the links connecting the nodes correspond to couplings between these elements. The objective is to understand how collective behavior like spatiotemporal patterns and synchronization appears and, moreover, how the underlying architecture of the system affects this behavior. Parallel to the first completely analysed topologies of biological networks the insight emerged that biochemical and metabolic processes are strongly influenced by the properties of the surrounding network (Fox and Hill 2001; Ravasz et al. 2002). Some methods of graph theory, especially the notion of hierarchical elements and the modular composition of real networks, have been related to concepts of cell biology (Hartwell et al. 1999). There, and in other articles, the importance of nonlinear dynamic elements and their connection to questions of network architecture is stressed (see also Guardiola et al. 2000). Understanding of many biological phenomena is much enhanced by applying the concept of coupled nonlinear oscillators. In such theoretical models the global behavior of many interacting dynamic elements can be discussed and then related to biological observations. Nonlinear dynamics and statistical physics provide the mathematical tools for this discussion. In the last few years, these tools have started to be combined with methods from graph theory, as the architecture of a network and its influence on pattern formation entered the focus of scientific interest. Linked systems of coupled oscillators have been under investigation for several years now. The underlying graphs, which have been taken into account so far, are regular arrangements like chains, grids or fully connected systems. With a statistical description of spontaneous synchronization in an ensemble of coupled phase oscillators beyond a critical coupling strength, Kuramoto (1984) formulated a model situation, which is still today the basis for nearly every discussion covering synchronization (see, e.g., Pikovsky et al. 2001). In an exemplary examination, namely, the numeric simulation of a chain of biochemically motivated nonlinear oscillators, Hütt et al. (2003) describe the induction of complex spatiotemporal patterns via biological variability. The relation of biological variability and connectivity in an ensemble of nonlinear oscillators – which is crucial for synchronization in networks – has been studied by Hütt and Lüttge (unpubl.). There, network variants of established spatiotemporal analysis tools have also been formulated and extended to a procedure for reconstructing connectivity from the time series of the individual oscillators. A simple and very successful mathematical model of gene regulation has been formulated by Kauffman (1969). It describes the interaction of binary elements in a random graph: In a network consisting of nodes ('genes') every single gene is regulated by other randomly selected genes via definite Boolean functions. Kauffman found that this model system undergoes a phase transition as soon as a critical value of the degree of linkage is exceeded. The behavior of the system changes from regular (ordered) to irregular (chaotic) at this point. Since experimental values are found to lie much above this critical value, it was for a long time an open question why deterministic chaos is almost absent in biological regulatory networks. Recent work shows ordered dynamics in a large parameter region if the random graph in Kauffman's model is replaced by a scale-free graph (Aldana 2003). This is a fundamental and novel example of the direct influence of architecture on possible forms of dynamics.

In a study of network dynamics, the vertices of a network are thought of as units processing information arriving from other vertices (input) and sending off information (output) based upon the input. In any investigation of network dynamics the individual element is less significant and an observed dynamical behavior arises due to the interplay between many elements. It is obvious that the spatial arrangement of these elements, i.e., the architecture of the system, plays a very important role. Figure 4 gives two simple examples of networks consisting of two (Fig. 4a) and three

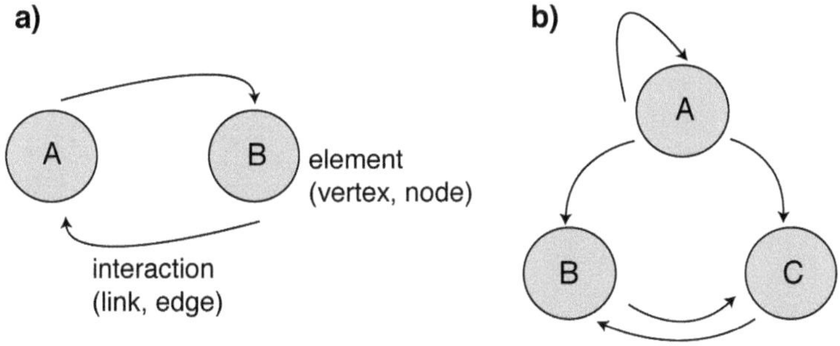

Fig. 4. Examples of simple networks consisting of two (**a**) and three (**b**) elements. (Adapted from Hütt 2001)

(Fig. 4b) elements, respectively. When a discretized time is used, such a picture can, from the point of view of network dynamics, also be thought of as a graphic representation of a set of *finite-difference equations*. Let A_t denote the state of element A at time t. Then Fig. 4b represents the following set of equations:

$$A_{t+1} = f_A(A_t), \quad B_{t+1} = f_B(A_t, C_t), \quad C_{t+1} = f_C(A_t, B_t) \tag{1}$$

If the state space Σ (i.e., the set Σ consisting of all possible vertex states) is restricted to 0 and 1, $\Sigma=\{0, 1\}$, one has *Boolean networks* (Aldana et al. 2003). When a regular lattice is used instead of an arbitrary network and the possible vertex states are discrete, one speaks of *cellular automata*. In the case of a continuous state space these models are called *coupled-map lattices*. If time is continuous rather than discrete, one has a network of nonlinear oscillators, each of which is formulated as a set of *differential equations*.

Figure 5 shows three trivial cases, namely, linear sequences, and open and closed (one-loop) networks. Even without knowledge of the vertex functions (or input–output relations) one can derive some dynamic properties of such simple networks. For the linear sequence from Fig. 5a, for example, one has

$$B_{t+1} = f_B(A_t), \quad C_{t+1} = f_C(B_t), \quad D_{t+1} = f_D(C_t) \tag{2}$$

with vertex functions f_B, f_C and f_D. Without knowing their mathematical forms, one can see how the value of the element A at time t propagates through the system:

$$B_{t+1} = f_B(A_t), \quad C_{t+2} = f_C(f_B(A_t)), \quad D_{t+3} = f_D(f_C(f_B(A_t))) \tag{3}$$

Thus, one finds simply from the fact that the element A does not have an input (open network condition) the state of each element after a few time

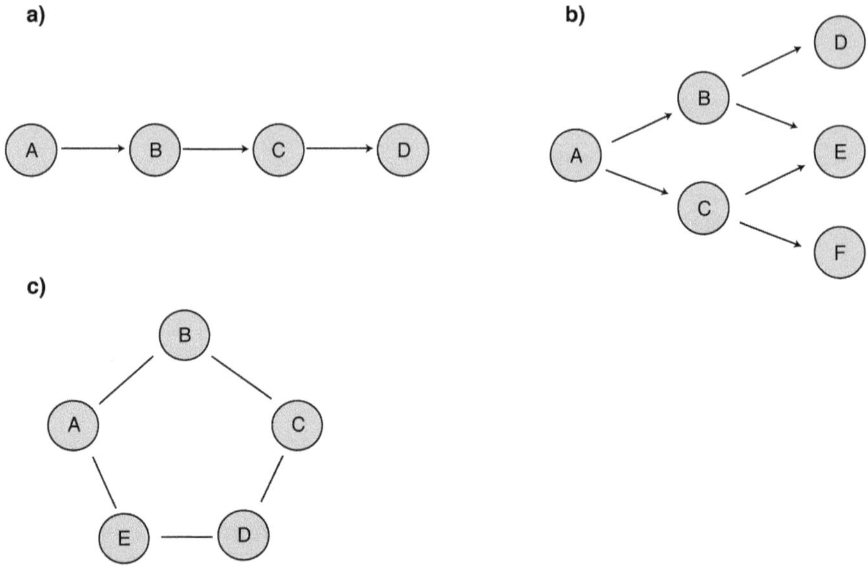

Fig. 5. Examples of open (**a**, **b**) and closed (**c**) networks. The first case of open networks (**a**) is a linear sequence, while the second (**b**) includes branching. The closed network (**c**) is a one-loop network. Arrows in this case have been omitted; they can be inserted in any direction, giving rise to different signaling lengths. It is clear that a variety of more complex closed network architectures are possible. In addition, one can have mixtures of the two basic types shown in **a** and **b**

steps, e.g., $A_t = A_0$ for all t and $D_t = D_3$ for $t > 3$. In general, an open network (or chain) of N elements reaches a steady state after $N-1$ time steps.

One major advantage of using networks lies in their capacity to efficiently translate complex activator–inhibitor scenarios or entangled signaling and metabolic systems into simple mathematical structures, which are then characterized by few observables. In this form, the analysis of large networks is a major task of bioinformatics. The key question for any study of network dynamics is to distinguish between those parts of observed dynamics coming from the individual units and those generated collectively by the network as a whole.

4 Historical Experience: Development of Metabolic Networks and Compartmentation

When individual biochemical enzyme reactions were unraveled, it was a major challenge for students to gain and retain an overview. However, gradually all details fell within pathways of reaction sequences and the great metabolic cycles of respiration, photosynthesis, photorespiration, the oxidative pentosephosphate cycle, nitrogen metabolism cycles, the glyoxylate cycle, etc. They are modules, sub-networks in the large network of metabolism. As promotion material, chemical firms often produce posters with the whole network of metabolism. With their lines, circles and names of metabolites, they look bewildering on the walls of laboratories. However, students quickly learn to find their way on these maps of pools and flows when they have the modules of metabolic pathways and cycles in their mind.

When techniques of biological sample preparation developed and electron microscopy could be applied to biological specimens, students were confronted with the confusing lines of membranes in the two-dimensional views of cells on the electron-micrographs and the bewildering variety of sights of different cell types. Conceptual order was established. One important step was Schnepf's eucyte hypothesis stating that each membrane separates an aqueous and a protoplasmic phase (Schnepf 1966). This provided orientation extracting the nucleo-cytoplasmic mixing phase, the mito- and plasto-plasm and the aqueous-phase cisternae of the Golgi and the endoplasmic reticulum. It explained intracellular membrane flow. It also explained the double membranes of mitochondria and plastids and the aqueous phase between them, and, thus, contributed much to reviving A.F.W. Schimper's original endosymbiont hypothesis of the origin of these organelles (Schimper 1883).

Compartmentation of metabolism became clear. Pathways and cycles in different compartments though separated must not operate in isolation. Studying regulation became an eminent task with coining the term 'transport metabolite' for molecules that can cross membrane borders of compartments as elements of metabolite shuttles and messengers for the conditions, such as energy charge, redox state, substrate availability and product output, in the various compartments (Klingenberg 1970; Heldt 1976a,b; Stocking and Heber 1976).

This very brief historical sketch of how large amounts of biochemical and cytological details have been put into perspective may be useful, because the situation then with new methodological breakthroughs was not principally different from the current situation of the era of 'omics' from genomics to channelomics described above (Sect. 1). History already illustrates that network concepts very much helped to advance progress.

5 Network Dynamics as a New View on Data Mountains

As examples to illustrate both how data mountains pile up rapidly and the current network fashion in plant biology, we can use three domains: (1)

signaling, (2) pattern formation in space (developmental biology) and (3) pattern formation in time (chronobiology). In all domains, we are confronted with a large diversity of control parameters, gene families, tissue-specific gene expression, signaling molecules (primary and secondary messengers) and the like, which can form modules and networks.

5.1 Signaling

In signaling systems, we have external control parameters, receptors and transducing primary and secondary messengers. External control parameters in plants are mainly light, blue and red light, and temperature. If we refrain from also listing all the resources as control parameters, e.g., substrates, nutrients, water, these are only few. However, diversity already appears with the receptors, e.g., the various phytochromes and cryptochromes for red and blue light, respectively.

With the transducing messengers in plants we have the established five types of phytohormones, auxins, gibberellins, cytokinins, abscisic acid (ABA) and ethylene. However, new members have successfully asked for admission, i.e., jasmonic acid, brassinolids (Mandava 1988) and polyamines (Walden et al. 1997; Antogoni et al. 1998). One of the latest triumphant arrivals is that of nitric oxide (NO) as a versatile and extensive signal molecule (Beligni and Lamattina 2000, 2001; Mata and Lamattina 2001; Stöhr et al. 2001; Lamattina et al. 2003). Recent accounts list and describe a plethora of NO effects and interactions abiotically and biotically involving hormones and other messengers, such as Ca^{2+}, cyclic GMP and also H_2O_2 (Lamattina et al. 2003; Neill et al. 2003). Secondary messengers are 1,4,5-inositol-tris-phosphate, diacylglycerol (Lehle 1990), Ca^{2+}, cyclic AMP and GMP, and even H^+ (pH) (Felle 2001) and K^+. Most recently, an increasing variety of elements of stress-reaction networks and anti-oxidant systems have also been considered as messengers, e.g., reactive oxygen species (ROS), especially the readily mobile H_2O_2 (Foyer et al. 1997; Bowler and Fluhr 2000; Basu et al. 2001). Furthermore, we must not forget the ubiquitous gas CO_2 and the internal signaling function of CO_2 in stomatal regulation and in Crassulacean acid metabolism (Cockburn and Patel 2001; Lüttge 2002a). An inflation of the term 'secondary messenger' is noticeable. One of the most well studied examples is stomatal regulation with an interaction of abscisic acid (ABA) and other phytohormones, metabolites, inorganic ions and CO_2 in a most complex signaling network (e.g., see Hunt et al. 2003).

Such enumerations are still far from precise formulations of network structures. Signaling systems of different external control parameters are also interconnected among themselves (Taylor and McAinsh 2004), e.g., for drought, salinity and low temperature (Chinnusamy et al. 2004). In monitoring expression profiles of rice genes in response to low temperature, drought, salinity and ABA, Rabbani et al. (2003) found 36, 62, 57 and 43 genes, respectively, induced by these four factors, where 15 genes responded to all four treatments.

It has been long known that transport and partitioning of substrates also contains information (Pitman 1972; Lüttge 1974). Sugar signaling (Koch 1996; Sheen et al. 1999; Gibson 2000; Smeekens 2000) and nitrogen signaling (Crawford 1997; Scheible et al. 1997; Morcuende et al. 1998; Stöhr 1999; Stöhr et al. 2001), however, became more recently coined terms and strongly developing fields of inquiry. The involvement of different types of modules in sugar signaling, such as sugar sensing enzymes (hexokinase) as well as sugar carriers (Lalonde et al. 1999), cytosolic pH and plasma membrane electrical potentials as affected by the sugar H^+ symporters, voltage-gated Ca^{2+} channels (Furuichi et al. 2001) and the phytohormone ABA (Laby et al. 2000), is a typical example of a regulation network (or a 'web': Gibson 2000).

Microarrays and screening techniques also reveal multiple interactions with inorganic ion cues, e.g., NO_3^-, P_i, S, Al, to quote but a few examples. In roots and shoots of *Arabidopsis*, 1,000 genes rapidly responding to NO_3^- with linkages to various domains of metabolism, viz. sugar, iron and sulphate, have been detected (Wang et al. 2003). In an array of 6,172 *Arabidopsis* genes, more than 1,800 were up- or downregulated by a factor of two or more within 72 h of P_i starvation (Wu et al. 2003). Transcriptome analysis of sulphur depletion in *Arabidopsis thaliana* reveals networks of gene regulation by metabolites with participation of phytohormones (Nikiforova et al. 2003). The Al-toxicity syndrome constitutes a network of interactions involving Ca^{2+} (Rengel and Zhang 2003). Transporters, i.e., pumps, carriers and channels, often represent large gene families of isoforms, just one arbitrarily selected example being P-type ATPases of membranes with 46 genes in *Arabidopsis* and 43 genes in rice (Baxter et al. 2003). In *Arabidopsis* the *cax1* gene appears to be a kind of transporter master-gene which alters the expression and/or activity of vacuolar Ca^{2+} transporters and the V-ATPase and thus regulates a myriad of plant processes (Cheng et al. 2003). Under stress of NaCl salinity and of K^+ and Ca^{2+} starvation, collective transcriptional responses of genes across many multigene transporter families are observed (Maathuis et al. 2003).

Distinguishing input parameters (external control parameters), primary (phytohormones) and secondary (as listed) messengers, we imply a hierarchy (i.e., a direction put forward by the architecture of the network, as depicted in Fig. 5a, b). However, with the inflation of messengers and other elements of signaling systems we can easily see that we do in fact move from linear signal transduction chains (Fig. 5a) to open networks (Fig. 5b) and then to closed networks (Fig. 5c) where hierarchy eventually is lost.

5.2 Development

From insect ontogeny we may adopt the conception of a hierarchy of genes, i.e., maternal master genes, gap genes, pair rule genes, homeotic genes and segment polarity genes regulating early development. However, as Theißen and Saedler (1997) write, the search for a central developmental control gene is just as frustrating as the search for a central circadian oscillator gene (see Sect. 5.3), and again we may expect to find complex networks.

An excellent example of development of a very complex super-network is the development of flowering in plants, because it perfectly illustrates the interlinking of sub-networks. There are (1) a day length-dependent pathway – or network as we would say – involving a circadian clock (see Sect. 5.3) in the input pathway of photoperiod sensing for flower induction, (2) a day length-independent pathway which can eventually support flowering in the absence of induction by photoperiod; and (3) an autonomous pathway modulating the other two pathways. A large number of genes are involved in the three pathways (Koorneef et al. 1998; Simpson et al. 1998; Blázquez and Weigel 2000; Devlin and Kay 2000; Samach et al. 2000).

5.3 Chronobiology

In chronobiology, i.e., the search for circadian clocks (as reviewed recently in this series; Lüttge 2002b), we distinguish input parameters, input pathways, oscillators, output pathways and overt output (Millar 1999; Somers 1999; Barak et al. 2000; McClung 2000; Staiger 2000; Golden and Strayer 2001). The two great questions are whether we have (1) one central oscillator or a diversity of independent oscillators, (2) a hierarchy of master genes or a diversity of independent clock-controlled genes (CCGs). In both cases the answer strongly tends toward the diversity (Lüttge 2002b).

Input pathways, the oscillator(s) and output pathways are clearly multifactorial networks (Lüttge 2002b). This begins to be worked out clearly on the molecular basis for input and output elements and the three-gene-product oscillator of the cyanobacterium *Synechococcus elongatus* (Ishiura et al. 1998; Katayama et al. 1999; Iwasaki and Kondo 2000; Schmitz et al. 2000). Specific roles of the large number of genes under circadian control (Harmer et al. 2000) and the precise allocation of the many genes involved in input, oscillator(s) and output in the higher plant *Arabidopsis thaliana* still remain open questions in many respects (Millar 1999). Although authors frequently still refer to the one 'central oscillator' when writing and central master genes are searched for (Lüttge 2002b; Maxwell et al. 2003), it is often implicitly understood that the central oscillator itself is a regulation network with different oscillating components and subordinate 'slave oscillators' (Kuno et al. 2003). Input, oscillator and output networks are connected to super networks (Fig. 3), where each of them may serve several others, e.g., an oscillator may have several inputs and create several outputs, an input pathway may feed into several oscillators, and so on.

6 A Case Study of Simple Network Dynamics

Topological analysis of the main classes of graphs has been performed with great success over the last years. A few studies of dynamics have also been carried out on those artificial graphs. Real technical, social and biological networks have been extensively described with respect to their topological properties on the basis of an enormous number of empirical findings (see, e.g., Strogatz 2001; Albert and Barabási 2002; Newman 2003). This chapter, apart from giving simple examples of dynamics on artificial graphs, outlines how to combine these two aspects (characterization of real graphs and network dynamics) by using dynamics as a quantitative criterion for information transport.

The first example is an immediate extension of viewing graphs as systems of finite-difference equations, as discussed in Sect. 3. This example, a rule-based dynamic 'game' describing a wave of excitation propagating through a graph, has been taken from Marr et al. (in prep.). Let us consider a system that is initialized by setting the states of all nodes to zero (quiescent state), except for one randomly selected node, the state of which is set to one (excited state). In the first time step this excitation is then transferred to all neighbors of this initially excited node, and a wave of excitation runs subsequently through the graph until all nodes are reached. Let $z_i(t)$ denote the state of the ith node at time t. The graph G is represented by its adjacency matrix A_{ij}, where the matrix element at position (ij) is one if there is a link between nodes i and j, and zero otherwise. For an undirected graph (a link from i to j implies a link from j to i) the adjacency matrix is symmetrical, $A_{ij}=A_{ji}$. Let τ_i be the time when node i becomes non-zero and τ^* the time when all nodes are excited and the game stops. This quantity τ^* depends on the initial node. In this particular game, we allow excitation to accumulate. Furthermore, excitation is passed on in units of one. The update rule for the game is then given by

$$z_i(t+1) = z_i(t) + \sum_j \delta_j(t) A_{ij} \tag{4}$$

where $\delta_i(t)$ is zero if $z_i(t)$ is zero, and one otherwise. Figures 6 and 7 compare the degree distribution $P(k)$ with the distribution of excitations $P(z)$, which has been averaged over all initial conditions (i.e., over different runs of the game, where the node carrying the initial excitation is varied). Two types of networks have been studied, namely scale-free graphs (Fig. 6) and small-world graphs obtained via link-adding (Fig. 7). Comparison of the two distributions for a scale-free graph in Fig. 6 gives a close agreement between $P(k)$ and $P(z)$. The distribution of excitations is fully determined by the topological features of the underlying graph. This result persists over a wide

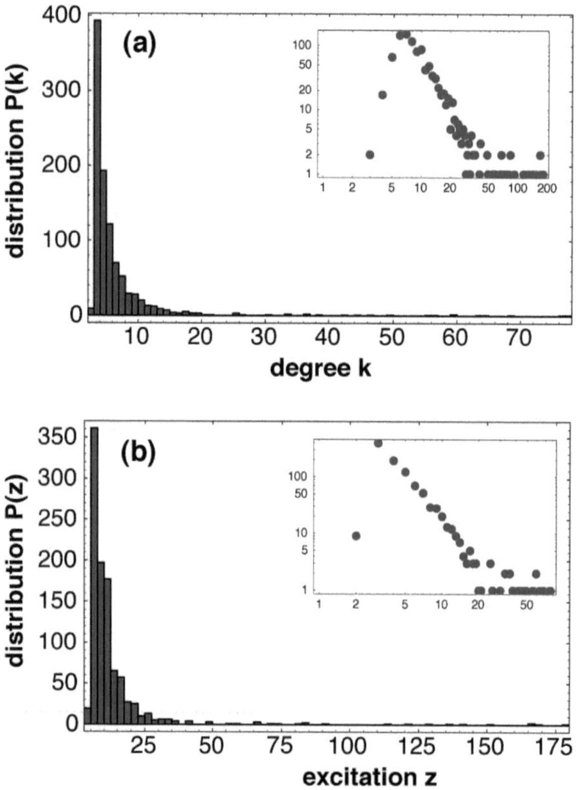

Fig. 6. Comparison of **a** the degree distribution $P(k)$ and **b** the excitation distribution $P(z)$ for the dynamic 'game' defined in Eq. (4) with the underlying architecture given by a scale-free graph. In both cases *inset* gives the corresponding distribution in a log-log plot. The scale-free graph has been obtained with the Barabási-Albert model described in the main text. One finds that the distribution of excitations (**b**) accurately reflects the degree distribution (**a**)

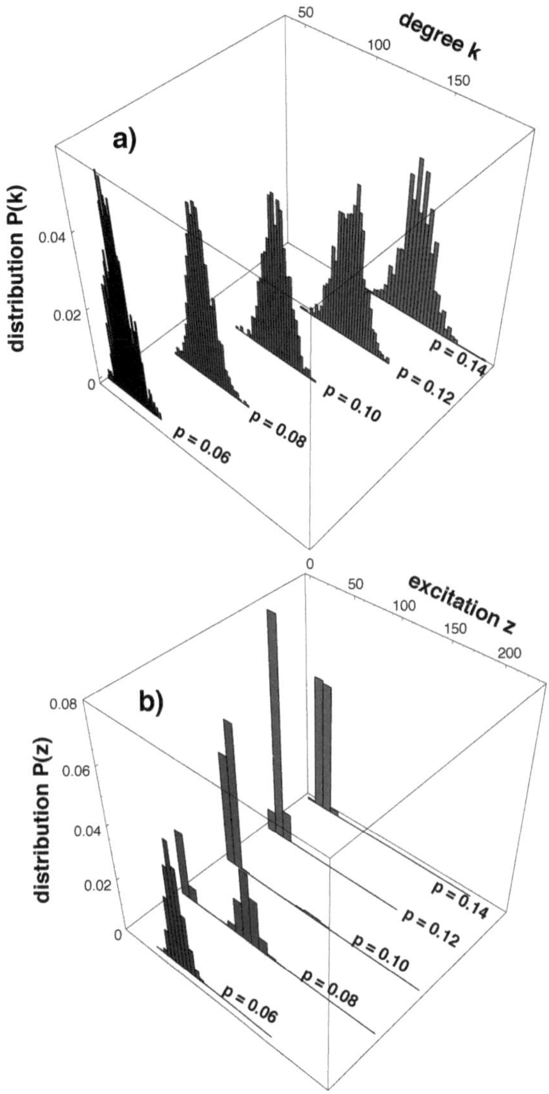

Fig. 7. Comparison of **a** the degree distribution $P(k)$ and **b** the excitation distribution $P(z)$ for the dynamic 'game' defined in Eq. (4), now with the underlying architecture given by a link-adding small-world graph for five different values of the link-adding probability p. One finds a transition in the distribution of excitation as p is increased: At low value of p the function $P(z)$ clearly reflects the degree distribution. When p is increased, a second peak appears at low excitation number, which then dominates the distribution as p is increased even further

range of topological parameters (e.g., the size of the initial graph, the number of links added with each iteration; cf. the prescription for generating scale-free graphs discussed in Sect. 3). For the small-world graph, the situation is not so simple. At low link-adding probability p one finds a similar agreement between the two distributions, as in the case of scale-free graphs. When p is increased, however, this relation is disrupted. The *diameter* of the graph (i.e., the maximum of all shortest paths between the nodes) decreases with increasing p and, consequently, the game time τ^* also decreases. At large τ^* the game is long enough for an excitation to reflect its degree. When τ^* becomes smaller, ever fewer nodes find their degree well represented by the excitation, while all other nodes are only weakly excited (corresponding to a low value of z). The transition between these two cases is marked by two distinct peaks appearing in the excitation distribution $P(z)$ representing the fully excited and weakly excited nodes, respectively, with the former peak disappearing when p is increased even further (Fig. 7). Note that the link-adding probability p continuously reshapes the graph from a regular graph at $p=0$ to a complete graph at $p=1$. Only at intermediate p (up to approx. $p=0.2$) do the corresponding graphs have the small-world property described in Section 3. Similarly, one passes from regular to random graphs in the rewiring procedure when the rewiring probability is changed.

The aim behind studying such simple dynamic rules is to understand by what mechanisms the topological features of a graph enter the dynamics on the graph. Such studies lead to an understanding of how information transport through a system depends on the underlying network (Marr et al., in prep.). Very recent findings show that in certain organisms the metabolite flux distribution also follows a power law (Almaas et al. 2004). This can be seen as one of the striking relations between network topology and network dynamics, similar to the outcome of the dynamic game shown in Fig. 6. It is beginning to emerge that, in addition to a simple reflection of topological features in the dynamics, a complex behavior can be found, which is triggered or enhanced by the network topology, but exceeds the purely static information in complexity. The idea is that some topologies may be optimal for transporting information through a network and it is precisely this line of thought that we expect to be most fruitful in the future.

Let us, as a second example of information propagating through a network, consider the simple epidemic model put forward by Watts and Strogatz (1998) to illustrate the dynamical implications of small-world architecture. Let G be a graph consisting of N interacting nodes. At time $t=0$ one randomly selected node is infected. In the next time step this node is removed from the system and with a probability r each neighbor is infected. This procedure (removal of all infected nodes, infection of a node linked to an infected node with probability r) is repeated iteratively. It is clear that either the spread of disease eventually dies out or the disease infects the whole system. Figure 8a shows the asymptotic system size (i.e., the portion of the system remaining unaffected by the disease when the dynamics has come to an end) as a function of the infection parameter r for different values of the link-adding probability p of the underlying small-world graph. As remarked by Watts and Strogatz (1998) the additional links strongly reduce the critical value r^* of the infection probability, above which the

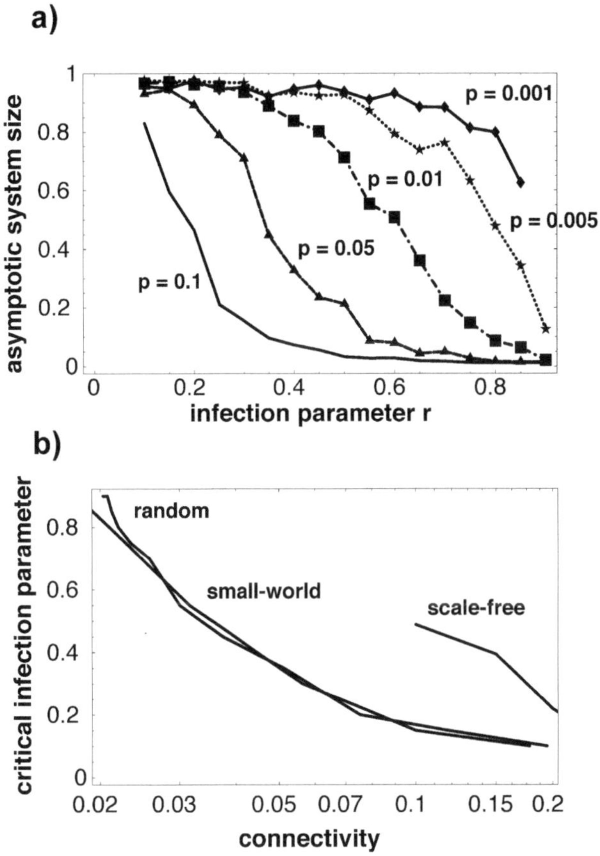

Fig. 8. Disease dynamics according to the simple model given by Watts and Strogatz (1998). **a** System size after spread of disease is shown as a function of the infection parameter *r* for small-world graphs generated with the link-adding procedure for different values of link-adding probability *p*. **b** Critical value r^* of the infection parameter as a function of connectivity π for different types of graphs. Numerical simulations were performed with graphs consisting of 300 nodes. Note that the range in connectivity accessible with the Barabási–Albert model of scale-free graphs is limited by construction (Marr et al., in prep.)

disease reaches at least half of the system. However, how does this effect depend on the class of graph? Is the reduction in r^* a characteristic feature of small-world graphs or is it also found in other types of graphs? Figure 8b shows this critical value as a function of connectivity for different types of graphs. Note that now the abscissa is given by the connectivity (i.e., the number of actual links divided by the number of possible links), rather than a probability regulating link insertion in a particular prescription for generating a graph. This is necessary for comparing different graphs. Remarkably, the scale-free graph turns out to be much more robust with respect to the spreading of the disease, while the random graph displays a signature almost identical to the case of a small-world graph.

Another example, again taken from Marr et al. (in prep.), considers a more refined version of epidemic spread, namely the so-called *forest fire model*, which is a stochastic cellular automata implementation of an excitable medium. An excitable medium is a spatial arrangement of identical elements, for which (at least) three states exist, namely 'quiescent' (excitable) (Q), 'excited' (E) and 'refractory' (R). A typical time sequence of states for a single cell is characterized by a switch from the quiescent state Q to the excited state E, when a certain condition is fulfilled; the falling into the refractory state R after one time step and the remaining in R for a fixed period of time (called the refractory time). This sequence immediately leads to the formation of propagating wave fronts and, when disturbed, to spiral waves (see, e.g., Hütt 2001 for more details). Several components of this cellular automaton, like the refractory time or the condition for a transition to the state E, can be replaced by probabilities, which leads to the concept of stochastic cellular automata, an example of which is the forest fire model (Drossel and Schwabl 1992; Jensen 1998) which is motivated by a phenomenon of ecological self-organization, namely the spreading of fires in a forest. Some aspects of pattern generation in this model have been useful to understand epidemic spreading (Jensen 1998). Transitions between the model's three states (tree T, fire F, empty site E) of each lattice point are governed by two probabilities, which are the control parameters of the system: the growth rate p and the lightning probability f. One has the following update rules leading to the state of the system at time t+1, when applied to each of the lattice sites (ij) at time t:

$$F \xrightarrow{\quad} E, \; E \xrightarrow{p} T, \; T \xrightarrow{F \in N_{ij}} F, \; T \xrightarrow{F \notin N_{ij} \wedge f} F \tag{5}$$

where the symbol above an arrow indicates the condition to be met in order to apply the corresponding rule. This condition is either a probability (p or f) or a neighborhood condition [e.g., a burning tree being in the neighborhood N_{ij} of the site (ij) under consideration]. Recently, the dynamics of the forest fire model have been investigated on small-world graphs, emphasizing the view as an epidemic model (Graham and Matthai 2003). Figure 9 shows the time-averaged number of fires (normalized to the number of nodes in the system) as a function of connectivity. This figure contains three astonishing results: (1) the number of fires changes with connectivity in a sigmoidal way, i.e., the system changes rather spontaneously from a low fire number to a high fire number, as connectivity is increased; (2) all three types of graphs, namely, random, small-world and scale-free graphs, obey

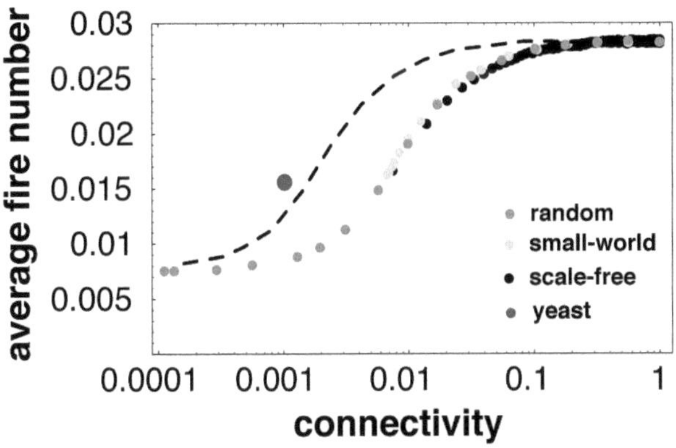

Fig. 9. Average number of fires in a forest fire model, Eq. (5), implemented on different types of graphs, namely random graphs, small-world graphs, scale-free graphs and the yeast protein network, as studied by Jeong et al. (2001). In all cases, graphs were characterized by their connectivity π. Again, graph size was set to N=300. When plotting the average number of fires as a function of the link-adding probability p, this curve is, indeed, independent of the number of nodes in the graph. However, as link-adding probability p and connectivity π are related via $\pi=p\ (1-\kappa)+\kappa$, where $\kappa=2/(N-1)$, increasing the graph size induces a shift of the *curve*. The corresponding *curve* shifted according to this relation is shown as a *dashed line* with N given by the size of the yeast protein network, namely N=1,870 (Marr et al., in prep.)

the same characteristic curve; (3) the average number of fires for a real biological network, namely the yeast protein network, is substantially higher (the value is indicated by the isolated dot in Fig. 9). As the number of fires, to a certain extent, reflects the information processed in the system, Fig. 9 provides evidence for an optimized information transport in the biological network, as compared to the artificial graphs.

In their recent review on the implications of network topology for biology, Barabási and Oltvai (2004) used the term 'network biology' for this remarkable interplay between theory and experimental efforts which has produced some astonishing results over the past 4 years. In particular, it seems that in spite of the obvious complexity of biological networks at all levels of cellular organization, some universal principles exist, making these different levels and observations comparable.

It is, however, precisely the idea of network dynamics that is as yet only poorly understood. One of the key questions is, under what conditions do network dynamics only reflect the topological properties of a network and when is network topology functionally exploited from the overlayed dynamics to produce some optimal transport of information through this network. The remarkable work by Kauffman et al. (2003), who implemented Boolean dynamics on the yeast transcriptional network, and the recent work by Almaas et al. (2004) are important steps in this direction.

7 Network Dynamics as a Tool for Interpreting Biological Systems

In a certain way networks of nonlinear dynamic elements provide a new language to account for complexity. The last few years have nourished the hope – as, e.g., in post-genomic views of protein–protein interaction networks (Jeong et al. 2001) – that the effects of network architecture on the dynamics of a system can be classified and understood by investigating networks of dynamically simple individual units. One then could establish basic organizational principles that provide a different perspective on the data mountains piling up in some branches of biology: The study of network dynamics is an attempt to interpret the system as a whole, rather than study (experimentally or theoretically) specific pathways of the system.

After presenting examples of networks rather theoretically (Sect. 3 and 6) and surveying actual examples of data mountains in signaling, development and chronobiology (Sect. 5), we may now project concrete biological relations into network models such as that of Fig. 4b. We chose to do this in analogy to circadian rhythmicity where essentially three sub-networks can be distinguished, an input pathway, an oscillator and an output pathway (Sect. 5.3; Lüttge 2003). The most general example is that of protein, gene and metabolite networks (Fig. 10A; Weckwerth 2003), where genes have proteins and metabolites as output and there is feedback from metabolites and proteins on genes, and proteins are receptors of input. In ecology (Fig. 10B), the receivers of environmental input are the phenotypes and their output is performance in shaping communities, where communities feed back to receivers directly or via environmental parameters (e.g., in Crassulacean acid metabolism; Lüttge 2004). In signaling systems the whole plant is transmitting input into the output of growth and development as depicted in the example of nitric oxide by Lamattina et al. (2003). Growth and development feed back to the plant via signals and also directly (Fig. 10C). In chronobiology (Fig. 10D), external signals feed into the oscillator network, which generates rhythmic output via the output network and all three sub-networks are feedback linked (Sect. 5.3; Lüttge 2003). Only the very major links are shown and mentioned here to underline the generic principle of such network structures. Clearly, these systems are highly non-linear (Thron 1991; Baier and Sahle 1998), and in system modeling even the contribution of stochastic noise must be taken into account (Beck et al. 2001; Hütt et al. 2002; Weckwerth 2003).

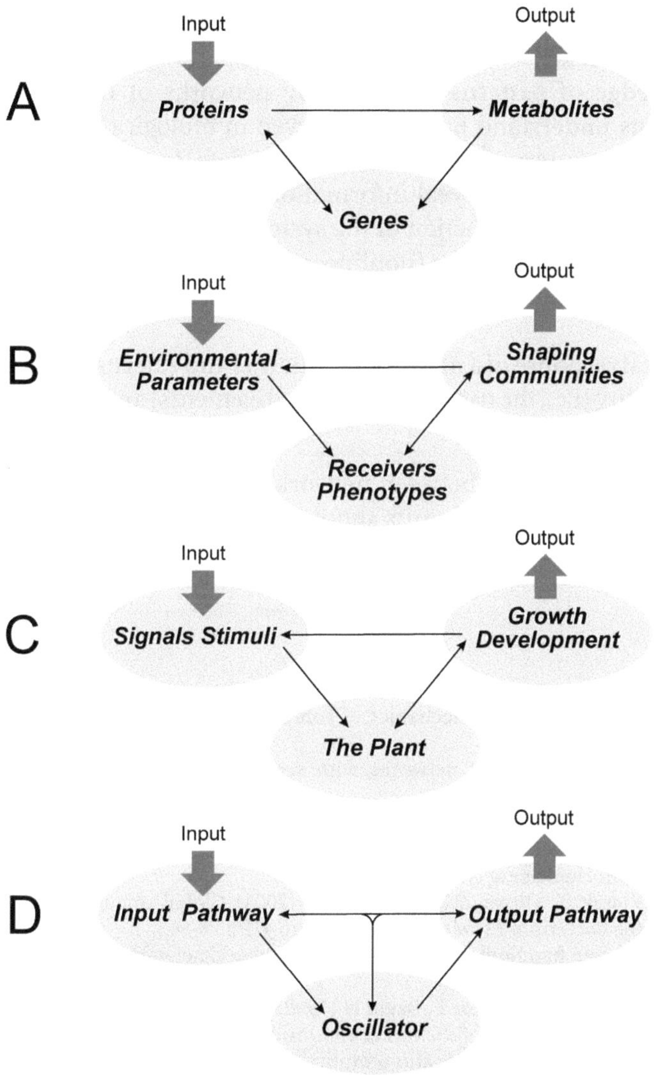

Fig. 10. Examples of networks relating three major elements to concrete biological systems. **A** Genomics, proteomics, metabolomics. **B** Ecology. **C** Signaling systems. **D** Chronobiology

In studying network dynamics, one is often led to alternative experiments and to different tools for the analysis of experimental data. From our point of view, the following questions and perspectives are put forward by studying and analyzing network dynamics:

- To what extent can a specific biological (e.g., genetic) network be thought of as a nonlinear system?
- Can our knowledge of patterns produced by networks of nonlinear oscillators help us understand patterns observed in biological systems (e.g., the expression patterns in cDNA microarray data)?
- Can nonlinear analysis tools reveal information about connectivity or the connection distribution function of the system?
- Can the forms and strengths of the (nonlinear) interactions at work in a biological network be determined from experimental data for a given network?
- What is the typical response of a network to (internal and external) noise, biological variability (i.e., the use of non-identical elements) and external perturbation?

To this end the intrinsic parallel between networks of nonlinear dynamic elements and interacting biological units should be exploited with special emphasis on topology and network architecture.

References

Albert R, Barabási AL (2002) Statistical mechanics of complex networks. Rev Mod Phys 74:47–97

Aldana M (2003) Boolean dynamics of networks with scale-free topology. Physica D 185:45–66

Aldana M, Coppersmith S, Kadanoff LP (2003) Boolean dynamics with random couplings. In: Kaplan E, Marsden JE, Sreenivasan KR (eds) Perspectives and problems in nonlinear science. Springer, Berlin Heidelberg New York

Almaas E, Kovács B, Vicsek T, Oltvai ZN, Barabási AL (2004) Global organization of metabolic fluxes in the bacterium *Escherichia coli*. Nature 427:839–843

Amit DJ (1989) Modeling brain function: the world of attractor neural networks. Cambridge University Press, Cambridge

Antogoni F, Fornale S, Grimmer C, Komor E, Bagni N (1998) Long-distance translocation of polyamines in phloem and xylem of *Ricinus communis* L. plants. Planta 204:520–527

Baier G, Sahle S (1998) Homogeneous and spatio-temporal chaos in biochemical reactions with feedback inhibition. J Theor Biol 193:233–242

Barabási AL (2002) Linked: the new science of networks. Perseus, London

Barabási AL, Albert R (1999) Emergence of scaling in random networks. Science 286:509–512

Barabási AL, Oltvai ZN (2004) Network biology: understanding the cell's functional organization. Nat Rev Genet 5:101–113

Barak S, Tobin EM, Andronis C, Sugano S, Green RM (2000) All in good time: the *Arabidopsis* circadian clock. Trends Plant Sci 5:517–522

Bassingthwaighte JB, Liebovitch LS, West BJ (1994) Fractal physiology. Oxford University Press, Oxford

Basu U, Good AG, Taylor GJ (2001) Transgenic *Brassica napus* plants overexpressing aluminium-induced mitochondrial manganese superoxide dismutase cDNA are resistant to aluminium. Plant Cell Environ 24:1269–1278

Baxter I, Tchieu J, Sussman MR, Boutry M, Palmgren MG, Gribskov M, Harper JF, Axelsen KB (2003) Genomic comparison of P-type ATPase ion pumps in *Arabidopsis* and rice. Plant Physiol 132:618–628

Beck F, Blasius B, Lüttge U, Neff R, Rascher U (2001) Stochastic noise interferes coherently with a model biological clock and produces specific dynamic behaviour. Proc R Soc B 268:1307–1313

Beligni MV, Lamattina L (2000) Nitric oxide stimulates seed germination and de-etiolation, and inhibits hypocotyl elongation, three light-inducible responses in plants. Planta 210:215–221

Beligni MV, Lamattina L (2001) Nitric oxide in plants: the history is just beginning. Plant Cell Environ 24:267–278

Blázquez MA, Weigel D (2000) Integration of floral inductive signals in *Arabidopsis*. Nature 404:889–892

Bornholdt S (2001) Modelling genetic networks and their evolution: a complex dynamical systems perspective. Biol Chem 382:1289–1299

Bornholdt S, Sneppen K (1998) Neutral mutations and punctuated equilibrium in evolving genetic networks. Phys Rev Lett 81:236–239

Bowler C, Fluhr R (2000) The role of calcium and activated oxygen as signals for controlling cross tolerance. Trends Plant Sci 5:241–246

Cheng N-H, Pittman JK, Barkla BJ, Shigaki T, Hirschi KD (2003) The *Arabidopsis cax1* mutant exhibits impaired ion homeostasis, development, and hormonal responses and reveals interplay among vacuolar transporters. Plant Cell 15:347–364

Chinnusamy V, Schumaker K, Zhu JK (2004) Molecular genetic perspectives on cross-talk and specificity in abiotic stress signalling in plants. J Exp Bot 55:225–236

Cockburn W, Patel K (2001) Control of internal carbon dioxide concentration during decarboxylation in CAM plants. In: Holtum JAM (ed) Proc CAM 2001, School of Tropical Biology, James Cook University, Townsville, p 213

Crawford NM (1997) Nitrate: nutrient and signal for plant growth. Plant Cell 7:859–868

De Perlinghi J (2003) Isabelle Stengers: 'Et si un jour les sciences devenaient civilisée...'. La Recherche 367:67–70

Devlin PF, Kay SA (2000) Flower arranging in *Arabidopsis*. Science 288:1600–1602

Drossel B, Schwabl F (1992) Self-organized critical forest-fire model. Phys Rev Lett 69:1629–1632

Erdös P, Rényi A (1960) On the evolution of random graphs. Publ Math Inst Hung Acad Sci 5:17–61

Erwin TL (1982) Tropical forests: their richness in coleoptera and other arthropod species. Coleopt Bull 36:74–75

Erwin TL (1995) Measuring arthropod diversity in the tropical forest canopy. In: Lowman MD, Nadkarni NM (eds) Forest canopies. Academic Press, San Diego, pp 109–127

Farkas IJ, Derényi I, Jeong H, Neda Z, Oltvai ZN, Ravasz E, Schurbert A, Barabási AL (2003) The topology of the transcription regulatory network in the yeast, *Saccharomyces cerevisiae*. Physica A381:601–612

Fell DA, Wagner A (2000) The small world of metabolism. Nat Biotechnol 18:1121–1122

Felle HH (2001) pH: signal and messenger in plant cells. Plant Biol 3:577–591

Fox JJ, Hill CC (2001) From topology to dynamics in biochemical networks. Chaos 11:809–815

Foyer CH, Lopez-Delgado L, Dat JF, Scott IM (1997) Hydrogen peroxide and glutathione associated mechanism of acclimatory stress tolerance and signaling. Physiol Plant 100:241–254

Furuichi T, Mori IC, Takahashi K, Muto S (2001) Sugar-induced increase in cytosolic Ca^{2+} in *Arabidopsis thaliana* whole plants. Plant Cell Physiol 42:1149–1155

Gibson SI (2000) Plant sugar-response-pathways. Part of a complex regulatory web. Plant Physiol 124:1532–1539

Giot L, Bader JS, Brouwer C, Chaudhuri A, Kuang B, Li YL, Hao YL, Ooi C E, Godwin B, Vitols E, Vijayadamodar G, Pochart P, Machineni H, Welsh M, Kong Y, Zerhusen B, Malcolm R, Varrone Z, Collis A, Minto M, Burgess S, McDaniel L, Stimpson E, Spriggs F, Williams J, Neurath K, Ioime N, Agee M, Voss E, Furtak K, Renzulli R, Aanensen N, Carrolla S, Bickelhaupt E, Lazovatsky Y, DaSilva A, Zhong J, Stanyon CA, Finley RL Jr, White KP, Braverman M, Jarvie T, Gold S, Leach M, Knight J, Shimkets RA, McKenna MP, Chant J, Rothberg JM (2003) A protein interaction map of *Drosophila melanogaster*. Science 3025:1727–1736
Girke T, Ozkan M, Carter D, Raikhel NV (2003) Towards a modeling infrastructure for studying plant cells. Plant Physiol 132:410–414
Golden SS, Strayer C (2001) Time for plants. Progress in plant chronobiology. Plant Physiol 125:98–101
Graham I, Matthai CC (2003) Investigation of the forest-fire model on a small-world network. Phys Rev E 68:036109
Guardiola X, Diaz-Guilera A, Lilas M, Pérez CJ (2000) Synchronization, diversity and topology of networks of integrate and fire oscillators. Phys Rev E 62:5565–5570
Guare J (1990) Six degrees of separation: a play. Vintage Books, New York
Harmer SL, Hogenesch JB, Straume M, Chang HS, Han B, Zhu T, Wang X, Kreps JA, Kay SA (2000) Orchestrated transcription of key pathways in *Arabidopsis* by the circadian clock. Science 290:2110–2113
Hartwell LH, Hopfield JJ, Leibler S, Murray AW (1999) From molecular to modular cell biology. Nature 402:C47–C52
Heldt HW (1976a) Metabolite carriers of chloroplasts. In: Stocking CR, Heber U (eds) Transport in plants III. Intracellular interactions and transport processes. Encyclopedia of plant physiology new series, vol 3. Springer, Berlin Heidelberg New York, pp 137–143
Heldt HW (1976b) Transport of metabolites between cytoplasm and the mitochondrial matrix. In: Stocking CR, Heber U (eds) Transport in plants III. Intracellular interactions and transport processes. Encyclopedia of plant physiology new series, vol 3. Springer, Berlin Heidelberg New York, pp 235–254
Hütt MT (2001) Datenanalyse in der Biologie. Springer, Berlin Heidelberg New York
Hütt MT, Lüttge U (2002) Nonlinear dynamics as a tool for data analysis and modeling in plant physiology. Plant Biol 4:281–297
Hütt MT, Neff R, Busch H, Kaiser F (2002) A method for detecting the signature of spatiotemporal stochastic resonance. Phys Rev E 66:026117
Hütt MT, Busch H, Kaiser F (2003) The effect of biological variability on spatiotemporal patterns: model simulations for a network of biochemical oscillators. Nova Acta Leopoldina 332:381–404
Hunt L, Mills LN, Pical C, Leckie CP, Aitken FL, Kopka J, Mueller-Roeber B, McAinsh MR, Hetherington AM, Gray JE (2003) Phospholipase C is required for the control of stomatal aperture by ABA. Plant Cell 34:47–55
Ishiura M, Kutsuna S, Aoki S, Iwasaki H, Andersson CR, Tanabe A, Golden SS, Johnson CH, Kondo T (1998) Expression of a gene cluster *kaiABC* as a circadian feedback process in cyanobacteria. Science 281:1519–1523
Iwasaki H, Kondo T (2000) The current state and problems of circadian clock studies in cyanobacteria. Plant Cell Physiol 41:1013–1020
Jensen HJ (1998) Self-organized criticality. Cambridge University Press, Cambridge
Jeong H, Tombor B, Albert R, Oltavi ZN, Barabási A-L (2000) The large scale organization of metabolic networks. Nature 407:651–654
Jeong H, Mason SP, Barabási A-L, Ottavi ZN (2001) Lethality and centrality in protein networks. Nature 411:41–42
Kanz KT (2003) Christian Gottfried Nees von Esenbeck. Briefwechsel mit Johann Wolfgang von Goethe nebst ergänzenden Schriften. Acta Historica Leopoldina, no. 40. Deutsche Akademie der Naturforscher Leopoldina, Halle and Wissenschaftliche Verlagsgesellschaft, Stuttgart

Katayama M, Tsinoremas NF, Kondo T, Golden SS (1999) *cpmA*, a gene involved in an output pathway of the cyanobacterial circadian system. J Bacteriol 181:3516–3524

Kauffman S (1969) Metabolic stability and epigenesis in randomly constructed genetic nets. J Theor Biol 22:437–467

Kauffman S, Peterson C, Samuelsson B, Troein C (2003) Random Boolean network models and the yeast transcriptional network. Proc Natl Acad Sci USA 100:14796–14799

Klingenberg M (1970) Metabolite transport in mitochondria: an example of intracellular membrane function. Essays Biochem 6:119–159

Koch KE (1996) Carbohydrate-modulated gene expression in plants. Annu Rev Plant Physiol Plant Mol Biol 47:509–540

Koorneef M, Alonso-Blanco C, Peeters AJM, Soppe W (1998) Genetic control of flowering in *Arabidopsis*. Annu Rev Plant Physiol Plant Mol Biol 49:345–370

Kratochwil A (1998) Biodiversity in ecosystems. Atti Convegni Lincei 145:23–62

Kuno N, Moller SG, Shinomura T, Xu XM, Chua NH, Furuya M (2003) The novel MYB protein EARLY-PHYTOCHROME-RESPONSIVE 1 is a component of a slave circadian oscillator in *Arabidopsis*. Plant Cell 15:2476–2483

Kuramoto Y (1984) Chemical oscillations, waves, and turbulence. Springer, Berlin Heidelberg New York

Laby RJ, Kincaid S, Kim D, Gibson SI (2000) The *Arabidopsis* sugar-insensitive mutants *sis4* and *sis5* are defective in abscisic acid synthesis and response. Plant J 23:587–596

Lalonde S, Boles E, Hellmann H, Barker L, Patrick JW, Frommer WB, Ward JM (1999) The dual function of sugar carriers: transport and sugar sensing. Plant Cell 11:707–726

Lamattina L, García-Mata C, Graziano M, Pagnussat G (2003) Nitric oxide: the versatility of an extensive signal molecule. Annu Rev Plant Biol 54:109–136

Lehle L (1990) Phosphatidyl inositol metabolism and its role in signal transduction in growing plants. Plant Mol Biol 15:647–658

Li S, Armstrong CM, Bertin N, Ge H, Milstein S, Boxem M, Vidalain PO, Han JDJ, Chesneau A, Hao T, Goldberg DS, Li N, Martinez M, Rual JF, Lamesch P, Xu L, Tewari M, Wong SL, Zhang LV, Berriz GF, Jacotot L, Vaglio P, Reboul J, Hirozane-Kishikawa T, Li Q, Gabel HW, Elewa A (2004) A map of the interactome network of the Metazoan *C. elegans*. Science 303:540–543

Linsenmair KE (1995) Biologische Vielfalt und ökologische Stabilität. In: Markl H, Geiler G, Großmann S, Oesterhelt D, Schmidbaur H, Quadbeck-Seeger H-J, Truscheit E (eds) Wissenschaft in der globalen Herausforderung. Edition Universitas. Hirzel Wissenschaftliche Verlagsgesellschaft, Stuttgart, pp 267–295

Lüttge U (1974) Co-operation of organs in intact higher plants: a review. In: Zimmermann U, Dainty J (eds) Membrane transport in plants. Springer, Berlin Heidelberg New York, pp 353–362

Lüttge U (2002a) CO_2-concentrating: consequences in Crassulacean acid metabolism. J Exp Bot 53:2131–2142

Lüttge U (2002b) Circadian rhythmicity: is the 'biological clock' hardware or software? Progress in Botany, vol 64. Springer, Berlin Heidelberg New York, pp 277–319

Lüttge U (2003) Circadian rhythms. In: Thomas B, Murphy DJ, Murray BG (eds) Encyclopedia of applied plant sciences. Elsevier, Academic Press, Amsterdam, pp 1084–1096

Lüttge U (2004) Ecophysiology of Crassulacean acid metabolism (CAM). Ann Bot 93:629–652

Maathuis FMJ, Filatov V, Herzyk P et al. (2003) Transcriptome analysis of root transporters reveals participation of multiple gene families in the response to cation stress. Plant J 35:675–692

Mandava NB (1988) Plant growth-promoting brassinosteroids. Annu Rev Plant Physiol 39:23–52

Maslov S, Sneppen K (2002) Specificity and stability in topology of protein networks. Science 296:910–913

Mata CG, Lamattina L (2001) Nitric oxide induces stomatal closure and enhances the adaptive plant responses against drought stress. Plant Physiol 126:1196–1204
Maxwell BB, Andersson CR, Poole DS, Kay SA, Chory J (2003) HY5, circadian clock-associated 1, and a cis-element, DET 1 dark response element, mediate DET1 regulation of *chlorophyll a/b-binding protein 2* expression. Plant Physiol 133:1565–1577
May RM (1988) How many species are there on earth? Science 241:1441–1449
McClung CR (2000) Circadian rhythms in plants: a millennial view. Physiol Plant 109:359–371
Milgram S (1967) The small world problem. Psychol Today 2:60–67
Millar AJ (1999) Biological clocks in *Arabidopsis thaliana*. New Phytol 141:175–197
Minorsky PV (2003) Achieving the in silico plant. Systems biology and the future of plant biological research. Plant Physiol 132:404–409
Morcuende R, Krapp A, Hurry V, Stitt M (1998) Sucrose-feeding leads to increased rates of nitrate assimilation, increased rates of α-oxoglutarate synthesis, and increased synthesis of a wide spectrum of amino acids in tobacco leaves. Planta 206:394–409
Neill SJ, Desikan R, Hancock JT (2003) Nitric oxide signalling in plants. New Phytol 159:11–35
Newman MEJ (2001) The structure of scientific collaboration networks. Proc Natl Acad Sci USA 98:404–409
Newman MEJ (2003) The structure and function of complex networks. SIAM Rev 45:167–256
Nikiforova V, Freitag J, Kempa S, Adamik M, Hesse H, Hoefgen R (2003) Transcriptome analysis of sulphur depletion in *Arabidopsis thaliana*: interlacing of biosynthetic pathways provides response specificity. Plant J 33:633–650
Pikovsky A, Rosenblum M, Kurths J (2001) Synchronization. A universal concept in nonlinear sciences. Cambridge University Press, Cambridge
Pitman MG (1972) Uptake and transport of ions in barley seedlings. III. Correlation of potassium transport to the shoot with plant growth. Aust J Biol Sci 25:905–919
Rabbani MA, Maruyama K, Abe H, Khan MA, Katsura K, Ito Y, Yoshiwara K, Seki M, Shinozaki K, Yamaguchi-Shinozaki K (2003) Monitoring expression profiles of rice genes under cold, drought, and high salinity stresses and abscisic acid and application using cDNA microarray and RNA gel-blot analyses. Plant Physiol 133:1755–1767
Raikhel NV, Coruzzi GM (2003) Plant systems biology. Plant Physiol 132:403
Ravasz E, Somera AL, Mongru DA, Oltvai ZN, Barábasi AL (2002) Hierarchical organization of modularity in metabolic networks. Science 297:1551–1555
Rengel Z, Zhang W-H (2003) Role of dynamics of intracellular calcium in aluminum-toxicity syndrome. Plant Physiol 159:295–314
Samach A, Onouchi H, Gold SE, Ditta GS, Schwarz-Sommer Z, Yanofsky MF, Coupland G (2000) Distinct roles of CONSTANs target genes in reproductive development of *Arabidopsis*. Science 288:1613–1616
Scheible W-R, Gonzales-Fontes A, Lauerer M, Müller-Röber B, Caboche M, Stitt M (1997) Nitrate acts as a signal to induce organic acid metabolism and repress starch metabolism in tobacco. Plant Cell 9:783–798
Schimper AFW (1883) Über die Entwicklung der Chlorophyllkörner und Farbkörner. Bot Zeit 41:105–114
Schmitz O, Katayama M, Williams SB, Kondo T, Golden SS (2000) CikA, a bacteriophytochrome that resets the cyanobacterial circadian clock. Science 289:765–768
Schnepf E (1966) Organellen-Reduplikation und Zellkompartimentierung. In: Sitte P (ed) Probleme der biologischen Reduplikation, 3. Wiss Konf Ges Dt Naturf Ärzte, Semmering 1965. Springer, Berlin Heidelberg New York, pp 371–393
Sheen J, Zhou L, Jang J-C (1999) Sugars as signaling molecules. Curr Opin Plant Biol 2:410–418
Simpson GG, Gendall AR, Dean C (1998) The control of flowering time and floral identity in *Arabidopsis*. Plant Physiol 117:1–8

Smeekens S (2000) Sugar-induced signal transduction in plants. Annu Rev Plant Physiol Plant Mol Biol 51:49–81

Solé RV, Manrubia SC, Luque B, Delgado J, Bascompte J (1996) Phase transitions and complex systems. Complexity 2:13

Somers DE (1999) The physiology and molecular bases of the plant circadian clock. Plant Physiol 121:9–19

Staiger D (2000) Biologische Zeitmessung bei Pflanzen. Biol Unserer Zeit 30:76–79

Stocking CR, Heber U (eds) (1976) Transport in plants III. Intracellular interactions and transport processes. Encyclopedia of plant physiology new series, vol 3. Springer, Berlin Heidelberg New York

Stöhr C (1999) Relationship of nitrate supply with growth rate, plasma membrane-bound and cytosolic nitrate reductase, and tissue nitrate content in tobacco plants. Plant Cell Environ 2:169–177

Stöhr C, Strube F, Marx G, Ullrich W, Rockel P (2001) A plasma membrane-bound enzyme of tobacco roots catalyses the formation of nitric oxide from nitrite. Planta 212:835–841

Strogatz SH (2001) Exploring complex networks. Nature 410:268–276

Sweetlove LJ, Last RL, Fernia AR (2003) Predictive metabolic engineering: a goal for systems biology. Plant Physiol 132:420–425

Taylor JE, McAinsh MR (2004) Signalling crosstalk in plants: emerging issues. J Exp Bot 55:147–149

Theißen G, Saedler H (1997) Molecular architects of plant body plans. Prog Bot 59:227–256

Thron CD (1991) Biochemical oscillations. Bull Math Biol 53:383

Tong AHY, Lesage G, Bader GD, Ding H, Xu H, Xin X, Young J, Berriz GF, Brost RL, Chang M, Chen YQ, Cheng X, Chua G, Friesen H, Goldberg DS, Haynes J, Humphries C, He G, Hussein S, Ke L, Krogan N, Li Z, Levinson JN, Lu H, Ménard P, Munyana C, Parsons AB, Ryan O, Tonikian R, Roberts T, Sdicu AM, Shapiro J, Sheikh B, Suter B, Wong SL, Zhang LV, Zhu H, Burd CG, Munro S, Sander C, Rine J, Greenblatt J, Peter M, Bretscher A, Bell G, Roth FP, Brown GW, Andrews B, Bussey H, Boone C (2004) Global mapping of the yeast genetic interaction network. Science 303:808–813

Wagner A, Fell DA (2001) The small world inside large metabolic networks. Proc R Soc Lond B 268:1803–1810

Walden R, Cordeiro A, Triburcio AF (1997) Polyamines: small molecules triggering pathways in plant growth and development. Plant Physiol 113:1009–1013

Wang R, Okamoto M, Xing X, Crawford NM (2003) Microarray analysis of the nitrate response in *Arabidopsis* roots and shoots reveals over 1,000 rapidly responding genes and new linkages to glucose, trehalose-6-phosphate, iron and sulphate metabolism. Plant Physiol 132:556–567

Watts DJ (1999) Small worlds: the dynamics of networks between order and randomness. Princeton University Press, Princeton

Watts DJ, Strogatz SH (1998) Collective dynamics of 'small-world' networks. Nature 393:440–442

Weckwerth W (2003) Metabolomics in systems biology. Annu Rev Plant Biol 54:669–689

Wu P, Ma L, Hou X, Wang M, Wu Y, Liu F, Deng XW (2003) Phosphate starvation triggers distinct alterations of genome expression in *Arabidopsis* roots and leaves. Plant Physiol 132:1260–1271

Marc-Thorsten Hütt
Ulrich Lüttge
Institut für Botanik
Technische Universität Darmstadt
Schnittspahnstrasse 3-5
64287 Darmstadt, Germany
Tel.: +49-6151-163700
Fax: +49-6151-164630
e-mail: luettge@bio.tu-darmstadt.de

Ecology

UV-B Radiation, Photomorphogenesis and Plant–Plant Interactions

Paul W. Barnes, James R. Shinkle, Stephan D. Flint, and Ronald J. Ryel

1 Introduction

Sunlight is an essential resource for autotrophic plants, and morphological traits that influence the placement and orientation of foliage are key determinants of a plant's success in a highly competitive environment. Although plant architecture influences light capture and photosynthetic carbon gain, solar radiation can also modify plant morphology in ways that influence this process (Aphalo and Ballaré 1995). Perhaps the most vivid example of how spectral quality can influence plant–plant interactions comes from studies examining phytochrome-mediated morphological responses of plants to changes in red:far red (R:FR) radiation that occur as a result of shading and neighborhood crowding (see reviews by Schmitt and Wulff 1993; Ballaré et al. 1997; Smith and Whitelam 1997). Although considerable attention has been given to understanding the physiological, adaptive and ecological aspects of the phytochrome system (Smith 1995; Schmitt 1997), other photoreceptor systems are known to exist in plants that also play a role in morphogenesis (Briggs and Olney 2001). These other photosensory systems include several receptors for blue (B)/ultraviolet-A (UV-A; 320–400 nm) radiation (Lin 2000), and an unknown ultraviolet-B (UV-B; 280–320 nm) photosensory system (Björn 1999). Relative to phytochrome, however, much less is known about the role that morphological alterations mediated by these other photoreceptor systems play in plant competition (Ballaré 1999).

In this chapter, we summarize what is known about morphological responses to UV-B, explore possible mechanisms and adaptive explanations for these responses, and evaluate the ecological consequences of these alterations for plant–plant interactions. We emphasize the role that UV-B plays in mediating competitive interactions between plants for light, though we recognize that UV-B also has the potential to influence belowground competition and allelopathic interactions (Caldwell 1997). As a working hypothesis, we propose that UV-B can induce specific photomorphogenic alterations in plant growth form and, under conditions that are

Progress in Botany, Vol. 66

ecologically relevant, subtle effects of UV-B on growth and morphology can be magnified by the competitive process when plants differ in their UV-B sensitivity. Because of these competitive interactions, the ecological consequences of UV-B-induced alterations in morphology are therefore likely to be of greater consequence for plants growing in dense stands than in isolation.

2 Overview of UV-B and Plants

Though UV-B comprises a relatively minor part of the solar spectrum, radiation in this waveband is known to elicit a variety of responses at the molecular, cellular and whole-organism level in higher plants (e.g., Day 2001; Jordan 2002; Tobin 2002). Evidence from molecular and physiological studies indicates that some of these UV-B responses represent general stress responses, while others are photomorphogenic in nature (Björn 1999; A-H-Mackerness 2000; Jansen 2002). The UV-B stress responses often involve some form of UV injury (e.g., DNA damage, oxidative stress, etc.) and appear to share gene activation and signal transduction pathways in common with other forms of abiotic (e.g., drought and ozone exposure) and biotic (e.g., herbivory and pathogen attack) stress (Holley et al. 2003; Izaguirre et al. 2003). By comparison, photomorphogenic responses involve one or several yet unknown UV-B receptors and result from molecular events and signal transduction pathways that are distinct from the more general stress responses (Brosché and Strid 2003; Casati and Walbot 2003). Examples of photomorphogenic responses to UV-B include (1) the induction of UV-absorbing compounds (flavonoids and other related phenolic compounds), which serve to protect sensitive tissue from UV-B exposure (Beggs and Wellmann 1994; Bornman et al. 1997; Mazza et al. 2000), and (2) the inhibition of stem elongation, which causes alterations in shoot morphology (Lercari et al. 1990; Ballaré et al. 1991, 1995a; Kim et al. 1998; Kobzar et al. 1998).

Over the past several decades, there has been concern over the ecological consequences of increases in solar UV-B associated with stratospheric ozone depletion resulting from anthropogenic releases of chlorofluorocarbons and related compounds. This concern has led to numerous studies evaluating the effects of elevated UV-B alone, or in combination with other elements of global change (e.g., CO_2 increase), on plants and terrestrial ecosystems (Ballaré et al. 1999; Björn et al. 1999; Caldwell et al. 2003; Flint et al. 2003; Kakani et al. 2003a). Even though the production and emission of these ozone-depleting substances have been greatly curtailed in recent years, stratospheric ozone depletion remains an issue in the near term, especially in light of recent reports predicting additional future reductions in stratospheric ozone due to release of hydrogen in the advent of widespread use of hydrogen fuel cell technology (Tromp et al. 2003).

Historically, considerable attention was given to identifying molecular targets of UV-B damage and assessing the potential negative effects of elevated UV-B on photosynthesis and production of important crop plants (Baker et al. 1997; Sullivan and Rozema 1999). However, while solar UV-B has been shown to reduce primary productivity in certain cases (Day and Neale 2002), the body of evidence indicates that the deleterious effects of increased UV-B on the production and photosynthesis of field-grown plants are likely to be less than previously expected (Searles et al. 2001). Increasingly, attention has shifted to understanding regulatory and photomorphogenic effects of UV-B on plants (Rozema et al. 1997; Jansen et al. 1998). In addition, there is growing interest in elucidating the fundamental role that solar UV-B plays in mediating important ecosystem processes such as competition, herbivory, disease, decomposition and biogeochemical cycling (Caldwell et al. 1999; Paul et al. 1999) even apart from the context of ozone depletion (Ballaré 2003; Paul and Gwynn-Jones 2003). Recent findings that increases in UV-absorbing tropospheric gases (e.g., ozone, SO_2 and NO_2) and aerosols can reduce the amount of solar UV reaching the earth's surface (McKenzie et al. 2001) further indicate that attention should also be given to understanding how *reductions* in solar UV-B impact ecosystems.

3 Morphological Responses of Plants to UV-B

Numerous studies have shown that exposure to UV-B can result in a wide variety of morphological alterations in higher plants (Table 1). These morphological changes can be observed under controlled environmental conditions in growth chambers or greenhouses, where UV-emitting lamps provide the sole source of UV-B, and also under a full solar spectrum in the field where ambient solar UV-B is supplemented using UV-emitting lamps or attenuated with UV-absorbing film. In some cases (e.g., reduced leaf area and shoot height), whole-plant changes in morphology are the result of an inhibition in the elongation or expansion of individual organs (leaves and stems), while in other cases (e.g., increased branching or leaf production) plant growth is stimulated by UV-B. Whereas the inhibitory effects of UV-B on stem and leaf elongation are consistently observed in response to ambient or enhanced levels of solar UV-B (both in greenhouse and field settings), the stimulatory effects on morphology appear to be more frequently observed under UV-B enhancement, especially under growth chamber or

Table 1. Summary of plant morphological responses to UV-B radiation. Direction of effect shows response under elevated or near-ambient UV-B relative to none or low UV-B (growth chamber/greenhouse), ambient UV-B (field enhancement) or sub-ambient UV-B (field attenuation). In cases where both positive and negative effects have been reported, studies illustrating these are followed by (+) and (–), respectively. Field enhancement includes studies conducted using square-wave and modulated UV-B systems. Reference list is not exhaustive but includes representative studies conducted on native and cultivated plants of various plant growth forms (grasses, herbaceous dicots, woody dicots and conifers) from a diversity of ecosystems

Morphological trait	Direction of effect	References[a] Growth chamber/ greenhouse	Field enhancement	Field attenuation
Whole-plant level				
Leaf area	–	1, 2, 3, 4	5, 6, 7, 8	9, 10, 11, 12
Number of leaves	+/–	1, 2, 13 (+)	14 (+)/ 6, 7 (–)	9, 11, 15, 16 (–)
Number of branches/tillers	+/–	3, 17, 18, 19 (+)	14, 20, 21 (+)/ 22, 23 (–)	9, 11, 12, 15 (–)
Shoot height	–	1, 2, 3, 4	5, 6, 24	9, 25, 26, 27
Root density (length/vol)	–			28
Root: shoot mass	+/–	17, 29, 30 (+)/3, 19, 31 (–)	20 (+)/14 (–)	27, 32, 33 (+)/33 (–)
Leaf: stem (shoot) mass	+	1, 17, 18	13	12
Organ-level				
Leaf size/ elongation	–	1, 12, 23, 30	6, 7, 13, 24	12, 25, 34, 35
Stem length/ elongation	–	4, 30, 36	13, 37, 38	9, 10, 39
Rhizome elongation	–			35
Flower/ petal size	–	4		40
Leaf thickness/ area: mass	+/–	1, 18, 19, 30 (+)/41 (–)	20, 21, 37, 38 (+)/6, 23 (–)	12, 25, 33, 39 (+)/42 (–)
Stem phototropism	+	43, 44		

Table 1. *Continued*

Morphological trait	Direction of effect	References[a] Growth chamber/ greenhouse	Field enhancement	Field attenuation
Cotyledon curling/opening	+	45, 46		
Tendril curling	+	47		

[a] *1* Barnes et al. (1990a); *2* Tosserams et al. (1997); *3* Bassman et al. (2001); *4* Kakani et al. (2003b); *5* Singh (1996); *6* Antonelli et al. (1997); *7* Correia et al. (1998); *8* Musil et al. (2002); *9* Krizek et al. (1997a); *10* Mark and Tevini (1997); *11* Ruhland and Day (2001); *12* Xiong et al. (2002); *13* Barnes et al. (1988); *14* Ziska et al. (1993); *15* Krizek et al. (1998); *16* Day et al. (2001); *17* Barnes et al. (1993); *18* Gwynn-Jones and Johanson (1996); *19* Furness and Upadhyaya (2002); *20* Sullivan et al. (1994); *21* Newsham et al. (1999); *22* Li et al. (2000); *23* Hakala et al. (2002); *24* Sullivan et al. (1996); *25* Searles et al. (1995); *26* Pal et al. (1997); *27* Pinto et al. (2002); *28* Zaller et al. (2002); *29* Furness et al. (1999); *30* Hofmann et al. (2001); *31* Krizek et al. (1997b); *32* Zavala and Botto (2002); *33* Deckmyn and Impens (1995); *34* Mazza et al. (1999); *35* Robson et al. (2003); *36* Ballaré et al. (1991); *37* Stephen et al. (1999); *38* Phoenix et al. (2001); *39* Ballaré et al. (1996); *40* Searles et al. (1999); *41* Stephanou and Manetas (1997); *42* Schumaker et al. (1997); *43* Baskin and Iino (1987); *44* Shinkle et al. (2004); *45* Wilson and Greenberg (1993); *46* Boccalandro et al. (2001); *47* Brosché and Strid (2000)

greenhouse conditions. Certain morphological effects of UV-B (phototropism and leaf curling) have only been documented under controlled laboratory conditions.

Relative to effects on shoot systems, fewer studies have examined impacts of UV-B on roots, and the emphasis in most of these studies has been on root biomass or root:shoot ratios in pot-grown plants. In general, results from these studies reveal no consistent effect of UV-B on root production or root:shoot allocation. Several recent field studies have shown, however, that near-ambient UV-B can reduce root length and rhizome internode elongation (Zaller et al. 2002; Robson et al. 2003).

Except under very high or unbalanced UV-B exposures, as generally occurs in growth chambers and greenhouses (Caldwell and Flint 1994), these morphological changes frequently occur without any detectable reduction in photosynthesis or total biomass production (Barnes et al. 1990a; Ziska et al. 1993; Searles et al. 1995; Newsham et al. 1999). When reductions in biomass accumulation are observed under realistic UV-B exposure, growth reductions are often the *result* of morphological changes (via reductions in leaf area) rather than the underlying *cause* of these alterations (Ballaré et al. 1996; González et al. 1998; Xiong and Day 2001).

The effects of UV-B on morphology are influenced by a number of other factors, such as the physiological and developmental status of the plant (Hunt and McNeil 1998; Jordan et al. 1998; Hofmann et al. 2003), and the amount and spectral composition of associated background radiation. Of particular importance is the balance between UV-B, UV-A and photosynthetically active radiation (PAR; 400–700 nm). Generally, exposure to UV-B under low levels of PAR accentuates UV damage and increases overall sensitivity to UV-B (Caldwell and Flint 1994; Gwynn-Jones et al. 1999). The effects of UV-A are more complex; in some cases UV-A interacts with UV-B to ameliorate UV-B responses (Caldwell et al. 1994; Flint and Caldwell 1996), whereas, in others, UV-A has a direct effect on morphology and the combination of UV-B + UV-A elicits greater responses than either waveband alone (see Flint and Caldwell 2003a for summary).

Because of the complexities associated with these responses, it has been difficult to identify and characterize the precise photosensory processes involved in the UV-B-induced morphological responses (Jansen et al. 1998; Brosché and Strid 2003), and little is known of the nature of the putative UV-B receptor(s) thought to mediate photomorphogenic responses (Jenkins et al. 1997; Björn 1999). Unlike the situation for UV-absorbing compounds (Landry et al. 1995; Bieza and Lois 2001) or DNA repair (Britt 1999; Tanaka et al. 2002), there is only one report of a genetic mutant identified that participates in signal transduction leading to specific morphological responses to UV-B (Suesslin and Frohnmeyer 2003). To date, attempts to elucidate mechanisms responsible for these responses have therefore largely employed more traditional photobiological and physiological approaches, though work with a variety of mutants has been conducted to explore involvement of other photoreceptor systems or specific cellular processes (e.g., DNA damage).

Indirect evidence from studies conducted under controlled-environmental conditions indicates that UV-B can induce morphological changes in plants that involve very specific sensory systems. For example, Ballaré et al. (1991) demonstrated that short-term UV-B exposure (6 h) equivalent to ambient UV-B in the tropics produced rapid inhibition (ca. 50%) in hypocotyl elongation in light-grown cucumber (*Cucumis sativus*) seedlings. This inhibition in hypocotyl elongation was reversible, occurred without any associated reductions in cotyledon area expansion or total plant biomass accumulation, and was due to UV-B perceived by the cotyledons and not by the stem or growing apex. Follow-up studies using de-etiolating seedlings of tomato (*Lycopersicon esculentum*) showed that, when exposed to near monochromatic UV-B against a background of visible radiation, the wavelength of maximum inhibition of hypocotyl elongation occurred near 300 nm (Ballaré et al. 1995a). This inhibitory effect of UV-B could be countered by treating seedlings with chemicals that interfere with flavin photochemistry, implicating a role for a flavin as a UV-B receptor. Kim et al. (1998) have shown that UV-B at low fluence rates inhibited hypocotyl cell elongation in *Arabidopsis thaliana*. This response was also interpreted as photomorphogenic because of the similarity in responses between wild types (WT) and mutants deficient in repair of DNA damage, accumulation of UV-absorbing compounds and biosynthesis of compounds responsible for protection against oxidative stress. Continuous high-resolution measurements with displacement transducers have further shown that there is an initial inhibition in hypocotyl elongation in dicot seedlings that is very rapid (within 15 min of initial exposure to UV-B) and inhibition can persist for several hours following exposure before full recovery occurs (Bertram and Lercari 2000; Shinkle et al. 2004).

Light-dependent inhibition in stem elongation is a morphological response that is influenced not only by UV-B but also by the phytochrome and B/UV-A receptors (Briggs and Olney 2001). Phytochrome absorbs into the UV-B region and UV-B can induce phytochrome photoconversion (Pratt and Butler 1970). Thus, it is conceivable that interactions between UV-B, phytochrome and/or other photoreceptors may be involved in the UV-B-induced elongation response, as is the case for some pigment responses (Mohr 1994). To date, studies with phytochrome-altered mutants have shown conflicting results. Using single phytochrome-deficient mutants (phyA or phyB) of tomato and cucumber, Ballaré et al. (1991, 1995a) found no differences in UV-B elongation responses between WT and mutants. Similarly, Bertram and Lercari (2000) found little difference in UV-B-induced inhibition in stem elongation between WT and mutants of tomatoes deficient in phytochrome (phyA or phyB) or overexpressing phytochrome responses. Responses of seedlings exposed to UV-B + R and UV-B + FR indicated additive rather than interactive effects of the UV-B and phytochrome photoreceptor systems on stem elongation. Kim et al. (1998) found similar results for single phytochrome mutants of *Arabidopsis*, but double phyA/phyB mutants were insensitive to UV-B-induced inhibition in hypocotyl elongation. Thus, these investigators argued that small amounts of phytochrome are necessary for expression of the UV-B morphological response. In contrast to the hypocotyl responses to UV-B, Boccalandro et al. (2001) found that phyB was required for photomorphogenic effects of UV-B on cotyledon opening in *Arabidopsis*. Studies with *Arabidopsis* mutants defective in blue-light responses have generally shown little requirement for functional B/UV-A receptors in UV-B effects on hypocotyl elongation, cotyledon opening or phototropism (Boccalandro et al. 2001; Shinkle et al. 2004).

Some of the morphological responses to UV-B need not involve a specific and dedicated photosensory system but may be a consequence of UV-B-induced changes in phytohormone metabolism and/or phenolic chemistry (Jansen 2002). Certainly, some of the effects of UV-B (e.g., stimulation of axillary branching) appear to be indicative of phytohormone disruption. It is known that UV-B can lead to the photodestruction of auxin, at least under in vitro conditions (Ros and Tevini 1995), and levels of indole-acetic acid (IAA) can be reduced by UV-B in greenhouse-grown plants (Huang et al. 1997). In addition, UV-B may alter the transport rates and distribution of auxin indirectly via effects on flavonoids (Brown et al. 2001) or phenol-oxidizing peroxidases (Jansen et al. 2001). In some cases, direct or indirect effects of UV-B on wall-bound phenolics, which influence cell wall expansion (Liu et al. 1995), have also been proposed as a mechanism to explain UV-B-induced inhibition in leaf elongation (Sullivan et al. 1996; Ruhland and Day 2000). A key unresolved issue regarding all of the above mechanisms, however, is whether they operate under ecologically realistic UV-B exposures under field conditions (Day 2001). In addition, it is unlikely that auxin-mediated morphological changes are sufficient to account for all UV-B induced changes in plant form. The morphology of the *tt4* mutant of *Arabidopsis*, which constitutively lacks the flavonoids that would be induced by UV-B, is quite similar to that expected of UV-B-exposed plants (Brown et al. 2001), and overexpression of the UV-B-induced peroxidase in tobacco causes thin leaves, not the expected leaf thickening (Jansen et al. 2001).

It is possible that some of the morphological responses are mediated, at least in part, by UV-B-induced damage to DNA (Jordan 2002). Absorption of DNA by UV-B can result in genome alterations and these changes can impact cellular metabolism (Jansen et al. 1998). Action spectra for DNA damage (specifically the induction of CPDs, see below) typically peak at wavelengths below 300 nm, but appreciable damage can be induced by wavelengths well into the UV-A (Quaite et al. 1992). The action spectrum developed by Flint and Caldwell (2003a) for several UV-induced morphological responses in light-grown *Avena sativa* seedlings shows increasing effectiveness with decreasing wavelengths in the UV (down to 275 nm, which was the lowest wavelength tested), and there is a significant tail into the UV-A. However, field tests conducted under varying UV-B and UV-A conditions indicated that the DNA damage action spectrum of Quaite et al. (1992) was inadequate in describing the spectral sensitivity of these morphological responses (Flint and Caldwell 2003b). Similarly, the action spectrum for UV-induced inhibition of stem elongation in *Lepidium sativus* reported by Steinmetz and Wellmann (1986) resembled that of DNA absorption, but an absence of photoreactivation (the light-driven repair of

DNA damage) suggested that DNA damage was not the cause of growth inhibition in this case. However, this conclusion has been made somewhat less certain by the finding that a key enzyme catalyzing photoreactivation (CPD photolyase) is not expressed in dark-grown *Arabidopsis* (Ahmad et al. 1997). Under low fluence rates in the laboratory, Boccalandro et al. (2001) found that the UV-B-induced inhibition in hypocotyl elongation was similar in WT and mutants of *Arabidopsis* deficient in the repair of the two most prominent types of DNA lesions [i.e., cyclobutane pyrimidine dimers (CPDs) and pyrimidine [6–4] pyrimidinone dimers (6–4 photoproducts)]. Mutants were, however, more responsive to UV-B under high fluence rates. Field UV-B attenuation studies by Britt and Fiscus (2003) have shown that ambient levels of solar UV-B reduced shoot height and rosette diameter more in these DNA repair mutants of *Arabidopsis* than in WT, but the responses of WT plants were also marked. In addition, plants carrying mutations in two DNA repair components did not show greater effects than the single mutant lines. Hence, this study does not necessarily implicate DNA damage as a regulator of morphological alterations in response to UV-B. In other field studies using plants with 'normal' DNA repair processes, morphological alterations have been found to be associated with DNA damage under near-ambient solar UV-B conditions (Ballaré et al. 1996; Mazza et al. 1999). Some general support for DNA damage as a regulatory process can be found in the identification of genes involved in regulating cell cycle in response to DNA damage (Garcia et al. 2003; Preuss and Britt 2003), but the genes identified to this point appear to sense double strand breaks rather than UV-induced lesions.

Given the diversity of morphological responses and the range of time-courses involved in these responses (e.g., rapid inhibition in stem elongation vs. longer-term stimulation in axillary branching) it would seem unlikely that a single mechanism exists that could account for all of the UV-B effects on morphology. Indeed, even different short-term responses, such as hypocotyl elongation and cotyledon opening, appear to involve distinctly different photosensory systems (e.g., Boccalandro et al. 2001). Moreover, what appears to be a single, overall morphological response to a certain light signal may actually consist of multiple, distinct growth responses (Rich et al. 1987; Folta and Spalding 2001). For example, results from laboratory studies by Shinkle et al. (2004) examining the kinetics of hypocotyl elongation in dim-red light grown cucumber seedlings (i.e., chlorophyll-containing but etiolated growth form) indicated that short and long wavelengths of UV-B elicited distinctly different morphological responses. In these studies, a 30-min exposure to a full UV-B (280–320 nm) + UV-A radiation treatment induced large (90%) and rapid inhibition in hypocotyl elongation with no recovery after 24 h. In contrast, a UV-B + UV-A treatment that lacked the short wavelengths of UV-B (<300 nm) induced only a 50% inhibition in elongation, and this effect was transient in nature. Both of these responses differed from that induced by UV-A, and the two UV-B responses differed in their adherence to the reciprocity principle (i.e., the response is proportional to the total dose of photons, independent of the rate at which they are delivered). The short wavelength UV-B treatment was more effective in inducing DNA damage (measured as CPD formation) than the long wavelength UV-B treatment and the amount of DNA damage was strongly correlated with the inhibition in hypocotyl elongation for the short wavelength UV-B treatment but not the long wavelength treatment (J. Shinkle, unpubl. data). Thus, these results suggest that, under these laboratory conditions, both damaging and non-damaging mechanisms of hypocotyl inhibition may be operating within the UV-B. Kim et al. (1998) reported similar findings for *Arabidopsis*, though they ascribed the two types of responses as resulting from different UV-B fluxes rather than different wavelengths. To what extent these two mechanisms contribute to the overall UV-B-induced inhibition in stem elongation observed in field-grown plants is unknown, but findings from these studies do suggest that differences in UV spectral quality and/or fluence rates may shift the balance between damaging and non-damaging effects of UV-B on morphology.

4 Are UV-B-Induced Changes in Morphology Adaptive?

While there is general agreement regarding the adaptive significance of photomorphogenic effects of UV-B on pigmentation, such is not the case for the UV-B-induced changes in morphology. Indeed, in cases where morphological responses are solely the result of stress responses to UV-B there may be no adaptive explanation for these responses (Dixon et al. 2001).

It has been hypothesized, however, that the UV-B receptor evolved as a mechanism to allow plants to obtain information on changes in the flux of solar UV-B which are not well coupled with those in wavebands (i.e., UV-A and visible radiation) where the other photoreceptors operate (Ballaré et al. 1995b; Caldwell 1997). For emerging seedlings, reductions in hypocotyl elongation could serve an important protective role by delaying emergence from the soil until the UV-protective mechanisms (i.e., UV-absorbing compounds) are in place (Ballaré et al. 1995b, 1996). Under laboratory conditions, the UV-B-induced inhibition in hypocotyl elongation usually occurs much faster than the accumulation of UV-absorbing compounds (Ballaré et al. 1995b; J. Shinkle, unpubl. data). In a field UV-B attenuation study, Ballaré et al. (1996) showed that ambient UV-B reduced the rate but not the total seedling emergence in *Datura ferox*. Zavala and Botto (2002) also reported that near-ambient UV-B delayed seedling emergence in radish (*Raphanus sativus*) by at least 1 day.

To what extent the above hypothesis would apply to leaf and stem elongation in established plants is less clear. In contrast to the rapid inhibition in hypocotyl elongation, which appears to result from transient reductions in cell elongation, UV-B reduces elongation in leaves by altering rates of both cell division and elongation (Hopkins et al. 2002). Beggs et al. (1986) argued that growth reduction could be viewed as a protective response to UV-B if cell division was delayed to allow for repair of UV-induced DNA damage. Normal cell division and growth would then proceed following cessation of the UV-B exposure, repair of DNA and/or acclimation to UV-B (i.e., flavonoid accumulation). If this is the case, it follows that an understanding of the *timing* as well as the *effectiveness* of UV-screening and/or DNA repair mechanisms may be critical to understanding what adaptive value, if any, these morphological responses represent. At present, many studies have shown that morphological sensitivity, the production of UV-absorbing compounds and UV-screening effectiveness vary both within and among species (e.g., Day et al. 1992; Hofmann et al. 2000; Li et al. 2000; Smith et al. 2000; Musil et al. 2002). However, clear relationships between the kinetics and effectiveness of accumulation of UV-absorbing compounds, DNA damage/repair and morphological change have yet to be established.

5 UV-B Radiation and Plant–Plant Interactions

For many field-grown plants, the effects of UV-B on shoot morphology under ecologically realistic conditions (i.e., attenuation of ambient solar UV-B or proportional enhancement of solar UV-B) are often quite subtle, and these alterations have minimal impact on the overall performance of isolated plants. Certainly, relative to other abiotic factors, such as moisture and temperature, UV-B is usually not considered a primary ecological factor influencing species abundances and distributions. Nevertheless, there are situations where UV-B-induced photomorphogenic effects can be ecologically important (Caldwell et al. 1999). Competition is one such arena where these subtle effects of UV-B on morphology can have considerable ecological consequences (Barnes et al. 1996).

To date, a surprisingly small number of studies have examined the influence of UV-B on plant competition (Table 2). Nearly all these studies have examined species pairs and most have been conducted in the field. The majority of studies have focused on herbaceous species and, of these, most

Table 2. Summary of UV-B effects on interspecific plant competition. When UV-B-induced shifts in competitive balance were detected, the species in *bold* increased in competitive status in response to UV-B (either near-ambient or elevated UV-B)

Competing species	Growth forms	Experimental conditions	Monocultures	Reference[a]
Agricultural crops				
Sinapis alba – ***Raphanus sativus***	Annual dicots	Greenhouse	No	1
Beta vulgaris – ***Spinacia oleracea***	Annual dicots	Greenhouse	No	1
Brassica chinensis – *Brassica oleracea*	Annual dicots	Greenhouse	No	1
Agricultural crops and weeds				
Oryza sativa var. *Lemont* – ***Echinochloa crusgalli***	Annual grasses	Greenhouse	Yes	2
Oryza sativa var. *IR36* – *Echinochloa crusgalli*	Annual grasses	Greenhouse	Yes	2
Brassica oleracea – *Chenopodium album*	Annual dicots	Greenhouse	Yes	3
Pisum sativum – *Alyssum alyssoides*	Annual dicots	Field enhancement	Yes	4
Medicago sativa – *Amaranthus retroflexus*	Perennial/annual dicots	Field enhancement	Yes	4
Medicago sativa – *Brassica nigra*	Perennial/annual dicots	Field enhancement	Yes	4
Allium cepa – *Amaranthus retroflexus*	Perennial monocot/annual dicot	Field enhancement	Yes	4
Trifolium pratense – *Setaria glauca*	Perennial dicot/annual grass	Field enhancement	Yes	4
Triticum aestivum – *Aegilops cylindrical*	Annual grasses	Field enhancement	Yes	5
Triticum aestivum – *Avena fatua*	Annual grasses	Field enhancement	Yes	5, 6
Triticum aestivum – ***Avena fatua***	Annual grasses	Field enhancement	Yes	7
Phaseolus vulgaris – *Avena fatua*	Annual dicot/annual grass	Field enhancement	Yes	8
Sinapsis alba – *Cirsium arvense*	Annual/perennial dicots	Field attenuation	No	9
Species of disturbed areas				
Setaria viridis – *Echinochloa crusgalli*	Annual grasses	Greenhouse	Yes	2
Bromus tectorum – *Alyssum alyssoides*	Annual grass/annual dicot	Field enhancement	Yes	4
Plantago patagonica – *Lepidium perfoliatum*	Annual dicots	Field enhancement	Yes	4
Verbascum phlomoides – ***Bellis perennis***	Biennial/perennial dicots	Field attenuation	Yes	10

Table 2. *Continued*

Competing species	Growth forms	Experimental conditions	Monocultures	Reference[a]
Taraxacum officinale – Bellis perennis	Perennial dicots	Field attenuation	Yes	9
Chrysanthemum vulgare *– Senecio sylvaticus*	Perennial/annual dicots	Field attenuation	Yes	10
Rumex obtusifolius – Rumex alpinus	Perennial dicots	Field attenuation	Yes	1
Verbascum phlomoides – ***Taraxacum officinale***	Biennial/perennial dicots	Field attenuation	Yes	10
Andropogon virginicus – Aster pilosus	Perennial grass/perennial dicot	Field attenuation	Yes	11
Forage/grassland species				
Sorghum halepense – Panicum virgatum	Perennial grasses	Greenhouse	Yes	2
Poa pratensis *– Geum macrophyllum*	Perennial grass/perennial dicot	Field enhancement	Yes	4
Dactylis glomerata – Trifolium repens	Perennial grass/perennial dicot	Field attenuation	Yes	12
Lolium perenne – Trifolium repens	Perennial grass/perennial dicot	Field attenuation	No	13
Lolium perenne – Poa trivialis – Dactylis glomerata – Phleum pratense – Cerastium holosteiodes – Plantago lanceolata	Perennial grasses (4 species)/ perennial dicots (2 species)	Field enhancement	No	14
Forest/woodland species				
Fraxinus excelsior – Acer pseudoplatanus	Woody dicot tree seedlings	Greenhouse	No	1
Fraxinus excelsior – ***Carpinus betulus***	Woody dicot tree seedlings	Field attenuation	No	1
Fagus sylvatica – Acer platanoides	Woody dicot tree seedlings	Field attenuation	No	9

[a] *1* B. Bruzek, unpubl. data, reported in Gold and Caldwell (1983); *2* Beck (1998); *3* Furness et al. (unpubl.); *4* Fox and Caldwell (1978); *5* Gold and Caldwell (1983); *6* Barnes et al. (1988); *7* Li et al. (1999); *8* R. Ryel, P. Barnes, S. Flint and M. Caldwell, unpubl. data; *9* Bogenrieder and Klein (1982); *10* S. Kiliani, unpubl. data, reported in Gold and Caldwell (1983); *11* Coon et al. (1997); 12 Cayenberghs et al. (2001); *13* Matthew et al. (1996); *14* Norton et al. (1999)

have involved crops and associated weeds, or co-occurring species of open or disturbed habitats. The most common experimental approach has been substitutive style designs where plants are grown in single-species stands (monocultures) and species mixtures (e.g., Fox and Caldwell 1978; Barnes et al. 1988; Li et al. 1999). Typically, plant performance has been assessed by measuring biomass production, though Li et al. (1999) examined both biomass and seed (grain) production. Interestingly, in this latter case the direction of UV-B-induced competitive shift appeared to be reversed when these different performance metrics were used. In several cases, only mixtures were examined. In the absence of monoculture data it is difficult to interpret the effects of UV-B on competition, as changes in mixture composition may be due to direct physiological or growth responses of the species to UV-B and not necessarily the result of UV-B-induced shifts in competitive balance (Gold and Caldwell 1983). Nonetheless, in Table 2, we report these changes in mixture composition as competitive shifts even when monoculture data are lacking, while recognizing the potential for this misinterpretation.

A total of 31 different species combinations have been examined, and UV-B-induced shifts in competitive balance have been reported in nearly half of the cases. Both near-ambient and elevated UV-B (in greenhouse and field settings) have been found to alter competitive balance. Studies with agricultural species and their associated weeds have shown that UV-B can often shift the balance of competition in favor of the crop, but this is not always the case. The UV-B-induced shifts in competitive balance have been found in mixtures of narrow-leaved grasses, broad-leaved dicots as well as grass + dicot combinations. In many (Fox and Caldwell 1978; Gold and Caldwell 1983; Barnes et al. 1988) but not all (Li et al. 1999) cases, shifts in competitive balance occur without any detectable changes in either monoculture or total mixture production.

In the most intensive study to date, Barnes et al. (1988) reported significant shifts in competitive balance between two grasses, wheat (*Triticum aestivum*) and a common weedy competitor, wild oat (*Avena fatua*), in response to supplemental UV-B under field conditions. This competitive shift occurred without any detectable effect of the UV-B treatment on single leaf photosynthesis (Beyschlag et al. 1988), total monoculture production (Gold and Caldwell 1983; Barnes et al. 1988) or simulated monoculture canopy photosynthesis (Barnes et al. 1995). Measurements of individual plants, however, revealed significant differential effects of the UV-B on shoot morphology (i.e., leaf length and leaf insertion height – a measure of the position of leaves in the canopy) with the weed, wild oat, being affected more than the crop, wheat. These plant-level morphological alterations then resulted in shifts in the relative positioning of leaf area for the two species in mixed canopies (Barnes et al. 1988). Using a multi-species canopy simulation model, the observed changes in canopy structure were found to be sufficient to alter light interception and canopy photosynthesis for these species (Ryel et al. 1990; Barnes et al. 1995). Thus, in this case, it was hypothesized that UV-B enhancement altered the competitive balance indirectly via changes in plant morphology, which then altered competition for light between the two species.

Since this long-term (6-year) field study, additional experiments have been conducted to test the generality of the above hypothesis of competitive change for other species combinations. In a field study with a broad-leaved crop, green bean (*Phaseolus vulgaris*), and wild oat, the UV-B treatment shifted the balance of competition in favor of bean and this change was again associated with differential morphological and canopy responses to the UV-B, and not changes in rates of single leaf photosynthesis (R. Ryel, P. Barnes, S. Flint and M. Caldwell, unpubl. data). Greenhouse studies conducted by Beck (1998) examined four mixtures of grasses and found that the UV-B altered competitive balance in one of these mixtures (*Oryza sativa* var. Lemont–*Echinochloa crusgalli*), and this was the only case where differential morphological responses to UV-B were observed. In this study, UV-B effects on leaf blade insertion height were more closely associated with the competitive shift than treatment effects on leaf blade length. More recently, UV-B-induced competitive shifts have been found for two dicot species (*Brassica oleracea* and *Chenopodium album*) under greenhouse conditions (Furness et al., unpubl.), and these changes were linked to morphological changes (Furness 2003). Thus, it appears that the above mechanism of UV-B-induced alterations in competitive balance mediated by alterations in shoot morphology is broadly applicable to various species mixtures representing different growth forms.

Within plant canopies, the supply and acquisition of light is largely unidirectional, coming primarily overhead from the direct beam of the sun and diffuse light of the sky. Consequently, competition for this resource is thought to be inherently asymmetrical between neighboring plants (Schwinning and Weiner 1998), such that even small differences in stature or leaf placement can result in large differences in light capture and photosynthesis (Barnes et al. 1990b). The quantitative importance of this phenomenon is illustrated in the following exercise, where changes in the relative positioning of leaf area were evaluated in hypothetical mixed-species canopies of herbaceous planophilic (i.e., many dicots) and erectophilic (grasses) growth forms (e.g., Monsi et al. 1973; Fig. 1A, B) using a multi-species light interception and photosynthesis canopy model (Ryel et al. 1990). Simulation models such as this can be effective tools for assessing the effects of changes in light climate in canopies where manipulations of

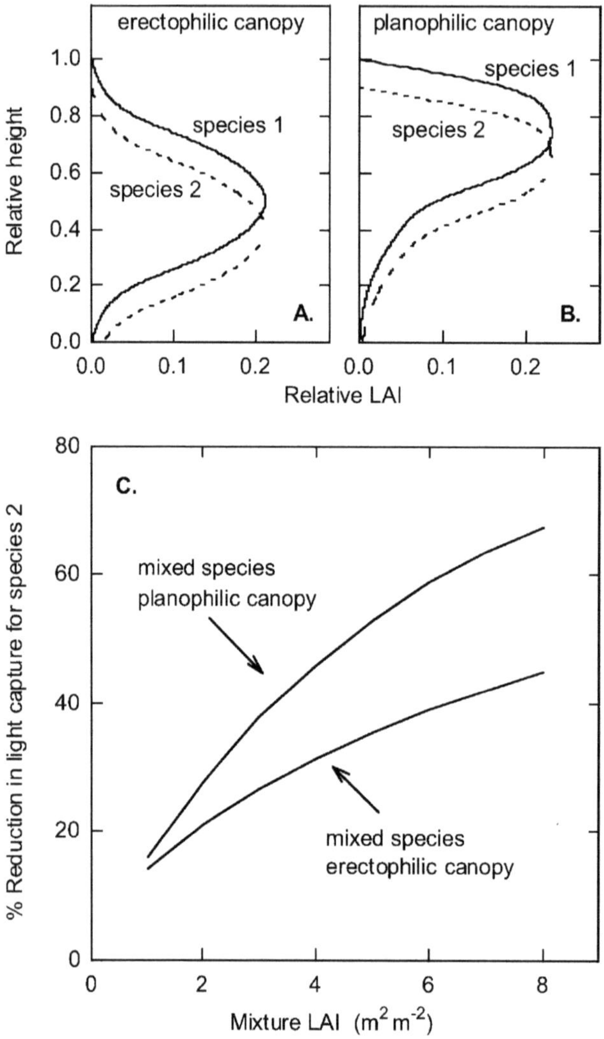

Fig. 1. Leaf area index (*LAI*; m^2 m^{-2}) profiles (**A, B**) and calculated changes in relative light capture (**C**) in hypothetical mixtures of erectophilic (grass; leaf angle 70°) and planophilic (forb; leaf angle 30°) species. For both canopy types, leaf angles are assumed uniform with depth, and leaf blades are the only foliage elements influencing light penetration (i.e., there is no stem or sheath material). Calculated changes in light capture are for integrated daily light capture of species 2 relative to species 1. Total LAI for species 1 and 2 in mixture is the same, but the height profile of LAI is reduced by 10% for species 2. Simulation conditions are for mid-latitude clear skies in summer. (Ryel et al. 1990)

canopy structure are difficult to do experimentally (Beyschlag et al. 1994; Ryel and Beyschlag 1999). The results from these simulations indicate that, even when the total leaf area index (LAI) of the two component species is identical and remains unchanged, a 10% reduction in the vertical distribution of leaf area for one of the species can reduce its relative light capture in mixture by 20–60% (Fig. 1C). The effects of these rather small differences in overtopping of one species by another increase with increasing LAI and are greater for the more horizontally inclined, planophilic canopies than the more vertical erectophilic canopies. These findings argue that mixtures of broad-leaved species (forbs) should be more susceptible to UV-B-induced competitive shifts than mixtures of grasses, which tend to have more steeply inclined foliage. This may not necessarily be the case, as these two growth forms appear to differ in their morphological sensitivity to UV-B (Barnes et al. 1990a; Tosserams et al. 1997).

Due to the asymmetrical nature of light competition, small initial differences in size and stature among competing individuals can become greatly magnified over time (Weiner et al. 2001). For monospecific stands, this asymmetric competition for light is thought to contribute to shifts in size distributions over time and the creation of size hierarchies (Weiner 1990). Similar mechanisms likely operate in mixed species stands (Freckleton and Watkinson 2001). Therefore, in a highly competitive environment, the potential exists for further amplification of the subtle effects of UV-B on growth and morphology. The degree to which this amplification occurs, however, will likely depend on the morphological responses to UV-B and other wavebands that occur within the canopy. For example, in some cases adjustments in morphology (i.e., increased stem elongation) in response to decreases in R:FR prior to and during canopy closure tend to counter the effects of asymmetrical competition and thereby dampen size variability in monocultures from that which would occur without these adjustments (Aphalo et al. 1999). Much less is known about how canopy structure and shading affect plant morphological responses to UV-B (Deckmyn et al. 2001). Observations do indicate that morphological responses to UV-B can often be quite different when plants are grown in mixtures and monocultures (Barnes et al. 1988), but whether these differences are the result of an altered competitive regime or spectral conditions is unknown.

In general, leaves transmit and reflect very little UV-B (Caldwell et al. 1983; Day et al. 1993), and, on average, UV-B penetrates less effectively in canopies than does UV-A, PAR, R or FR (Grant 1997). Canopy architecture can influence UV-B penetration, with erectophilic canopies having greater transmission of UV-B than planophilic canopies (Deckmyn et al. 2001). Due to the rapid attenuation of UV-B in canopies, shorter, more shaded plants will receive less *absolute* UV-B exposure and therefore may exhibit reduced

Table 3. Measured irradiances for ultraviolet and visible wavebands in full sunlight and below mature tropical forest canopies. To allow for comparisons with other studies, irradiance is reported in units most commonly reported [i.e., photon flux density for photosynthetically active radiation (PAR; 400–700 nm) and energy units for ultraviolet-B (UV-B; 280–320 nm) and ultraviolet-A (UV-A; 320–400 nm) radiation]. The UV-B irradiances are unweighted except UV-B_{BE}, which is weighted according to the generalized plant action spectrum of Caldwell (1971) and normalized to unity at 300 nm. Ratios of red:far-red (R:FR) are for photon flux densities from 658–662 nm (R) and 728–732 nm (FR)

Waveband	Full sunlight	Below canopy		Reference
		Sunlit gaps	Shade	
PAR ($\mu mol\ m^{-2}\ s^{-1}$)	2,000	1,350	70	Flint and Caldwell (1998)
UV-A ($W\ m^{-2}$)	52.8	–	9.4	Parisi and Kimlin (1999)
UV-B ($W\ m^{-2}$)	3.6	–	0.84	Parisi and Kimlin (1999)
UV-B_{BE} ($mW\ m^{-2}$)	350	128	19	Flint and Caldwell (1998)
UV-B_{305}:PAR (ca.)[a]	0.8	0.5	1.5	Flint and Caldwell (1998)
UV-B:UV-A	0.068	–	0.089	Parisi and Kimlin (1999)
R:FR	1.33	1.15	0.35	Lee (1987)

[a] Ratio of UV-B at 305 nm to PAR (UV-B_{305}:PAR) is multiplied by 10^4 for ease of presentation

inhibition in stem and leaf elongation to UV-B relative to taller plants. These effects would tend to counteract the development of strong competitive asymmetries in both monocultures and mixtures. On the other hand, morphological responses within the canopy may be affected by *relative* UV-B fluxes in ways that could enhance competitive asymmetry. For example, with respect to radiation of longer wavelengths (especially the PAR), a much greater proportion of solar UV-B is present in the diffuse component than the direct beam. Consequently, appreciable differences in the ratios of UV-B to other wavebands exist in sunlit (direct + diffuse radiation) and shaded (diffuse radiation only) areas in canopies (Grant 1999). In forest canopies, where it is easy to precisely position sensors in sunlit and shaded areas, the ratio of UV-B:PAR can be two- to threefold higher in the shade than in sunflecks or above the canopy (Table 3). By comparison, gaps are somewhat depleted in UV-B relative to PAR. It is conceivable, therefore, that morphological responses to UV-B could be accentuated in shaded regions (higher UV-B:PAR) as compared to the canopy gaps (lower UV-B:PAR). Given the large fraction of shaded foliage in canopies, this effect could be appreciable. However, apart from the well-documented effects of maintaining proper UV-B:PAR balance under high irradiance conditions simulating ozone depletion (Caldwell and Flint 1994), very little is known of the significance of absolute vs. relative changes in UV-B on morphology (Deckmyn et al. 1994). In addition to the altered UV-B:PAR ratios, shaded areas of a canopy may also exhibit slightly elevated ratios of UV-B:UV-A. Thus, shade created by vegetation not only exhibits low R:FR ratios, which can influence phytochrome responses, but also may have altered ratios of UV-B:UV-A:PAR. To what extent these combined spectral changes within the canopy influence morphology and interactions between the UV-B, B/UV-A and phytochrome photoreceptor systems is largely unknown (Ballaré 1999), but they may be important in mediating competitive interactions under ambient as well as enhanced or reduced UV-B conditions.

6 Summary and Conclusions

Ultraviolet-B radiation comprises a relatively small amount of the solar spectrum at the earth's surface, yet radiation in this waveband can induce a variety of growth and morphological responses in higher plants. Although the precise photosensory mechanisms responsible for these morphological responses have yet to be fully elucidated, it does appear that some of these responses represent specific photomorphogenic effects of UV-B, while others are the result of more general stress responses to UV-B. The degree to which these specific and non-specific mechanisms contribute to the overall morphological responses to UV-B in field-grown plants has received little attention. While the majority of studies to date have focused on the potential responses to elevated UV-B that would occur with stratospheric ozone depletion, it is clear from a number of recent studies that current levels of solar UV-B, even in regions experiencing minimal stratospheric ozone depletion, can influence plant morphology. Thus, solar UV-B is presently an important abiotic factor influencing plant growth. The morphological alterations induced by ambient or realistic UV-B enhancements, however, are typically subtle and may be of minor consequence for the overall performance and success of isolated plants. These alterations have the potential to be ecologically important in highly productive, crowded environments where subtle differences in morphology and leaf placement in canopies can be magnified by competitive interactions for light. The extent to which asymmetric competition for light further amplifies size differences will likely be determined by the degree to which plant morphology is affected by the changes in UV-B, UV-A, PAR and R:FR that occur within plant canopies.

Acknowledgements. We thank the US Department of Agriculture (NRICGP Grant 96-35100-3166 to Texas State University and Trinity University, and UV-B Radiation Monitoring Program subcontract G 1502-2 to Utah State University) for financial support of some of the research and ideas presented here. We are grateful for the support and encouragement of M. Caldwell.

References

A-H-Mackerness S (2000) Plant responses to ultraviolet-B (UV-B: 280–320 nm) stress: what are the key regulators? Plant Growth Reg 32:27–39

Ahmad M, Jarillo JA, Klimczak LJ, Landry LG, Peng T, Last RL, Cashmore AR (1997) An enzyme similar to animal type II photolyases mediates photoreactivation in *Arabidopsis*. Plant Cell 9:199–207

Antonelli F, Grifoni D, Sabatini F, Zipoli G (1997) Morphological and physiological responses of bean plants to supplemental UV radiation in a Mediterranean climate. Plant Ecol 128:127–136

Aphalo PJ, Ballaré CL (1995) On the importance of information-acquiring systems in plant–plant interactions. Funct Ecol 9:5–14

Aphalo PJ, Ballaré CL, Scopel AL (1999) Plant–plant signalling, the shade-avoidance response and competition. J Exp Bot 50:1629–1634

Baker NR, Nogués S, Allen DJ (1997) Photosynthesis and photoinhibition. In: Lumsden PJ (ed) Plants and UV-B: responses to environmental change. Cambridge University Press, Cambridge, pp 95–111

Ballaré CL (1999) Keeping up with the neighbors: phytochrome sensing and other signalling mechanisms. Trends Plant Sci 4:97–102

Ballaré CL (2003) Stress under the sun: spotlight on ultraviolet-B responses. Plant Physiol 132:1725–1727

Ballaré CL, Barnes PW, Kendrick RE (1991) Photomorphogenic effects of UV-B radiation on hypocotyl elongation in wild type and stable-phytochrome-deficient mutant seedlings of cucumber. Physiol Plant 83:652–658

Ballaré CL, Barnes PW, Flint SD (1995a) Inhibition of hypocotyl elongation by ultraviolet-B radiation in de-etiolating tomato seedlings. I. The photoreceptor. Physiol Plant 93:584–592

Ballaré CL, Barnes PW, Flint SD, Price S (1995b) Inhibition of hypocotyl elongation by ultraviolet-B radiation in de-etiolating tomato seedlings. II. Time course, comparison with flavonoid responses and adaptive significance. Physiol Plant 93:593–601

Ballaré CL, Scopel AL, Stapleton AE, Yanovsky MJ (1996) Solar ultraviolet-B radiation affects seedling emergence, DNA integrity, plant morphology, growth rate, and attractiveness to herbivore insects in *Datura ferox*. Plant Physiol 112:161–170

Ballaré CL, Scopel AL, Sánchez RA (1997) Foraging for light: photosensory ecology and agricultural implications. Plant Cell Environ 20:820–825

Ballaré CL, Scopel AL, Mazza CA (1999) Effects of solar UV-B radiation on terrestrial ecosystems: case studies from southern South America. In: Rozema J (ed) Stratospheric ozone depletion: the effects of enhanced UV-B radiation on terrestrial ecosystems. Backhuys, Leiden, pp 293–311

Barnes PW, Flint SD, Caldwell MM (1990a) Morphological responses of crop and weed species of different growth forms to ultraviolet-B radiation. Am J Bot 77:1354–1360

Barnes PW, Beyschlag W, Ryel R, Flint SD, Caldwell MM (1990b) Plant competition for light analyzed with a multispecies canopy model. III. Influence of canopy structure in mixtures and monocultures of wheat and wild oat. Oecologia 82:560–566

Barnes PW, Flint SD, Caldwell MM (1995) Early-season effects of supplemented solar UV-B radiation on seedling emergence, canopy structure, simulated stand photosynthesis and competition for light. Global Change Biol 1:43–53

Barnes PW, Ballaré CL, Caldwell MM (1996) Photomorphogenic effects of UV-B radiation on plants: consequences for light competition. J Plant Physiol 148:15–20

Barnes PW, Jordan PW, Gold WG, Flint SD, Caldwell MM (1988) Competition, morphology and canopy structure in wheat (*Triticum aestivum* L.) and wild oat (*Avena fatua* L.) exposed to enhanced ultraviolet-B radiation. Funct Ecol 2:319–330

Barnes PW, Maggard S, Holman SR, Vergara BS (1993) Intraspecific variation in sensitivity to UV-B radiation in rice. Crop Sci 33:1041–1046

Baskin TI, Iino M (1987) An action spectrum in the blue and ultraviolet for phototropism in alfalfa. Photochem Photobiol 46:127–136

Bassman JH, Robberecht R, Edwards GE (2001) Effects of enhanced UV-B radiation on growth and gas exchange in *Populus deltoides* Bartr. ex Marsh. Int J Plant Sci 162:103–110

Beck J (1998) Effects of ultraviolet-B radiation on grass morphology and competition. MS Thesis, Biology Department, Texas State University, San Marcos, Texas, p 75

Beggs CJ, Wellmann E (1994) Photocontrol of flavonoid biosynthesis. In: Kendrick RE, Kronenberg GHM (eds) Photomorphogenesis in plants, 2nd edn. Kluwer, Dordrecht, pp 733–751

Beggs CJ, Schneider-Ziebert U, Wellmann E (1986) UV-B radiation and adaptive mechanisms in plants. In: Worrest RC, Caldwell MM (eds) Stratospheric ozone reduction, solar ultraviolet radiation and plant life. Springer, Berlin Heidelberg New York, pp 235–250

Bertram L, Lercari B (2000) Evidence against the involvement of phytochrome in UVB-induced inhibition of stem growth in green tomato plants. Photosyn Res 64:107–117

Beyschlag W, Barnes PW, Flint SD, Caldwell MM (1988) Enhanced UV-B irradiation has no effect on photosynthetic characteristics of wheat (*Triticum aestivum* L.) and wild oat (*Avena fatua* L.) under greenhouse and field conditions. Photosynthetica 22:516–525

Beyschlag W, Ryel RJ, Caldwell MM (1994) Photosynthesis of vascular plants; assessing canopy photosynthesis by means of simulation models. In: Caldwell MM, Schulze D-E (eds) Ecophysiology of photosynthesis. Ecological studies 100. Springer, Berlin Heidelberg New York, pp 409–430

Bieza K, Lois R (2001) An *Arabidopsis* mutant tolerant to lethal ultraviolet-B levels shows constitutively elevated accumulation of flavonoids and other phenolics. Plant Physiol 126:1105–1115

Björn LO (1999) UV-B effects: receptors and targets. In: Singhal GS, Renger G, Sopory SK, Irrgang K-D, Govindjee (eds) Concepts in photobiology: photosynthesis and photomorphogenesis. Narosa, New Delhi, pp 821–832

Björn LO, Callaghan TV, Gehrke C, Gwynn-Jones D, Lee JA, Johanson U, Sonesson M, Buck ND (1999) Effects of ozone depletion and increased ultraviolet-B radiation on northern vegetation. Polar Res 18:331–337

Boccalandro HE, Mazza CA, Mazzella MA, Casal JJ, Ballaré CL (2001) Ultraviolet B radiation enhances a phytochrome-B-mediated photomorphogenic response in *Arabidopsis*. Plant Physiol 126:780–788

Bogenrieder A, Klein R (1982) Does solar UV influence the competitive relationship in higher plants? In: Calkins J (ed) The role of solar ultraviolet in marine ecosystems. Plenum Press, New York, pp 641–649

Bornman JF, Reuber S, Cen Y-P, Weissenböck G (1997) Ultraviolet radiation as a stress factor and the role of protective pigments. In: Lumsden PJ (ed) Plants and UV-B: responses to environmental change. Cambridge University Press, Cambridge, pp 157–168

Briggs WR, Olney MA (2001) Photoreceptors in plant photomorphogenesis to date. Five phytochromes, two cryptochromes, one phototropin, and one superchrome. Plant Physiol 125:85–88

Britt AB (1999) Molecular genetics of DNA repair in higher plants. Trends Plant Sci 4:20–25

Britt A, Fiscus EL (2003) Growth responses of *Arabidopsis* DNA repair mutants to solar irradiation. Physiol Plant 118:183–192

Brosché M, Strid Å (2000) Ultraviolet-B radiation causes tendril coiling in *Pisum sativum*. Plant Cell Physiol 41:1077–1079

Brosché M, Strid Å (2003) Molecular events following perception of ultraviolet-B radiation by plants. Physiol Plant 117:1–10

Brown DE, Rashotte AM, Murphy AS, Normanly J, Tague BW, Peer WA, Taiz L, Muday GK (2001) Flavonoids act as negative regulators of auxin transport in vivo in *Arabidopsis*. Plant Physiol 126:524–535

Caldwell MM (1971) Solar UV irradiation and the growth and development of higher plants. Photophysiology 6:131–177

Caldwell MM (1997) Alterations in competitive balance. In: Lumsden PJ (ed) Plants and UV-B: responses to environmental change. Cambridge University Press, Cambridge, pp 305–315

Caldwell MM, Ballaré CL, Bornman JE, Flint SD, Björn LO, Teramura AH, Kulandaivelu G, Tevini M (2003) Terrestrial ecosystems, increased solar ultraviolet radiation and interactions with other climatic change factors. Photochem Photobiol Sci 2:29–38

Caldwell MM, Flint SD (1994) Stratospheric ozone reduction, solar UV-B radiation and terrestrial ecosystems. Climatic Change 28:375–394

Caldwell MM, Flint SD, Searles PS (1994) Spectral balance and UV-B sensitivity of soybean: a field experiment. Plant Cell Environ 17:267–276

Caldwell MM, Robberecht R, Flint SD (1983) Internal filters: prospects for UV-acclimation in higher plants. Physiol Plant 58:445–450

Caldwell MM, Searles PS, Flint SD, Barnes PW (1999) Terrestrial ecosystem responses to solar UV-B radiation mediated by vegetation, microbes and abiotic photochemistry. In: Press MC, Scholes JD, Barker MG (eds) Physiological plant ecology. Blackwell, Oxford, pp 241–262

Casati P, Walbot V (2003) Gene expression profiling in response to ultraviolet radiation in maize genotypes with varying flavonoid content. Plant Physiol 132:1739–1754

Cayenberghs E, Deckmyn G, Ceulemans R (2001) Decreased ultraviolet-B radiation alters the vertical biomass distribution in cocksfoot. Biol Plant 44:385–389

Coon CR, Forseth IN, Sullivan JH (1997) Ambient UV-B radiation effects on competition between two early successional species. Bull Ecol Soc Am Suppl 78:236

Correia CM, Areal ELV, Torres-Pereira MS, Torres-Pereira JMG (1998) Intraspecific variation in sensitivity to ultraviolet-B radiation in maize grown under field conditions. I. Growth and morphological aspects. Field Crop Res 59:81–89

Day TA (2001) Ultraviolet radiation and plant ecosystems. In: Cockell CS, Blaustein AR (eds) Ecosystems, evolution, and ultraviolet radiation. Springer, Berlin Heidelberg New York, pp 80–117

Day TA, Neale PJ (2002) Effects of UV-B radiation on terrestrial and aquatic primary producers. Annu Rev Ecol Syst 33:371–396

Day TA, Vogelmann TC, DeLucia EH (1992) Are some plant life forms more effective than others in screening out ultraviolet-B radiation? Oecologia 92:513–519

Day TA, Martin G, Vogelmann TC (1993) Penetration of UV-B radiation in foliage: evidence that the epidermis behaves as a non-uniform filter. Plant Cell Environ 16:735–741

Day TA, Ruhland CT, Xiong FS (2001) Influence of solar ultraviolet-B radiation on Antarctic terrestrial plants: results from a 4-year field study. J Photochem Photobiol B Biol 62:78–87

Deckmyn G, Impens I (1995) UV-B increases the harvest index of bean (*Phaseolus vulgaris* L). Plant Cell Environ 18:1426–1433

Deckmyn G, Martens C, Impens I (1994) The importance of the ratio UV-B/photosynthetic active radiation (PAR) during leaf development as determining factor of plant sensitivity to increased UV-B irradiance: effects on growth, gas exchange, and pigmentation of bean plants (*Phaseolus vulgaris* cv. Label). Plant Cell Environ 17:295–301

Deckmyn G, Cayenberghs E, Ceulemans R (2001) UV-B and PAR in single and mixed canopies grown under different UV-B exclusions in the field. Plant Ecol 154:125–133

Dixon P, Weinig C, Schmitt J (2001) Susceptibility to UV damage in *Impatiens capensis* (Balsaminaceae): testing for opportunity costs to shade-avoidance and population differentiation. Am J Bot 88:1401–1408

Flint SD, Caldwell MM (1996) Scaling plant ultraviolet spectral responses from laboratory action spectra to field spectral weighting factors. J Plant Physiol 148:107–114

Flint SD, Caldwell MM (1998) Solar UV-B and visible radiation in tropical forest gaps: measurements partitioning direct and diffuse radiation. Global Change Biol 4:863–870

Flint SD, Caldwell MM (2003a) A biological spectral weighting function for ozone depletion research with higher plants. Physiol Plant 117:137–144

Flint SD, Caldwell MM (2003b) Field testing of UV biological spectral weighting functions for higher plants. Physiol Plant 117:145–153

Flint SD, Ryel RJ, Caldwell MM (2003) Ecosystem UV-B experiments in terrestrial communities: a review of recent findings and methodologies. Agric For Meteorol 120:177–189

Folta KM, Spalding EP (2001) Unexpected roles for cryptochrome 2 and phototropin revealed by high-resolution analysis of blue light-mediated hypocotyl growth inhibition. Plant J 26:471–478

Fox FM, Caldwell MM (1978) Competitive interaction in plant populations exposed to supplementary ultraviolet-B radiation. Oecologia 36:173–190

Freckleton RP, Watkinson AR (2001) Asymmetric competition between plant species. Funct Ecol 15:615–623

Furness NH (2003) Influence of ultraviolet-B radiation on crop-weed competitive interactions. PhD Diss, Faculty of Agricultural Sciences, University of British Columbia, Vancouver, British Columbia, p 243

Furness NH, Upadhyaya MK (2002) Differential susceptibility of agricultural weeds to ultraviolet-B radiation. Can J Plant Sci 82:789–796

Furness NH, Upadhyaya MK, Ormrod DP (1999) Seedling growth and leaf surface morphological responses of three rangeland weeds to ultraviolet-B radiation. Weed Sci 47:427–434

Garcia V, Bruchet H, Camescasse D, Granier F, Bouchez D, Tissier A (2003) *AtATM* is essential for meiosis and the somatic response to DNA damage in plants. Plant Cell 15:119–132

Gold WG, Caldwell MM (1983) The effects of ultraviolet-B radiation on plant competition in terrestrial ecosystems. Physiol Plant 58:435–444

González R, Mepsted R, Wellburn AR, Paul ND (1998) Non-photosynthetic mechanisms of growth reduction in pea (*Pisum sativum* L.) exposed to UV-B radiation. Plant Cell Environ 21:23–32

Grant RH (1997) Partitioning of biologically active radiation in plant canopies. Int J Biometeorol 40:26–40

Grant RH (1999) Ultraviolet-B and photosynthetically active radiation environment of inclined leaf surfaces in a maize canopy and implications for modeling. Agric Forest Meteorol 95:187–201

Gwynn-Jones D, Johanson U (1996) Growth and pigment production in two subarctic grass species under four different UV-B irradiation levels. Physiol Plant 97:701–707

Gwynn-Jones D, Lee JA, Johanson U, Phoenix GK, Callaghan TV, Sonesson M (1999) The responses of plant functional types to enhanced UV-B radiation. In: Rozema J (ed) Stratospheric ozone depletion: the effects of enhanced UV-B radiation on terrestrial ecosystems. Backhuys, Leiden, pp 173–185

Hakala K, Jauhiainen L, Koskela T, Kayhko P, Vorne V (2002) Sensitivity of crops to increased ultraviolet radiation in northern growing conditions. J Agron Crop Sci 188:8–18

Hofmann RW, Swinny EE, Bloor SJ, Markham KR, Ryan KG, Campbell BD, Jordan BR, Fountain DW (2000) Responses of nine *Trifolium repens* L. populations to ultraviolet-B radiation: differential flavonol glycoside accumulation and biomass production. Ann Bot 86:527–537

Hofmann RW, Campbell BD, Fountain DW, Jordan BR, Greer DH, Hunt DY, Hunt CL (2001) Multivariate analysis of intraspecific responses to UV-B radiation in white clover (*Trifolium repens* L.). Plant Cell Environ 24:917–927

Hofmann RW, Campbell BD, Fountain DF (2003) Sensitivity of white clover to UV-B radiation depends on water availability, plant productivity and duration of stress. Global Change Biol 9:473–477

Holley SR, Yalamanchili RD, Moura DS, Ryan CA, Stratmann JW (2003) Convergence of signaling pathways induced by systemin, oligosaccharide elicitors, and ultraviolet-B radiation at the level of mitogen-activated protein kinases in *Lycopersicon peruvianum* suspension-cultured cells. Plant Physiol 132:1728–1738

Hopkins L, Bond MA, Tobin AK (2002) Ultraviolet-B radiation reduces the rates of cell division and elongation in the primary leaf of wheat (*Triticum aestivum* L. cv Maris Huntsman). Plant Cell Environ 25:617–624

Huang S, Dai Q, Peng S, Chavez AQ, Miranda LL, Visperas RM, Vergara BS (1997) Influence of supplemental ultraviolet-B on indoleacetic acid and calmodulin in the leaves of rice (*Oryza sativa* L). Plant Growth Reg 21:59–64

Hunt JE, McNeil DL (1998) Nitrogen status affects UV-B sensitivity of cucumber. Aust J Plant Physiol 25:79–86

Izaguirre MM, Scopel AL, Baldwin IT, Ballaré CL (2003) Convergent responses to stress. Solar ultraviolet-B radiation and *Manduca sexta* herbivory elicit overlapping transcriptional responses in field-grown plants of *Nicotiana longiflora*. Plant Physiol 132:1755–1767

Jansen MAK (2002) Ultraviolet-B radiation effects on plants: induction of morphogenic responses. Physiol Plant 116:423–429

Jansen MAK, Gaba V, Greenberg BM (1998) Higher plants and UV-B radiation: balancing damage, repair and acclimation. Trends Plant Sci 3:131–135

Jansen MAK, van den Noort RE, Tan MYA, Prinsen E, Lagrimini LM, Thorneley RNF (2001) Phenol-oxidizing peroxidases contribute to the protection of plants from ultraviolet radiation stress. Plant Physiol 126:1012–1023

Jenkins GI, Fuglevand G, Christie JM (1997) UV-B perception and signal transduction. In: Lumsden PJ (ed) Plants and UV-B: responses to environmental change. Cambridge University Press, Cambridge, pp 135–156

Jordan BR (2002) Molecular response of plant cells to UV-B stress. Funct Plant Biol 29:909–916

Jordan BR, James PE, A-H-Mackerness S (1998) Factors affecting UV-B-induced changes in *Arabidopsis thaliana* L. gene expression: the role of development, protective pigments and the chloroplast signal. Plant Cell Physiol 39:769–778

Kakani VG, Reddy KR, Zhao D, Sailaja K (2003a) Field crop responses to ultraviolet-B radiation: a review. Agric Forest Meteorol 120:191–218

Kakani VG, Reddy KR, Zhao D, Mohammed AR (2003b) Effects of ultraviolet-B radiation on cotton (*Gossypium hirsutum* L.) morphology and anatomy. Ann Bot 91:817–826

Kim BC, Tennessen DJ, Last RL (1998) UV-B-induced photomorphogenesis in *Arabidopsis thaliana*. Plant J 15:667–674

Kobzar EF, Kreslavski VD, Muzafarov EN (1998) Photomorphogenetic responses to UV radiation and short-term red light in lettuce seedlings. Plant Growth Reg 26:73–76

Krizek DT, Mirecki RM, Britz SJ (1997a) Inhibitory effects of ambient levels of solar UV-A and UV-B radiation on growth of cucumber. Physiol Plant 100:886–893

Krizek DT, Kramer GF, Mirecki RM (1997b) Influence of UV-B radiation and putrescine on shoot and root growth of cucumber seedlings grown in nutrient solution. J Plant Nutr 20:613–623

Krizek DT, Britz SJ, Mirecki RM (1998) Inhibitory effects of ambient levels of solar UV-A and UV-B radiation on growth of cv. New Red Fire lettuce. Physiol Plant 103:1–7

Landry LG, Chapple CCS, Last RL (1995) *Arabidopsis* mutants lacking phenolic sunscreens exhibit enhanced ultraviolet-B injury and oxidative damage. Plant Physiol 109:1159–1166

Lee DW (1987) The spectral distribution of radiation in two neotropical rainforests. Biotropica 19:161–166

Lercari B, Sodi F, di Paola ML (1990) Photomorphogenic responses to UV radiation: involvement of phytochrome and UV photoreceptors in the control of hypocotyl elongation in *Lycopersicon esculentum*. Physiol Plant 79:668–672

Li Y, Yue M, Wang X-L, Hu Z-D (1999) Competition and sensitivity of wheat and wild oat exposed to enhanced UV-B radiation at different densities under field conditions. Environ Exp Bot 41:47–55

Li Y, Zu Y, Chen H, Chen J, Yang J, Hu Z (2000) Intraspecific responses in crop growth and yield of 20 wheat cultivars to enhanced ultraviolet-B radiation under field conditions. Field Crop Res 67:25–33

Lin C (2000) Plant blue-light receptors. Trends Plant Sci 5:337–342

Liu L, Gitz DC, McClure JW (1995) Effects of UV-B on flavonoids, ferulic acid, growth and photosynthesis in barley primary leaves. Physiol Plant 93:725–733

Mark U, Tevini M (1997) Effects of solar ultraviolet-B radiation, temperature and CO_2 on growth and physiology of sunflower and maize seedlings. Plant Ecol 128:224–234

Matthew C, Hofmann A, Rapson GL, McKenzie RL, Kemp PD, Osborne MA (1996) Growth of ryegrass and white clover under canopies with contrasting transmission of ultraviolet-B radiation. Proc Agron Soc N Z 26:23–30

Mazza CA, Battista D, Zima AM, Szwarcberg-Bracchitta M, Giordano CV, Acevedo A, Scopel AL, Ballaré CL (1999) The effects of solar ultraviolet-B radiation on the growth and yield of barley are accompanied by increased DNA damage and antioxidant responses. Plant Cell Environ 22:61–70

Mazza CA, Boccalandro HE, Giordano CV, Battista D, Scopel AL, Ballaré CL (2000) Functional significance and induction by solar radiation of ultraviolet-absorbing sunscreens in field-grown soybean crops. Plant Physiol 122:117–125

McKenzie RL, Seckmeyer G, Bais AF, Kerr JB, Madronich S (2001) Satellite retrievals of erythemal UV dose compared with ground-based measurements at northern and southern midlatitudes. J Geophys Res 106:24,051–24,062

Mohr H (1994) Coaction between pigment systems. In: Kendrick RE, Kronenberg GHM (eds) Photomorphogenesis in plants, 2nd edn. Kluwer, Dordrecht, pp 353–373

Monsi M, Uchijima Z, Oikawa T (1973) Structure of foliage canopies and photosynthesis. Annu Rev Ecol Syst 4:301–327

Musil CF, Chimphango SBM, Dakora FD (2002) Effects of elevated ultraviolet-B radiation on native and cultivated plants of southern Africa. Ann Bot 90:127–137

Newsham KK, Greenslade PD, McLeod AR (1999) Effects of elevated ultraviolet radiation on *Quercus robur* and its insect and ectomycorrhizal associates. Global Change Biol 5:881–890

Norton LR, McLeod AR, Greenslade PD, Firbank LG, Watkinson AR (1999) Elevated UV-B radiation effects on experimental grassland communities. Global Change Biol 5:601–608

Pal M, Sharma A, Abrol YP, Sengupta UK (1997) Exclusion of UV-B radiation from normal solar spectrum on the growth of mung bean and maize. Agric Ecosyst Environ 61:29–34

Parisi AV, Kimlin MG (1999) Comparison of spectral biologically effective solar ultraviolet in adjacent tree and sun. Phys Med Biol 44:2071–2080

Paul ND, Gwynn-Jones D (2003) Ecological roles of solar UV radiation: towards an integrated approach. Trend Ecol Evol 18:48–55

Paul ND, Callaghan T, Moody S, Gwynn-Jones D, Johanson U, Gehrke C (1999) UV-B impacts on decomposition and biogeochemical cycling. In: Rozema J (ed) Stratospheric ozone depletion: the effects of enhanced UV-B radiation. Backhuys, Leiden, pp 117–133

Phoenix GK, Gwynn-Jones D, Callaghan TV, Sleep D, Lee JA (2001) Effects of global change on a sub-arctic heath: effects of enhanced UV-B radiation and increased summer precipitation. J Ecol 89:256–267

Pinto ME, Edwards GE, Riquelme AA, Ku MSB (2002) Enhancement of nodulation in bean (*Phaseolus vulgaris*) by UV-B irradiation. Funct Plant Biol 29:1189–1196

Pratt LH, Butler WL (1970) Phytochrome conversion by ultraviolet light. Photochem Photobiol 11:503–509

Preuss SB, Britt AB (2003) A DNA-damage-induced cell cycle checkpoint in *Arabidopsis*. Genetics 164:323–334

Quaite FE, Sutherland BM, Sutherland JC (1992) Action spectrum for DNA damage in alfalfa lowers predicted impact of ozone depletion. Nature 358:576–578

Rich TCG, Whitelam GC, Smith H (1987) Analysis of growth rates during phototropism: modifications by separate light-growth responses. Plant Cell Environ 10:303–311

Robson TM, Pancotto VA, Flint SD, Ballaré CL, Sala OE, Scopel AL, Caldwell MM (2003) Six years of solar UV-B manipulations affect growth of *Sphagnum* and vascular plants in a Tierra del Fuego peatland. New Phytol 160:379–389

Ros J, Tevini M (1995) Interaction of UV-radiation and IAA during growth of seedlings and hypocotyl segments of sunflower. J Plant Physiol 146:295–302

Rozema J, van de Staaij J, Björn LO, Caldwell M (1997) UV-B as an environmental factor in plant life: stress and regulation. Trend Ecol Evol 12:22–28

Ruhland CT, Day TA (2000) Effects of ultraviolet-B radiation on leaf elongation, production and phenylpropanoid concentrations of *Deschampsia antarctica* and *Colobanthus quitensis* in Antarctica. Physiol Plant 109:244–251

Ruhland CT, Day TA (2001) Size and longevity of seed banks in Antarctica and the influence of ultraviolet-B radiation on survivorship, growth and pigment concentrations of *Colobanthus quitensis* seedlings. Environ Exp Bot 45:143–154

Ryel RJ, Beyschlag W (1999) Canopy modeling. In: Pugnaire FI, Valladares F (eds) Handbook of functional plant ecology. Dekker, New York, pp 771–804

Ryel RJ, Barnes PW, Beyschlag W, Caldwell MM, Flint SD (1990) Plant competition for light analyzed with a multispecies canopy model. I. Model development and influence of enhanced UV-B conditions on photosynthesis in mixed wheat and wild oat canopies. Oecologia 82:304–310

Schmitt J (1997) Is photomorphogenic shade avoidance adaptive? Perspectives from population biology. Plant Cell Environ 20:826–830

Schmitt J, Wulff RD (1993) Light spectral quality, phytochrome and plant competition. Trend Ecol Evol 8:47–51

Schumaker MA, Bassman JH, Robberecht R, Radamaker GK (1997) Growth, leaf anatomy, and physiology of *Populus* clones in response to solar ultraviolet-B radiation. Tree Physiol 17:617–626

Schwinning S, Weiner J (1998) Mechanisms determining the degree of size asymmetry in competition among plants. Oecologia 113:447–455

Searles PS, Caldwell MM, Winter K (1995) The response of five tropical dicotyledon species to solar ultraviolet-B radiation. Am J Bot 82:445–453

Searles PS, Flint SD, Díaz SB, Rousseaux MC, Ballaré CL, Caldwell MM (1999) Solar ultraviolet-B radiation influence on *Sphagnum* bog and *Carex* fen ecosystems: first field season findings in Tierra del Fuego, Argentina. Global Change Biol 5:225–234

Searles PS, Flint SD, Caldwell MM (2001) A meta-analysis of plant field studies simulating stratospheric ozone depletion. Oecologia 127:1–10

Shinkle JR, Atkins AK, Humphrey EE, Rodgers CW, Wheeler SL, Barnes PW (2004) Growth and morphological responses to different UV wavebands in cucumber (*Cucumis sativum*) and other dicotyledonous seedlings. Physiol Plant 120:240–248

Singh A (1996) Growth, physiological, and biochemical responses of three tropical legumes to enhanced UV-B radiation. Can J Bot 74:135–139

Smith H (1995) Physiological and ecological function within the phytochrome family. Annu Rev Plant Physiol Plant Mol Biol 46:289–315

Smith H, Whitelam GC (1997) The shade avoidance syndrome: multiple responses mediated by multiple phytochromes. Plant Cell Environ 20:840–844

Smith JL, Burritt DJ, Bannister P (2000) Shoot dry weight, chlorophyll and UV-B-absorbing compounds as indicators of a plant's sensitivity to UV-B radiation. Ann Bot 86:1057–1063

Steinmetz V, Wellmann E (1986) The role of solar UV-B in growth regulation of cress (*Lepidium sativum* L.) seedlings. Photochem Photobiol 43:189–193

Stephanou M, Manetas Y (1997) Ultraviolet-B radiation effects on the Mediterranean ruderal *Dittrichia viscosa*. Plant Ecol 128:109–112

Stephen J, Woodfin R, Corlett JE, Paul ND, Jones HG, Ayres PG (1999) Response of barley and pea crops to supplementary UV-B radiation. J Agric Sci 132:253–261

Suesslin C, Frohnmeyer H (2003) An *Arabidopsis* mutant defective in UV-B light-mediated responses. Plant J 33:591–601

Sullivan JH, Rozema J (1999) UV-B effects on terrestrial plant growth and photosynthesis. In: Rozema J (ed) Stratospheric ozone depletion: the effects of enhanced UV-B radiation on terrestrial ecosystems. Backhuys, Leiden, pp 39–57

Sullivan JH, Teramura AH, Dillenburg LR (1994) Growth and photosynthetic responses of field-grown sweetgum (*Liquidambar styraciflua*; Hamamelidaceae) seedlings to UV-B radiation. Am J Bot 81:826–832

Sullivan JH, Howells BW, Ruhland CT, Day TA (1996) Changes in leaf expansion and epidermal screening effectiveness in *Liquidambar styraciflua* and *Pinus taeda* in response to UV-B radiation. Physiol Plant 98:349–357

Tanaka A, Sakamoto A, Ishigaki Y, Nikaido O, Sun G, Hase Y, Shikazono N, Tano S, Watanabe H (2002) An ultraviolet-B-resistant mutant with enhanced DNA repair in *Arabidopsis*. Plant Physiol 129:64–71

Tobin AK (2002) The effects of ultraviolet B radiation on crop plants. In: Ambasht RS, Ambasht NK (eds) Modern trends in applied terrestrial ecology. Kluwer, Dordrecht, pp 183–193

Tosserams M, Magendans E, Rozema J (1997) Differential effects of elevated ultraviolet-B radiation on plant species of a dune grassland ecosystem. Plant Ecol 128:266–281

Tromp TK, Shia R-L, Allen M, Eiler JM, Yung YL (2003) Potential environmental impact of a hydrogen economy on the stratosphere. Science 300:1740–1742

Weiner J (1990) Asymmetric competition in plant populations. Trend Ecol Evol 5:360–364

Weiner J, Stoll P, Muller-Landau H, Jasentuliyana A (2001) The effects of density, spatial pattern, and competitive symmetry on size variation in simulated plant populations. Am Nat 158:438–450

Wilson MI, Greenberg BM (1993) Specificity and photomorphogenic nature of ultraviolet-B-induced cotyledon curling in *Brassica napus* L. Plant Physiol 102:671–677

Xiong FS, Day TA (2001) Effect of solar ultraviolet-B radiation during springtime ozone depletion on photosynthesis and biomass production of Antarctic vascular plants. Plant Physiol 125:738–751

Xiong FS, Ruhland CT, Day TA (2002) Effect of springtime solar ultraviolet-B radiation on growth of *Colobanthus quitensis* at Palmer Station, Antarctica. Global Change Biol 8:1146–1155

Zaller JG, Caldwell MM, Flint SD, Scopel AL, Sala OE, Ballaré CL (2002) Solar UV-B radiation affects below-ground parameters in a fen ecosystem in Tierra del Fuego, Argentina: implications of stratospheric ozone depletion. Global Change Biol 8:867–871

Zavala JA, Botto JF (2002) Impact of solar UV-B radiation on seedling emergence, chlorophyll fluorescence, and growth and yield of radish (*Raphanus sativus*). Funct Plant Biol 29:797–804

Ziska LH, Teramura AH, Sullivan JH, McCoy A (1993) Influence of ultraviolet-B (UV-B) radiation on photosynthetic and growth characteristics in field-grown cassava (*Manihot esculentum* Crantz). Plant Cell Environ 16:73–79

Paul W. Barnes
Department of Biological Sciences
Loyola University
New Orleans, Louisiana 70118, USA
Tel.: +1-504-8652008
Fax: +1-504-8652920
e-mail: pwbarnes@loyno.edu

James R. Shinkle
Department of Biology
Trinity University
San Antonio, Texas 78212, USA

Stephan D. Flint
Ronald J. Ryel
Department of Forest, Range and Wildlife Sciences
Utah State University
Logan, Utah 84322, USA

Vertical Vegetation Structure Below Ground: Scaling from Root to Globe

H. Jochen Schenk

So sind wohl manche Sachen, die wir getrost belachen,
weil unsre Augen sie nicht sehn.
('Abendlied', poem by Matthias Claudius)[*]

1 Introduction

The structure of vegetation that is below ground truly is the hidden half (Waisel et al. 2002), hidden from view and too often hidden from biological research. Over the last 30 years, the Biological Abstracts database has recorded almost twice as many publications in plant ecology that address the ecology of leaves than studies of root ecology. Biologists do not even have a term to refer to the subterranean parts of vegetation. In the English language, biologists use the term canopy to refer to the collective arrangement of all leaves in a plant community, but lack such a term for roots. (The old-fashioned word rootage suggests itself.) Despite this unequal attention to the visible and the hidden halves of vegetation, the two parts are equally important for the function of terrestrial plants and ecosystems and have general functions in common: Both halves are structured to enable resource capture. They are made up of organs that are specialized for resource acquisition, leaves above ground and the distal parts of roots below ground, and of the axes that support and connect these sites of resource acquisition within individual plants. Both also contain structures that are devoted to mechanical support, to storage of resources, and to reproduction; these are, however, beyond the scope of this chapter. Both exhibit a vertical structure that responds to resources that have a strong vertical dimension, light above ground, water, nutrients, and physical soil structure below ground.

* Translation: So there are many things, which we scoff at without thought because our eyes do not see them. ('Evening song', poem by Matthias Claudius)

Progress in Botany, Vol. 66

Because vegetation structure is largely determined by the spatial distribution of resource-acquiring organs, there ought to be predictable relationships between vertical vegetation structure and the vertical distribution of plant resources in space and time. Indeed, the structure of plant canopies is strongly related to the temporal and spatial distribution of light intensity, even though physical factors that act as stressors also play a major role in structuring vegetation above ground (Russell et al. 1989; Ehleringer and Field 1993). Understanding such relationships has enabled ecologists to use information about environmental conditions to successfully predict the spatial structure of plant canopies at scales ranging from individual plants to whole communities (Horn 1971; Botkin et al. 1972; Schieving and Poorter 1999; Anten and Hirose 2001) and even to global vegetation (Haxeltine and Prentice 1996; Bonan et al. 2003; Sitch et al. 2003).

Recent innovations in the methodology of root research (Smit et al. 2000) have enabled much progress toward understanding relationships between roots, soil resources, and other physical and biotic soil factors, a fact attested to by two recent edited volumes on root biology and ecology (Waisel et al. 2002; de Kroon and Visser 2003). However, until very recently there have been relatively few attempts to synthesize and integrate such information in order to develop predictive models of spatial root distributions. Such models are now being developed and are discussed at the end of this chapter.

This chapter explores the current state of information about quantitative relationships between root growth and physical factors, such as soil characteristics and water availability, and assesses the usefulness of this information for developing predictive models of vertical vegetation structure below ground at scales ranging from root systems to global vegetation. Such models would greatly help us to quantify interactions between climate, soil, and plants, to better predict patterns of water and nutrient uptake by plants, and thus to better predict changes in the structure and function of vegetation above and below ground in response to global change. In the following, we will discuss the factors that influence the spatial distribution of roots, beginning at the scale of individual roots and parts of root systems, then scale up to root systems of individual plants, to plant populations and communities, and finally to global vegetation. This is followed by a review of theoretical and empirical models that may be used to predict the vertical vegetation structure below ground and scale up from individual roots and root systems to the globe.

2 Spatial Distribution of Individual Roots in Response to Soil Conditions

Individual roots can respond to environmental stimuli by curving, changing elongation rates, growing or abscising branches, or increasing or decreasing root diameter and specific root length. Together these responses make up what is usually referred to as root growth. Other responses to environmental factors include changes in root life span (i.e., root turnover), in root anatomy, or in physiology, and such changes may in turn affect root growth and therefore the architecture of root systems. In the following some examples of responses that directly affect spatial distributions of roots are highlighted.

2.1 Responses of Root Distributions to Soil Structure

The physical structure of the rooting medium may be more important for determining small-scale distributions of roots than any other factor. Rocks in the soil, boundaries between a litter layer and mineral soil, or between mineral soil and bedrock will obviously affect root distributions. As strong effects of tillage on vertical root distributions demonstrate, soil strength and the distribution and quantity of macropores also greatly affect root growth (Unger and Kaspar 1994; Dwyer et al. 1996; Ball-Coelho et al. 1998). Root elongation rates typically decrease exponentially with increasing soil strength (Bengough 2003; Bingham and Bengough 2003). The higher the soil strength of the bulk soil, the more likely are roots to be confined to soil macropores, such as fissures in clay soils (Dardanelli et al. 2003), animal burrows (Pitkanen and Nuutinen 1997; Springett and Gray 1997), ant nests (Carvalheiro and Nepstad 1996; Moutinho et al. 2003), and root channels (Dell et al. 1983; Gish and Jury 1983; Johnston et al. 1983; van Noordwijk et al. 1991; Rasse and Smucker 1998). In bedrock, roots are always confined to fissures, where they may play an important role in enlarging rock fissures and mechanically contributing to weathering (Matthes-Sears and Larson 1995; Zwieniecki and Newton 1995; Sternberg et al. 1996; Jackson et al. 1999; Werner and Luepnitz 1999; Hubbert et al. 2001; Witty et al. 2003). In general, roots strongly respond to the physical structure of the rooting medium, but they also create and modify physical structure, thus enabling future roots to take advantage of such root-mediated soil modifications.

2.2 Root Curvature: Tropic Responses to Environmental Stimuli

Roots respond to a large variety of environmental stimuli by curving; stimuli include gravity (Knight 1811; Chen et al. 1999), water (Hooker 1915; Takahashi 1994), electrical stimuli (Schrank 1959; Wolverton et al. 2000), temperature (Wortmann 1885; Fortin and Poff 1991), various chemicals (Lilienfeld 1905; Konarzewski and Guminski 1975), oxygen (Porterfield and Musgrave 1998), and mechanical stimuli (Darwin 1880; Ishikawa and Evans 1997). Such tropic responses enable roots to grow towards soil regions with favorable conditions for growth or away from regions with unfavorable conditions. However, the available data suggest that such plastic responses differ greatly between species and depend on the specific combination of multiple environmental factors encountered by the roots. Gravitropism is by far the most important of these responses, because it tends to override other tropic responses (Takahashi 1994). Gravitropism enables root systems to direct root growth of main root axes and lateral branches at genetically determined angles from the vertical, thus increasing the efficiency at which the soil volume is explored and reducing the probability of root overlap within a root system.

2.3 Root Proliferation: Local Increases in Root Length in Response to Environmental Stimuli

In response to resource gradients or resource patches, roots often show proliferation, i.e., increases in local length density (root length per volume of soil), either through increased root elongation, increased branching, or both (Hutchings and de Kroon 1994). Ecological experiments that document root proliferation typically do not distinguish between these two forms of root growth (but see Mou et al. 1997) and simply report measurements of root length density.

2.3.1 Responses of Root Distributions to Soil Water Heterogeneity

Root elongation rates typically are highest at soil water potentials close to the field capacity of the soil, and decrease linearly or exponentially with decreasing soil water potential (Teskey and Hinckley 1981; Kuhns et al. 1985; Torreano and Morris 1998; Joslin et al. 2001; Zou et al. 2001; Siegel-Issem 2002). This effect is due in part to the increase in soil strength that is associated with a decrease in water potential (Wraith and Wright 1998). The minimum water potential at which roots elongate varies greatly between species, and is often lower than −1.5 MPa, the limit usually assumed to be the 'permanent wilting point' for plants. Roots of *Arabidopsis thaliana* seedlings elongated at water potentials of −1.6 MPa in agar (van der Weele et al. 2000). Roots of various forest tree seedlings also elongate at water potentials below −1.5 MPa (Siegel-Issem 2002). Not surprisingly, roots of desert plants can elongate at lower soil water potential. Desert succulents produced roots at water potentials as low as −2.6 MPa and showed 50% of maximum elongation rates at −1.4 MPa (Jordan and Nobel 1984). Roots of some desert shrubs elongate at water potentials lower than −6 MPa (Fernandez and Caldwell 1975). Root elongation rates strongly decrease with decreasing water potentials, but, on average, the limiting water potential is likely to be much lower in dry and seasonally dry environments than in mesic ones.

Root proliferation in moist parts of the soil may be caused in part by growth towards gradients of increasing relative humidity (Takahashi 1994; Takano et al. 1995; Tsutsumi et al. 2003a,b). Until recently, hydrotropism was thought to be largely restricted to dry soil where humidity gradients are steeper than in moist soils (Wraith and Wright 1998). However, recent experiments designed to decouple hydrotropism from gravitropism suggest that hydrotropism may be one of the dominant forces shaping root system development (Tsutsumi et al. 2003a).

Due to changes in elongation, branching, and hydrotropic growth, vertical root distributions can change greatly in response to shifts in vertical distributions of soil water (Wraith and Wright 1998). In response to drying of the soil surface, root distributions may even change from having the highest density close to the surface in well-watered soil to having the highest density at depth (Klepper et al. 1973). In general, frequent shallow irrigation causes shallow root distributions and localized drip irrigation causes local root proliferation around the irrigation source (Fabiao et al. 1990, 1995; Meyer and Barrs 1991; Benasher and Silberbush 1992; Carmi et al. 1992, 1993). Root responses to soil water heterogeneity tend to be highly plastic (Klepper 1991; Hook and Lauenroth 1994; Tsutsumi et al. 2003a), but only if roots can reach moist soil layers without being impeded by dry layers with high soil strength (Plaut et al. 1996). Vertical root distributions are most likely to shift downward in response to drying when some deep roots are already present at depth at the onset of the drought (Klepper et al. 1973).

2.3.2 Responses of Root Distributions to Soil Nutrient Heterogeneity

There is great variation between plant species in their response to nutrient patchiness, but on average root length density increases in nutrient-rich patches when compared to a uniformly nutrient-deficient soil. Local proliferation tends to be strongest in patches containing nutrients most limiting to plant growth, such as nitrogen and phosphorus (Robinson 1994; Robinson and van Vuuren 1998; Forde and Lorenzo 2001). Proliferation in one part of the root system is typically compensated for by reduced root growth elsewhere, resulting in asymmetric root distributions (Robinson and van Vuuren 1998). Increased root growth and branching are usually assumed to be the main causes for root proliferation in patches, but roots in such patches may also have longer life spans (Pregitzer et al. 1993), further contributing to the persistence of high root length densities in such patches. Whatever the underlying mechanisms, proliferation in itself does not necessarily lead to increased nutrient uptake; instead it appears to confer advantages to plants competing with neighbors for nutrients (Robinson et al. 1999; Robinson 2001).

Root length densities are almost always highest in the most fertile soil horizons, such as organic and A horizons. Soils with permanent litter layers and other organic horizons tend to have a high percentage of all roots in these horizons (Fig. 1; Schenk and Jackson 2002a). Roots tend to proliferate in patches of organic matter (St. John et al. 1983; Hodge 2003) and in some

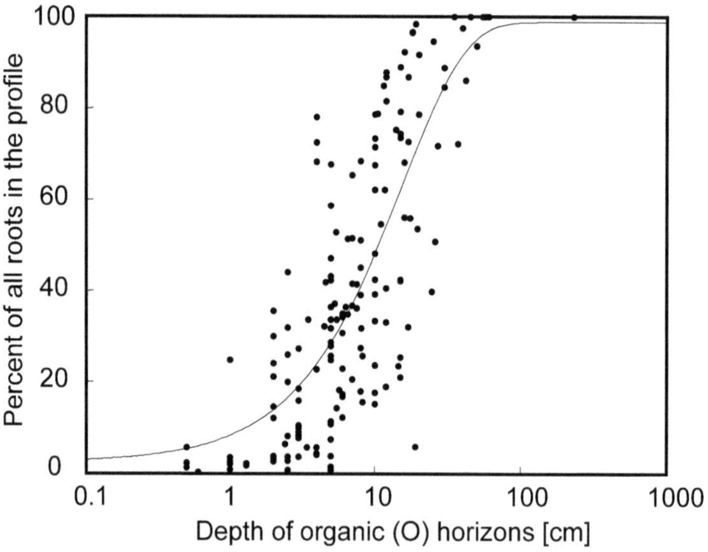

Fig. 1. Percentage of all roots in the soil profile that are located in organic soil horizons as a function of the depth of the organic horizons. Data are from 514 soil profiles from many global vegetation types (Schenk and Jackson 2002a). The fitted *curve* is a sigmoid function with the equation $y=2440.8+2539.5/(1+\exp(-(x-49.4)/15.3))$ ($r^2=0.637$)

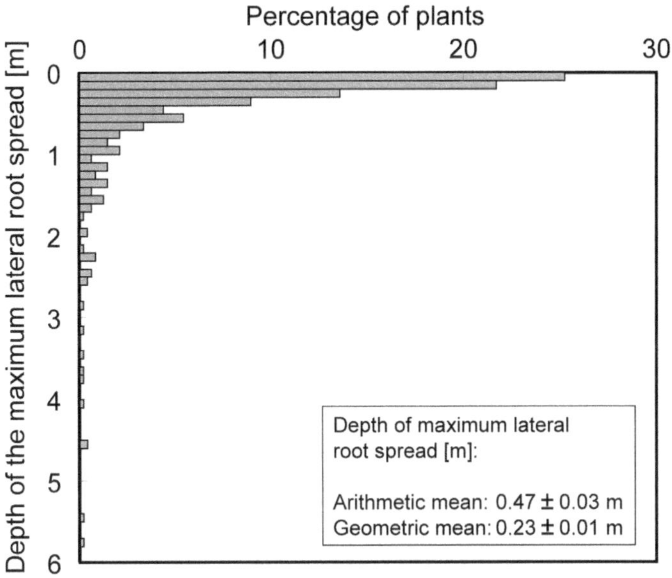

Fig. 2. Soil depth at which maximum lateral root spread was observed for 479 plants of different growth forms in global, water-limited environments (ratio of mean annual precipitation/mean annual potential evapotranspiration <1). Data were determined from detailed drawings of root systems compiled by Schenk and Jackson (2002b)

cases also in coarse organic debris (Vogt et al. 1995), presumably in response to the high fertility resulting from microbial breakdown of organic matter and also in response to its high water holding capacity. Seventy four percent of all plants in water-limited environments have their largest lateral root spread in the nutrient-rich upper 0.5 m of the soil profile (Fig. 2). Globally, the upper 0.2 m of the soil profile contains on average almost 60% of all roots in the profile (Fig. 3), and two thirds of all roots in soils with permanent litter layers (Schenk and Jackson, unpubl. data). Compare this with average nutrient contents of the upper 0.2 m of global soils: >40% of all plant-available phosphorus, >30% of total nitrogen, and about 30% of potassium (Jobbágy and Jackson 2001). Clearly, nutrient availability is strongly related to root densities, but on its own does not fully explain the high concentration of roots in the upper layers of the soil.

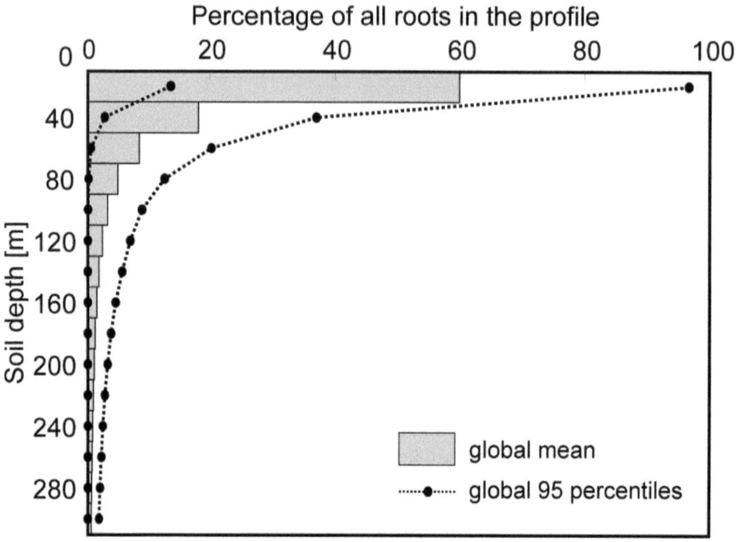

Fig. 3. Global average vertical root distribution and 95 percentiles calculated from 564 soil profiles from many global vegetation types (Schenk and Jackson 2002a). Data are available online at http://www.daac.ornl.gov. (Schenk and Jackson 2003a)

2.3.3 Responses of Root Distributions to Unfavorable Soil Conditions

Shallow rooting depths in boreal, alpine, and tundra vegetation may at least in part be due to low temperatures of the subsoil (Bonan and Shugart 1989; Kutschera and Lichtenegger 1997). Anaerobic soil conditions (Crawford 1992) and soil toxicity, such as high heavy-metal or aluminum concentrations (Baligar et al. 1998) can also restrict root growth, and may be locally important for causing shallow rooting. Vertical root distributions can change seasonally as soil conditions change (e.g., Fernandez and Caldwell 1975). Roots in soil horizons that develop unfavorable conditions may stop growing, may go dormant (Leshem 1970), or die, in some cases due to controlled abscission of branches (Pregitzer 2002).

2.4 Effects of Roots on the Soil Environment: Potential Feedbacks into Spatial Patterns

Soil formation and development is greatly affected by plants, especially by roots and their associated microbial organisms (Jenny 1980). Roots and their rhizoflora modify their soil environment by altering its structure, texture, carbon content, water contents, nutrient contents, pH, and by adding organic root exudates and organic matter, as well as by the effects of all of these factors on microbial communities and the soil fauna. Russian ecologists have called such plant-generated soil alterations phytogenic fields when they are associated with individual plants (Uranov 1965; Kryshen 2000). More generally, one may refer to such phenomena as phytogenic soil modifications. Little research has been done to address the effects of phytogenic modifications on resulting spatial patterns in root distributions. They could potentially result either in positive feedbacks if roots persist or even proliferate in soil space modified by roots, or in negative feedbacks if roots avoid or grow poorly in such modified soil space.

Examples of negative feedback effects in roots are well documented for many annual crops. A lack of crop rotation often leads to poor root growth, usually due to an accumulation of soil pathogens (Olsson and Alstrom 1996). This negative feedback effect tends to be species-specific and may increase species diversity by increasing species turnover during succession (Mills and Bever 1998). Root-mediated allelopathy is another example of such species-specific negative feedback effects (Schenk et al. 1999; Bais et al. 2003). Vertical root distributions will be affected by negative feedback if different species replacing each other differ markedly in their root distributions, but the quantitative significance of such effects for the persistence and dynamics of vertical vegetation structure below ground is unclear.

More important for the development of the vertical vegetation structure below ground are probably positive feedback effects, which promote the proliferation of roots where roots had been previously established. I have already mentioned that roots often grow in macropores created by other roots that may still be alive or may have been decomposed. Furthermore, roots may ameliorate the soil through exudates and components of mucilage which increase the availability of inorganic nutrients (Inderjit 2003; Read et al. 2003), affect mineral weathering (Bormann et al. 1998), and create and stabilize soil aggregates (Graham et al. 1995). Roots can also redistribute water in the soil environment and transport it into dry parts of soils, thereby enabling roots to extend into soil horizons that may not often be reached by infiltration or may rapidly dry due to surface evaporation (Hunter and Kelley 1946; Schulze et al. 1998; Burgess et al. 2001). Water may also reach dry parts of the soil due to gravitational flow along existing roots and root channels (Johnston et al. 1983; Karpachevskiy et al. 1995; Whitford et al. 1995; Martinez-Meza and Whitford 1996; Devitt and Smith 2002). Hydraulic redistribution and preferential water flow along roots can rapidly recharge deep soil layers (Burgess et al. 2001; Ryel et al. 2003) and may account for the occurrence of deep roots in habitats where water infiltration is typically shallow. Redistribution of water in the soil allows roots to maintain a relatively stable distribution in the soil rather than constantly shifting distributions with every change in soil water status.

Effects of roots on soil fertility may involve both positive and negative feedback. Roots and their associated microorganisms and soil fauna are the main source for soil nitrogen, which means that soils previously occupied by roots contain more nitrogen than those that never contained roots. However, roots also locally deplete nutrients, which may temporarily inhibit further root growth in the depletion zones. In the long term, plant activities redistribute nutrients in the soil profile by locally reducing concentrations of those nutrients that are typically most limiting to growth, nitrogen, phosphorus, and potassium, and by adding nutrients to surface layers via litterfall (Jobbágy and Jackson 2001). In some cases, root activities can drastically and rapidly change soil nutrient status and pH, as has been observed for *Eucalyptus* trees planted in the Argentinean Pampas (Jobbágy and Jackson 2003).

In general, negative feedback effects are likely to largely affect individual species and thus species composition, while positive feedback effects will enable roots of many species to grow in soil previously occupied by roots. Such positive feedback effects may even persist after disturbances and affect the rooting patterns of successional vegetation. Where deep roots occurred before a major disturbance removed much of the vegetation, the subsequent successional vegetation may grow roots down existing root channels and thus quickly reestablish deep roots (Sommer et al. 2000, 2003).

3 Spatial Structure of Root Systems of Individual Plants

The architecture of individual root systems can be analyzed quantitatively by topological analysis of branching patterns (Fitter et al. 1991; Smilauerova and Smilauer 2002; Robinson et al. 2003). This approach can be used to characterize the efficiency of root systems in exploring soil space, to quantitatively compare root systems, and to quantify the phenotypic plasticity of root growth (Fitter and Stickland 1991, 1992). Analyses of root system architecture and branching angles have also been used to determine the degree of competition within an individual root system (Ge et al. 2000). Recent findings suggest that individual roots encountering another root may be able to distinguish whether that root belongs to the same plant or a different plant and alter growth patterns accordingly (Mahall and Callaway 1991, 1992; de Kroon et al. 2003; Falik et al. 2003; Holzapfel and Alpert 2003). The ability to discriminate between self and non-self roots could potentially decrease intra-plant competition.

Many other studies that address variables structuring root systems of individual species could be mentioned here, but reviewing such studies would be of limited value in the context of this chapter because the species is not a useful unit for scaling up from individuals to communities and biomes. Theoretically, phylogeny may play a role in structuring communities, but there is currently a lack of evidence to support such a role (Silvertown et al. 2001), and therefore no way to scale up from species-specific traits to properties of communities or ecosystems (Naeem and Wright 2003). Of course, if one had data on typical morphologies and sizes of root systems for all plant species within a given region one could probably use this information to predict vertical root distributions at the plant community level. Such detailed information on multiple species is actually available, but only for very few regions. Over the last century a handful of researchers have excavated hundreds of root systems and documented their morphologies in detailed drawings, for example John Weaver in the North American Midwest (1919, 1920), M.S. Shalyt in the Ukraine and in Kazakhstan (1950, 1952), Isa Baitulin in Kazakhstan and Mongolia (1979, 1993), and Lore Kutschera, Erwin Lichtenegger, and colleagues in central Europe (many volumes, from Kutschera 1960 to Kutschera and Lichtenegger 2002). The importance of this ground- and backbreaking work for our understanding of the variability of root system morphology and species-specific differences can hardly be overstated. However, in order to use this massive amount of information for predictions at the community level it first has to be converted into quantitative data.

Scanned drawings of root systems have been used to quantify root system sizes and shapes (Sun et al. 1997) and to analyze differences between root systems of different plant growth forms and quantify relationships between root system sizes, climate, and soil characteristics (Schenk and Jackson 2002b), as well as for analyses of plant allometry (Casper et al. 2003). Strong allometric relationships exist between the volumes occupied by individual plant canopies and the volumes occupied by their root systems, and these relationships span taxonomically and ecologically diverse species across ten orders of magnitude of size (Karl J. Niklas, H.J. Schenk, and R.B. Jackson, unpubl. data). Variance in aboveground sizes statistically accounts for a large proportion of the variance in root system sizes between plant growth forms (Fig. 4; Schenk and Jackson 2002b). As a case in point, no differences in rooting depths or lateral root spreads were found in an analysis of large numbers of species of perennial grasses and perennial forbs, two growth forms that typically differ substantially in above- and belowground plant architecture, but have similar size ranges (Schenk and Jackson 2002b). These growth forms may differ in the architecture of roots involved in anchorage, but the volume of a root system is defined by the

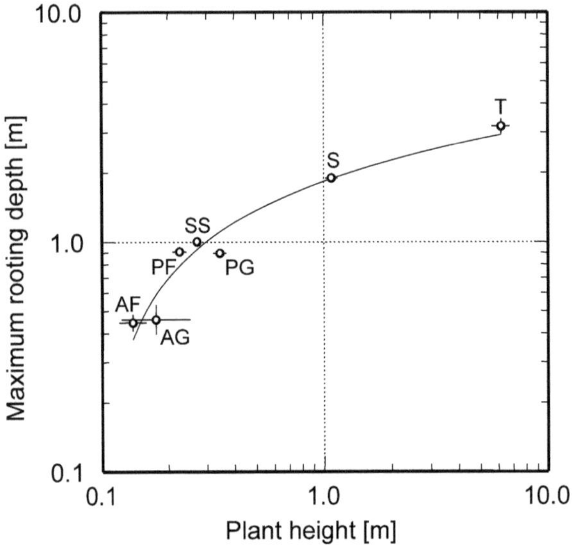

Fig. 4. Relationship between plant height and maximum rooting depth for seven plant growth forms in global, water-limited environments (ratio of mean annual precipitation/mean annual potential evapotranspiration <1). Data shown for each growth form are the geometric means of plant height and maximum rooting depth. Growth forms include annual forbs (*AF*, n=81), annual grasses (*AG*, n=9), perennial forbs (*PF*, n=212), perennial grasses (*PG*, n=176), semi-shrubs (*SS*, n=259), shrubs (*S*, n=154), and trees (*T*, n=88). The equation for the regression line is $y=0.7244 \ln x+1.8419$ ($r^2=0.986$). (Data from Schenk and Jackson 2002b)

most distal roots, and these are typically not involved in anchorage (Ennos and Fitter 1992; Ennos 1993).

Because soil resources are most limiting to plant growth in dry environments, it is hardly surprising that root system sizes and root–shoot allometries change along climatic gradients. In dry or seasonally dry environments with less than 1,000 mm annual precipitation, maximum rooting depths tend to increase with increasing precipitation (Fig. 5), while root/shoot size ratios decrease (Schenk and Jackson 2002b). Rooting depths of herbaceous plants are more strongly correlated with precipitation and potential evapotranspiration than rooting depths of woody plants. Climate also affects the shape of root systems, as plants in arid environments tend to have larger lateral root spreads and larger ratios of lateral root spread to rooting depth than plants of similar aboveground size in more mesic environments (Schenk and Jackson 2002b; Casper et al. 2003). Contrary to a common assumption, plants in arid environments are generally not the most deeply rooted plants. The notable exception to this rule is provided

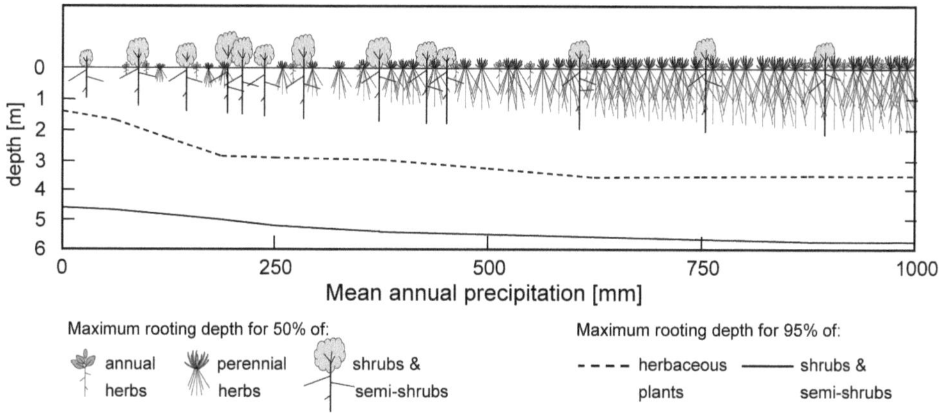

Fig. 5. Schematic diagram of the relationship between mean annual precipitation (MAP) and plant rooting depths. The *drawings of individual plants* approximate the median rooting depths for three broad growth form categories along the MAP gradient. The maximum rooting depths depicted were those for 95% of all plants in the respective precipitation range, i.e., only 5% of plants in that growth form category had deeper roots than indicated by these *lines*. Also shown are the changing plant density and changing proportions of annual herbs, perennial herbs, and semi-shrubs/shrubs along the gradient, from deserts and shrublands dominated by woody plants to grasslands dominated by herbaceous plants. (Schenk and Jackson 2002b, reproduced with permission)

by phreatophytic trees in arid lands, which tend to be the most deeply rooted plants on Earth (Table 1).

4 Spatial Structure of Root Systems in Plant Populations and Communities

Studies of root system architecture in soil typically involve plants growing in monospecific stands or in more or less diverse plant communities, which means that these root systems are grown in competition with roots from other plants. Studies on root systems of single plants in the absence of competition are typically done in pots, and observations from such studies may be difficult to extrapolate to natural settings. Until quite recently, few studies had addressed the effects of root competition on spatial root distributions. In a review of such studies, Fitter (1987) found little evidence for significant effects of root competition on spatial root distributions. Such evidence began to accumulate in the 1990s, beginning with a seminal study by Mahall and Callaway (1991) on roots of two desert shrub species, *Larrea tridentata* and *Ambrosia dumosa*. Root elongation rates of both species were found to significantly decrease when roots encountered *Larrea* roots,

Table 1. Plant rooting depths greater than 30 m that have been reported in the scientific literature. Other accounts of deep roots exist, but most of these are anecdotal. The world record for deepest roots according to the Guinness Book of Records (2003) is said to be about 120 m for a fig tree (*Ficus* sp.) near Ohrigstad, South Africa

Species	Maximum rooting depth (m)	Geographic location	Climate	Source
Acacia erioloba	45	Kalahari, Botswana	Arid	Leistner (1967)
Acacia erioloba	60	Kalahari, Botswana	Arid	Jennings (1974)
Acacia tortilis ssp. *raddiana*	35	Niger	Arid	Fagg (1991)
Boscia albitrunca	68	Kalahari, Botswana	Arid	Jennings (1974)
Eucalyptus calophylla	36.5	SW Australia	Humid	Campion (1926)
Eucalyptus marginata	40	S Western Australia	Humid	Dell et al. (1983)
Faidherbia albida	40	SW Niger	Semi-arid	Leduc et al. (2001)
Juniperus monosperma	61	New Mexico, USA	Arid	Cannon (1960)
Prosopis juliflora	35	Sahel region	Arid	Breman and Kessler (1995)
Prosopis velutina	58	Arizona, USA	Arid	Phillips (1963)
Prosopis cineraria	60	Oman	Arid	Brown (1992)

and, for *Ambrosia*, also when its roots encountered other *Ambrosia* roots. Several studies published since 1991 have addressed the effects of root competition on spatial root distributions, and their results have been reviewed extensively by Schenk et al. (1999), de Kroon et al. (2003), and Hutchings and John (2003). To briefly summarize the findings to date: (1) root growth in many species reacts to the presence of other roots; (2) in at least some cases such reactions are triggered by signals between roots or by allelochemicals; (3) types of reactions differ depending on the species involved; and (4) reactions appear to be different when roots encounter roots from the same plant or from a separate plant. The overall ecological effects of these mechanisms seem to be that they allow plants to shift root growth to soil space unoccupied by roots, or, if such space is unavailable, to soil space occupied by roots of competitors, and to reduce intra-plant root overlap.

The overall effect at the community level of shifts in root distributions in response to competition is that soil space is partitioned more efficiently by roots than it would be without these mechanisms. For example, vertical root distributions of some species may shift in response to interspecific root competition and thus exhibit vertical niche partitioning (D'Antonio and Mahall 1991; Mou et al. 1997; Leuschner et al. 2001; Schmid and Kazda 2001). Niche partitioning by multiple species has been found to increase resource use efficiency at the community level and thereby increase primary productivity (Loreau and Hector 2001). Partitioning of the soil space by roots of multiple species is likely to be complex and the patterns will be species-specific. It should not be assumed that different plant growth forms or functional types tend to partition the soil profile in a similar fashion in different ecosystems. For example, Heinrich Walter's (1939) classic two-layer hypothesis of soil depth partitioning between woody and herbaceous plants, originally proposed for African savannas, does not appear to be universally applicable and may be largely restricted to relatively dry climates (Fig. 5; Schenk and Jackson 2002b).

Regardless of whether root growth reacts to the presence of other roots, all terrestrial plant species have evolved under conditions of root competition and are therefore likely to have evolved a root system architecture and root growth patterns that allow them to grow and survive in the presence of competitors. As mentioned above, roots tend to proliferate in nutrient-rich patches of the soil, even though such proliferation may not in itself increase nutrient uptake rates. Instead it confers advantages by pre-empting or reducing the use of such patches by competitors (Robinson et al. 1999; Robinson 2001). The effect of such root proliferation in fertile patches or soil layers will be that, within a soil profile, root densities will increase more than linearly with nutrient concentrations. This may in part explain the observations discussed above that the upper 0.2 m of soil profiles contains on average > 40% of all plant-available phosphorus (Jobbágy and Jackson 2001) but almost 60% of all roots (Fig. 3).

Because most soil profiles receive most of their water inputs from above, the upper soil layers will be most likely to contain plant-available water (except the surface layer which is directly affected by soil evaporation). Root competition will favor placement of roots into those soil layers that have the most reliable access to water, and shallow roots will generally confer competitive advantages over deeper roots in soils that receive most water by infiltration from above (van Wijk and Bouten 2001).

Because of the diversity of root system architectures in diverse plant communities, niche partitioning, and competition, vertical root distributions in communities are more likely to be closely associated with the distribution of soil resources than the root distributions of individual plants or mono-specific stands. The overall effect of plant diversity on vegetation structure is that it decreases the proportion of variance explained by phylogenetic components and thus increases the proportion explained by environmental variance, including the ecological history of the community (e.g., successional status) and environmental factors. The reduction of the phylogenetic component means that spatial community structure can be predicted, at least to some degree, if its history and environmental variables are known.

5 Belowground Spatial Structure of Global Vegetation

Maximum plant rooting depths and vertical root distributions for global biomes were first compiled and analyzed by Robert B. Jackson and coworkers (Canadell et al. 1996; Jackson et al. 1996). Schenk and Jackson (2002a,b, 2003a) further expanded these data sets, analyzed relationships of rooting depths with climate, soil texture, and vegetation types, and used these analyses to develop models and maps of global plant rooting depths (Schenk and Jackson 2003b; Schenk and Jackson 2004). Three variables were used to characterize vertical root distributions: the soil depth above which 50% of all roots are located (D_{50}), the depth above which 95% of all roots are located (D_{95}), and, for individual plants, the maximum rooting depth (D_{max}). Globally, D_{50} was less or equal to 0.3 m and D_{95} was less or equal to 2 m for 90% of all soil profiles. Deeper rooting depths were mainly found in seasonally water-limited ecosystems and in warm-temperate to tropical environments (Schenk and Jackson 2002a). Climatic variables most strongly and positively correlated with rooting depths were annual potential evapotranspiration (PET) and precipitation. Globally, the dominant plant growth forms and soil texture explained only minor proportions of the total variance in rooting depths. Responding to the need for root parameters for global climate models, Schenk and Jackson (2003b) developed global maps of average rooting depths using nonlinear regression models relating D_{50} and D_{95} to PET, precipitation, and vegetation type.

Predicting average rooting depths is problematic and associated with substantial errors because of the enormous spatial variability of vertical root distributions. In recognition of this problem, Schenk and Jackson (2004) developed global maps of the probability of deep rooting (rather than mean rooting depths) based on two global data sets of rooting depths, long-term means of monthly precipitation, and potential evapotranspiration, and soil texture (Figs. 6 and 7). According to these analyses, deep roots (defined as $D_{95} > 2$ m) are most likely to occur in seasonally dry, semi-arid to humid tropical regions under savanna or thorn-scrub vegetation or under seasonally dry semi-deciduous to evergreen forests. Deep roots are

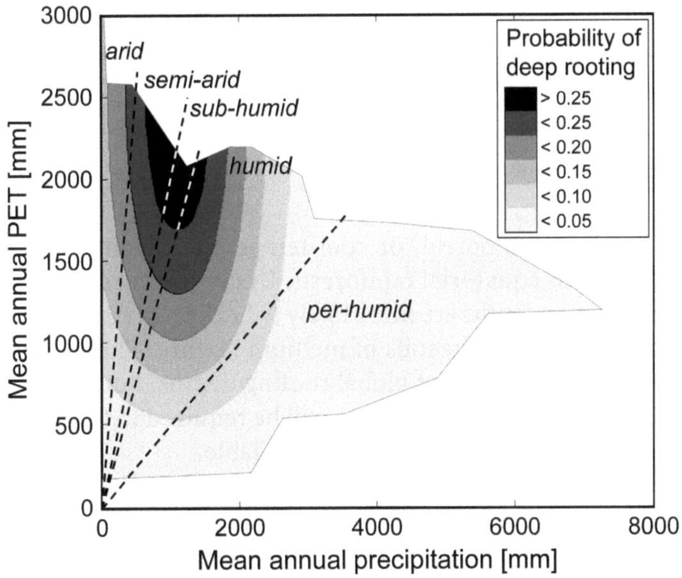

Fig. 6. Global probability of deep rooting (defined as > 5% of all roots below 2 m) as a function of mean annual precipitation and mean annual potential evapotranspiration (*PET*). *Area shaded in different tones of gray* represents the regional climatic averages of terrestrial 0.5° latitudinal-longitudinal grid cells for the late 20th century. Climates in the *white part of the graph* did not globally exist during this period, at least not as grid cell averages. The climatic continuum is divided into humidity zones according to UNEP (1992). Probabilities of deep rooting were calculated from the empirical distribution of records for deep roots within the climatic continuum, using an analysis described in Schenk and Jackson (2004)

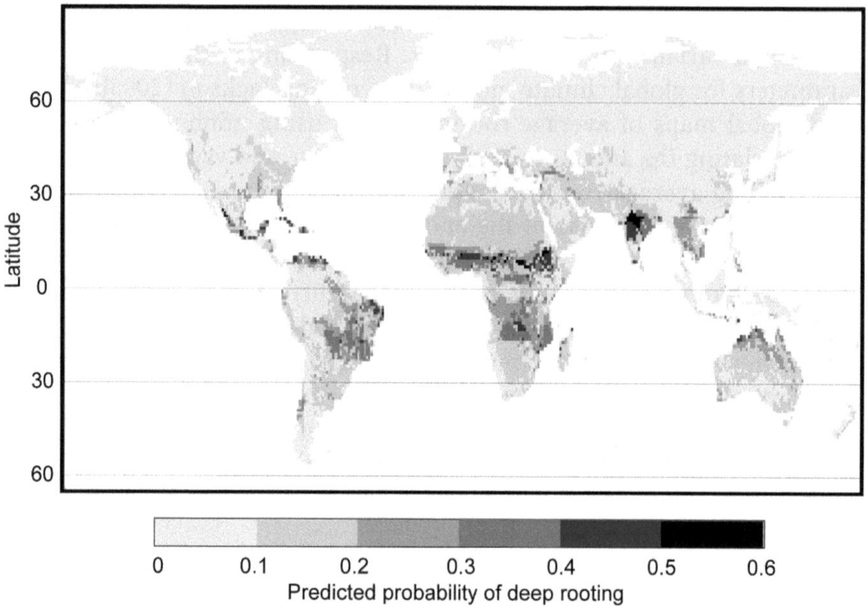

Fig. 7. Global distribution of the predicted probability of deep rooting (defined as > 5% of all roots below 2 m) for 1° latitudinal-longitudinal grid cells, calculated by a non-linear regression model relating the probability of deep rooting to precipitation, potential evapotranspiration, and soil texture. (Modified from a figure in Schenk and Jackson 2004)

least likely to occur in arctic, boreal, or cool-temperate regions and in per-humid climates such as equatorial rainforests. Under warm-temperate to tropical climates, rooting depths are more likely to be deep in coarse-textured and fine-textured soil than in soils of medium texture. These maps are the first empirically based maps of global rooting depths, but they are based on limited data, so systematic sampling will be required to test these predictions for areas for which no root data are available.

6 Scaling from Root to Globe: Predictive Models

A large number of variables affect root growth and vertical root distributions (see summary in Table 2). If the responses of root growth and root demography to all these variables could be incorporated into a mathematical model, then such a model could theoretically be used to predict vertical vegetation structure below ground for individual plants, plant communities, or global biomes. No single mathematical model published to date incorporates all of the variables listed in Table 2 explicitly, but several

Table 2. Important variables to be considered in the development of models used for predicting belowground structure of vegetation at the scale of plant communities to global biomes. Predictions for individual plants or plant populations may require consideration of additional variables, such as species-specific root system architecture. Depending on the spatial and temporal scale of the model, not all of these variables will have to be modeled explicitly, and/or additional variables may have to be included. See text for further discussions of all variables listed

General categories	Specific variables	Potential effects on roots
Mechanical soil properties	1. Vertical distribution of soil strength	1. With increasing soil strength root growth declines, and roots are increasingly likely to preferentially grow in soil macropores
	2. Vertical distribution of soil texture, including rocks	2. Affects the relationship between soil water potential and soil water content and thereby root growth
	3. Vertical distribution of soil structure	3. Affects presence and abundance of macropores which can affect patterns of root growth and infiltration
Chemical soil properties	1. Vertical distribution of organic matter	1. Affects soil water holding capacity and fertility. Roots typically proliferate in soil that is rich in organic matter
	2. Vertical distribution of plant-available nutrients	2. Roots proliferate in soil volumes with high contents of plant-available nitrogen and phosphorus
	3. Vertical distribution of toxins, especially aluminum concentrations	3. Toxins may negatively affect root growth
Soil water balance	1. Inputs: precipitation, run-on, groundwater	1. Affect infiltration depth, vertical distribution of soil water contents, and thereby root growth
	2. Outputs: runoff and evapotranspiration	2. Affect vertical distribution of soil water contents and thereby root growth
	3. Vertical distribution of hydraulic conductivity	3. Affects water flow rates, root water uptake, and potentially hydrotropic root growth
	4. Dynamic vertical distribution of soil water resulting from 1—3	4. Affects vertical distribution of root growth, including due to hydrotropism

Table 2. *Continued*

General categories	Specific variables	Potential effects on roots
Biotic variables	1. Plant sizes above ground	1. Root system sizes are positively correlated with shoot system sizes
	2. Costs (construction, maintenance) and benefits (resource uptake) of root growth	2. Root distributions generally (but not always) will reflect efficient use of energy (carbon) and nutrients by plants
	3. Root competition	3. Shifts root distributions to parts of the profile with higher resource availability (water and nutrients)
	4. 'Permanent wilting points' of root growth	4. The lowest soil water potential for root growth is lower for plants in dry environments than in wet environments
Phytogenic soil modifications	1. Positive feedback effects	1. Root-mediated soil modifications allow roots to grow more easily where previous roots have grown before
	2. Negative feedback effects	2. 'Lack of crop rotation' effects. Roots may 'avoid' strong overlap with roots from the same plant. Some species have root systems that exclude other roots through allelochemicals

process-based models have been developed in recent years that incorporate at least some of the more important variables required to predict vertical root distributions. Following Wraith and Wright (1998), modeling approaches may be divided into models of root system architecture, which simulate root branching patterns, and continuum models that calculate root mass or root length density for given locations in the soil.

6.1 Models of Root System Architecture

Recent models used to calculate effects of nutrient distributions on three-dimensional root distributions include those by Somma et al. (1998) and Dunbabin et al. (2002). The latter also incorporates effects of soil strength and water content on root growth. Ge et al. (2000) modeled three-dimensional root growth in response to heterogeneous distributions of phosphorus in the soil and also analyzed the effects of different branching patterns on competition among roots of the same plant. Ho et al. (2004) added responses to heterogeneous distributions of soil water to this model. Root distributions resulting from hydrotropic growth were simulated successfully in two dimensions by a model developed by Tsutsumi et al. (2003a,b). In two other recent modeling studies (Grant 1998; Biondini 2001), simulations of three-dimensional root growth were incorporated into highly complex plant growth models that simulate numerous aspects of plant growth, development, and resource allocation in response to environmental conditions. Root growth in these models responds to heterogeneous distributions of nutrients and soil water.

6.1.1 Continuum Models of Root Growth

Recent two-dimensional continuum models that incorporate responses of root densities to soil conditions include the model of Chen and Lieth (1993), in which root densities respond to soil water potential and temperature, and the model of Heinen et al. (2003), in which roots respond to nutrient contents. In the one-dimensional models of Adiku et al. (1996) and van Wijk and Bouten (2001), root densities respond to soil water content. Both of these models simulate vertical root distributions using optimization methods. Such methods are discussed below. Continuum models hold promise for application at the community level, where explicit simulations of the root system architectures of multiple plants would be impractical.

6.1.2 Optimization Approaches to Predict Vertical Root Distributions

Optimization schemes that distribute vertical root distributions in ways that minimize plant energy expenditure and/or maximize water and nutrient uptake have been used in several models, at scales ranging from the level of individual root systems (Ge et al. 2000; Ho et al. 2004) to the globe (Kleidon and Heimann 1998). The continuum model by Adiku et al. (1996) used an optimization scheme to partition roots within the soil profile in order to maximize water uptake while minimizing plant energy use. A related approach was used in a model developed by van Wijk and Bouten (2001), in which roots were distributed vertically so as to maximize water uptake in the presence or absence of root competition. In a model by Schwinning and Ehleringer (2001) a carbon-gain optimization scheme was used to partition roots of different functional types of desert plants between two soil layers in response to fluctuating rainfall regimes. Kleidon and Heimann (1998) used a global biosphere model to predict optimized rooting depths that would maximize net primary production (NPP). Rooting depth in that model was represented by the depth of a soil compartment comprised of a single soil layer.

Optimization approaches hold promise for application at all hierarchical levels ranging from the individual to communities and the biosphere, but great care has to be taken when choosing variables for optimization, because these variables will differ substantially depending on the hierarchical scale. For example, high resource use efficiency or high productivity of plant species may be favored by natural selection under some circumstances (e.g., Dudley 1996), but wasteful resource use that pre-empts competitors from accessing resources may actually be favored in competitive environments (Cohen 1970; DeLucia and Schlesinger 1991; Gersani et al. 2001). Highly productive species are not necessarily better competitors than plants of low productivity (Loreau and Hector 2001). Thus species should not be expected to maximize resource use efficiency or productivity. 'Optimal' root growth traits for individual species are those that maximize fitness in a given environment, including its competitive regime, and predicting such traits for individual species in specific places may be of little help in understanding community structure in general.

Plant communities are more likely to optimize resource use efficiency than individual species, because community structure responds to environmental factors by changing species composition in addition to the processes of phenotypic plasticity, gene–environment interactions, and natural selection operating at the species level. Species composition matters greatly for ecosystem function at the local scale (Lavorel and Garnier 2002), but it can change over time and allow additional species to take advantage of resources not used by previously established species. On average, such changes over time would tend to increase resource-use efficiency at the community level. NPP of plant communities tends to increase linearly with the supply of both solar energy and water (Rosenzweig 1968; Chong et al. 1993), and this would not be the case if resource-use efficiencies at the community level fluctuated widely and randomly. Recent studies of the effects of biodiversity on NPP suggest that, on average, plant interactions and resulting niche partitioning in diverse communities tend to increase resource use efficiency at the community level (Loreau and Hector 2001) and thus shift NPP towards the optimum level that can be achieved given the physical constraints of the environment and the phylogenetic constraints of the species in the community, which is not to suggest that the optimum level is ever attained.

Based on this line of reasoning, it may be postulated that vertical root distributions of communities will tend to approach shapes that maximize water and nutrient use efficiency at the community scale. This development will be driven by plastic root growth in response to environmental conditions, positive and negative species interactions, the resulting niche partitioning of the soil profile, successional changes in species composition, and by positive feedback effects. One may further predict that root distributions after major disturbances and in early successional or species-poor communities are less likely to approach optimal resource use efficiency and are more likely to be restricted to shallow soil layers which usually offer the greatest amount of resources at the lowest cost to plants (Gale and Grigal 1987).

These ideas about resource use optimization at the community level were the basis for a conceptual model of global rooting depths (Schenk and Jackson 2002b), which was based on the observation that many of the variables shaping vertical root distributions (see Table 2) tend to favor shallow roots over deep roots: (1) energy costs for construction, maintenance, and resource uptake are lower for shallow roots; (2) water in most soils enters the profile predominantly from above; (3) nutrient concentrations are typically highest in the upper soil layers; (4) shallow soil layers are usually less likely to be oxygen-deficient; and (5), because of reasons (1)–(4), shallow roots generally have competitive advantages over deeper roots. Based on this conceptual model, rooting depths should only increase if (1) there is transpirational demand for water that cannot be fulfilled by taking up water stored closer to the soil surface and (2) water is available at depth. Therefore, average rooting depths are predicted to be shallow in cold environments due to low transpirational demand, shallow in dry environments regardless of the thermal regime because of a general lack of water at depth, shallow in per-humid environments because deep roots are not needed there to fulfill transpirational demands, and deep in warm and seasonally dry environments because of high transpirational demands and the availability of deep soil water. Analyses of large data sets of rooting depths from dry environments (Schenk and Jackson 2002b) and from a broad range of global environments (Schenk and Jackson 2004) strongly support these general trends (Fig. 6), thereby allowing the predictions for the global distribution of deep roots depicted in Fig. 7. Much further research will be required to test the validity of the assumptions underlying the conceptual model of Schenk and Jackson (2002b), to test the accuracy of the global map in Fig. 7, and to develop process-based mathematical models that can increase the accuracy of predictions.

7 Conclusions and Future Challenges

Roots experience a resource environment that has a strong vertical dimension. Just as leaves respond to the vertical distribution of light in a canopy, roots will respond to vertical gradients of nutrients, water, and other soil properties. Just as the top of the canopy is generally the optimal location for leaves, so the upper soil layers are the optimal location for roots. Some plants do not compete for the best spot in the canopy; they are shade-adapted and specialized to live at the lower edge of the canopy. Similarly, some plants will be adapted to be deeply rooted and will not compete with other species for the abundant resources in upper soil layers. However, unlike the top of the canopy which may be out of reach for any but the tallest species, the upper soil layers are reachable by all plants, and therefore vertical stratification of the soil profile by different species is less likely to be as pronounced as vertical stratification of the canopy.

Roots live in a heterogeneous and complex environment, but, as this chapter shows, their average responses to soil conditions are predictable to a large degree. Moreover, roots modify the soil environment, and the positive feedback effects of root-mediated soil modifications make root distributions more stable and predictable than they would be without such effects.

Root research faces many methodological challenges, but it now has access to a large variety of tools and methods (Smit et al. 2000). Over recent decades, it has moved from description to explanation and, by taking advantage of descriptive work accumulated over many decades, is now well on its way towards prediction. Because of the large number of variables affecting root growth, our predictions about root behavior will never be very precise, but the increasing number of ecological models that successfully predict general patterns of root distributions demonstrates that our understanding of root ecology has greatly progressed in recent years.

Obviously, being able to predict root distributions is still a long way from understanding the spatial distribution of root function. We are only beginning to understand the functional diversity of roots within root systems (Pregitzer 2002). There is mounting evidence that water and nutrient uptake from a soil volume are not always closely related to root length density or root surface area within that volume (Adiku et al. 2000; Robinson 2001), proving that roots, just like leaves, are not all alike and are capable of adjusting their physiologies in response to external or internal signals (Forde and Lorenzo 2001; Pregitzer 2002). Root dormancy is another neglected issue that will have large effects on the distribution of functions within root systems (North and Nobel 1998). Studies of root function require a different set of tools from root distribution studies, such as tracers of nutrient and water uptake (Casper et al. 2003) and inverse modeling approaches that calculate root function from changes in the distribution of soil resources (Adiku et al. 2000; Hupet et al. 2003). The greatest challenge in root research remain mycorrhizae and their effects on root distribution and function. The vertical distribution of mycorrhizal hyphae below the upper soil layers is an open research question for most terrestrial ecosystems (Rosling et al. 2003).

Ecologists cannot be accused to ever 'scoff at what their eyes cannot see' – to paraphrase the German poet Matthias Claudius – but too often in the past have they ignored what their eyes did not see. Too often the soil remains the black box of studies in terrestrial ecology (Hammer 1998). The visible and hidden halves of plants, of vegetation, and of ecosystems deserve and need equal attention.

Acknowledgements. The author wishes to thank Susan Schwinning and David Robinson for helpful comments on the manuscript and Laura Middlebrooke for help with data processing. This research was made possible in part by a postdoctoral fellowship from the National Center for Ecological Analysis and Synthesis [a Center funded by NSF (DEB-94-21535), the University of California at Santa Barbara, and the State of California] and by a grant from the Andrew W. Mellon Foundation.

References

Adiku SGK, Braddock RD, Rose CW (1996) Modelling the effect of varying soil water on root growth dynamics of annual crops. Plant Soil 185:125–135

Adiku SGK, Rose CW, Braddock RD, Ozier-Lafontaine H (2000) On the simulation of root water extraction: examination of a minimum energy hypothesis. Soil Sci 165:226–236

Anten NPR, Hirose T (2001) Limitations on photosynthesis of competing individuals in stands and the consequences for canopy structure. Oecologia 129:186–196

Bais HP, Vepachedu R, Gilroy S, Callaway RM, Vivanco JM (2003) Allelopathy and exotic plant invasion: from molecules and genes to species interactions. Science 301:1377–1380

Baitulin IO (1979) Kornevaja sistema rastenij aridnoj zony Kazakhstana (in Russian). (Root systems of plants of the arid zone of Kazakhstan) Nauka, Alma-Ata

Baitulin IO (ed) (1993) Fitoékologicheskie issledovanija v juzhnoj Gobi (in Russian with English summary). (Phytoecological investigations in southern Gobi) Gylim, Alma-Ata

Baligar VC, Fageria NK, Elrashidi MA (1998) Toxicity and nutrient constraints on root growth. HortScience 33:960–965

Ball-Coelho BR, Roy RC, Swanton CJ (1998) Tillage alters corn root distribution in coarse-textured soil. Soil Till Res 45:237–249

Benasher J, Silberbush M (1992) Root distribution under trickle irrigation – factors affecting distribution and comparison among methods of determination. J Plant Nutr 15:783–794

Bengough AG (2003) Root growth and function in relation to soil structure, composition, and strength. In: de Kroon H, Visser EJW (eds) Root ecology, vol 168. Springer, Berlin Heidelberg New York, pp 151–171

Bingham IJ, Bengough AG (2003) Morphological plasticity of wheat and barley roots in response to spatial variation in soil strength. Plant Soil 250:273–282

Biondini M (2001) A three-dimensional spatial model for plant competition in an heterogeneous soil environment. Ecol Model 142:189–225

Bonan GB, Shugart HH (1989) Environmental factors and ecological processes in boreal forests. Annu Rev Ecol Syst 20:1–28
Bonan GB, Levis S, Sitch S, Vertenstein M, Oleson KW (2003) A dynamic global vegetation model for use with climate models: concepts and description of simulated vegetation dynamics. Global Change Biol 9:1543–1566
Bormann BT, Wang D, Bormann FH, Benoit G, April R, Snyder MC (1998) Rapid, plant-induced weathering in an aggrading experimental ecosystem. Biogeochemistry 43:129–155
Botkin DB, Janak JF, Wallis JR (1972) Some ecological consequences of a computer model of forest growth. J Ecol 60:849–872
Breman H, Kessler J-J (1995) Woody plants in agro-ecosystems of semi-arid regions. Springer, Berlin Heidelberg New York
Brown K (1992) *Prosopis cineraria* woodlands in Oman, past, present and future. In: Dutton RW, Powell M, Ridley RJ (eds) *Prosopis* species, aspects of their value, research and development. FAO, Rome, pp 131–144
Burgess SSO, Adams MA, Turner NC, White DA, Ong CK (2001) Tree roots: conduits for deep recharge of soil water. Oecologia 126:158–165
Campion WE (1926) The depth attained by roots. Aust For J 9:128
Canadell J, Jackson RB, Ehleringer JR, Mooney HA, Sala OE, Schulze ED (1996) Maximum rooting depth of vegetation types at the global scale. Oecologia 108:583–595
Cannon HL (1960) The development of botanical methods of prospecting for uranium on the Colorado Plateau. US Geological Survey Bulletin, vol 1085-A. US Government Printing Office, Washington, DC
Carmi A, Plautz Z, Heuer B, Grava A (1992) Establishment of shallow and restricted systems in cotton and its impact on plant response to irrigation. Irrig Sci 13:87–91
Carmi A, Plaut Z, Sinai M (1993) Cotton root growth as affected by changes in soil water distribution and their impact on plant tolerance to drought. Irrig Sci 13:177–182
Carvalheiro KO, Nepstad DC (1996) Deep soil heterogeneity and fine root distribution in forests and pastures of eastern Amazonia. Plant Soil 182:279–285
Casper BB, Schenk HJ, Jackson RB (2003) Defining a plant's belowground zone of influence. Ecology 84:2313–2321
Chen DX, Lieth JH (1993) A 2-dimensional, dynamic-model for root-growth distribution of potted plants. J Am Soc Hortic Sci 118:181–187
Chen RJ, Rosen E, Masson PH (1999) Gravitropism in higher plants. Plant Physiol 120:343–350
Chong DLS, Mougin E, Gastellu-Etchegorry JP (1993) Relating the Global Vegetation Index to net primary productivity and actual evapotranspiration over Africa. Int J Remote Sensing 14:1517–1546
Cohen D (1970) The expected efficiency of water utilization in plants under different competition and selection regimes. Isr J Bot 19:50–54
Crawford RMM (1992) Oxygen availability as an ecological limit to plant distribution. Adv Ecol Res 23:93–185
D'Antonio CM, Mahall BE (1991) Root profiles and competition between the invasive, exotic perennial, *Carpobrotus edulis*, and two native shrub species in California coastal scrub. Am J Bot 78:885–894
Dardanelli JL, Calmon MA, Jones JW, Andriani JM, Diaz MP, Collino DJ (2003) Use of a crop model to evaluate soil impedance and root clumping effects on soil water extraction in three Argentine soils. Trans Am Soc Agric Eng 46:1265–1275
Darwin C (1880) The power of movement in plants. Murray, London
De Kroon H, Visser EJW (eds) (2003) Root ecology. Springer, Berlin Heidelberg New York
De Kroon H, Mommer L, Nishiwaki A (2003) Root competition: towards a mechanistic understanding. In: de Kroon H, Visser EJW (eds) Root ecology, vol 168. Springer, Berlin Heidelberg New York, pp 215–234
Dell B, Bartle JR, Tacey WH (1983) Root occupation and root channels of jarrah forest subsoils. Aust J Bot 31:615–627

DeLucia EH, Schlesinger WH (1991) Resource-use efficiency and drought tolerance in adjacent Great-Basin and Sierran plants. Ecology 72:51–58
Devitt DA, Smith SD (2002) Root channel macropores enhance downward movement of water in a Mojave Desert ecosystem. J Arid Environ 50:99–108
Dudley SA (1996) Differing selection on plant physiological traits in response to environmental water availability: a test of adaptive hypotheses. Evolution 50:92–102
Dunbabin VM, Diggle AJ, Rengel Z, van Hugten R (2002) Modelling the interactions between water and nutrient uptake and root growth. Plant Soil 239:19–38
Dwyer LM, Ma BL, Stewart DW, Hayhoe HN, Balchin D, Culley JLB, McGovern M (1996) Root mass distribution under conventional and conservation tillage. Can J Soil Sci 76:23–28
Ehleringer J, Field CB (eds) (1993) Scaling physiological processes: leaf to globe. Academic Press, San Diego
Ennos AR (1993) The scaling of root anchorage. J Theor Biol 161:61–75
Ennos AR, Fitter AH (1992) Comparative functional morphology of the anchorage systems of annual dicots. Funct Ecol 6:71–78
Fabiao A, Madeira M, Steen E (1990) Effect of water and nutrient supply on root distribution in an *Eucalyptus globulus* plantation. Water Air Soil Pollut 54:635–640
Fabiao A, Madeira M, Steen E, Katterer T, Ribeiro C, Araujo C (1995) Development of root biomass in an *Eucalyptus globulus* plantation under different water and nutrient regimes. Plant Soil 168/169:215–223
Fagg C (1991) *Acacia tortilis*: fodder tree for desert sands. Nitrogen Fixing Tree Association, Hawaii
Falik O, Reides P, Gersani M, Novoplansky A (2003) Self/non-self discrimination in roots. J Ecol 91:525–531
Fernandez OA, Caldwell MM (1975) Phenology and dynamics of root growth of three cool semi-desert shrubs under field conditions. J Ecol 63:703–714
Fitter AH (1987) Spatial and temporal separation of activity in plants communities: prerequisite or consequence of coexistence. In: Gee JHR, Giller PS (eds) Proc 27th Symp of the British Ecological Society, Organization of Communities: Past and Present, Aberystwyth, Wales, UK, 1986. Blackwell, Palo Alto, California, pp 119–139
Fitter AH, Stickland TR (1991) Architectural analysis of plant root systems. 2. Influence of nutrient supply on architecture in contrasting plant species. New Phytol 118:383–389
Fitter AH, Stickland TR (1992) Architectural analysis of plant root systems III. Studies on plants under field conditions. New Phytol 121:243–248
Fitter AH, Stickland TR, Harvey ML, Wilson GW (1991) Architectural analysis of plant root systems 1. Architectural correlates of root exploitation efficiency. New Phytol 118:375–382
Forde B, Lorenzo H (2001) The nutritional control of root development. Plant Soil 232:51–68
Fortin M-C, Poff KL (1991) Characterization of thermotropism in primary roots of maize: dependence on temperature and temperature gradient, and interaction with gravitropism. Planta 184:410–414
Gale MR, Grigal DF (1987) Vertical root distribution of northern tree species in relation to successional status. Can J For Res 17:829–834
Ge ZY, Rubio G, Lynch JP (2000) The importance of root gravitropism for inter-root competition and phosphorus acquisition efficiency: results from a geometric simulation model. Plant Soil 218:159–171
Gersani M, Brown JS, O'Brien EE, Maina GM, Abramsky Z (2001) Tragedy of the commons as a result of root competition. J Ecol 89:660–669
Gish TJ, Jury WA (1983) Effect of plant roots and root channels on solute transport. Trans Am Soc Agric Eng 26:440–441, 451
Graham RC, Ervin JO, Wood HB (1995) Aggregate stability under oak and pine after four decades of soil development. Soil Sci Soc Am J 59:1740–1744

Grant RF (1998) Simulation in *ecosys* of root growth response to contrasting soil water and nitrogen. Ecol Model 107:237–264
Guinness (2003) Guinness world records 2004. Guinness World Records, New York
Hammer RD (1998) Space and time in the soil landscape: the ill-defined ecological universe. In: Peterson DL, Parker VT (eds) Ecological scale: theory and applications. Columbia University Press, New York, pp 105–140
Haxeltine A, Prentice IC (1996) BIOME3: an equilibrium terrestrial biosphere model based on ecophysiological constraints, resource availability, and competition among plant functional types. Global Biogeochem Cycl 10:693–709
Heinen M, Mollier A, de Willigen P (2003) Growth of a root system described as diffusion. II. Numerical model and application. Plant Soil 252:251–265
Ho MD, McCannon BC, Lynch JP (2004) Optimization modeling of plant root architecture for water and phosphorus acquisition. J Theor Biol 226:331–340
Hodge A (2003) Plant nitrogen capture from organic matter as affected by spatial dispersion, interspecific competition and mycorrhizal colonization. New Phytol 157:303–314
Holzapfel C, Alpert P (2003) Root cooperation in a clonal plant: connected strawberries segregate roots. Oecologia 134:72–77
Hook PB, Lauenroth WK (1994) Root system response of a perennial bunchgrass to neighbourhood-scale soil water heterogeneity. Funct Ecol 8:738–745
Hooker HD Jr (1915) Hydrotropism in roots of *Lupinus albus*. Ann Bot 29:265–283
Horn HS (1971) The adaptive geometry of trees. Princeton University Press, Princeton, New Jersey
Hubbert KR, Graham RC, Anderson MA (2001) Soil and weathered bedrock: components of a Jeffrey pine plantation substrate. Soil Sci Soc Am J 65:1255–1262
Hunter AS, Kelley OJ (1946) The extension of plant roots into dry soil. Plant Physiol 21:445–451
Hupet F, Lambot S, Feddes RA, van Dam JC, Vanclooster M (2003) Estimation of root water uptake parameters by inverse modeling with soil water content data. Water Resour Res 39(11), 1312, doi: 10.1029/2003WR002046
Hutchings MJ, de Kroon H (1994) Foraging in plants: the role of morphological plasticity in resource acquisition. Adv Ecol Res 25:160–238
Hutchings MJ, John EA (2003) Distribution of roots in soil and root foraging activity. In: de Kroon H, Visser EJW (eds) Root ecology, vol 168. Springer, Berlin Heidelberg New York, pp 33–60
Inderjit (2003) Root exudates: an overview. In: de Kroon H, Visser EJW (eds) Root ecology, vol 168. Springer, Berlin Heidelberg New York, pp 235–255
Ishikawa H, Evans ML (1997) Induction of curvature in maize roots by calcium or thigmostimulation: role of the postmitotic isodiametric growth zone. Plant Physiol 100:762–768
Jackson RB, Canadell J, Ehleringer JR, Mooney HA, Sala OE, Schulze ED (1996) A global analysis of root distributions for terrestrial biomes. Oecologia 108:389–411
Jackson RB, Moore LA, Hoffmann WA, Pockman WT, Linder CR (1999) Ecosystem rooting depth determined with caves and DNA. Proc Natl Acad Sci USA 96:11387–11392
Jennings CMH (1974) The hydrology of Botswana. PhD Thesis, University of Natal, Pietermaritzburg, South Africa
Jenny H (1980) The soil resource. Origin and behavior. Springer, Berlin Heidelberg New York
Jobbágy EG, Jackson RB (2001) The distribution of soil nutrients with depth: global patterns and the imprint of plants. Biogeochemistry 53:51–77
Jobbágy EG, Jackson RB (2003) Patterns and mechanisms of soil acidification in the conversion of grasslands to forests. Biogeochemistry 64:205–229
Johnston CD, Hurle DH, Hudson DR, Height MI (1983) Water movement through preferred paths in lateritic profiles of the Darling Plateau, Western Australia. CSIRO Groundwater Res Tech Pap 1:1–34

Jordan PW, Nobel PS (1984) Thermal and water relations of roots of desert succulents. Ann Bot 54:705–718

Joslin JD, Wolfe MH, Hanson PJ (2001) Factors controlling the timing of root elongation intensity in a mature upland oak stand. Plant Soil 228:201–212

Karpachevskiy LO, Borovinskaya LB, Khaidapova DD (1995) Importance of root systems for soil formation in dry steppes. Euras Soil Sci 27:68–78

Kleidon A, Heimann M (1998) A method of determining rooting depth from a terrestrial biosphere model and its impacts on the global water and carbon cycle. Global Change Biol 4:275–286

Klepper B (1991) Crop root-system response to irrigation. Irrig Sci 12:105–108

Klepper B, Taylor HM, Fiscus EL (1973) Water relations and growth of cotton in drying soil. Agron J 65:307–310

Knight TA (1811) On the causes which influence the direction of the growth of roots. Philos Trans R Soc Lond 1811:209–219

Konarzewski Z, Guminski S (1975) Chemotropic effect of metal cations on *Sinapis alba* roots as affected by some factors. Biol Plant 17:458–467

Kryshen AM (2000) Fitogennoe pole: teoriia i projavleniia v prirode (in Russian). (The phytogenic field: theory and manifestations in nature) Izvest Akad Nauk Ser Biol 4:437–443

Kuhns MR, Garrett HE, Teskey RO, Hinckley TM (1985) Root-growth of black walnut trees related to soil-temperature, soil-water potential, and leaf water potential. For Sci 31:617–629

Kutschera L (1960) Wurzelatlas mitteleuropäischer Ackerunkräuter und Kulturpflanzen. DLG, Frankfurt am Main

Kutschera L, Lichtenegger E (1997) Bewurzelung von Pflanzen in verschiedenen Lebensräumen. Land Oberösterreich, OÖ. Landesmuseum, Linz

Kutschera L, Lichtenegger E (2002) Wurzelatlas mitteleuropäischer Waldbäume and Sträucher. Stocker, Graz

Lavorel S, Garnier E (2002) Predicting changes in community composition and ecosystem functioning from plant traits: revisiting the Holy Grail. Funct Ecol 16:545–556

Leduc G, Favreau G, Schroeter P (2001) Long-term rise in a Sahelian water-table: the continental Terminal in South-West Niger. J Hydrol 24:43–54

Leistner OA (1967) The plant ecology of the southern Kalahari. Bot Surv S Afr Mem 38:172 pp

Leshem B (1970) Resting roots of *Pinus halepensis*: structure, function, and reaction to water stress. Bot Gaz 131:99–104

Leuschner C, Hertel D, Coners H, Büttner V (2001) Root competition between beech and oak: a hypothesis. Oecologia 126:276–284

Lilienfeld M (1905) Über den Chemotropismus der Wurzel. Thieme, Leipzig

Loreau M, Hector A (2001) Partitioning selection and complementarity in biodiversity experiments. Nature 412:72–76

Mahall BE, Callaway RM (1991) Root communication among desert shrubs. Proc Natl Acad Sci USA 88:874–876

Mahall BE, Callaway RM (1992) Root communication mechanisms and intracommunity distributions of two Mojave Desert shrubs. Ecology 73:2145–2151

Martinez-Meza E, Whitford WG (1996) Stemflow, throughfall and channelization of stemflow by roots in three Chihuahuan desert shrubs. J Arid Environ 32:271–287

Matthes-Sears U, Larson DW (1995) Rooting characteristics of trees in rock: a study of *Thuja occidentalis* on cliff faces. Int J Plant Sci 156:679–686

Meyer WS, Barrs HD (1991) Roots in irrigated clay soils: measurement techniques and responses to root zone conditions. Irrig Sci 12:125–134

Mills KE, Bever JD (1998) Maintenance of diversity within plant communities: soil pathogens as agents of negative feedback. Ecology 79:1595–1601

Mou P, Mitchell RJ, Jones RH (1997) Root distribution of two tree species under a heterogeneous nutrient environment. J Appl Ecol 34:645–656

Moutinho P, Nepstad DC, Davidson EA (2003) Influence of leaf-cutting ant nests on secondary forest growth and soil properties in Amazonia. Ecology 84:1265–1276
Naeem S, Wright JP (2003) Disentangling biodiversity effects on ecosystem functioning: deriving solutions to a seemingly insurmountable problem. Ecol Lett 6:567–579
North GB, Nobel PS (1998) Water uptake and structural plasticity along roots of a desert succulent during prolonged drought. Plant Cell Environ 21:705–713
Olsson S, Alstrom S (1996) Plant-affecting streptomycin-sensitive microorganisms in barley monoculture soils. New Phytol 133:245–252
Phillips WS (1963) Depth of roots in soil. Ecology 44:424
Pitkanen J, Nuutinen V (1997) Distribution and abundance of burrows formed by *Lumbricus terrestris* L and *Aporrectodea caliginosa* Sav in the soil profile. Soil Biol Biochem 29:463–467
Plaut Z, Carmi A, Grava A (1996) Cotton root and shoot responses to subsurface drip irrigation and partial wetting of the upper soil profile. Irrig Sci 16:107–113
Porterfield DM, Musgrave ME (1998) The tropic response of plant roots to oxygen: oxytropism in *Pisum sativum* L. Planta 206:1–6
Pregitzer KS (2002) Fine roots of trees – a new perspective. New Phytol 154:267–270
Pregitzer KS, Hendrick RL, Fogel R (1993) The demography of fine roots in response to patches of water and nitrogen. New Phytol 125:575–580
Rasse DP, Smucker AJM (1998) Root recolonization of previous root channels in corn and alfalfa rotations. Plant Soil 204:203–212
Read DB, Bengough AG, Gregory PJ, Crawford JW, Robinson D, Scrimgeour CM, Young IM, Zhang K, Zhang X (2003) Plant roots release phospholipid surfactants that modify the physical and chemical properties of the soil. New Phytol 157:315–326
Robinson D (1994) The responses of plants to non-uniform supplies of nutrients. New Phytol 127:635–674
Robinson D (2001) Root proliferation, nitrate inflow and their carbon costs during nitrogen capture by competing plants in patchy soil. Plant Soil 232:41–50
Robinson D, van Vuuren MMI (1998) Responses of wild plants to nutrient patches in relation to growth rate and life form. In: Lambers H, Poorter H, Van Vuuren MMI (eds) Inherent variation in plant growth. Physiological mechanisms and ecological consequences. Backhuys, Leiden, pp 237–257
Robinson D, Hodge A, Griffiths BS, Fitter AH (1999) Plant root proliferation in nitrogen-rich patches confers competitive advantage. Proc R Soc Lond B 266:431–435
Robinson D, Hodge A, Fitter A (2003) Constraints on the form and function of root systems. In: de Kroon H, Visser EJW (eds) Root ecology, vol 168. Springer, Berlin Heidelberg New York, pp 1–31
Rosenzweig ML (1968) Net primary productivity of terrestrial communities: prediction from climatological data. Am Nat 102:67–74
Rosling A, Landeweert R, Lindahl BD, Larsson KH, Kuyper TW. Taylor AFS, Finlay RD (2003) Vertical distribution of ectomycorrhizal fungal taxa in a podzol soil profile. New Phytol 159:775–783
Russell G, Marshall B, Jarvis PG (eds) (1989) Plant canopies: their growth, form and function. Cambridge University Press, Cambridge
Ryel RJ, Caldwell MM, Leffler AJ, Yoder CK (2003) Rapid soil moisture recharge to depth by roots in a stand of *Artemisia tridentata*. Ecology 84:757–764
Schenk HJ, Jackson RB (2002a) The global biogeography of roots. Ecol Monogr 72:311–328
Schenk HJ, Jackson RB (2002b) Rooting depths, lateral spreads, and below-ground/above-ground allometries of plants in water-limited ecosystems. J Ecol 90:480–494
Schenk HJ, Jackson RB (2003a) Global distribution of root profiles in terrestrial ecosystems. Online data set with documentation, available at http://www.daac.ornl.gov. Oak Ridge National Laboratory Distributed Active Archive Center, Oak Ridge, Tennessee
Schenk HJ, Jackson RB (2003b) Global maps of ecosystem rooting depths. NASA ISLSCP Initiative 2. Online data set with documentation, available at http://islscp2.sesda.com/

Schenk HJ, Jackson RB (2004) Mapping the global distribution of deep roots in relation to climate and soil characteristics. Geoderma (in press)
Schenk HJ, Callaway RM, Mahall BE (1999) Spatial root segregation: are plants territorial? Adv Ecol Res 28:145–180
Schieving F, Poorter H (1999) Carbon gain in a multispecies canopy: the role of specific leaf area and photosynthetic nitrogen-use efficiency in the tragedy of the commons. New Phytol 143:201–211
Schmid I, Kazda M (2001) Vertical distribution and radial growth of coarse roots in pure and mixed stands of *Fagus sylvatica* and *Picea abies*. Can J For Res 31:539–548
Schrank AR (1959) Electronasty and electrotropism. In: Ruhland W (ed) Movements due to mechanical and electrical stimuli and to radiation, vol 17(1). Springer, Berlin Heidelberg New York, pp 148–163
Schulze ED, Caldwell MM, Canadell J, Mooney HA, Jackson RB, Parson D, Scholes R, Sala OE, Trimborn P (1998) Downward flux of water through roots (i.e. inverse hydraulic lift) in dry Kalahari sands. Oecologia 115:460–462
Schwinning S, Ehleringer JR (2001) Water use trade-offs and optimal adaptations to pulse-driven arid ecosystems. J Ecol 89:464–480
Shalyt MS (1950) Podzemnaja cast' nekotorykh lugovykh, stepnykh i pustynnykh rastenyi i fitocenozov. C. I. Travjanistye i polukustarnigkovye rastenija i fitocenozy lesnoj (luga) i stepnoj zon (in Russian). (Belowground parts of some meadow, steppe, and desert plants and plant communities, part 1. Herbaceous plants and subshrubs and plant communities of forest and steppe zones) Trudy Botanicheskogo Instituta im VL Komarova Akademii nauk SSSR Seriia III. Geobotanika 6:205–442
Shalyt MS (1952) Podzemnaja cast' nekotorykh lugovykh, stepnykh i pustynnykh rastenyi i fitocenozov. C. 2. Travjanistye, polukustarnigkovye i kustarnigkovye rastenija i fitocenozy pustynnoj zony (in Russian). (Belowground parts of some meadow, steppe, and desert plants and plant communities, part 2. Herbaceous plants, subshrubs, and shrubs, and plant communities of the desert zone) Trudy Botanicheskogo Instituta im VL Komarova Akademii nauk SSSR Seriia III. Geobotanika 8:71–139
Siegel-Issem CM (2002) Forest productivity as a function of root growth opportunity. MSc Thesis, Virginia Polytechnic Institute and State University, Blacksburg, Virginia
Silvertown J, Dodd M, Gowing D (2001) Phylogeny and the niche structure of meadow plant communities. J Ecol 89:428–435
Sitch S, Smith B, Prentice IC, Arneth A, Bondeau A, Cramer W, Kaplan JO, Levis S, Lucht W, Sykes MT,Thonicke K, Venevsky S (2003) Evaluation of ecosystem dynamics, plant geography and terrestrial carbon cycling in the LPJ dynamic global vegetation model. Global Change Biol 9:161–185
Smilauerova M, Smilauer P (2002) Morphological responses of plant roots to heterogeneity of soil resources. New Phytol 154:703–715
Smit AL, Bengough AG, Engels C, van Noordwijk M, Pellerin S, van de Geijn SC (eds) (2000) Root methods: a handbook. Springer, Berlin Heidelberg New York
Somma F, Hopmans JW, Clausnitzer V (1998) Transient three-dimensional modeling of soil water and solute transport with simultaneous root growth, root water and nutrient uptake. Plant Soil 202:281–293
Sommer R, Denich M, Vlek PLG (2000) Carbon storage and root penetration in deep soils under small-farmer land-use systems in the eastern Amazon region, Brazil. Plant Soil 210:231–241
Sommer R, Fölster H, Vielhauer K, Maklouf Carvalho EJ, Vlek PLG (2003) Deep soil water dynamics and depletion by secondary vegetation in the eastern Amazon. Soil Sci Soc Am J 67:1672–1686
Springett J, Gray R (1997) The interaction between plant roots and earthworm burrows in pasture. Soil Biol Biochem 29:621–625
St John TV, Coleman DC, Reid CPP (1983) Growth and spatial distribution of nutrient-absorbing organs: selective exploitation of soil heterogeneity. Plant Soil 71:487–493

Sternberg PD, Anderson MA, Graham RC, Beyers JL, Tice KR (1996) Root distribution and seasonal water status in weathered granitic bedrock under chaparral. Geoderma 72:89–98

Sun G, Coffin DP, Lauenroth WK (1997) Comparison of root distributions of species in North American grasslands using GIS. J Veg Sci 8:587–596

Takahashi H (1994) Hydrotropism and its interaction with gravitropism in roots. Plant Soil 165:301–308

Takano M, Takahashi H, Hirasawa T, Suge H (1995) Hydrotropism in roots - sensing of a gradient in water potential by the root cap. Planta 197:410–413

Teskey RO, Hinckley TM (1981) Influence of temperature and water potential on root-growth of white oak. Physiol Plant 52:363–369

Torreano SJ, Morris LA (1998) Loblolly pine root growth and distribution under water stress. Soil Sci Soc Am J 62:818–827

Tsutsumi D, Kosugi K, Mizuyama T (2003a) Effect of hydrotropism on root system development in soybean (*Glycine max*): growth experiments and a model simulation. J Plant Growth Regul 21:441–458

Tsutsumi D, Kosugi K, Mizuyama T (2003b) Root-system development and water-extraction model considering hydrotropism. Soil Sci Soc Am J 67:387–401

UNEP (1992) World atlas of desertification. United Nations Environment Programme. Edward Arnold, London

Unger PW, Kaspar TC (1994) Soil compaction and root growth - a review. Agron J 86:759–766

Uranov AA (1965) Fitogennoe pole (in Russian) (The phytogenic field). In: Lavrenko EM (ed) Problemy sovremennoj botaniki, vol 1. Nauka, Moscow, pp 251–254

Van der Weele CM, Spollen WG, Sharp RE, Baskin TI (2000) Growth of *Arabidopsis thaliana* seedlings under water deficit studied by control of water potential in nutrient-agar media. J Exp Bot 51:1555–1562

Van Noordwijk M, Widianto MH, Hairiah K (1991) Old tree root channels in acid soils in the humid tropics: important for crop root penetration, water infiltration and nitrogen management. Plant Soil 134:37–44

Van Wijk MT, Bouten W (2001) Towards understanding tree root profiles: simulating hydrologically optimal strategies for root distribution. Hydrol Earth Syst Sci 5:629–644

Vogt KA, Vogt DJ, Asbjorsen H, Dahlgren RA (1995) Roots, nutrients and their relationship to spatial patterns. Plant Soil 168–169:113–123

Waisel Y, Eshel A, Kafkafi U (eds) (2002) Plant roots: the hidden half. Dekker, New York

Walter H (1939) Grasland, Savanne und Busch der arideren Teile Afrikas in ihrer ökologischen Bedingtheit. Jahrb Wissensch Bot 87:750–860

Weaver JE (1919) The ecological relations of roots. Carnegie Institution of Washington, Washington, DC

Weaver JE (1920) Root development in the grassland formation. Carnegie Institution of Washington, Washington, DC

Werner M, Luepnitz D (1999) The root-systems of endemic *Echium* species on Tenerife (Canary Islands) in relation to habitat conditions. Feddes Repert 110:31–36

Whitford WG, Martinez-Turanzas G, Martinez-Meza E (1995) Persistence of desertified ecosystems: explanations and implications. Environ Monit Assess 37:1–3

Witty JH, Graham RC, Hubbert KR, Doolittle JA, Wald JA (2003) Contributions of water supply from the weathered bedrock zone to forest soil quality. Geoderma 114:389–400

Wolverton C, Mullen JL, Ishikawa H, Evans ML (2000) Two distinct regions of response drive differential growth in *Vigna* root electrotropism. Plant Cell Environ 23:1275–1289

Wortmann J (1885) Über den Thermotropismus der Wurzel. Bot Zh 43:225–235

Wraith JM, Wright CK (1998) Soil water and root growth. HortScience 33:951–959

Zou C, Penfold C, Sands R, Misra RK, Hudson I (2001) Effects of soil air-filled porosity, soil matric potential and soil strength on primary root growth of radiata pine seedlings. Plant Soil 236:105–115

Zwieniecki MA, Newton M (1995) Roots growing in rock fissures: their morphological adaptations. Plant Soil 172:181–187

H. Jochen Schenk
Department of Biological Science
California State University Fullerton
P.O. Box 6850
Fullerton, California 92834-6850, USA
Tel.: +1-714-2783678
Fax: +1-714-2783426
e-mail: jschenk@fullerton.edu

Ectomycorrhizal Community Structure: Linking Biodiversity to Function

Douglas L. Godbold

1 Introduction

Mycorrhizal fungal diversity has been shown to influence both plant community structure and production (van der Heijden et al. 1998; Horton et al. 1999). Similarly, plant communities have strong effects on soil microbial diversity (Yin et al. 2000), with subsequent possible feedbacks on ecosystem functioning. There have been considerable developments in methods to identify both ectomycorrhizas and vesicular arbuscular mycorrhizas, and hence gain estimates of the community structure. Advances in techniques to investigate the physiological function of mycorrhizas have shown considerable differences between species in both ectomycorrhizas and vesicular arbuscular mycorrhizas. These differences are particularly obvious for nutrient capture (see Chalot et al. 2002), but also include factors such as water uptake (see Auge 2001) and protection of the host plant against pathogens (Duchesne et al. 1988). As our understanding of the composition of mycorrhizal communities advances, the possibilities of applying commonly used ideas from plant community ecology and our understanding of the function of mycorrhizas in natural and managed ecosystems increase.

In plant ecology it has been assumed that biodiversity influences ecosystem stability and productivity. The concept that increases in biodiversity increase ecosystem stability has been supported by studies on a number of plant ecosystems, particularly grasslands (Naeem et al. 1994; Tilman and Downing 1994). However, recently, this has been questioned. Pfisterer and Schmid (2002) showed an inverse relationship between diversity and ecosystem functioning. Stability may be more related to the number of guilds with a functional significance than species number per se; thus when assessing biodiversity not only the number and distribution of species must be considered, but also functional diversity and redundancy. Functional diversity is defined as the number of distinct processes or functions that are carried out by a community. Functional redundancy is an estimate of the number of different species within the various functional groups or guilds

Progress in Botany, Vol. 66

(Gaston 1996). These concepts have recently been extended to soil bacterial communities (Yin et al. 2000). In the work of Yin et al. (2000) along a vegetation gradient, bacterial function redundancy increased in relation to plant community regrowth. These authors suggest that bacterial functional redundancy may be a useful indication of soil quality and ecosystem functioning.

Most forest trees are colonised by a number of ectomycorrhizal species (Dahlberg 2001) In an investigation using vesicular-arbuscular mycorrhizas, van der Heijden et al. (1998) showed that aboveground plant production increased with addition of up to eight mycorrhizal species, above which no further increase was found. These results indicate that for the beneficial effects of mycorrhizas, functional redundancy also occurs. As both ectomycorrhizas and vesicular-arbuscular mycorrhizas have similar physiological functions (Smith and Read 1997), it may be assumed that more diverse ectomycorrhizal communities are of greater benefit to trees than pauperate ectomycorrhizal communities.

2 Ectomycorrhizal Community Structure

Ectomycorrhizal community structure has been investigated using inventories of aboveground sporocarps, and more recently using belowground ectomycorrhizas on fine roots. The last 10 years has seen an explosion in methods to identify ectomycorrhizas, and thus estimate ectomycorrhizal community structure. Much of the pioneering work to allow identification of species of fungi-forming ectomycorrhizal root tips was carried out by Agerer and his group (Agerer 1987–1996). This group developed methods for identification of ectomycorrhizal root tips based on the hyphal mantel colour and structure, and ramification. These investigations are extremely time consuming and require a high level of expertise and experience, but still form the basis for almost all investigations of community structure. The easily seen difference in ectomycorrhizas such as colour and ramification allow ectomycorrhizas to be sorted into groups of morphotypes. Indeed several large-scale investigations have been carried out using only morphotype techniques (Sittig 1999). The pre-sorting of ectomycorrhizal root tips decreases the number of roots that are then subsequently identified using molecular methods. The last 5 years has seen great advances in non-morphological methods to identify ectomycorrhizal species (see Horton and Bruns 2001). These methods are based on different applications of molecular tools, and have mainly been applied to determine which species of ectomycorrhizal fungi are colonising root tips (Pritsch et al. 1997; Horton and Bruns 1998; Jonsson et al. 1999a; Peter et al. 2001; Nara et al. 2003;

Rosling et al. 2003). However, as a large part of the biomass of ectomycorrhizas is formed by the extramatrical mycelium, methods to determine the extent and species of the extramatrical mycelium in soils have become critically important, not only to link biodiversity to functionality, but also in defining the extent of individuals, and thus true estimates of biodiversity.

Molecular methods to identify ectomycorrhizas have been used in a large number of investigations of population structure (Pritsch et al. 1997; Horton and Bruns 1998; Jonsson et al. 1999a; Peter et al. 2001; Nara et al. 2003; Rosling et al. 2003). The most important advance in these methods has been the application of the polymerase chain reaction (PCR) in the identification methods. The PCR reaction forms the basis for all subsequent variations in the methods used for final identification. The primary targets to be amplified have been ribosomal genes and spacers. For amplification a number of primers have been developed. The advantages of using ribosomal genes and spacers is that these regions have a high copy number, they have highly conserved sequence tracks which can serve as sites for primer design, and they also have variable regions between the priming sites (Horton and Bruns 2001). In most cases, molecular identification of ectomycorrhizal fungi has used restriction analysis of the internal transcribed spacer region (Gardes and Bruns 1996a; Kårén and Nylund 1997; Jonsson et al. 1999a). The internal transcribed spacer (ITS) region is a nuclear region lying between the small subunit and large subunit ribosomal genes and contains two noncoding spacing regions separated by the 5.8s rRNA gene. The size of the region is in the range of 650–900 base pairs, and is usually amplified by universal, fungal-specific or basidomycete-specific primers. However, although most of the more specific primers have been designed to show specificity for a particular group, these must be treated with caution (Horton and Bruns 2001). Not only does co-amplification of non-target sequences occur, but also primers may not amplify DNA sequences of every species within the intended group (Horton and Bruns 2001). This failure of these primers to amplify some DNA sequences has considerable implications for investigations of species diversity.

Characterisation of the PCR products is normally carried out by restriction fragment length polymorphism (RFLP) analysis or by sequencing. RFLP analysis of the ITS region can be carried out typically using two or three restriction enzyme digests. Relatively few enzyme digests are needed as sequence differences between taxa are usually the result of indels, insertions or deletions of nucleotides that cause length variations, decreasing the need to successively cleave the fragment to produce different size fractions (Horton and Bruns 2001). The RFLP approach allows a high degree of success in identification of species if RFLP patterns from sporocarps and ectomycorrhizal fungi colonising root tips are compared. The technique also allows easy estimation of the number of ectomycorrhizal species colonising a root system even if complete identification is not possible. The inability to match sequences to the ITS-RFLP database of known sporocarp species may be for a number of reasons. The most obvious is that RFLP databases tend to contain patterns of species-forming conspicuous sporocarps. In an analysis of sporocarps from a restricted region of 7 km, Horton (2002) could show that from 44 species, 38 produced single species-specific RFLP types. Polymorphic ITS-RFLP types were observed in only six species. This suggests that within a small range, polymorphism should not greatly limit species identification. However, intraspecific variation may occur across wide geographic ranges (Kårén and Nylund 1997; Selosse et al. 1998; Methvyn et al. 2000), making exact matches difficult. Many of the problems of matching ITS-RFLP types can be overcome by determining and comparing DNA sequences. This has been used in a number of investigations of ectomycorrhizal fungal communities (Peter et al. 2001; Rosling et al. 2003).

Using DNA sequences of the ITS rDNA region, Rosling et al. (2003) were able to identify over 95% of root tips that could be successfully amplified and sequenced. Of the original 247 root tips, 75% were successfully sequenced. The root tips were identified using published sequences in the GenBank database using the BLAST program (Altschul et al. 1997); a number of *Piloderma* morphotypes were identified using unpublished sequences. Similarly, in the work of Nara et al. (2003) looking at belowground primary succession, these authors identified the majority of the ectomycorrhizas using RFLP patterns, and sequenced PCR products of unidentified types. The sequences were then compared with sequences of known types in the DDBJ/EMBL GenBank database. In both of these examples (Nara et al. 2003; Rosling et al. 2003) and in many recent studies (Erland et al. 1999; Peter et al. 2001), pre-sorting of the ectomycorrhizal root tips into morphotypes was carried out before molecular identification. In the study of Rosling et al. (2003), 8,275 root tips were sorted into 39 morphological groups. Each time a morphological group was distinguished five representative root tips were selected. When the root tips from the same group were subsequently distinguished, one or two representative samples were taken. Of the 39 morphological groups, in 33 the ITS rDNA region could be amplified and sequenced. Using the sequences, the 33 morphological groups could be rearranged into 22 taxa. This emphasises the importance of final molecular identification.

3 Estimation of the Extramatrical Mycelium

The extramatrical mycelium constitutes the largest part of the biomass of most species of ectomycorrhizas. As the extramatrical mycelium is instrumental in the functioning of ectomycorrhizas in terms of nutrient acquisition, estimations of the extent of the extramatrical mycelium and also the species composition are fundamental to understanding the functionality of ectomycorrhizal communities. Above and beyond this, the extramatrical mycelium may be important in a number of other ecosystem processes, including transfer of C to the soil C pool. The total amount of mycelium (hyphal mantle mycelium and extramatrical mycelium) was estimated to have a biomass of between 500 and 700 kg ha^{-1} in the humus layer of a forest soil (Wallander et al. 2001). Based on only few empirical data, the turnover of ectomycorrhizal and arbuscular mycorrhizal hyphae is assumed to be rapid, with a life span of only 5–7 days (Friese and Allen 1991). Recently, using ^{14}C labelling techniques with ^{14}C-depleted CO_2 and determination using accelerator mass spectrometry, the average life span of the extraradical hyphal biomass of arbuscular mycorrhizas has been shown to be 6 days (Staddon et al. 2003). Assuming that in forests the hyphae of the extramatrical mycelium of ectomycorrhizas have a similar turnover rate, the turnover of the extramatrical mycelium must be important for C input into forest soils and nutrient dynamics. To estimate the biomass of the extramatrical mycelium, a number of biochemical markers have been used. Chitin is a fungal cell wall compound, which has been used to measure biomass of the fungal components of ectomycorrhizas and the biomass of the extramatrical mycelium (Ekblad et al. 1998). Chitin is a relatively recalcitrant compound and persists after the death of the fungi, and thus gives a measure of both living and dead hyphae. However, chitin is not specific to fungi; in addition chitin is a component found in both insects and arthropods, and thus background levels of chitin in soils can be high (Olsson 1999). Ergosterol is a constituent of fungal membranes but not plant membranes, and has been used to estimate the biomass of the living extramatrical mycelium (Wallander and Nylund 1992). In addition to chitin and ergosterol, Olsson (1999) has shown that in most fungi phospholipid fatty acid 18:2ω6,9 makes up a large proportion of the phospholipid fatty acids in fungal membranes, and can be used for detecting the growth of the extramatrical mycelium of ectomycorrhizas (Olsson and Wallander 1998). The amount of phospholipid fatty acid 18:2ω6,9 correlates with the concentration of ergosterol in soil samples (Frostegård and Bååth 1996). Wallander et al. (2001) used both chitin and the phospholipid fatty acid 18:2o6,9 to estimate the mycelia from ectomycorrhizas in a forest soil humus layer and determined 900 kg ha^{-1} using ergosterol and 700 kg ha^{-1} using PLFA

18:2ω6,9, showing that the methods are comparable. However, again both ergosterol and PLFA 18:2ω6,9 are not specific to ectomycorrhizal fungi, and the hyphal biomass from ectomycorrhizal fungi can only be estimated if the hyphal biomass from saprophytic fungi can be subtracted. The background biomass of hyphae from saprophytic fungi is normally measured by root trenching, which inhibits growth of hyphae from mycorrhizal roots. Methods using phospholipid fatty acids and ergosterol give an estimate of the total biomass of the extramatrical mycelium, but give no indication of which species of ectomycorrhiza contributes to the biomass. In a study in which the ectomycorrhizal root tips were sorted by morphological characteristics and then samples of root tips identified by PCR and denaturing gradient gel electrophoresis, the extramatrical mycelium in soil was also identified using the same method (Landeweert et al. 2003a). Using this technique, a total of 14 fungal taxa could be detected in soils, and seven of the taxa could be identified as ectomycorrhizal fungi. All of the seven ectomycorrhizal taxa identified from soils could be found on the root tips. In soils this technique does not distinguish between DNA from hyphae, spores or sclerotia. However, Landeweert et al. (2003a) in their study suggested that co-extraction of DNA from spores and sclerotia was not important. Although these investigations show that the extramatrical mycelium can be identified, the quantification of the extramatrical mycelium by species can only be advanced by development of quantitative PCR methods. Competitive PCR is a popular method in bacterial community profiling (Brüggemann et al. 2000). In competitive PCR, a competitive sequence is added to the PCR reaction mixture. As both the target sequence and the competitor sequence are amplified simultaneously with the same primer pair, the ratio of these sequences in the final PCR reaction mixture is proportional to the initial known amount of the competitor sequence and the unknown amount of the target sequence. This method also has the advantage that false negatives due to failed amplification can be immediately detected. Using competitive PCR Guidot et al. (2002) detected hyphae from *Hebeloma cylindrosporum* in sandy soils low in organic matter, and *Hebeloma cylindrosporum* DNA in soil decreased rapidly with increasing distance from sporocarps. No *Hebeloma cylindrosporum* DNA was detected more than 50 cm from the sporocarps, and even within the vicinity of the sporocarps the concentration of DNA varied greatly. Guidot et al. (2002) used primers (IGS2a and IGS2b) that were specific for *Hebeloma cylindrosporum*, and amplified a 533-base pair fragment within the nuclear rDNA untranscribed intergenic spacer 2, which spans between the 3-end of the 5S and 5-end of the 18S rDNA genes. In bacterial studies using competitive PCR, the amount of DNA determined is usually converted into cell numbers, to allow quantification of numbers of cells per gram of soil (Lleó et al. 1999; Phillips et al. 2000). However, filamentous fungi may have variable numbers of nuclei per cell, making conversion of the determined amount of DNA to hyphal mass difficult (Guidot et al. 2002). Similarly, it has been shown that when investigating bacterial populations with multiple-competitive and hence multi-template PCR, some primer template combinations do not accurately reflect the initial ratio of template molecules in the PCR mixture (Hansen et al. 1998; Polz and Cavanaugh et al. 1998). Clearly, before competitive PCR can be used to estimate the relative contribution of ectomycorrhizal species to the total extramatrical mycelium, considerable development is still needed.

4 Structure of Communities and Sampling

Much of the knowledge of community structure and the size of genets of ectomycorrhizal fungi has been based on aboveground inventories of sporocarps (Dahlberg 2001). It has, however, become increasingly apparent that rarely if ever do estimates of sporocarp numbers match estimates of belowground abundance (Dahlberg 2001; Erland and Taylor 2002; Taylor 2002). In a study in central Sweden (Taylor 2002), *Cortinarius* was the most abundant genus, forming 42.3% of sporocarps and representing 25 species. However, only 1.6% of the mycorrhizal root tips were found to be colonised by *Cortinarius* species. There are numerous examples of this (Dahlberg et al. 1997; Kårén and Nylund 1997; Gehring et al. 1998; Jonsson et al. 1999a), and there are numerous explanations (Erland and Taylor 2002). One of the simplest is that some species of ectomycorrhizas fruit frequently and abundantly above ground, although they appear infrequently below ground. Alternatively some species may appear frequently below ground, investing strongly in vegetative growth, but invest little in reproductive growth and thus fruit rarely. However, central to this discussion is the role of sampling and the size of genets below ground, and thus definition of individuals. Both these aspects have been discussed in some detail by Horton and Bruns (2001) and Taylor (2002). The common pattern that emerges in studies of ectomycorrhizal communities is that they are very diverse, with a large number of rare types (Horton and Bruns 2001). This results in an inverse relationship between abundance and rarity (Figs. 1 and 2). This relationship can be seen in Fig. 1 for belowground populations (Horton and Bruns 2001), but also for estimates of sporocarps and belowground populations (Fig. 2) at the same site (Taylor 2002). Thus in all cases the community comprises a few common species and a large number of rare species.

In an analysis of a number of studies, Horton and Bruns (2001) reported that most studies are based on 30 or fewer soil samples and cover less that

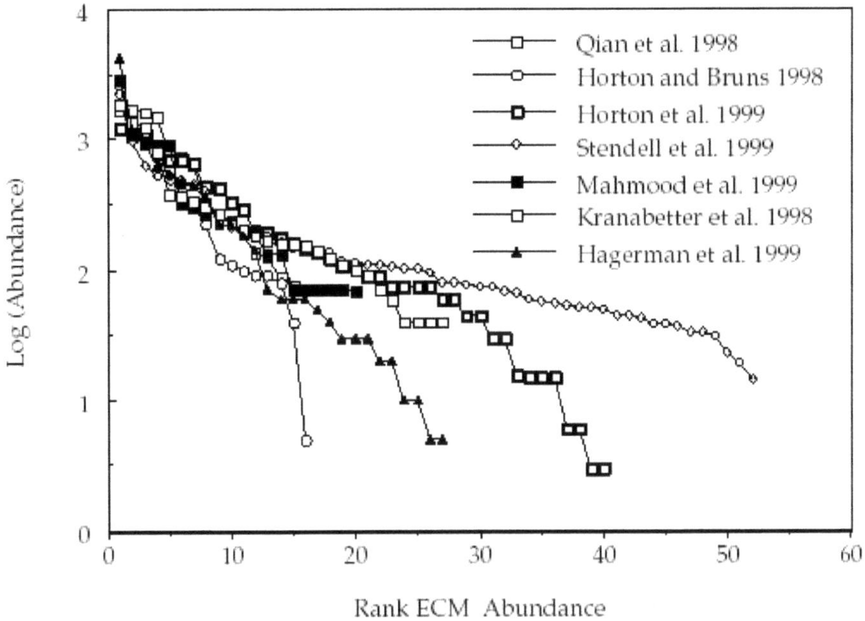

Fig. 1. Log normal distribution of species abundance from a number of studies. *ECM* Ectomycorrhizal. (Horton and Bruns 2001)

1 ha. In these studies (Horton and Bruns 2001, their Table 2), 50 or fewer species of ectomycorrhizal fungi were detected. In a study with a higher sampling effort (198 soil samples) and a larger area covered (2.1 ha) (Luoma et al. 1997), over 200 species were determined. This study, however, was carried out in a mature stand dominated by Douglas fir which is known to have a high number of associated ectomycorrhizal species. In studies reported by both Horton and Bruns (2001; Fig. 3) and Taylor (2002; Fig. 4), the number of species determined increases with the number of soil samples taken. Only in one study (Horton and Bruns 1998) was an asymptote reached, suggesting that in many cases an insufficient number of soil samples were taken to fully assess the number of species. In addition, the distribution of species may be patchy. Horton and Bruns (2001) suggest that most species frequently occur in less than 10% of soil samples taken, and that individual soil cores are usually dominated by one or two species even though several species may be found. However, this idea has been refuted by Taylor (2002), who could show that although in his study 84% of the morphotypes occurred in three or fewer soil samples out of a total of 30, due to the small area of the site sampled, this occurrence does not support a non-random distribution of mycelia of a species across the site investi-

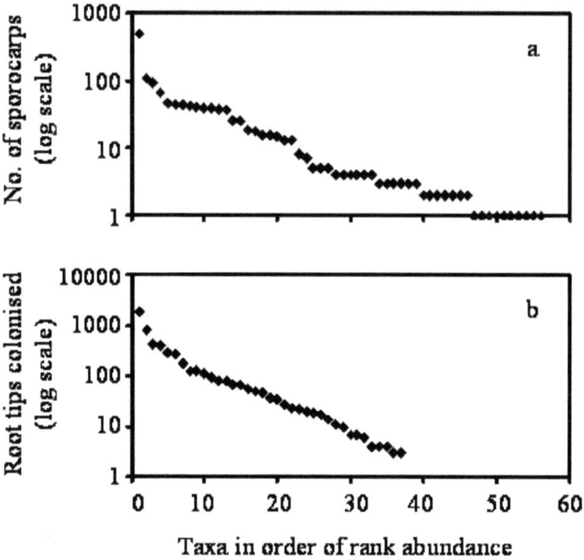

Fig. 2. Rank abundance patterns of ectomycorrhizal sporocarps (**a**) and morphotypes (**b**) using log abundance plotted against rank abundance. Data from a 50-year-old pine stand in central Sweden. (Taylor 2002)

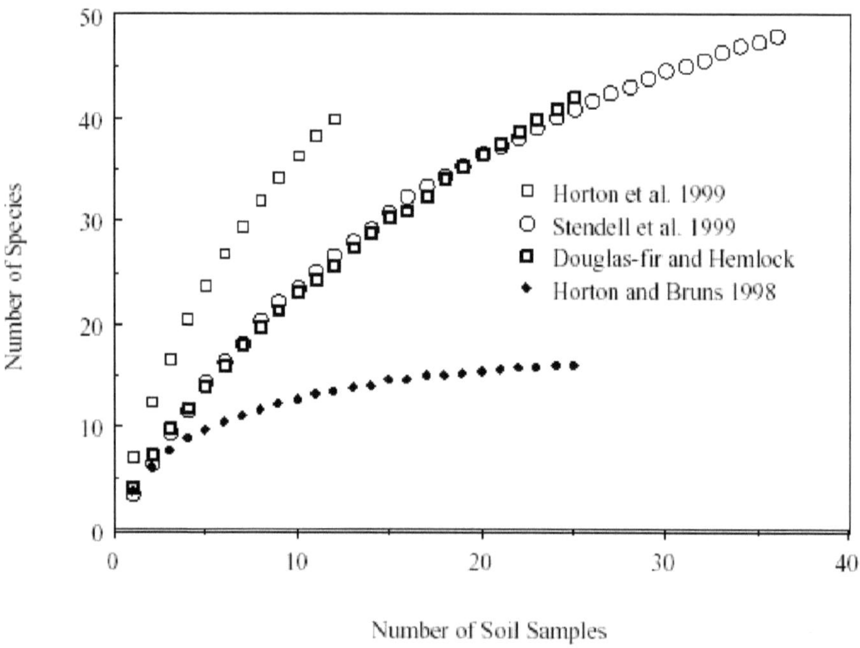

Fig. 3. Species area curves from four studies, with number of soil samples taken used as a substitute for the area sampled. (Horton and Bruns 2001)

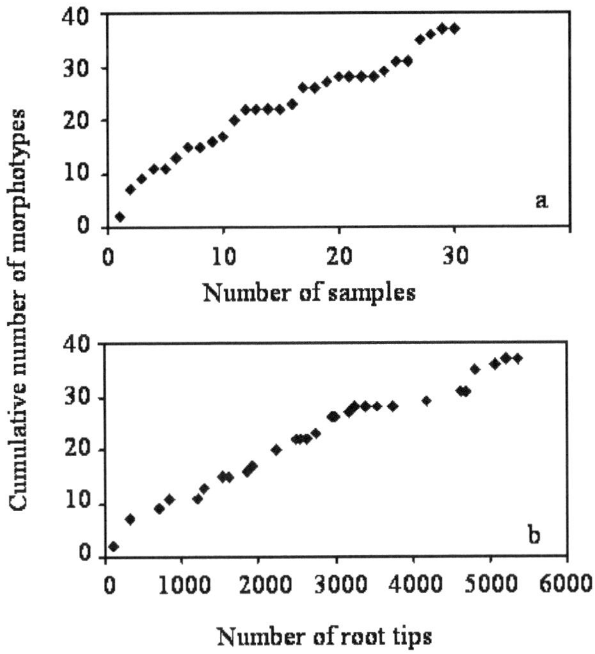

Fig. 4. Species area plots for ectomycorrhizal morphotypes found in a study of the mycorrhizal community in a 50-year-old pine stand in central Sweden. Cumulative number of morphotypes is plotted against the number of samples (**a**) and cumulative number of root tips (**b**) examined. (Taylor 2002)

gated. The size of genets of different ectomycorrhizal fungi has been found to vary from more than 100 per 100 m^2 in *Laccaria amethystine* (Gherbi et al. 1999) to single genets of more than 10 m^2, for example *Suillus bovinus* (Dahlberg and Stenlid 1990) and *Cortinarius rotundisporus* (Sawyer et al. 1999). As within a cluster of root tips colonised by the same species of ectomycorrhizal fungus, it is highly likely that they are colonised by the same individual, the extent of genets and the structure of populations is not trivial to our understanding of biodiversity. For large genets even samples taken within a few meters of one another may represent repeated sampling of the same individual.

In addition to the problems discussed above, it is becoming apparent that there are vertical gradients in the distribution of ectomycorrhizal fungi, and also seasonal fluctuations in belowground abundance. In forests, the density (units per volume of soil) of root biomass and root tips is normally highest in organic soil layers compared to the mineral soil (Godbold et al. 2003). A number of investigations have shown that the distribution of ectomycorrhizal species differs between soil layers (Stendell et al.

1999; Taylor and Bruns 1999). Fransson et al. (2000) could show that *Cenococcum geophilum* mycorrhizas were preferentially found in organic layers whereas *Tylospora fibillosa* was found preferentially in the mineral soil. In a study of the vertical distribution of ectomycorrhizal species in a boreal forest podzol, Rosling et al. (2003) could show clear vertical gradients in the distribution of species. At this study site, in a spruce/pine forest dominated by 60- to 80-year-old *Picea abies*, seven horizons were sampled, and root tips were identified by morphological sorting and sequencing of the rDNA ITS region. Twenty-two taxa were identified. An *Inocybe* spp. and *Piloderma byssinum* were restricted to the organic layer, and *Tomentellopsis submollis, Piloderma fallax, Hygrophorus olivacelalbus, Russula decolorans* and a *Dermocybe* spp. were found in the organic layer and the E horizons. *Suillus luteus, Lactarius utilis* and three unidentified *Piloderma* species were associated with the mineral horizons. More importantly, two thirds of the root tips were found in the mineral soil and half the taxa were restricted to the mineral soil. This strongly shows the importance of including the mineral horizons in belowground estimates of ectomycorrhizal community structure. Also of note is the distribution of pine-specific species *Suillus luteus* which was only found in the mineral soil. This distribution probably reflects the greater rooting depth of pine compared to spruce, and emphasises the importance of the tree root distribution for the vertical distribution of ectomycorrhizal fungi.

The population structure of ectomycorrhizas also varies throughout the vegetation period and probably between years. Differences were observed in species assemblage and dominance in samples taken from two successive years in a ponderosa pine forest, even though the samples were only taken 25 cm apart between years (Stendell et al. 1999). Horton and Bruns (2001) attributed this difference to patchiness. In a study using horizontal and vertical rhizotrones (Sittig 1999), clear changes in dominance in the ectomycorrhizal community structure in two beech forests over the vegetation period could be shown. In this study, the fate of 1,306 root tips in the calcareous site (Göttinger Wald) and 447 root tips in the acid site (Solling) was followed. In the calcareous beech forest the percentage of *Lactarius subdulcis* in the total population of root tips observed increased from 12 to ca. 60% over the vegetation period (Fig. 5). Similarly, in the acidic beech forest, the percentage of *Xerocomus chrysenteron* decreased and that of *Cenococcum geophilum* increased over the vegetation period (Fig. 6). These results reflect the colonisation rates of the fungi on root systems observed in the rhizotrones, but also the longevity of the different species and the underpinning fine root turnover. *Lactarius subdulcis* mycorrhizas had an average life span of 83 days compared to a 39-day average for the other mycorrhizas. Similarly, the life span of *Cenococcum geophilum* mycorrhi-

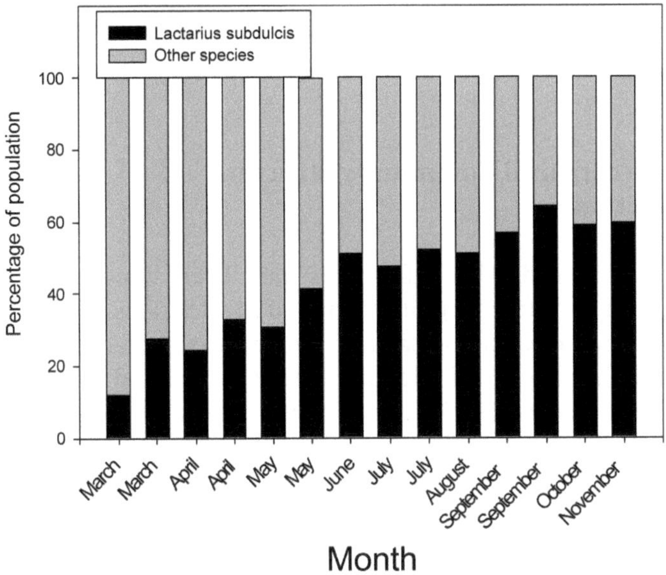

Fig. 5. Proportion of *Lactarius subdulcis* and other species in the ectomycorrhizal population observed using horizontal rhizotrons, in a calcareous beech forest over the vegetation period March to November 1997. In April, May, July and September two measurements were made. (Sittig 1999)

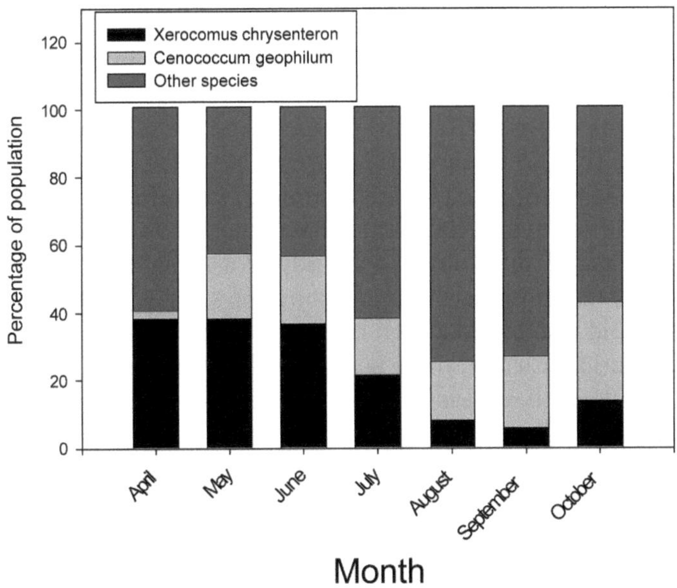

Fig. 6. Proportion of *Xerocomus chrysenteron*, *Cenococcum geophilum* and other species in the ectomycorrhizal population observed using horizontal rhizotrons, in an acidic beech forest over the vegetation period April to October 1997. (Sittig 1999)

zas was 94 days, that of *Xerocomus chrysenteron* 76 days and the other mycorrhizas 72 days.

5 Relating Ectomycorrhizal Community Structure to Ecosystem Function

As mentioned above, considerable differences have been shown in the physiological function of different ectomycorrhizal species. This has been demonstrated for a range of physiological functions such as nutrient uptake (see Chalot et al. 2002), heavy metal tolerance (see Jentschke and Godbold 2000), direct mobilisation of nutrients from minerals (see Landeweert et al. 2001) and utilisation of organic nitrogen and phosphorus forms (Bending and Read 1995; Perez-Mereno and Read 2000). All of these processes have important implications for forest ecosystem function, in particular mineral nutrition and soil development. Much of the information about physiological function of ectomycorrhizas is limited to a relatively small number of species, such as *Paxillus involutus*, *Laccaria laccata*, *Pisolithus tinctorius*, *Hebeloma cylindrosporum*, *Suillus bovinus* and *Telephora terrestris*. These species are, in the main, early successional fungi and can be cultured with a range of host tree seedlings.

In conifer forests, the dominant taxa on tree roots have been shown to be members of the Russulaceae, Thelephoraceae and non-thelephoroid resupinates (Horton and Bruns 2001). In a *Picea abies* forest in southern Sweden the ectomycorrhizal community was dominated by *Cenococcum geophilum*, *Telephora terrestris*, *Tylospora fibrillose* and *Tylophilus felleus* (Erland et al. 1999). The dominance of *Tylospora* has been shown in a number of studies (Dahlberg et al. 1997; Flynn et al. 1998; Jonsson et al. 1999b). The taxa identified in the Thelephoraceae are members of the genus *Tomentella* (Horton and Bruns 2001). In a study of a *Picea abies* forest, Dahlberg et al. (1997) found the second most common species to be *Cenococcum geophilum*, and also found a high presence of *Piloderma croceum* in an uneven distribution. Clearly, there is a large dichotomy between the group of species about which most is known physiologically and those that occur most commonly in natural ecosystems. Many of the species most common in natural ecosystems are species that do not form an extensive extramatrical mycelium, such as those of *Tylospora* and *Russula*. However, hyphae from genera such as *Russula* and *Lactarius*, both of which have a smooth hyphal mantle, have been detected on soils (Dickie et al. 2002; Landeweert et al. 2003b; Smit et al. 2003). The ability of ectomycorrhizas to capture and transport nutrients is believed to be strongly related to the explorative ability and function of the extramatrical mycelium. Clearly many of the ectomycorrhizal species that do not produce an extensive extramatrical mycelium may have another function.

As described above, the ectomycorrhizal community often comprises a few common species and a large number of rare species, and the ability to detect rare species may be determined by sampling effort (Horton and Bruns 2001; Taylor 2002). In terms of community function, as shown for bacteria communities (Yin et al. 2000), as the number of species increases, functional redundancy increases. Although detecting all species of ectomycorrhizas present may be important for understanding community structure, it may be less important for understanding community function. Similarly, although patchy distributions may complicate assessments of belowground biodiversity, they may have a functional significance. As discussed by Taylor (2002); Gherbi et al. (1999) suggested that *Laccaria amethysina* recolonises root systems of beech more or less annually. The main source of inoculum is spores from sporocarps, giving a clumped distribution pattern of genets. A similar distribution pattern was reported for *Russula cremoricolor* (Redecker et al. 2001). Both *Laccaria amethysina* and *Russula cremoricolor* are suggested to produce short-lived genets. Within the soil environment resources may also be patchy, with new areas forming and others becoming depleted; thus colonisation of different soil areas between seasons may aid resource capture. We can speculate that the most long-lived and explorative genets should be found in the most stable environments, such as the mineral soil.

Clearly, our understanding of ectomycorrhizal community composition has greatly progressed with the development of molecular and other methods. Our understanding of community ecology, however, will only increase if we increase our understanding of the function of species in the community.

References

Agerer R (1987–1996) Colour atlas of ectomycorrhizae. Einhorn, Schwäbisch Gmünd

Altschul SF, Madden TL, Schaffer AA, Zhang J, Zhang Z, Miller W, Lipman DJ (1997) Gapped BLAST and PSI-BLAST: a new generation of protein database search programs. Nucleic Acids Res 25:3389–3402

Auge RM (2001) Water relations, drought and vesicular-arbuscular mycorrhizal symbiosis. Mycorrhiza 11:3–42

Bending GD, Read DJ (1995) The structure and function of the vegetative mycelium of ectomycorrhizal plants. V. Foraging behaviour and translocation of nutrients from exploited litter. New Phytol 130:401–409

Brüggemann J, Stephen JR, Yun-Juah C, Macnaughton SJ, Kowalchuk GA, Kline E, White DC (2000) Competitive PCR-DGGE analysis of bacterial mixtures: an internal standard and an appraisal of template enumeration accuracy. J Microbiol Meth 40:111–123

Chalot M, Javelle A, Blaudez D, Lambilliote R, Cooke R, Sentenac H, Wipf D, Botton B (2002) An update on nutrient transport processes in ectomycorrhizas. Plant Soil 244:165–175
Dahlberg A (2001) Community ecology of ectomycorrhizal fungi: an advancing interdisciplinary field. New Phytol 150:555–562
Dahlberg A, Stenlid J (1990) Size, distribution and biomass of genets in populations of *Suillus bovinus* (L.:Fr.) Ruossel revealed by somatic incompatibility. New Phytol 115:487–493
Dahlberg A, Jonsson L, Nylund JE (1997) Species diversity and distribution of biomass above and below ground among ectomycorrhizal fungi in an old-growth Norway spruce forest in south Sweden. Can J Bot 75:1323–1335
Dickie IA, Xu B, Koide RT (2002) Vertical niche differentiation of ectomycorrhizal hyphae in soil as shown by T-RFLP analysis. New Phytol 156:527–535
Duchesne LC, Peterson RL, Ellis BE (1988). Interaction between the ectomycorrhizal fungus *Paxillus involutus* and *Pinus resinosa* induces resistance to *Fusarium oxysporum*. Can J Bot 66:558–562
Ekblad A, Wallander H, Näsholm T (1998) Chitin and ergosterol combined to measure total and living fungal biomass in ectomycorrhizas. New Phytol 138:143–149
Erland S, Taylor AFS (2002) Diversity of ectomycorrhizal communities in relation to the abiotic environment. In: van der Heijden MGA, Sanders IR (eds) Mycorrhizal ecology. Ecological studies, vol 157. Springer, Berlin Heidelberg New York, pp 470–485
Erland S, Jonsson T, Mahmood S, Finlay RD (1999) Below-ground ectomycorrhizal community structure in two *Picea abies* forests in southern Sweden. Scand J For Res 14:209–217
Flynn D, Newton AC, Ingleby K (1998) Ectomycorrhizal colonization of Sitka spruce (*Picea sitchensis* (Bong.) Carr) seedlings in a Scottish plantation forest. Mycorrhiza 7:313–317
Fransson PMA, Taylor AFS, Finlay RD (2000) Effects of continuous optimal fertilisation upon a Norway spruce ectomycorrhizal community. Tree Physiol 20:599–606
Friese CF, Allen MF (1991) The spread of VA mycorrhizal fungal hyphae in the soil – inoculum types and external hyphal architecture. Mycologia 83:409–418
Frostegård Å, Bååth E (1996) The use of phospholipids fatty acid analysis to estimate bacterial and fungal biomass in soil. Biol Fertil Soil 22:59–65
Gardes M, Bruns T D (1996a) Community structure of ectomycorrhizal fungi in a *Pinus muricata* forest: above- and below-ground views. Can J Bot 74:1572–1583
Gardes M, Bruns TD (1996b) ITS-RFLP matching for identification of fungi. Meth Mol Biol 50:177–186
Gaston KJ (1996) Biodiversity: a biology of numbers and difference. Blackwell, Oxford
Gehring CA, Theimer TC, Whitham TG, Keim P (1998) Ectomycorrhizal fungal community structure of pinyon pines growing in two environmental extremes. Ecology 79:1562–1572
Gherbi H, Delaruelle C, Selosse MA, Martin F (1999) High genetic diversity in a population of the ectomycorrhizal basidiomycete *Laccaria amethystina* in a 150-year-old beech forest. Mol Ecol 8:2003–2013
Godbold DL, Fritz HW, Jentschke G, Meesenburg H, Rademacher P. (2003) Root turnover and root necromass accumulation of Norway spruce (*Picea abies*) are affected by soil acidity. Tree Physiol 23:915–921
Guidot A, Debaud JC, Marmeisse R (2002) Spatial distribution of the below-ground mycelia of an ectomycorrhizal fungus inferred from specific quantification of its DNA in soil samples. FEMS Microbiol Ecol 42:477–486
Hagerman SM, Jones MD, Bradfield GE, Gillespie M, Durall DM (1999) Effects of clear-cut logging on the diversity and persistence of ectomycorrhizae at a sub alpine forest. Can J For Res 29:124–134
Hansen MC, Tolker-Nielsen T, Givskov M, Molin S (1998) Biased 16S rDNA PCR amplification caused by interference from DNA flanking the template region. FEMS Microbiol Ecol 26:141–149
Horton TR (2002) Molecular approaches to ectomycorrhizal diversity studies: variation in ITS at a local scale. Plant Soil 244:29–39

Horton TR, Bruns TD (1998) Multiple host fungi are the most frequent and abundant ectomycorrhizal types in a mixed stand of Douglas fir (*Pseudotsuga menziesii* (Mirb.) Franco) and bishop pine (*Pinus muricata* D. Don). New Phytol 139:331–339

Horton TR, Bruns TD (2001) The molecular revolution in ectomycorrhizal ecology: peeking into the black box. Mol Ecol 10:1855–1871

Horton TR, Bruns TD, Parker VT (1999) Ectomycorrhizal fungi associated with *Arctostaphylos* contribute to *Pseudotsuga menziesii* establishment. Can J Bot 77:93–102

Jentschke G, Godbold DL (2000) Metal toxicity and ectomycorrhizas. Physiol Plant 109:107–116

Jonsson L, Dahlberg A, Nilsson MC, Zackrisson O, Kårén O (1999a) Ectomycorrhizal fungal communities in late-successional Swedish boreal forests, and their composition following wildfire. Mol Ecol 8:205–215

Jonsson T, Kokalj S, Finlay R, Erland S (1999b) Ectomycorrhizal community structure in a limed spruce forest. Mycol Res 103:501–508

Kårén O, Nylund JE (1997) Effects of ammonium sulphate on the community structure and biomass of ectomycorrhizal fungi in a Norway spruce stand in southwestern Sweden. Can J Bot 75:1628–1642

Kranabetter JM, Wylie T (1998) Ectomycorrhizal community structure across forest openings on naturally regenerated western hemlock seedlings. Can J Bot 76:189–196

Landeweert R, Hoffland F, Finlay RD, Kuyper TW, van Breemen N (2001) Linking plants to rocks: ectomycorrhizal fungi mobilize nutrients from minerals. Trends Ecol Evol 16:248–254

Landeweert R, Leeflang P, Smit E, Kuyper T (2003a) Diversity of an ectomycorrhizal fungal community studied by a root tip and total soil DNA approach. Mycorrhiza. Springer, Berlin Heidelberg New York, www.springer-link.com

Landeweert R, Leeflang P, Kuyper TW, Hoffland E, Rosling A, Wernars K, Smit E (2003b) Molecular identification of ectomycorrhizal mycelium in soil horizons. Appl Environ Microbiol 69:327–333

Lleó MM, Tafi MC, Signoretto C, Dal Cero C, Canepari P (1999) Competitive polymerase chain reaction for quantification of nonculturable *Enterococus faecalis* cells in lake water. FEMS Microbiol Ecol 30:345–353

Luoma DL, Eberhart JL, Amaranthus MP (1997) Biodiversity of ectomycorrhizal types from southwest Oregon. In: Kaye TN, Liston A, Love RM, Luoma DL, Meinke RJ, Wilson MV (eds) Conservation and management of native plants and fungi. Native Plant Society of Oregon, Corvallis, Oregon, pp 249–253

Mahmood S, Finlay R, Erland S (1999) Effects of repeated harvesting of forest residues on the ectomycorrhizal community in a Swedish spruce forest. New Phytol 142:557–585

Methvyn A, Hughes KW, Petersen RH (2000) *Flammulina* RFLP patterns identify species and show biogeographical patterns within species. Mycologia 92:1064–1070

Naeem S (1998) Species redundancy and ecosystem reliability. Conserv Biol 12:39–45

Naeem S, Thompson LJ, Lawler SP, Lawton JH, Woodfin RM (1994) Declining biodiversity can alter the performance of ecosystems. Nature 368:734–737

Nara K, Nakaya H, Wu B, Zhou Z, Hogetsu T (2003) Underground primary succession of ectomycorrhizal fungi in a volcanic desert on Mount Fuji. New Phytol 159:743–756

Olsson PA (1999) Signature fatty acids provide tools for determination of the distribution and interactions of mycorrhizal fungi in soil. FEMS Microbiol Ecol 29:303–310

Olsson PA, Wallander H (1998) Interactions between ectomycorrhizal fungi and the bacterial community in soils amended with various primary minerals. FEMS Microbiol Ecol 27:95–205

Peter M, Ayer F, Egli S (2001) Nitrogen addition in Norway spruce stand altered macromycete sporocarp production and belowground ectomycorrhizal species composition measured by PCR-RFLP analysis of the ribosomal ITS-region. New Phytol 149:311–324

Perez-Moreno J, Read DJ (2000) Mobilization and transfer of nutrients from litter to tree seedlings via the vegetative mycelium of ectomycorrhizal plants. New Phytol 145:301–309
Pfisterer AB, Schmid B (2002) Diversity-dependent production can increase the stability of ecosystem functioning. Nature 416:84–86
Phillips CJ, Paul EA, Prosser JI (2000) Quantitative analysis of ammonia oxidising bacteria using competitive PCR. FEMS Microbiol Ecol 32:167–175
Polz MF, Cavanaugh CM (1998) Bias in template-to-product ratios in multitemplate PCR. Appl Environ Microbiol 64:3724–3730
Pritsch K, Boyle H, Munch JC, Buscot F (1997) Characterization and identification of black alder ectomycorrhizas by PCR-RFLP analyses of the rDNA internal transcribed spacer (ITS). New Phytol 137:357–369
Qian XM, Kottke I, Oberwinkler F (1998) Influence of liming and acidification on the activity of the mycorrhizal communities in a *Picea abies* (L.) Karst. stand. Plant Soil 199:99–109
Redecker D, Szaro TM, Bowman RJ, Bruns TD (2001) Small genets of *Lactarius xanthogalactus*, *Russula cremoricolor* and *Amanita francheti* in late stage ectomycorrhizal successions. Mol Ecol 10:1025–1034
Rosling A, Landeweert R, Lindahl BD, Larsson KH, Kyper TW, Taylor AFS, Finlay RD (2003) Vertical distribution of ectomycorrhizal fungal taxa in a podzol soil profile. New Phytol 159:775–783
Sawyer NA, Chambers SM, Cairney JWG (1999) Molecular investigation of genet distribution and genetic variation of *Cortinarius rotundisporus* in eastern Australian sclerophyll forests. New Phytol 142:561–568
Selosse MA, Jacquot D, Bouchard D, Martin F, Le Tacon F (1998) Temporal persistence and spatial distribution of an American inoculant strain of the ectomycorrhizal basidiomycete *Laccaria bicolor* in a French forest plantation. Mol Ecol 7:561–573
Sittig U (1999) Zur saisonalen Dynamik von Ektomycorrhizen der Buche (*Fagus sylvatica* L.). PhD Thesis, University of Göttingen, Göttingen, Germany
Smit E, Veenman C, Baar J (2003) Molecular analysis of ectomycorrhizal basidiomycete communities in a *Pinus sylvestris* L. stand reveals long-term increased diversity after removal of litter and humus layers. FEMS Microbiol Ecol 45:49–57
Smith SE, Read DJ (1997) Mycorrhizal symbiosis. Academic Press, San Diego
Staddon PL, Ramsey CB, Ostle N, Ineson P, Fitter AH (2003) Rapid turnover of hyphae of mycorrhizal fungi determined by AMS Microanalysis of ^{14}C. Science 300:1138–1139
Stendell ER, Horton TR, Bruns TD (1999) Early effects of prescribed fire on the structure of the ectomycorrhizal fungus community in a Sierra Nevada ponderosa pine forest. Mycol Res 103:1353–1359
Taylor AFS (2002) Fungal diversity in ectomycorrhizal communities: sampling effort and species detection. Plant Soil 244:19–28
Taylor DL, Bruns TB (1999) Community structure of ectomycorrhizal fungi in a *Pinus muricata* forest: minimal overlap between the mature forest and resistant propagule communities. Mol Ecol 8:1837–1850
Tilman D, Downing JA (1994) Biodiversity and stability in grasslands. Nature 367:363–365
Van der Heijden MGA, Klironomos JN, Ursic M (1998) Mycorrhizal fungal diversity determines plant biodiversity, ecosystem variability and productivity. Nature 396:69–72
Wallander H, Nylund JE (1992) Effects of excess nitrogen and phosphorus starvation on the extramatrical mycelium of ectomycorrhizas of *Pinus sylvestris* L. New Phytol 120:495–503
Wallander H, Nilsson LO, Hagerberg D, Bååth E (2001) Estimation of the biomass and production of external mycelium of ectomycorrhizal fungi in the field. New Phytol 151:53–760
Yin B, Crowley D, Sparovek G, de Melo WJ, Borneman J (2000) Bacterial functional redundancy along a soil reclamation gradient. Appl Environ Microbiol 66:4361–4365

Douglas L. Godbold
School of Agricultural and Forest Sciences
University of Wales
Bangor, Gwynedd LL57 2UW, UK
e-mail: d.l.godbold@bangor.ac.uk

Tracing the Behaviour of Plants in Ecosystems: How Can Molecular Ecology Help?

Michael Kaldorf, Carsten Renker, and François Buscot

1 Introduction

Analyzing interactions between plants themselves or plants and other organisms is a key issue in plant ecology. With typology-based identification criteria, the species composition of plant communities can be easily characterized. This provides a basis for investigating interactions between two or more plant species. In contrast, working on intraspecific interactions often requires molecular tools, as classical phenotypic markers are in most cases inappropriate for resolving the fine population structure of a plant species. This intraspecific level addresses not only relationships between individuals within given populations, but also comparative studies on population structure in different species, which allows understanding of the fundamental aspects of ecology such as reproductive strategies and dispersal success, speciation and evolutionary dynamics, or responses of plant populations to disturbance.

A second important application field for molecular ecological methods concerns interactions between plants and microorganisms, particularly those that take place in soils, for example with rhizosphere bacteria or mycorrhizal fungi. In this context, neutral molecular markers can be employed for species identification and the characterization of microbial biodiversity. Genes related to ecologically significant processes, such as the *nifH* gene used as a marker for the ability of bacteria to fix nitrogen, are of increasing importance in functional analyses of the ecological dimensions of plant–microbe interactions.

The aim of this chapter is not to give a full review of recent results in molecular plant ecology, but to outline a wide range of applications for molecular methods in ecological research. Only a few examples are given in detail, highlighting new findings in plant ecology achieved by the use of molecular tools. In the Conclusion, possible directions for the future per-

Progress in Botany, Vol. 66

spectives of molecular plant ecology, using techniques for the tracing of functions in situ, are discussed.

2 Tracing Intraspecific Population Structure – From Plants in Wild Communities to Cultivated Plants in Agriculture and Forestry

2.1 Molecular Ecological Tools for Population Analysis

The variability of coding target regions within plant genomes is often not sufficient to detect intraspecific genetic diversity suitable for population analysis. Therefore neutral molecular markers are necessary to characterize individuals or populations within a given plant species (Bachmann 1994). The first techniques for DNA fingerprinting in plants were developed in the 1980s, based on the detection of restriction fragment length polymorphisms (RFLPs) in genomic DNA (Saghai-Maroof et al. 1984; Rogstad et al. 1988; Ryskov et al. 1988). After the invention of the PCR, faster and more efficient methods were developed, making large-scale studies on population genetics possible.

2.1.1 Random Amplified Polymorphic DNA

Random amplified polymorphic DNA (RAPD, Williams et al. 1990) markers are obtained by the amplification of multiple genomic DNA sequences with short (9–12 bases) arbitrary primers. The resulting complex banding profiles are detected by electrophoresis, usually on denaturing polyacrylamide gels. RAPD was the first PCR-based technique for DNA fingerprinting to be introduced into plant science (Welsh and McClelland 1990; Hu and Quiros 1991). The major weakness of this method is that, due to the short primer length, low and less stringent annealing temperatures are required, which reduces the specificity of the PCR. As a result, banding patterns are only reproducible when all PCR conditions (e.g. the temperature profile, including the heating and cooling rates) are exactly identical, making it nearly impossible to compare RAPDs from different laboratories.

2.1.2 Amplified Fragment Length Polymorphism

The amplified fragment length polymorphism (AFLP, Vos et al. 1995) technique is based on the selective PCR amplification of restriction fragments from a total digest of genomic DNA. After restriction, oligonucleotide adapters are ligated to the ends of the restriction fragments, followed by PCR amplification. The adapter-restriction site sequence serves as a target for primer annealing. Specificity is obtained by the use of primers extending into the restriction fragments. Amplified restriction fragments are separated and detected on denaturing polyacrylamide gels (Vos et al. 1995).

2.1.3 Microsatellite Markers: Simple Sequence Repeats and Inter-Simple Sequence Repeats

As in animals and fungi, microsatellites (Morgante and Olivieri 1993) in plants are highly polymorphic as a consequence of variations in the number of repeats. To amplify simple sequence repeats (SSR) themselves by PCR, it is necessary to know the sequences of the flanking DNA. The cloning and sequencing effort for the development of SSR primers can be avoided by using the inter-simple sequence repeats (ISSR) technique. ISSR primers anneal within the repetitive sequences, extending only a few base pairs into the flanking regions. In PCR, only one ISSR primer is used, allowing the amplification of genomic DNA between two identical microsatellites with inverted orientation. The utilization of long primers for both SSR and ISSR allows more stringent annealing temperatures as for RAPD primers, leading to a high reproducibility. Especially for ISSR, the high number of polymorphic fragments produced is an additional advantage compared to RAPD (Salimath et al. 1995).

2.2 Examples of the Application of DNA Fingerprinting Techniques in Recent Plant Population Analyses

Detailed genetic characterization of population structures is often helpful to analyze the efficiency of vegetative versus generative plant dispersal. AFLP (Wang et al. 2003) and RAPD (Shimizu et al. 2002) have been used to determine the number and size of plant genets within natural populations of clonally growing trees, allowing conclusions about the impact of disturbance or different habitats. Likewise, based on microsatellite markers, Bockelmann et al. (2003) investigated the genetic differentiation among populations of the clonal grass *Elymus athericus* in order to characterize the influence of habitat adaptation on the gene flow between populations. Polymorphic DNA markers are also used to determine the relative importance of dispersal mechanisms, e.g. pollen flow versus seed dispersal (Ziegenhagen et al. 2003), to follow the spread of invasive species (Bleeker 2003; Saltonstall 2003), the colonization of open habitats by founder populations after disturbance (Tremetsberger et al. 2003) or to perform paternity analyses (Oddou-Muratorio et al. 2003, and references therein).

In conservation genetics, the above-mentioned molecular techniques are important tools to understand the level at which populations interact and to determine the degree of local habitat adaptation in order to set up adequate conservation strategies for endangered species (Bonnin et al. 2002). For example, facilitation of new colonization might be considered to protect endangered species characterized by a low intraspecific genetic variation, while the preservation of existing populations is important when a high degree of microgeographical genetic variation indicates local habitat adaptation (Gustafsson and Sjörge-Gulve 2002; Lemes et al. 2003).

2.3 Examples of the Use of Molecular Markers in Agriculture

In the case of agriculturally important plants, molecular analyses are used for the characterization and identification of ecotypes, landraces, cultivars and even clones (Pupilli et al. 2000; Devarumath et al. 2002; and references therein). A second important research area is the determination of the genetic diversity between populations, forming a basis for DNA marker-assisted breeding programs (Kantety et al. 1995; Ude et al. 2003). As Rebourg et al. (2001) pointed out, the combination of molecular and morphological descriptions might give a more accurate classification of genetic resources compared to pure molecular approaches in many cases. Research in this context is not only focused on crops with worldwide importance like maize, potato, soybean and wheat, but also embraces applications in the non-food plant sector such as the use of the grass *Miscanthus* for biomass production (Hodkinson et al. 2002) and breeding programs for flowers like *Primula* in horticulture (Nan et al. 2003) or *Eucalyptus* in forestry (Jones et al. 2002).

3 Tracing Interactions of Genetically Modified Plants with Other Plants and Microorganisms

Molecular ecological methods are major tools for the risk assessment of releasing genetically modified plants (GMPs). Two important fields of current research are the analysis of the dispersal capacity of GMPs and studies investigating the impact of GMPs on soil microbes (Dale et al. 2002).

3.1 Monitoring the Spread of Genetically Modified Plants

In order to avoid an uncontrolled spread of GMPs, knowledge of their dispersal capacities is essential. Intraspecific gene flow by pollination of wild type by transgenic cultivars is easily detectable if the GMP carries a dominant selectable marker gene, e.g. the *npt*II gene conferring resistance to kanamycin (Ritala et al. 2002) or herbicide resistance genes (Scheffler et al. 1995; Reboud 2003). In such cases, the progenies of wild-type plants can be screened for the expression of marker genes to detect an eventual gene flow from transgenic plants. Recently, highly specific methods for the PCR-based detection of several transgenic crops and even of single transgenic events have been developed (Rønning et al. 2003), which are independent of the presence of marker genes. These methods might allow tracing of the dispersal of transgenic pollen under field conditions, although their use is currently limited to the detection of GMPs in food. Transgenic *Nicotiana tabacum* plants, carrying a gene for a green fluorescent protein under the control of a pollen-active promoter, have been proposed as a model system for the fast monitoring of the spatial distribution patterns of transgenic pollen under field conditions (Hudson et al. 2001).

Most studies assessing the spread of GMPs are focused on pollen dispersal, but at least two further mechanisms of dispersal have been recognized as relevant for the risk assessment of GMPs. Arnaud et al. (2003) analysed the gene flow from transgenic sugar beet (*Beta vulgaris* ssp. *vulgaris*) to coastal wild populations of *Beta vulgaris* ssp. *maritima*, using a chloroplastic marker and several nuclear microsatellite loci. The study gave clear evidence of an important contribution by seed dispersal in addition to pollen dispersal. Besides reproductive dispersal, even vegetative growth might be a factor that should be included in risk assessment studies in some cases. Vegetative propagation, e.g. of several tree species like *Populus* or *Salix*, can be important for the local distribution of plants. In a release experiment with wild type and *rolC* transgenic *Populus*, planted in a field of only 1,500 m^2, more than 200 root sucker plants were detected within four growth seasons after planting. In a PCR with *rolC*-specific primers, the presence of the *rolC* gene was demonstrated for more than 100 of the root suckers, growing at a distance up to 7 m from the next transgenic tree (Fladung et al. 2003).

3.2 Interactions Between GMPs and Microorganisms

An increasing number of studies are aimed at the influence of GMPs on both target and non-target soil microorganisms. Communities of several groups such as protozoa, micro-arthropods, nematodes and saprophytic fungi in the rhizosphere of GMPs have been analysed by classical microscopic and microbiological methods (Donegan et al. 1996, 1999; Griffiths et al. 2000). In contrast, many studies on rhizosphere bacterial communities used molecular tools, either alone or in combination with cultivation and biochemical profiling methods (Bruinsma et al. 2003). In most cases, parts of bacterial 16S rRNA genes were amplified in a first step by PCR with either universal primers for eubacteria or group-specific primers (e.g. for α- or β-Proteobacteria). Different methods were then used for the separation of PCR products and, thus, for the generation of DNA fingerprints representative of bacterial communities. Denaturing gradient gel electrophoresis (DGGE, Fischer and Lerman 1983), separating DNA fragments of the same length with single base pair substitutions based on differences in denaturation kinetics, was applied for bacterial community analysis in the cases of T4-lysozyme producing transgenic potatoes (Heuer and Smalla 1999; Heuer et al. 2002) and glufosinate-tolerant transgenic oilseed rape (Gyamfi et al. 2002). As a further method, single-strand conformation polymorphism (SSCP, Orita et al. 1989) was used for genetic profiling of rhizosphere bacteria from transgenic herbicide-resistant sugar beet and maize (Schmalenberger and Tebbe 2002, 2003a). Other alternatives are RFLP (Donegan et al. 1995) or terminal-restriction fragment length polymorphism (T-RFLP, Liu et al. 1997), as suggested by Lukow et al. (2000). In all the studies cited above, the methods were appropriate to detect treatment-specific differences between bacterial communities, but in all cases transgene-specific changes were minor compared to those specific to plant line or seasonal variation.

Based on the set of these publications, the conclusion is reasonable that the influence of GMPs on soil bacterial communities is generally negligible. This interpretation seems to be at least rash, as recent studies have clearly demonstrated that bacterial communities in soils are far more diverse than detected by SSCP or DGGE (Schmalenberger and Tebbe 2003b). In this context, the results of di Giovanni et al. (1999) are remarkable. In their study, DNA located between enterobacterial repetitive intergenic consensus sequences (ERIC-PCR) was amplified from enrichment cultures to generate fingerprints of cultivable rhizosphere bacteria from wild type and transgenic alfalfa lines. The ERIC-PCR fingerprints detected clear differences between the rhizospheres of the wild type and an α-amylase-producing line, while PCR-RFLP analyses of the same lines revealed no differences

(Donegan et al. 1999). Likewise, Sessitsch et al. (2003) observed significant effects of transgene expression in potatoes, producing the lytic peptide cecropin B, when analysing *Bacillus* isolates from the rhizospheres by 16S rDNA PCR-RFLP. Thus, restricting the analyses to a rather small part of the total bacterial community (e.g. cultivatable Gram-negative bacteria or *Bacillus* ssp.) allows the detection of potentially relevant differences between bacterial communities, which are not obvious when more general genetic profiling methods are used.

These results cause doubts whether the sensitivity of the available methods for the separation of 16S rDNA PCR products is sufficient to compare total bacterial communities and if they are detailed enough to draw conclusions on the real microbiological impact of GMPs. Additionally, the complete lack of data concerning functional gene diversity in the rhizosphere of GMPs is an important flaw. In conclusion, although the existing studies indicate no alarming effects of GMPs on soil microorganisms, much more effort is needed to characterize the rhizosphere of GMPs before risks of GMPs can be considered as excluded.

4 Tracing Mycorrhizas as a Biological Link That Connects Plants Together and to the Soil Medium

Most terrestrial plants are associated with soil fungi forming mycorrhizal symbioses with roots (Smith and Read 1997). There is fossil evidence that even the first plants colonizing soil substrates in the Ordovician about 460 million years ago contained mycorrhizal fungi similar to the arbuscular mycorrhizal fungi (AMF) of today (Redecker et al. 2000). All AMF belong to a monophyletic systematic group which has been recently elevated to the rank of a distinct phylum, the Glomeromycota (Schüßler et al. 2001). Arbuscular mycorrhiza (AM) is still the most common type of mycorrhiza, being found in up to 80% of all plant species and in almost all ecosystems (Smith and Read 1997).

Additionally, many trees and shrubs are dependent on ectomycorrhizas (EM). In contrast to AM, the ability to form EM was achieved several times during evolution by a wide range of at least 5,000 fungal species of the Basidiomycota and Ascomycota, and even a few Zygomycota (Smith and Read 1997). This diversity provides a link to sparse and heterogeneously distributed resources under a wide range of forms in organic soil horizons (Read 1993). Other types of mycorrhizas are the ericoid, arbutoid, monotropoid and orchid mycorrhizas, which show some different morphological features, corresponding to the special ecophysiological situations in the habitats of their associated plants and are therefore found in a small range of plant families (Smith and Read 1997).

From the plant's point of view, the key function of mycorrhizas is the mobilization of mineral nutrients, especially phosphate and nitrogen, from soil. In exchange, the fungal carbon demand is covered by plant photoassimilates. Additionally, mycorrhizas mitigate different kinds of plant stresses such as drought or heavy metal toxicity (e.g. Hildebrandt et al. 1999; Augé et al. 2001) and protect plants against root herbivory (Gange 2001). An extreme example of plant dependency on mycorrhizal fungi is that of mycoheterotrophic plants like *Arachnitis uniflora* and *Voyria* sp., which live epiparasitically on members of a narrow lineage within the genus *Glomus* and obtain most nutrients, including carbon, via these AMF (Bidartondo et al. 2002).

4.1 Analysis of Ectomycorrhizal Communities

In EMF, features of the fungal sheath and the Hartig net allow characterization of mycorrhizal morphotypes which roughly correspond to species (Agerer 1987–1996). To achieve the identification of these morphotypes, their molecular biological characterization with reference to identified fruiting bodies or to published sequences is a powerful tool (Dahlberg et al. 1997; Horton and Bruns 2001). A common target for the molecular characterization of morphotypes are the internal transcribed spacer (ITS) regions, which are usually analyzed by PCR/RFLP or sequencing (Buscot et al. 2000). Since the groundbreaking work of Gardes et al. (1991), EMF community structures have been studied in many different ecosystems (see also Godbold, this Vol.). A decrease in EMF diversity was, for example, related to a less adequate ecological context in a part of an alder stand (Pritsch et al. 1997). Further examples are evaluations of the impact on EM communities of environmental stress (Gehring et al. 1998), disturbance by fire (Jonsson et al. 1999), nitrogen deposition (Lilleskov et al. 2002) or the planting of transgenic trees (Kaldorf et al. 2002).

4.2 Analysis of AMF Communities

A major goal of current plant ecological research is the characterization of mechanisms that determine plant biodiversity, ecosystem variability and productivity. Based on model experiments with introduced AMF species, van der Heijden et al. (1998) proposed that AMF species composition might be a key factor determining plant biodiversity in natural ecosystems. To validate this hypothesis under realistic field conditions, tools for the analysis of AMF communities within roots are necessary. As a morphological determination of AMF is only possible based on spore characteristics (Morton and Benny 1990), molecular methods are essential. Extraction of DNA from mycorrhizas always results in a mixture of plant and fungal DNA, and even DNA from many saprophytic, endophytic and pathogenic fungi is likely to be included to a similar extent as DNA from AMF. Therefore primer sets specific for the Glomeromycota are necessary to amplify mainly AMF DNA from mycorrhizal roots. White et al. (1990) proposed several sets of primers for the amplification of fungal rDNA. The use of these universal primers allowed the amplification of AMF rDNA from spores (e.g. Hildebrandt et al. 1999), but the results were not satisfactory when field-collected roots were used as the source for AMF DNA (Renker et al. 2003).

The adoption of the nested PCR technique in AMF research in combination with specific primer sets increased the progress of community studies (van Tuinen et al. 1998). The first broad-scale field experiments on AMF were performed in England using 18S rDNA as the molecular marker (Helgason et al. 1998, 1999). As a fine population analysis of single taxa is not possible with this marker, Redecker (2000) developed a set of different primers targeting the more variable ITS region of several groups within the Glomeromycota. These different primers were used as a mix in PCR on field roots (Redecker et al. 2003). However, the disadvantage of this method is its specificity, limiting the range of detection to certain species groups within the Glomeromycota, with the loss of several non-compatible species. In a further approach, we developed a primer set for the first step of a nested PCR being as specific as possible for all AMF taxa. As it was impossible to totally achieve this requisite, we performed a restriction digest with the enzyme *Alu*I after the first step of the nested PCR because most AMF do not contain any restriction site for this enzyme while many contaminating fungi do. Therefore, it was possible to increase the amount of mycorrhizal fungi detected on root material from the field (Renker et al. 2003).

As the identification of AMF in root samples is currently based on sequence data, reliable sequence databases are inevitably necessary to directly address taxa. Mislabelling of available sequences in databases like GenBank or EMBL has been recently discussed as a source of confusion in mycorrhizal research (Clapp et al. 2002; Pringle et al. 2003; Schüßler et al. 2003). An evaluation process of the databases is urgently needed to remove erroneous data, but this seems unlikely in view of the amount of new sequence data produced daily. Thus, phylogenetic analysis including sequences of possible contaminants should be performed in order to identify an organism based on its sequence (Schüßler 1999).

Another specific problem for the analysis of AM communities is the fact that genetically heterogeneous nuclei may coexist in mycelium, and even the multicopy-regions of the rDNA of single nuclei seem to be heterogeneous (Kuhn et al. 2001; Sanders 2002). Therefore, a cost-intensive cloning of PCR products is inevitable to separate heterogeneous PCR products, especially when working with field material. Approaches to obtain an overview of AMF diversity in the field should therefore combine the sequencing of target regions for taxon identification with DNA fingerprinting techniques to increase the number of analyzed samples.

The first DNA fingerprinting techniques applied to AMF analysis were RAPD (Wyss and Bonfante 1993) and RFLP (Sanders et al. 1996). T-RFLP, being originally developed to analyze bacterial communities (Liu et al. 1997), has been recently introduced into AMF community analysis (Tonin et al. 2001; Johnson et al. 2003; Vandenkoornhuyse et al. 2003). T-RFLP is based on a PCR in which one of the primers is fluorescently labeled. The obtained PCR products are digested with one single or a set of restriction enzyme(s). The restriction bands obtained at the labeled end of the amplification product can be detected by laser-induced fluorescence.

There are some other possible DNA fingerprinting techniques recently introduced in mycorrhizal research. Douhan and Rizzo (2003), for example, proposed the use of amplified fragment length microsatellites (AFLM) to develop microsatellite markers especially in organisms with limited amounts of DNA, e.g. AMF. Kowalchuk et al. (2002) and Öpik et al. (2003) used DGGE to separate 18S rDNA for fingerprinting. The approach was found to be efficient for the separation of AMF-like sequences down to genus and possibly species level. The results indicate that DGGE banding patterns rather provide an underestimate of total diversity than an overestimate as often proposed for direct estimations of AMF community structure. Solutions such as the use of other, more variable DNA regions to improve the resolution of the method were discussed by Kowalchuk et al. (2002).

5 Conclusions and Future Perspectives

Molecular tools have contributed to a major part of the progress in plant ecology over recent years. Nevertheless, there are still many open questions, especially well exemplified in the ecological aspects of the interactions between plants and associated microorganisms like AMF. In most studies on AMF communities, the sampling design is in major part directed by the rather small number of samples that can be processed with the available molecular methods. To achieve a real community survey, factors like soil heterogeneity (Ettema and Wardle 2002; Carvalho et al. 2003) and seasonality (Husband et al. 2002) should be considered in a sampling design based on geostatistics.

For ecological studies in the future, development of microarrays or microchips might be very helpful to reduce the time needed for such monitoring and to extend analyses to the detection of functional gene expression in the field. By hybridization of PCR products from the field to given DNA fragments on the arrays or chips, a fast and proper identification might be possible for large numbers of samples. Arrays and chips are based on slides or membranes on which oligonucleotides of known DNA or cDNA fragments are spotted. Fluorescently labeled DNA or RNA extracts can be hybridized to these oligonucleotides. Screening of arrays or chips with a laser detects the hybridized oligonucleotides (de Benedetti et al. 2000). This technology allows a screening of DNA polymorphisms or an expression analysis of up to several thousand genes (Xiang and Chen 2000).

Microchips have already been used to detect at the gene expression level mechanisms involved in adaptation and response of plants to environmental stresses such as drought (Watkinson et al 2003) or wounding, low temperatures and heavy metals (Sugano 2003). Peter et al. (2003) introduced microchips into mycorrhizal research to analyze functional aspects, while using this technique for the identification of mycorrhizal fungi was considered by Martin (2001). The effects of cold stress have also recently been investigated by proteomics (Bae et al. 2003). In this approach, proteins are isolated and analyzed by two-dimensional gel electrophoresis. After desorption, those proteins corresponding to stress-specific spots can be characterized by mass spectrometry. Actually the applications of both chip and proteomic technologies are limited to plants cultivated under controlled laboratory conditions. However, due to the rapid progress, they should soon be extendable to field studies. This will open a new chapter of plant ecophysiology. Besides validating the ecological efficiency of regulative pathways presently detected on model plants such as *Arabidopsis*, these new techniques will allow us to trace the behaviour of plants involved in complex interactions with microbial consortia, that can mediate adaptation to the ecological niche or on the contrary be adverse. It is expected that metabolic adaptations not adopted under simplified culture conditions might be identified in that way.

References

Agerer R (1987–1996) Colour atlas of ectomycorrhizae. Einhorn, Schwäbisch Gmünd

Arnaud J-F, Viard F, Delescluse M, Cuguen J (2003) Evidence for gene flow via seed dispersal from crop to wild relatives in *Beta vulgaris* (Chenopodiaceae): consequences for the release of genetically modified crop species with weedy lineages. Proc R Soc Lond B 270:1565–1571

Augé RM, Stodola AJW, Tims JE, Saxton AM (2001) Moisture retention properties of a mycorrhizal soil. Plant Soil 230:87–97

Bachmann K (1994) Tansley review no. 63: molecular markers in plant ecology. New Phytol 126:403–418

Bae MS, Cho EJ, Choi E-Y, Park OK (2003) Analysis of the *Arabidopsis* nuclear proteome and its response to cold stress. Plant J 36:652–663

Bidartondo MI, Redecker D, Hijri I, Wiemken A, Bruns TD, Domínguez L, Sérsic A, Leake JR, Read DJ (2002) Epiparasitic plants specialized on arbuscular mycorrhizal fungi. Nature 419:389–392

Bleeker W (2003) Hybridization and *Rorippa austriaca* (Brassicaceae) invasion in Germany. Mol Ecol 12:1831–1841

Bockelmann A-C, Reusch TBH, Bijlsma R, Bakker JP (2003) Habitat differentiation vs. isolation-by-distance: the genetic population structure of *Elymus athericus* in European salt marshes. Mol Ecol 12:505–515

Bonnin I, Colas B, Bacles C, Holl A-C, Hendoux F, Destiné B, Viard F (2002) Population structure of an endangered species living in contrasted habitats: *Parnassia palustris* (Saxifragaceae). Mol Ecol 11:979–990

Bruinsma M, Kowalchuk GA, van Veen JA (2003) Effects of genetically modified plants on microbial communities and processes in soil. Biol Fertil Soils 37:329–337

Buscot F, Munch JC, Charcosset JY, Gardes M, Nehls U, Hampp R (2000) Recent advances in exploring physiology and biodiversity of ectomycorrhizas highlight the functioning of these symbioses in ecosystems. FEMS Microbiol Rev 24:601–614

Carvalho LM, Correia PM, Ryel RJ, Martins-Loução MA (2003) Spatial variability of arbuscular mycorrhizal fungal spores in two natural plant communities. Plant Soil 251:227–236

Clapp JP, Helgason T, Daniell TJ, Young JPW (2002) Genetic studies of the structure and diversity of arbuscular mycorrhizal fungal communities. In: van der Heijden MGA, Sanders IR (eds) Mycorrhizal ecology. Ecological studies, vol 157. Springer, Berlin Heidelberg New York, pp 201–224

Dahlberg A, Jonsson L, Nylund J-E (1997) Species diversity and distribution of biomass above and below ground among ectomycorrhizal fungi in an old-growth Norway spruce forest in south Sweden. Can J Bot 75:1323–1335

Dale PJ, Clarke B, Fontes EMG (2002) Potential for the environmental impact of transgenic crops. Nat Biotechnol 20:567–574

De Benedetti VMG, Biglia N, Sismondi P, de Bortoli M (2000) DNA chips: the future of biomarkers. Int J Biol Markers 15:1–9

Devarumath RM, Nandy S, Rani V, Marimuthu S, Muraleedharan N, Raina SN (2002) RAPD, ISSR and RFLP fingerprints as useful markers to evaluate genetic integrity of micropropagated plants of three diploid and triploid elite tea clones representing *Camellia sinensis* (China type) and *C. assamica* ssp. *assamica* (Assam-India type). Plant Cell Rep 21:166–173

Di Giovanni GD, Wartrud LS, Seidler RJ, Widmer F (1999) Comparison of parental and transgenic alfalfa rhizosphere bacterial communities using Biolog GN metabolic fingerprinting and enterobacterial repetitive intergenic consensus sequence-PCR (ERIC-PCR). Microb Ecol 37:129–139

Donegan KK, Palm CJ, Fieland VJ, Porteous LA, Ganio LM, Schaller DL, Bucao LQ, Seidler RJ (1995) Changes in levels, species and DNA fingerprints of soil microorganisms associated with cotton expressing the *Bacillus thuringiensis* var. *kurstaki* endotoxin. Appl Soil Ecol 2:111–124

Donegan KK, Schaller DL, Stone JK, Ganio LM, Reed G, Hamm PB, Seidler RJ (1996) Microbial populations, fungal species diversity and plant pathogen levels in field plots of potato plants expressing the *Bacillus thuringiensis* var. *tenebrionis* endotoxin. Transgen Res 5:25–35

Donegan KK, Seidler RJ, Doyle JD, Porteous LA, di Giovanni G, Widmer F, Watrud LS (1999) A field study with genetically engineered alfalfa with recombinant *Sinorhizobium meliloti*: effects on the soil ecosystem. J Appl Ecol 36:920–936

Douhan GW, Rizzo DM (2003) Amplified fragment length microsatellites (AFLM) might be used to develop microsatellite markers in organisms with limited amounts of DNA applied to arbuscular mycorrhizal (AM) fungi. Mycologia 95:368–373

Ettema CH, Wardle DA (2002) Spatial soil ecology. Trends Ecol Evol 17:177–183

Fischer SG, Lerman LS (1983) DNA fragments differing by single base-pair substitutions are separated in denaturating gradient gels: correspondence with melting theory. Proc Natl Acad Sci USA 80:1579–1583

Fladung M, Nowitzki O, Ziegenhagen B, Kumar S (2003) Vegetative and generative dispersal capacity of field released transgenic aspen trees. Trees 17:412–416

Gange AC (2001) Species-specific responses of a root- and shoot-feeding insect to arbuscular mycorrhizal colonization of its host plant. New Phytol 150:611–618

Gardes M, White TJ, Fortin JA, Bruns TD, Taylor JW (1991) Identification of indigenous and introduced symbiotic fungi in ectomycorrhizae by amplification of nuclear and mitochondrial ribosomal DNA. Can J Bot 69:180–190

Gehring CA, Theimer TC, Whitham TG, Keim P (1998) Ectomycorrhizal fungal community structure of pinyon pines growing in two environmental extremes. Ecology 79:1562–1572

Griffiths BS, Geoghegan IE, Robertson WM (2000) Testing genetically engineered potato, producing the lectins GNA and Con A, on non-target soil organisms and processes. J Appl Ecol 37:159–170

Gustafsson S, Sjögren-Gulve P (2002) Genetic diversity in the rare orchid, *Gymnadenia odoratissima* and a comparison with the more common congener, *G. conopsea*. Cons Genet 3:225–234

Gyamfi S, Pfeifer U, Stierschneider M, Sessitsch A (2002) Effects of transgenic glufosinate-tolerant oilseed rape (*Brassica napus*) and the associated herbicide application on eubacterial and *Pseudomonas* communities in the rhizosphere. FEMS Microbiol Ecol 41:181–190

Helgason T, Daniell TJ, Husband R, Fitter AH, Young JPW (1998) Ploughing up the wood-wide web? Nature 394:431

Helgason T, Fitter AH, Young JPW (1999) Molecular diversity of arbuscular mycorrhizal fungi colonising *Hyacinthoides non-scripta* (bluebell) in a seminatural woodland. Mol Ecol 8:659–666

Heuer H, Smalla K (1999) Bacterial phyllosphere communities of *Solanum tuberosum* L. and T4-lysozyme-producing transgenic variants. FEMS Microbiol Ecol 28:357–371

Heuer H, Kroppenstedt RM, Lottmann J, Berg G, Smalla K (2002) Effects of T4 lysozyme release from transgenic potato roots on bacterial rhizosphere communities are negligible relative to natural factors. Appl Environ Microbiol 68:1325–1335

Hildebrandt U, Kaldorf M, Bothe H (1999) The zinc violet and its colonization by arbuscular mycorrhizal fungi. J Plant Physiol 154:709–717

Hodkinson TR, Chase MW, Renvoize SA (2002) Characterization of a genetic resource collection for *Miscanthus* (Saccharinae, Andropogoneae, Poaceae) using AFLP and ISSR PCR. Ann Bot 89:627–636

Horton TR, Bruns TD (2001) The molecular revolution in ectomycorrhizal ecology: peeking into the black-box. Mol Ecol 10:1855–1871

Hu J, Quiros CF (1991) Identification of broccoli and cauliflower cultivars with RAPD markers. Plant Cell Rep 10:505–511

Hudson LC, Chamberlain D, Stewart CN (2001) GFP-tagged pollen to monitor pollen flow of transgenic plants. Mol Ecol Notes 1:321–324

Husband R, Herre EA, Young JPW (2002) Temporal variation in the arbuscular mycorrhizal communities colonising seedlings in a tropical forest. FEMS Microbiol Ecol 42:131–136

Johnson D, Vandenkoornhuyse PJ, Leake JR, Gilbert L, Booth RE, Grime JP, Young JPW, Read DJ (2003) Plant communities affect arbuscular mycorrhizal fungal diversity and community composition in grassland microcosms. New Phytol 161:503–515

Jones RC, Steane DA, Potts BM, Vaillancourt RE (2002) Microsatellite and morphological analysis of *Eucalyptus globulus* populations. Can J For Res 32:59–66

Jonsson L, Dahlberg A, Nilsson M-C, Zackrisson O, Kårén O (1999) Ectomycorrhizal fungal communities in late-successional Swedish boreal forests, and their composition following wildfire. Mol Ecol 8:205–215

Kaldorf M, Fladung M, Muhs H-J, Buscot F (2002) Mycorrhizal colonization of transgenic aspen in a field trial. Planta 214:653–660

Kantety RV, Zeng XP, Bennetzen JL, Zehr BE (1995) Assessment of genetic diversity in dent and popcorn (*Zea mays* L.) inbred lines using inter-simple sequence repeat (ISSR) amplification. Mol Breed 1:365–373

Kowalchuk GA, de Souza FA, van Veen JA (2002) Community analysis of arbuscular mycorrhizal fungi associated with *Ammophila arenaria* in Dutch coastal sand dunes. Mol Ecol 11:571–581

Kuhn G, Hijri M, Sanders IR (2001) Evidence for the evolution of multiple genomes in arbuscular mycorrhizal fungi. Nature 414:745–748

Lemes MR, Gribel R, Proctor J, Grattapaglia D (2003) Population genetic structure of mahogany (*Swietenia macrophylla* King, Meliaceae) across the Brazilian Amazon, based on variation at microsatellite loci: implications for conservation. Mol Ecol 12:2875–2883

Lilleskov EA, Fahey TJ, Horton TR, Lovett GM (2002) Belowground ectomycorrhizal fungal community change over a nitrogen deposition gradient in Alaska. Ecology 83:104–115

Liu W-T, Marsh TL, Cheng H, Forney LJ (1997) Characterization of microbial diversity by determining terminal restriction fragment length polymorphisms of genes encoding 16S rRNA. Appl Environ Microbiol 63:4516–4522

Lukow T, Dunfield PF, Liesack W (2000) Use of the T-RFLP technique to assess spatial and temporal changes in the bacterial community structure within an agricultural soil planted with transgenic and non-transgenic potato plants. FEMS Microbiol Ecol 32:241–247

Martin F (2001) Frontiers in molecular mycorrhizal research – genes, loci, dots and spins. New Phytol 150:499–505

Morgante M, Olivieri AM (1993) PCR-amplified microsatellites as markers in plant genetics. Plant J 3:175–182

Morton JB, Benny GL (1990) Revised classification of arbuscular mycorrhizal fungi (Zygomycetes): a new order, Glomales, two new suborders, Glomineae and Gigasporineae, and two new families, Acaulosporaceae and Gigasporaceae, with an emendation of Glomaceae. Mycotaxon 37:471–491

Nan P, Shi S, Peng S, Tian C, Zhong Y (2003) Genetic diversity in *Primula obconica* (Primulaceae) from central and south-west China as revealed by ISSR markers. Ann Bot 91:329–333

Oddou-Muratorio S, Houot M-L, Demesure-Musch B, Austerlitz F (2003) Pollen flow in the wildservice tree, *Sorbus torminalis* (L.) Crantz. I. Evaluating the paternity analysis procedure in continuous populations. Mol Ecol 12:3427–3439

Öpik M, Moora M, Liira J, Kõljalg U, Zobel M, Sen R (2003) Divergent arbuscular mycorrhizal fungal communities colonize roots of *Pulsatilla* spp. in boreal Scots pine forest and grassland soils. New Phytol 160:581–593

Orita M, Suzuki Y, Sekiya T, Hayashi K (1989) Rapid and sensitive detection of point mutations and DNA polymorphisms using the polymerase chain reaction. Genomics 5:874–879

Peter M, Courty P-E, Kohler A, Delaruelle C, Martin D, Tagu D, Frey-Klett P, Duplessis S, Chalot M, Podila G, Martin F (2003) Analysis of expressed sequence tags from the ectomycorrhizal basidiomycetes *Laccaria bicolor* and *Pisolithus microcarpus*. New Phytol 159:117–129

Pringle A, Moncalvo J-M, Vilgalys R (2003) Revisiting the rDNA sequence diversity of a natural population of the arbuscular mycorrhizal fungus *Acaulospora colossica*. Mycorrhiza 13:227–231

Pritsch K, Boyle H, Munch J-C, Buscot F (1997) Characterization and identification of black alder ectomycorrhizas by PCR/RFLP analyses of the rDNA internal transcribed spacer (ITS). New Phytol 137:357–369

Pupilli F, Labombarda P, Scotti C, Arcioni S (2000) RFLP analysis allows for the identification of alfalfa ecotypes. Plant Breed 119:271–276

Read DJ (1993) Mycorrhiza in plant communities. In: Tommerup IC (ed) Advances in plant pathology 9. Mycorrhizal synthesis. Academic Press, London, pp 1–31

Reboud X (2003) Effect of a gap on gene flow between otherwise adjacent transgenic *Brassica napus* crops. Theor Appl Genet 106:1048–1058

Rebourg C, Gouesnard B, Charcosset A (2001) Large scale molecular analysis of traditional European maize populations. Relationships with morphological variation. Heredity 86:574–587

Redecker D (2000) Specific PCR primers to identify arbuscular mycorrhizal fungi within colonized roots. Mycorrhiza 10:73–80

Redecker D, Kodner R, Graham LE (2000) Glomalean fungi from the Ordovician. Science 289:1920–1921

Redecker D, Hijri I, Wiemken A (2003) Molecular identification of arbuscular mycorrhizal fungi in roots: perspectives and problems. Folia Geobot 38:113–124

Renker C, Heinrichs J, Kaldorf M, Buscot F (2003) Combining nested PCR and restriction digest of the internal transcribed spacer region to characterize arbuscular mycorrhizal fungi on roots from the field. Mycorrhiza 13:191–198

Ritala A, Nuutila AM, Aikasalo R, Kauppinen V, Tammisola J (2002) Measuring gene flow in the cultivation of transgenic barley. Crop Sci 42:278–285

Rogstad SH, Patton JC, Schaal BA (1988) M13 repeat probe detects DNA minisatellite-like sequences in gymnosperms and angiosperms. Proc Natl Acad Sci USA 85:9176–9178

Rønning SB, Vaïlingom M, Berdal KG, Holst-Jensen A (2003) Event specific real-time quantitative PCR for genetically modified Bt11 maize (*Zea mays*). Eur Food Res Technol 216:347–354

Ryskov AP, Jincharadze AG, Prosnyak MI, Ivanov PL, Limborska SA (1988) M13 phage DNA as a universal marker for DNA fingerprinting of animals, plants and microorganisms. FEBS Lett 233:388–392

Saghai-Maroof MA, Soliman KM, Jorgensen RA, Allard RW (1984) Ribosomal DNA spacer-length polymorphism in barley: Mendelian inheritance, chromosomal location, and population-dynamics. Proc Natl Acad Sci USA 81:8014–8018

Salimath SS, de Oliveira AC, Godwin ID, Bennetzen JL (1995) Assessment of genome origins and genetic diversity in the genus *Eleusine* with DNA markers. Genome 38:757–763

Saltonstall K (2003) Microsatellite variation within and among North American lineages of *Phragmites australis*. Mol Ecol 12:1689–1702

Sanders IR (2002) Ecology and evolution of multigenomic arbuscular mycorrhizal fungi. Am Nat 160 (Suppl):S128–S141

Sanders IR, Clapp JP, Wiemken A (1996) The genetic diversity of arbuscular mycorrhizal fungi in natural ecosystems – a key to understanding the ecology and functioning of the mycorrhizal symbiosis. New Phytol 133:123–134

Scheffler JA, Parkinson R, Dale PJ (1995) Evaluating the effectiveness of isolation distances for field plots of oilseed rape (*Brassica napus*) using a herbicide-resistance transgene as a selectable marker. Plant Breed 114:317–321
Schmalenberger A, Tebbe CC (2002) Bacterial community composition in the rhizosphere of a transgenic, herbicide-resistant maize (*Zea mays*) and comparison to its non-transgenic cultivar *Bosphore*. FEMS Microbiol Ecol 40:29–37
Schmalenberger A, Tebbe CC (2003a) Genetic profiling of noncultivated bacteria from the rhizospheres of sugar beet (*Beta vulgaris*) reveal field and annual variability but no effect of a transgenic herbicide resistance. Can J Microbiol 49:1–8
Schmalenberger A, Tebbe CC (2003b) Bacterial diversity in maize rhizospheres: conclusions on the use of genetic profiles based on PCR-amplified partial small subunit rRNA genes in ecological studies. Mol Ecol 12:251–262
Schüßler A (1999) Glomales SSU rRNA gene diversity. New Phytol 144:205–207
Schüßler A, Schwarzott D, Walker C (2001) A new fungal phylum, the Glomeromycota: phylogeny and evolution. Mycol Res 105:1413–1421
Schüßler A, Schwarzott D, Walker C (2003) *Glomeromycota* rRNA genes – the diversity of myths? Mycorrhiza 13:233–236
Sessitsch A, Kan F-Y, Pfeifer U (2003) Diversity and community structure of culturable *Bacillus* spp. populations in the rhizospheres of transgenic potatoes expressing the lytic peptide cecropin B. Appl Soil Ecol 22:149–158
Shimizu Y, Ando M, Sakai F (2002) Clonal structure of natural populations of *Cryptomeria japonica* growing at different positions of slopes, detected using RAPD markers. Biochem Syst Ecol 30:733–748
Smith SE, Read DJ (1997) Mycorrhizal symbiosis, 2nd edn. Academic Press, San Diego
Sugano S, Kaminaka H, Rybka Z, Catala R, Salinas J, Matsui K, Ohme-Takagi M, Takatsuji H (2003) Stress-responsive zinc finger gene *ZPT2-3* plays a role in drought tolerance in petunia. Plant J 36:830–841
Tonin C, Vandenkoornhuyse P, Joner EJ, Straczek J, Leyval C (2001) Assessment of arbuscular mycorrhizal fungi diversity in the rhizosphere of *Viola calaminaria* and effect of these fungi on heavy metal uptake by clover. Mycorrhiza 10:161–168
Tremetsberger K, Stuessy TF, Samuel RM, Baeza CM, Fay MF (2003) Genetics of colonization in *Hypochaeris tenuifolia* (Asteraceae, Lactuceae) on Volcán Lonquimay, Chile. Mol Ecol 12:2649–2659
Ude GN, Kenworthy WJ, Costa JM, Cregan PB, Alvernaz J (2003) genetic diversity of soybean cultivars from China, Japan, North America, and North American ancestral lines determined by amplified fragment length polymorphism. Crop Sci 43:1858–1867
Van der Heijden MGA, Klironomos JN, Ursic M, Moutoglis P, Streitwolf-Engel R, Boller T, Wiemken A, Sanders IR (1998) Mycorrhizal fungal diversity determines plant biodiversity, ecosystem variability and productivity. Nature 396:69–72
Van Tuinen D, Jacquot E, Zhao B, Gollotte A, Gianinazzi-Pearson V (1998) Characterization of root colonization profiles by a microcosm community of arbuscular mycorrhizal fungi using 25S rDNA-targeted nested PCR. Mol Ecol 7:879–887
Vandenkoornhuyse P, Ridgway KP, Watson IJ, Fitter AH, Young JPW (2003) Co-existing grass species have distinctive arbuscular mycorrhizal communities. Mol Ecol 12:3085–3095
Vos P, Hogers R, Bleeker M, Reijans M, van de Lee T, Hornes M, Frijters A, Pot J, Peleman J, Kuiper M, Zabeau M (1995) AFLP: a new technique for DNA fingerprinting. Nucleic Acid Res 23:4407–4414
Wang X-R, Chhatre VE, Nilsson M-C, Song W, Zackrisson O, Szmidt AE (2003) Island population structure of Norway spruce (*Picea abies*) in northern Sweden. Int J Plant Sci 164:711–717
Watkinson JI, Sioson AA, Vasquez-Robinet C, Shukla M, Kumar D, Ellis M, Heath LS, Ramakrishnan N, Chevone B, Watson LT, van Zyl L, Egertsdotter U, Sederoff RR, Grene

R (2003) Photosynthetic acclimation is reflected in specific patterns of gene expression in drought-stressed loblolly pine. Plant Physiol 133:1702–1716

Welsh J, McClelland M (1990) Fingerprinting genomes using PCR with arbitrary primers. Nucleic Acid Res 18:7213–7218

White TJ, Bruns T, Lee S, Taylor J (1990) Amplification and direct sequencing of fungal ribosomal RNA genes for phylogenetics. In: Innis MA, Gelfand DH, Sninsky JJ, White TJ (eds) PCR protocols: a guide to methods and applications. Academic Press, San Diego, pp 315–322

Williams JG, Kubelik AR, Livak KJ, Rafalski JA, Tingey SV (1990) DNA polymorphisms amplified by arbitrary primers are useful as genetic markers. Nucleic Acid Res 18:6531–6535

Wyss P, Bonfante P (1993) Amplification of genomic DNA of arbuscular-mycorrhizal (AM) fungi by PCR using short arbitrary primers. Mycol Res 97:1351–1357

Xiang CC, Chen Y (2000) cDNA microarray technology and its application. Biotech Adv 18:35–46

Ziegenhagen B, Bialozyt R, Kuhlenkamp V, Schulze I, Ulrich A, Wulf M (2003) Spatial patterns of maternal lineages and clones of *Galium odoratum* in a large ancient woodland: inferences about seedling recruitment. J Ecol 91:578–586

Michael Kaldorf
Carsten Renker
François Buscot
Terrestrial Ecology
Institute of Botany
University of Leipzig
Johannisallee 21
04103 Leipzig, Germany
Tel.: +49-341-9738581
Fax: +49-341-9738599
e-mail: buscot@uni-leipzig.de

History of Flora and Vegetation During the Quaternary North America

Burkhard Frenzel

The 1998 and 2002 papers of this series (Frenzel 1998, 2002) were devoted to paleoecological problems of central and northern Asia, the article of the year 2000 to those of the tropical and subtropical zones. Thus, many very interesting papers on Quaternary paleoecology and vegetation history of the Americas have accumulated meanwhile. Only some of them can be dealt with here. Others will be discussed in the forthcoming issue of this series.

1 The Problem of Reliably Dating Paleoecological Events

A consequence of the relatively long duration of the Quaternary (ca. 2.6 million years) is that for different parts of this epoch different dating methods have to be used. All of them have drawbacks, and this is a well-known fact. To overcome these difficulties critically carried out methodological studies are needed. Ingram and Ingle (1998), when studying the marine Merced Formation off San Francisco, showed that strontium isotope ages of the foraminifer *Elphidiella hannai* can be used with some confidence in sediments that are some hundred thousand years old (standard deviation of ±0,1 million years). In older sediments the uncertainties increase considerably, so that the base of this formation gave ages of between 2.4 and 4.8 million years. Lamphere (2000) studied possible differences between the results of the conventional K-Ar dating method and those of the $^{40}Ar/^{39}Ar$ method used for dating young mafic volcanic rocks. Both these methods give the same results up to an age of ca. 700,000 years, but the data of the $^{40}Ar/^{39}Ar$ method repeatedly seem to be more precise. Swanson and Caffee (2001) made a very profound analysis of the ^{36}Cl production-rate method on rock surfaces that were freshly exposed during deglaciation. The results are said to be valid for the zone between 46.5 and 51° northern latitude. Differences to the results of other research teams are carefully discussed.

Concerning younger sediments, the radiocarbon method is of prime importance, yet it is a well-known fact that this method, too, is rich in traps.

Progress in Botany, Vol. 66

Andrews et al. (1999) investigated the AMS-age data of marine bottom-surface sediments of the Ross Sea, Antarctica. They found age differences ranging from 2,000 to 21,000 years. These differences may have been caused by the incorporation of older sediments into the surface samples, methodological difficulties, the geographical position of the sites studied, differing sedimentation rates or differences in the amount of reworked material, etc.

Colman et al. (2002), when studying accumulation rates of Holocene estuarine sediments from Chesapeake Bay, pointed out the fact that redeposition of older sediments causes series problems concerning the dating accuracy of near-shore marine sediments. To avoid these difficulties only the deepest parts of lakes or estuaries should be analysed.

Another problem was discussed by Abbott et al. (2000) when comparing AMS data of individual botanical samples with those of nearby sediment bulk samples in Birch Lake, central Alaska. It turns out that the bulk samples, which include material of different ages, repeatedly give older ages than the AMS data (from 3853 to 230 years).

Hall (1999) studied ^{14}C ages of modern soils and of the underlying parent material (loess). They discovered that the samples from near the soil surface do not show any correct age relations with those of the underlying material. It is thought that these discrepancies are caused by wind action and cryoturbation. Regrettably, this problem has not been studied at the same sites concerning well-developed fossil soils. Clapp et al. (2001) found that reliable ^{10}Be and ^{26}Al age data on rock surfaces can only be obtained if erosion and accumulation within the geomorphic basin are in equilibrium with each other. Thus, before dating land surfaces by these methods, comprehensive geomorphological research should be carried out.

The problem of the so-called reservoir effect when dating lake or marine sediments will be dealt with later on. In general, this effect is considered to be constant with respect to a certain water-filled basin. When studying an archaeological shell mound on the seashore of northern California, Ingram (1998) found that this reservoir effect had changed there over time. Remarkably, this was evidently caused by changing wind directions and wind strengths, thus causing differences in the strength of marine upwelling and river activities.

The problems discussed are a warning against a too uncritical comparison of absolute age data of different regions when, e.g., trying to synchronize deep-sea data with those of loess or glacial regions. Of course, these correlations are necessary to understand paleoecological consequences of atmospheric or oceanic driving factors, i.e. the problem of lead and lag in paleoecology. However, I feel that one should exercise much more caution when doing so than is being applied at present.

2 The Paleoclimatological Background of Changes in the Paleoecological Setting

In line with these sceptical remarks are the very interesting papers by Struck et al. (1993), Tzedakis et al. (1997), Sarnthein et al. (2000) and Shackleton et al. (2002). It is stressed that the dating quality of marine and terrestrial sediment sequences is very often poor. Moreover, it would be better to focus on the reasons for divergent age data or processes than on trying to synchronize past processes observed in the oceans and on the continents without taking into consideration that differences in the timing of important paleoecological events in the oceans compared to those on the continents (Tzedakis et al. 1997), or even within different regions of the oceans, may have existed (Struck et al. 1993). Besides all the difficulties in producing exact paleoclimatological data (Sarnthein et al. 2000), it might be dangerous to believe too uncritically in the correctness of the time scale given by Martinson et al. (1987), especially if it is not taken into consideration that the definition of the marine isotope stage 5e (132,000 to 115,000 b.p.) strongly differs from that used for the paleobotanical definition of the Last Interglacial on land (Shackleton et al. 2002). In this respect, it is interesting, too, that according to Herterich et al. (1992) the 3-D modelling approaches for describing the modern global climate system at present cannot be used successfully to better understand past climates. Smart and Richards (1992), relying globally on 300 age data of fossil coral reefs, stated that statistically age data for high sea stands can be established, however, the standard deviations of those of the Last Interglacial (129,000±33,000 and 123,000±13,000) are very long.

Toscano and York (1992), using amino acid racemization techniques for dating fossil sea levels along the mid-Atlantic coast of the United States, showed that during deep-sea stage 5e this level was some 5 to 6 m higher than it is at present (that of 5c is at the present level, that of 5a is at about 8 m below that of today). These data and those of the Last Glacial stadials and interstadials are helpful in trying to reconstruct the volumes of former inland ice masses. A vivid discussion has continued on the credibility of paleoclimatological reconstructions done on the Last Interglacial parts of the GRIP ice core in central Greenland, which are characterized by repeatedly occurring abrupt changes in ice structure and in stable isotopes. Johnsen et al. (1995) stated that the instability of the Last Interglacial climate deduced from these changes "is likely to be real, though no conclusive evidence is available". It is thought that meridional changes in the North Atlantic marine circulation might have been responsible, so that beyond the Arctic, traces of these changes will probably not be found. In this respect, it is interesting to see that neither Knudsen et al. (2002) nor

Cortijo et al. (1994), Bauch et al. (1999) or McManus et al. (2002) could observe any traces of these abrupt climatic changes when studying sediment cores of northern Denmark and the North Atlantic, respectively. On the contrary, sea-surface temperatures (SST) remained high in the North Atlantic during the Last Interglacial even when inland ice masses had already begun to grow. In this respect, it should be mentioned that Adkins et al. (1997) showed that the SST at Bermuda Rise experienced no remarkable changes during the Last Interglacial. These relatively stable conditions ended within about 400 years when very abruptly cold deep-sea waters moved to lower latitudes (35 to 36°N). Berger et al. (1992) used a 2-D coupled climate model to study the onset of the Last Glaciation. They found that changes in the solar radiation intensity, calculated for June and July between 60 and 70°N, preceded changes in the volume of inland ice masses by about 4,000 years, but that the model used was incapable of reconstructing the strong cooling in climate of those times. This could be improved if changes in the CO_2 content of the atmosphere were also taken into consideration. Raynaud et al. (1992) pointed out the diverging development in the CO_2 and CH_4 contents of air bubbles enclosed in Antarctic inland ice dating from the transition from the Last Interglacial to the Last Glaciation. CH_4 is thought to have originated mainly in the continents, whereas CO_2 is believed to have been produced by the oceans (see another interpretation regarding CO_2 in Frenzel 2001). Kukla et al. (2002), when using climate modelling techniques, are of the opinion that during the beginning of the Last Glaciation the number of ENSO (El Niño-Southern Oscillation) events in time might have been twofold greater than those of the Last Interglacial and it is speculated that this fact might have strengthened the growth of ice masses at higher latitudes. Yet, I feel that up till now nobody has been able to show either the frequency or the intensity of these events during the Last Interglacial.

According to Clark et al. (2002), abrupt climatic changes during the Last Glaciation are thought to have been caused by changes in the Atlantic thermohaline circulation, which in turn was influenced by small changes in the hydrological cycle and by several feedback mechanisms, which strengthened the impact. Also, Broecker (2000) stresses that the chronological sequences of changes in inland ice budgets and in sedimentation rates were influenced by changes in the seasonality of climate, which in turn depended on changes in the orbital parameters. Yet these factors were faint, only. So strengthening, yet still unknown, mechanisms must have existed, which should be considered. Stuiver and Grotes (2000) comprehensively reported on the amplitude and velocity of changes in the oxygen isotope ratios in Greenland ice cores. An ^{18}O curve with a high resolution in time is given for the last 90,000 years. This curve shows the velocity (some years to some decades, only) of past climatic changes. Blunier et al. (1998) thought that the Dansgaard-Oeschger events, so characteristic in the northern Atlantic Ocean, were triggered by changes in temperatures of the Antarctic region. It is stressed that between 45,000 and 23,000 b.p., changes in temperature in Greenland were delayed compared with those of the Antarctic by about 1,000 to 2,500 years. However, is the dating quality sufficient for drawing this conclusion?

Hesse and McTainsh (1999) investigated the contents in eolian sediments in two deep-sea cores of the Tasman Sea. During the Last Glacial Maximum (LGM), wind strengths evidently had not increased there compared to modern times, but much more sediment dust could be transported, because over 40% of the land surface of Australia active dunes were migrating at those times. Bush (2001), according to climate modelling, found that changing SSTs as well as changing orbital parameters influenced the Southeast Asian monsoon equally intensively. This does not seem to be an unexpected result. Haberle and Ledru (2001) investigated the amount of charcoal in sediment cores from ten localities situated between Indonesia in the west and Central and South America in the east. It is thought that a broad synchronicity of fire events existed in these regions during times of rapid climatic change since about 16 000 calibrated (cal) years b.p. due to the Walker circulation of the tropical to subtropical atmosphere. The data reported show higher fire intensities at the sites studied at about 13,500, 11,500 and about 18,000 to 15,000 cal years b.p. and during the Upper Holocene. Several papers, which are dealt with later, show the same, at least with respect to the Late Glacial and the Early Holocene fire frequencies. Evidently, these fires have to be taken into consideration when studying paleoecological processes of those times.

A very interesting point is made by Cowling and Sykes (1999) when discussing the paleoclimatological interpretation of Last Glacial to Holocene pollen diagrams. The authors feel that the changing shares of C_3 and C_4 plants should be taken more into consideration than has been done up till now. For discussion, see Cowling and Sykes (2000) and Williams et al. (2000). I feel that it is much more important when reconstructing quantitatively past climates to use many independent indicators of climate, rather than to rely on types of vegetation or on past distribution patterns of systematically poorly defined taxa only (see Frenzel 1980).

3 Timing and Extent of Former Glaciations

A knowledge of timing and extent of past glaciations is crucial to understand well the history of various plant taxa and vegetation types in different parts of the Americas (see Table 1 for references).

It is worthwhile noting that in the Americas clear traces of Early and Middle Wisconsinan Glaciations could be repeatedly found in quite different regions of this huge continent, though in Europe these traces generally seem to have been overlooked or misinterpreted (for references of comparable observations in Europe, see Frenzel 1991, 2004).

4 Modern Eolian Pollen Transport

Of crucial importance for the correct interpretation of pollen diagrams is the knowledge of recent pollen transport and pollen deposition in various environments. This problem has already been dealt with in this series repeatedly. Some more recent observations may be mentioned.

According to Linskens and Jorde (1986), the pollen flora that was entrapped during the trip of a gas balloon over northwestern Germany did not reflect the terrestrial vegetation of the region passed over. Moreover, Linskens investigated the modern pollen flora of northwestern Greenland (Linskens 1995) and Severnaya Zemlya (Linskens 1996) used surface samples. As already known (see Frenzel 1969), the share of long-distance transport there is very high in general, with arboreal pollen transported to more than 2,100 km. It is interesting that driftwood, which in several regions of the Arctic is very common, is repeatedly accompanied by a relatively high amount of exotic pollen grains. Thus one should be careful when studying the pollen flora of driftwood there, because it not only contains sediments of Arctic or glacial climates. Comparable findings were made in various parts of Antarctica and adjacent islands (Linskens et al. 1991, 1993; Linskens 1992). These studies showed that there even quite strange *mixturae mirabiles* can contain pollen from remote areas. Kvavadze (2000) investigated recent pollen spectra between 1,356 and 2,001 m a.s.l. in the Chernogora (Ukrainian Carpathians). Although the profiles studied crossed the timberline, this phytogeographically important borderline could not be observed in the arboreal pollen flora (AP) nor in the sum of the nonarboreal pollen flora (NAP). According to my own experience, this can only be done by scrupulously analysing a great number of NAP. The same difficulties were encountered by Bush (2000) when studying modern pollen rain at 80 localities in Panama and Costa Rica, situated between 1,500 and 4,300 m a.s.l. The aim was to establish climatological response matrices

Table 1. Papers dealing with timing and extent of mountain and low-land glaciations in the Americas. *LGM* Last Glacial Maximum

	Reference	Region	Remarks
1. Early and Middle Pleistocene	Westgate et al. (2001)	Central Yukon	Oldest glacigene sediments ca. 1,500,000 years b.p., most of them since about 250,000 years b.p.
2. Last Glaciation	Briner and Kaufman (2000)	Southwest Alaska	Maps of glaciated areas during an important Early Wisconsinan glaciation and during Late Wisconsinan glaciations
	Briner et al. (2001)	Southwestern Alaska	The same
	Lamoureux and England (2000)	Central sector of the Canadian High Arctic	LGM
	Winograd (2001)	Canada	Ice-sheet growth: timing, volumes, causes
	Karrow et al. (2000)	Great Lakes region	Early, Middle and Late Wisconsinan glaciations
	Hemming et al. (2000)	Canada	Iceberg discharges 22,000–10,500 years b.p.
	Barrie and Conway (1999)	Canadian northern Pacific margin	Map of glaciated areas during the Late Wisconsinan, even in regions that were formerly believed to have been glacial refuge areas of various plant and animal taxa
	Porter and Swanson (1998)	Puget Sound area near Seattle	Maps of glaciated areas during various phases of the Last Glaciation, ice volumes, velocities of retreat

Table 1. *Continued*

Reference	Region	Remarks
Thackray (2001)	Olympic Peninsula	Strong Early and Middle Wisconsinan glaciations; during the Late Wisconsinan climate appreciably dry
Benson et al. (1998b)	Sierra Nevada	Several glacier advances between 52,600 and 12,500 years b.p.
James et al. (2002)	Sierra Nevada	Strong Early and Late Wisconsinan glacier advances
Bischoff and Cummins (2001)	Sierra Nevada	Several glacier advances during the Early, Middle and Late Wisconsinan
Orvis and Horn (2000)	Cerro Chirripó, Costa Rica	Maps of the Late Wisconsinan and Late Glacial glaciated areas. Carefully done calculations of position of the 'snowline' (ELA=equilibrium line altitude) during various stades of the Last Glaciation
Goodman et al. (2001)	Southeastern Peru	Most important Wisconsinan glaciation at 70,000 years b.p.; strong LGM glacial advance
Mark et al. (2002)	Southeastern Peru	Calculation of the ELA for various stades of the Last Glaciation, since the LGM
Benn and Clapperton (2000)	Strait of Magellan	Maps of present-day glaciation and of those of the LGM. Timing and temperatures of the Late Pleistocene glaciations

Table 1. *Continued*

	Reference	Region	Remarks
	Marden and Clapperton (1995)	South Patagonia	Maps of glaciated areas during the LGM, Late Glacial and during various phases of the Holocene. Younger Dryas glacier advance
3. The Holocene	Begét and Motyka (1998)	Southeastern Alaska	Dating of a Late Glacial volcanic eruption, the tephra of which can serve as an age level
	Leonard and Reasoner (1999)	Banff National Park, Alberta	Late Glacial and Holocene glacier history together with forecasts concerning development if temperature increase
	Hallett et al. (2001)	Southern British Columbia	Maps of glaciated areas during the Holocene
	Orvis and Horn (2000)	Cerro Chirripó, Costa Rica	History of Holocene glacier advances
	Strelin and Malagnino (2000)	Argentina	Maps of Late Glacial and Holocene glacier advances in the Lago Argentino region
	Goodman et al. (2001)	Southeastern Peru	Holocene glacier history
	Mark et al. (2002)	Southeastern Peru	Holocene glacier history, equilibrium line altitudes as indicators of climate
	Wenzens (1999)	Southern Andes of Argentina	Maps of glaciated areas during Late Glacial and Holocene times (eight Holocene glacier advances)

from modern pollen floras. In the end, only 17 pollen types could be used for this purpose.

5 The History of Flora and Vegetation

5.1 Alaska and Northernmost Canada

Important new findings on the vegetation history at the transition from the Last Interglacial to the Last Glaciation in southwestern Alaska (surroundings of the Ahklun Mountains) were reported by Kaufman et al. (2001b). During the Last Interglacial, when climate was warmer than it is today, vegetation of the forest-tundra ecotone was thriving there with an appreciably higher share of *Picea* than is found today. With the onset of the Last Glaciation, the sea level retreated and a tundra-like vegetation spread, followed by a forest tundra and finally a tundra vegetation, before between 70,000±10,000 and 53,600±2,000 a glacier covered the sediments. At about 35,000 to 30,000 years b.p. a warming of the climate was felt in eastern Beringia, but it was not intensive enough to permit conifer forests or stands to spread from the Yukon Territory to the west. Fossil beetle assemblages show the ensuing harsh climatic conditions during the LGM (Elias 2000). At this time (about 21,500 cal years b.p.) pedogenesis was very faint on the northern Seward Peninsula, where permafrost had been very active. It has been reported that the vegetation here in those times resembled neither modern cold-climate steppes nor polar deserts, but may have been related to the vegetation of dry sites of the arctic coast of northernmost Yakoutia (Höfle et al. 2000). At the end of the LGM or at the very beginning of Late Glacial, according to lake-level reconstructions of Birch Lake, central Alaska, the climate was evidently extremely dry and cold, because the lake dried up either seasonally or even completely for longer times (Abbott et al. 2000). History of the strongly oscillating level of Birch Lake has also been investigated carefully, and it is stressed that for the Holocene it is very often impossible to differentiate between influences of climate and of vegetation on the paleohydrology of the lake. Concerning the westernmost Yukon Territory, Pienitz et al. (2000)have described mesosaline lake conditions at about 11,000 to 8,100 ^{14}C years b.p. This fits quite well into the picture just given. It is interesting to note that during the Holocene the salinity of the lake seems to have changed repeatedly. This does not always correspond to changes in hydrological conditions of other parts of this zone. Yet probably local conditions have to be taken into consideration. Comparable studies were undertaken by Hu et al. (1998) concerning a lake in the northwestern Alaska Range. Here, pollen analytical data are combined with a trace-element analysis of ostracod shells. The time between 11,000 and 9,000 ^{14}C years b.p. is said to have been cold and dry. Yet, according to former publications, this time was considered to have been a climatic optimum, because *Populus* cf. *balsamifera* had spread simultaneously. It can be shown now that according to the contents of trace elements in ostracod shells the time between 8,500 and 8,000 ^{14}C years b.p. was the warmest, followed at about 6,500 years b.p. by increasing moisture. *Betula* followed *Populus* cf. *balsamifera* and *Salix*, when synchronously the share of Poaceae and *Artemisia* had decreased. This phase in the development of vegetation was followed by a decrease in *Betula* and increases in the share of *Alnus* and *Picea* pollen: this sequence strongly resembles a normal succession from heliophytes to sciadophytes. Yet, the time needed for this development (from about 11,000 to ca. 6,500 years b.p.) was much longer than a normal succession would need. Thus climate and migrating facilities as well as the positions of Last-Glacial refuge areas must have been of prime importance. In the Central Brooks Range, Alaska, lake history was analysed by Anderson et al. (2001) using oxygen isotope ratios in lake sediments. Only the time from about 8,500 cal years b.p. onwards could be studied.

At first, climate was evidently drier than it is nowadays. Yet, from 6,000 cal years b.p. onwards, moisture increased when temperatures decreased. It is said that sometimes during colder episodes loesses accumulated. Thus the vegetation cover must have been very open. On southwestern Victoria Island, Canadian Arctic Archipelago, driftwood seems to have accumulated from about 4,700 years b.p., and from about 2,000 years b.p. its quantity increased. Evidently, larch trunks came from northern Siberia, *Picea* from the Mackenzie and Yukon areas. It may be possible that even older driftwood had already been used by paleo-Eskimos (Dyke and Savelle 2000). As to the history of the Bering Sea Bowhead Whale there, see Dyke and Savelle (2001). The glacier and forest history of southeastern Alaska was studied dendrochronologically by Wiles et al. (1999) and a very comprehensive report on the dendroclimatology of the northern hemisphere together with a wealth of most interesting maps are presented by Briffa et al. (2002a,b). The maps seem to show the migrations to and fro of the Rossby waves (see Frenzel 1975). Lake-level changes at the boreal to subarctic ecotone in northern Québec were analyzed dendroclimatologically by Bégin (2001).

Zdanowicz et al. (2000) investigated the atmospheric dust deposition on Penny Ice Cap, Baffin Island. Evidently, here, too, Late-Glacial climate was much drier and it was much windier than at present. Because the grain sizes of the minerogenic particles in Penny Ice Cap are finer than those in the GISP2 ice core of central Greenland, different dust sources must have existed, which regrettably cannot be located more precisely for that time.

According to Bourgeois et al. (2000), the pollen contents and pollen influx rates in the Agassiz Ice Cap on Ellesmere Island experienced comparable changes to those that have just now been described concerning the deposition of dust on Penny Ice Cap. Although the dating quality is not equally good in all parts of a 127-m-long ice core, it seems that the pollen concentration between 11,500 and 6,000 cal years b.p., mostly of *Pinus*, *Picea*, *Betula*, Ericaceae, Cyperaceae and Poaceae, has been remarkably high, evidently indicating a very intensive long-distance pollen transport, which was much less from 6,000 years b.p. up till about 3,000 years b.p. It seems to have increased again remarkably between 1,600 years AD and modern times. Regrettably, the reasons for the changes in pollen concentration after 6,000 years b.p. are not discussed thoroughly.

5.2 The Rocky Mountains

Bennett et al. (2001) investigated the diatom and pollen floras together with some aspects of the former chemistry of Big Lake in British Columbia. It is interesting to note that neither the diatom nor the pollen assemblages of between 12,900 and 8,400 cal years b.p. have modern analogues. By about 8,400 years b.p. the fresh-water lake became saline when a steppe vegetation spread. Only at about 7,500 cal years b.p. *Abies* seemed to have immigrated into the area. The ensuing changing hydrological and geobotanical conditions have been studied thoroughly. On the eastern flanks of the Canadian Rockies, Pine Lake seems to reflect changes in eolian dust and sand import since about 3,540 years b.p., which shows the changing position of the forest-steppe ecotone there during the Late Holocene (Campbell 1998).

The Rocky Mountains are strongly influenced by tectonic movements. So it is of some interest to learn from Benson et al. (2001) how forest vegetation in an estuarine position reacted to these processes.

Research has strongly concentrated on paleohydrological and paleoecological conditions of the intramontane dry basins of the US Rockies. Here repeatedly in the past huge lakes evidently existed, the timing of which seems to be equivocal. Formerly it was held that during the last 780 000 years former lake Bonneville experienced 17 phases of deep lakes and ensuing shallow lakes and of pedogenesis. Yet, Oviatt et al. (1999) could find traces of four lake phases only, which are considered to have corresponded to the marine isotope stages (MIS) 2, 6, 12 and 16. Ku et al. (1998) and Yang et al. (1999) investigated the same deep geological boring in Death Valley, California. Here, the depth of the fossil lake has changed repeatedly over the last 200,000 years. Yet, these changes evidently did not occur in line with those of neighbouring lakes. The reason for this discrepancy is unknown. In Death Valley, a deep fossil lake is said to have existed between 186,000 and ca. 128,000 years b.p., which is believed to have become shallow until about 120,000 years b.p., i.e. during the Last Interglacial (Hooke 1999). Enzel et al. (2002) stress that in this case lake levels of quite different ages, even of Late Wisconsinan age, have been confused with each other (see Hookes reply: 2002). On the other hand, four erosion-accumulation phases were reconstructed in the western Grand Canyon, Arizona, dating from more than ca. 603,000±8,000 to between 250,000 and 100,000 years b.p., pointing to the remarkable tectonic instability of this mountain area (Lucchitta et al. 2000). According to Reheis (1999), who had mapped the former lakes between 38 and 42°N and 115 and 120°W, the size of the fossil lakes has decreased continuously from Middle to Late Pleistocene times. However, because comparable tendencies cannot be found in deep-sea sediments of the eastern Pacific, it is thought that within the lake region the effective moisture must have decreased, caused by changes in local climate, tectonics and/or configuration of the catchment areas. According to stable isotopes in pocket gopher teeth (Microtinae) found in the southern Rocky Mountains, during the Late Matuyama period (before ca. 740,000 years b.p.), several changes existed between phases characterized by pine pollen (MIS 21, 23 and 25) or by *Artemisia* (MIS 20, 22, 24 and 27). This means that the glacial periods there were characterized by a steppe- or desert steppe-like vegetation. Earliest traces of a stronger mountain glaciation were found in MIS 22, i.e. at about 820,000 years b.p. (Rogers and Wang 2002).

In these regions, much paleoecological interest has focused on the Upper Quaternary. Kaufman et al. (2001a) described traces of a deep fossil lake along the Bear River, Utah, which existed at about 59,000±5,000 years b.p. (TL and IRSL data). It is said that also the Gulf of Mexico experienced at about 60,000 years b.p. a strong meltwater influx, which evidently had originated in inland ice masses of the upper Mississippi region. It is believed that the ice volumes of that time were between two-thirds and three-fourths of those of the LGM. This lake phase was followed repeatedly during the Last Glaciation by younger lake phases (Anderson et al. 2002). Much interest has concentrated for a long time on the history of fossil lakes Bonneville, Utah, Mono, California and Lahontan. Liddicoat and Coe (1998) suggest that the sediments of Lake Bonneville had begun to accumulate at about 18,000 or 19,000 years b.p., yet Sack (1999) found that the history of Lake Bonneville was much more complicated than was thought previously, and Waters and Ravesloot (2000) describe the strongly changing activity of the Gila River and pedogenesis there. The general consequence is that this modern semidesert region has experienced much moister climates during the Last Glaciation, though the amount of moisture available has changed repeatedly during those times. The climatological backgrounds of these changes in moisture are discussed by Benson et al. (1998a) and Lin et al. (1998). It is tempting to interpret these changes in terms of Dansgaard-Oeschger cycles in Greenland or by the former position of the jet stream. Yet, it may be questioned whether these correlations are well founded simply in view of the difficulties in exact datings. Finally, Broecker et al. (1998) compared these paleoclimatological changes with those of east-central Africa.

The fish fauna of Lake Bonneville was thoroughly analyzed by Broughton et al. (2000), based on 14,866 fish remains. At 11,300 years b.p. the remains point to repeatedly occurring periods of very low lake levels. Yet, the entire Late Pleistocene fish fauna there reappeared during the equivalent of the Younger Dryas Period. At about 10,400 years b.p. fishes disappeared completely. At the same time, the richness in small rodent species decreased remarkably (Grayson 1998). According to Smith and Betancourt (1998), the body size of the wood rat *Neotoma cinerea* seems nearly everywhere to have been reduced prior to the change in climate to drier conditions. It is discussed whether this was caused by wood rats being more temperature-sensitive than most of the plants thriving there or whether winter temperatures had increased earlier than the summer temperatures: wood rats are very sensitive to winter temperatures.

According to Vierling (1998), the vegetation history of central Colorado was divided during Late Glacial and Holocene times into three phases: between 11,800 and 9,100 years b.p. an *Artemisia* steppe with stands of *Pinus*, *Picea* and cf. *Juniperus* followed an earlier *Artemisia* steppe rich in various herb taxa, yet nearly without any trees. The period from 9,100 to 1,800 years b.p. was characterized by small areas of *Artemisia*Poaceae communities, but with appreciably dense *Pinus-Picea* forests, which finally were replaced by a steppe vegetation rich in *Artemisia* and with a decreasing share of Poaceae. This is interpreted by the climate having become colder and drier, yet did man not exist there?

In general, the Late Pleistocene to Holocene vegetation history seems to have been quite different in the different tectonical basins of this mountain system. Doerner and Carrara (2001) studied the vegetation history of one of these tectonic basins, that of Long Valley, west-central Idaho. The peat bog analyzed is situated within the end moraine systems of the Pinedale Glaciation, which according to the thickness of weathering rinds on erratic boulders is thought to date from about 20,000 to 14,000 years b.p.. Beyond these end moraines loesses are situated. To me this indicateds a very open steppe vegetation for that time. This view is corroborated by the pollen flora of sediments dating from ca. 16,500 to 12.200 ^{14}C years b.p.: nearly no AP was found, yet pollen of *Artemisia*, together with Chenopodiaceae/Amaranthaceae, Rosaceae, *Salix* and *Selaginella densa*, was at 60 to 70%. Between

12,200 and 9,750 years b.p. the oldest peat layers were formed when AP was about 90%, most of all *Pinus*, together with some *Picea* and even with *Abies*. Between 9,750 and 3,200 years b.p. tree pollen decreased somewhat (from 90 to 73%). *Pinus* still dominated together with *Picea* and *Abies*. The amount of NAP, most of all of *Artemisia*, *Ambrosia*, *Sarcobatus* and Ranunculaceae, increased. The share of water plants was relatively high. Finally, AP increased again remarkably; even that of *Picea* did so, yet Cyperaceae and water plants retreated. All these changes were interpreted climatologically, only. Another sequence of events is described from the surroundings of Owens Lake in eastern California (Mensing 2001). At present, the lake is surrounded by a desert-steppe vegetation, yet on the flanks of the nearby mountains pine and fir forests very rich in various species of these genera are characteristic. From the palynological data it is concluded that between 16,200 and 11,750 cal years b.p. at first *Juniperus* was quite important (30%), which since about 15,000 cal years b.p. retreated rapidly, having been replaced by *Artemisia* and Chenopodiaceae, some *Ambrosia* and Poaceae. Between 11,750 and 11,200 years b.p. the share of desert vegetation increased remarkably, and steppe and forest plants retreated strongly. A short phase with increasing *Juniperus* pollen (11,200 to 11,000) was followed from 11,000 to 7,850 cal years b.p. by Chenopodiaceae-*Artemisia* steppe vegetation. Traces of a Late Glacial to Early Holocene amelioration of climate like those in the Long Valley could not be found here.

On the other hand, Davis (1999) investigated the vegetation history of the hypersaline Mono Lake in California. Here, since Late Glacial times, the pollen spectra were nearly always dominated by *Pinus* (up to 90%), *Artemisia* (10–20%) and by a relatively high percentage of Cupressaceae pollen, together with some *Tsuga mertensiana*, *Quercus* and *Sequoiadendron*. The share of NAP was nearly always very low. Since about 2,000 years b.p., modern pollen flora has become established. According to modern pollen spectra, past precipitation rates and temperatures were estimated. Yet, to me, it seems that the pollen flora mentioned does not reflect past vegetation, but rather it is the consequence of pollen-preserving conditions in a hypersaline lake, the surroundings of which do not produce much local pollen so that long-distance pollen is overrepresented. This very strange situation caused by highly salty soils or sediments has already been described very comprehensively in the Russian literatur (see Frenzel 1973, pp. 194, 195). These paleoecological conditions, which are so difficult to reconstruct correctly, were investigated by Connin et al. (1998) in the south-western United States using paleontological data ($\delta^{13}C$ of dental enamel of *Mammuthus* spp., *Bison* spp., *Equus* spp. and *Camelops* spp.). Nearly no fossils there date from the LGM. This seems to favour the view that the bioproduction from that time must have been very low. Yet, the number of these animals increases considerably in Late Glacial sediments. It seems that at that time the amount of grazed/browsed C_4 plants increased in an eastern direction. The antilopides are said to have always preferred C_3 plants. Hockett (2000) analyzed nearly 70,000 bone remains of vertebrates and lizards found in Pintwater Cave, southern Nevada. They date from between 32,000 and 7,350 years b.p. It is said that the fauna from between 32,000 and 10,100 years b.p. points to a mixture of xeric and cool-mesic taxa, yet the dating quality is very poor (only one ^{14}C date). It is thought that from 10,100 years b.p. onwards summer precipitation and temperatures increased. In the Southern Great Plains at sites where springs are active, in sediment profiles black mats can be found which originate in just these springs. The ostracod fauna as well as the geochemistry and sedimentology were studied (Quade et al. 1998).

According to the $\delta^{13}C$ values of the organic material dating from between 11,800 and 6,300 years b.p., both C_3 and C_4 plants seem to have lived there, with C_3 plants dominating. Between 6,300 and 2,300 years b.p., due to an arid climate springs do not seem to have been active. The oldest black mat dates from 11,800 to 11,600 ^{14}C years b.p.; most were formed at about 10,000 years b.p. It is thought that the aquifers were filled with water from at least 11,800 years b.p. Thus, this process must have begun at least 400 years in advance of the onset of the Younger Dryas Period. In Long Valley, north-central Nevada, during parts of the last glaciation, a pluvial lake existed, which was drained during Late Glacial times, followed between 11,000 and 9,800 ^{14}C years b.p. by river aggradation, then by a drier climate, which caused dunes to migrate there from 8,000 years b.p. The rich fauna of all these times is analyzed in a very detailed way by Huckleberry et al. (2001). Regarding Late Holocene activity phases of debris flows as indicators of climate in the Grand Canyon, Arizona, see Hereford et al. (1998); for climate variability of the last 1,000 years in east-central California, analyzed by using geochemistry data of sediments of Owens Lake, see Li et al. (2000); and for the Holocene fire history in Montane Meadows, Sierra Nevada, see Anderson and Smith (1997). Here, charcoal maxima occurred between 9,200 and 8,700, at 6,000, and since 4,500 to 3,800 years b.p with very short interruptions only. This sequence of events is mostly interpreted in a climatological way. To me it is relatively difficult, often, to follow these interpretations. However, regarding the last 4,500 years, the authors state that "human influence cannot be ruled out". Sharpe (2002) very comprehensively studied the plant remains in packrat (*Neotoma* spp.) middens in Dinosaur National Monument, Colorado. The aim of this investigation was to decipher climate history there quantitatively; 93 botanical taxa were identified. From 9,870 to 8,460 years b.p., remains of *Picea pungens, Juniperus communis, J. scopulorum, Pinus flexilis* and *Pseudotsuga menziesii* were found. The problem seems to be to differentiate here between the prevailing vegetation, on the one hand, and some scattered relic taxa, on the other. Holliday (2000) investigated climate history, bioproduction of the spontaneous vegetation and dune activity in the Southern High Plains, in the time of the Clovis Culture (ca. 11,200 to 10,900 years b.p.) and of the ensuing Folsom Culture (10,900 to 10,200 years b.p.). During the Clovis Culture, perennial rivers occurred in the study area, yet during the Folsom Culture lakes shrank, sands spread in the uplands and dunes formed.

Taking all the above-mentioned data into consideration, it seems that different regions of this geomorphologically very diverse mountain area behaved climatologically in quite a different way, provided paleoclimate had been reconstructed correctly, without overlooking local conditions.

Within an evidently very narrow belt between this dry region and the Pacific Ocean, during the LGM and the MIS 3, forest communities of various types existed (western Oregon: Grigg and Whitlock 1998; Grigg et al. 2001). Evidently, in general, these forests strongly resembled modern natural forest communities of different altitudinal zones of the western Rockies. Only the forest types that had thrived there during the LGM have had no modern equivalents. Since macrofossils of the tree taxa involved generally lack taxonomic data, they are given according to the paleoecological situation met with. This may turn out to be somewhat dangerous, especially since even here the amount of available moisture seems to have been strongly reduced in comparison to modern conditions (Dugas 1998). Due to the difficulties in exact dating of Last Glacial phases of climate and vegetation history, new age data on volcanic tephra layers, which might serve as additional stratigraphic markers, are welcome (King et al. 2001).

Various aspects of Holocene paleoclimatology and paleohydrology in the coastal regions of California were investigated by Ingram (1998), Goman and Wells (2000) and Byrne et al. (2001). Investigating the amount of charcoal particles in the Santa Barbara Basin, California, Mensing et al. (1999) convincingly showed that fire has been a naturally occurring factor in the ecology of the chaparral vegetation there (since 1,425 years AD there have been 20 large fires, the periodicity of which did not change remarkably during this time). Yet, here again, the possible activity of the former Indians has not been taken into consideration. The Late Pleistocene history of juniper woodlands and chaparral shows quite remarkable differences to the modern geobotanical situation there (Rhode 2002). According to plant remains in *Neotoma* middens, dating from the final parts of the Younger Dryas Period in central Baja California, geobotanical conditions must have existed there which occur at present some hundred kilometres farther to the north. Evidently, at that time, woods and chaparral thrived where nowadays the sarcophyllous and sarcocaulescent Sonoran desert vegetation is found. Thus annual precipitation must have been some 125 mm higher than it is in modern times and the mean annual temperatures were some 5 to 6 °C lower than they are today. According to Socorro-Lozano-García et al. (2002), chaparral seems to have begun to occur in Baja California at about 34,000 to 22,000 years b.p., following pine-juniper woodlands, which seem to have been widespread between about 44,000 to 34,000 years b.p. Chaparral and open woodland, mostly composed of *Juniperus* and *Pinus*, yet with appreciably low NAP values, was a very characteristic element in the San Felipe Basin between 22,000 and ca. 13,000 years b.p. This fits quite well with the data just mentioned (Rhode 2002). The paleoclimatological background of the occurrence of woodland and chaparral there and in New Mexico was investigated by Krider (1998). However, it seems that Early Holocene pluvial conditions cannot be proven.

The Late Holocene history of *Pinus torreyana*, endemic to Torrey Pines State Reserve, California, was studied by Cole and Wahl (2000). Evidently, this pine spread remarkably from about 2,100 years b.p. For the Late Holocene frequency and paleoecological consequences of earthquakes along the North American Pacific coast, see Polenz and Kelsey (1999), Sherrod et al. (2000) and Williams and Hutchinson (2000). For the Late Holocene East Pacific sea surface temperatures, precipitation and river runoff, see Jones and Kennett (1999) as well as Schimmelmann et al. (1998).

5.3 Central North America

During deglaciation of the Last Glaciation, huge ice-dammed lakes were formed in northern North America, which should have been of great paleoecological importance. The largest of these lakes was Lake Agassiz. Leverington et al. (2000, 2002) studied its configuration, bathymetry and volume in the times 11,000, 10,700, 10,300, 10,200, 9,900, 9,400, 9,300, 9,200, 8,900, 8,500, 8,200 and 7,700 years b.p. The results are given in a number of most interesting maps. On the other hand, Marshall and Clarke (1999) modelled North American fresh-water runoff during the Last Glacial cycle. The results are also given in maps. It is said that, although a net reduction in precipitation occurred during the times studied, the continental runoff was two to three times greater than that of today. This is strange, because a wealth of data exists showing that during the stadials of the Last Glaciation precipitation and soil moisture were much less in huge areas of North America than they are at present. Regrettably, these data were not discussed by the authors. Campbell et al. (2000) studied the chronological sequence of the beginning of Holocene paludification in 71 paludified peatlands of western interior Canada. It seems that most data concentrate on 9,500,

8,500, 7,000, 5,200 and 3,700 cal years b.p. From these and other data, a periodicity of ca. 1,450 years is calculated, which is compared with age data of CO_2 maxima at Taylor Dome, Antarctica. The authors state: "This demonstrates that global carbon-budgets are sensitive to small climatic fluctuations". However, to test this assumption, would it not be better to study the history of all of the major peat-bog areas of the globe? As to younger ground- and surface waters in thick Pleistocene sediments of the Lake Huron/Lake Erie region, see Cumming and Al-Aasm (1999).

In order to understand the timing and the distribution pattern of Last Glacial dry climates, loesses and continental dunes are very significant. There is a widely accepted view that the Last Glacial North American loesses originated in the huge meltwater plains coming from the inland ice masses. However, Mason (2001) very convincingly showed that the Peoria loess of Nebraska (ca. 25,000 to 11,000 years b.p.) was blown in from the northwest, because the thickness of this loess layer strongly decreases from the northwest to the southeast, independently of the big meltwater rivers. The same is shown by Muhs and Bettis (2000) when investigating grain sizes and geochemistry of the Peoria loess in western Iowa. The authors stress that these winds coming from the west contradict the results of general circulation models, which for that time had reconstructed winds coming from the east or even northeast. This west-northwesterly wind direction of Last Glacial times was reconstructed by Marsh (1998), too, studying the former wind relief in Pennsylvania, where huge and long earth ridges were formed by these winds. Thus vegetation must have been very open at those times, and it is thought that migrating dunes blocked the North and South Platte Rivers in Nebraska at about 12,000 years b.p. (Muhs et al. 2000). Even in the Gulf Coast region, southern Louisiana, widely spread loesses were formed at about 27,500 years b.p. and most of them between 17,200 and 15,600 years b.p. Yet, when during Late Glacial times the Mississippi River drained huge meltwater masses, no loesses were formed there (Otvos and Price 2001). As to the magneto-stratigraphic division of the Peoria loess, see Grimley et al. (1998).

Curry (1998), studying the history of the Mississippi River from about 40,000 years b.p. to the LGM interprets the then existing ostracod fauna occurring at about 91°W and between 39 and 43°N as having existed under climatic conditions that can be found at present in the eastern Canadian prairies, where evaporation is higher than precipitation. These conditions seem to have ameliorated between 19,300 and 17,200 years b.p., yet even then annual precipitation was definitely less than it is there today. This conclusion was corroborated by Gobetz and Bozarth (2001), who reported that the opal phytoliths found in mastodon teeth of that time in Kansas indicate that these huge animals grazed on grasses only, which very often surrounded moist sites. Yet, it seems that already during Late Glacial and Early Holocene times the groundwater recharge rates of the Central Great Plains had strongly improved, being twice as high as those of the Late Holocene (MacFarlane et al. 2000).

The observations and interpretations mentioned seem to show that during the stadials of the Last Glaciation an open, sparse vegetation of dry habitats was widespread to the south of the North American inland ice masses. Yet, very often, it is believed that at those times coniferous forests of various types occurred already very near to the southern margin of the ice masses. In this respect, some new observations are quite interesting: Russel and Stanford (2000) describe the Upper Pleistocene to Late Glacial vegetation which had thrived there at two localities in Delaware and New Jersey, i.e. very close to the Appalachians, and which are considered to have been important Last Glacial forest-refuge areas. At both these sites the NAP content was relatively high (30 to 60%), although pollen of exacting trees did occur. It is thought that the vegetation of that time strongly resembled that of the modern park and grassland of central-western Canada and that this vegetation was far away from forests. However, the vegetation was not a tundra. It seems to have occurred there for quite a long time. A comparable situation was encountered by Maher et al. (1998) at Valders, Wisconsin. Most of the flora investigated dates from about 14,200 cal years b.p., yet some fossils seem to have been redeposited from ca. 17,000-year-old sediments. Although the tree-pollen flora (50 to 70%) contains, besides the dominating *Picea*, traces of *Fraxinus*, *Betula*, *Alnus*, *Abies*, *Pinus*, *Ulmus*, *Quercus*, *Ostrya*, *Acer*, *Carya*, *Juglans*, *Populus* and *Salix*, always together with *Juniperus*, the macrofossils are only represented by *Salix arctica* type, *Salix herbacea*, *Salix uva-ursi*, *Arenaria rubella*, *Cerastium alpinum* type, *Sibbaldia procumbens*, *Potentilla* cf. *crantzii*, *Dryas integrifolia*, *Vaccinium uliginosum* var. *alpina* and *Armeria maritima*, and no macrofossils of the arboreal flora mentioned could be found. Thus, it is concluded that the vegetation was very open and far away from the forests. A very similar macro- and microfossil flora was described by Baker et al. (1999) from southeastern Minnesota, dating from about 18,700±180 years b.p. It is stressed that pine pollen seems to have originated from long-distance transport; if *Picea* and *Larix* were found, they could have occurred in isolated stands only. The low NAP values are to me quite correctly interpreted as due to a flora the pollen production of which has always been very low. A comparable situation was described by Shuman et al. (2001) from southeastern Massachusetts, dating from about 15,000 to 11,100 cal years b.p.

Taking all the geological and paleoecological observations previously mentioned together, it seems that the problem of determining the full-glacial vegetation in central North America needs much more critical studies than are currently available.

As to the Last Interglacial and the beginning of the Last Glaciation, very comprehensively conducted pollen analytical observations were reported from the Pittsburg Basin, Illinois (Teed 2000). Here, within the modern distribution area of *Quercus-Carya* forests poor in species, which are thriving along the rivers, together with prairies on the drier upland sites, the sequence from Late Glacial times of the Illinoian Glaciation to the beginning of the Wisconsinan Glaciation was investigated. At the beginning of the Last Interglacial, broad-leaved forests, very rich in tree taxa, formed and dominated the region. During Mid-Interglacial times, a forest steppe had spread, which was followed still during the Interglacial by a second phase of species-rich broad-leaved forests. The transition of these forests into a prairie vegetation at the onset of the Last Glaciation can also be very clearly shown. The pedological situation of these Interglacial times was investigated in south-central Indiana by Jacobs (1998). Evidently, the soils of that time were much better drained there than they are at present.

For dating fossil sediments repeatedly paleontological data are used, if exact physical datings are not available. In this respect, the body sizes of black bear (*Ursus americanus*) have repeatedly been used in the mid-western United States. It is thought that continuously over time the black bear has become much smaller so that animals of large size are considered to date from Last Glacial times. Yet, Wolverton and Lymen (1998) show that this is not correct. Bones of big animals of this species gave ^{14}C ages of ca. 200 years only, and were found together with modern fauna elements. So body sizes of this animal can no longer be used there for dating fossil sediments.

According to U/Th age data, combined with $\delta^{13}C$ and $\delta^{18}O$ values, vertebrate stratigraphy and some cursory pollen data, Denniston et al. (1999) construed that in the southern mid-western United States steppe vegetation dominated between 9,000 and 1,500 cal years b.p. Between 4500 and 3000 years b.p. climate seems to have slightly changed from warm and dry to cooler and dry conditions. The responses of man to Middle Holocene climates of the North American Great Plains have been intensively discussed by Meltzer (1999), and Holocene pedogenesis on different substrates in Michigan and Wisconsin has been analyzed by Ewing and Nater (2002). As to Late Holocene hydrological variations in the Northern Great Plains, see Fritz et al. (2000), and for phases of dune formation and pedogenesis at the eastern shore of Lake Michigan, see Loope and Arbogast (2000).

5.4 The Appalachians and the Atlantic Coastal Region

The history of inland dune formation and pedogenesis on the coastal plain of Georgia, comprising nearly the entire Last Glaciation, was investigated by Ivester et al. (2001). For paleoecological reconstruction of important storm events on the coasts of northwestern Florida, see Liu and Fearn (2000), as well as Liu and Fearn (2002) and Otvos (2002).

A paleoecologically very important process during the Holocene in eastern North America was the so-called *Tsuga* decline. During Mid- to Late Holocene times the share of *Tsuga* pollen diminished there drastically, yet the reasons for this process are still unknown. Bennett and Fuller (2002) investigated this event by analysing pollen at 60 sites. The critically carried out dating of the *Tsuga* decline seems to concentrate on about 4,750 ^{14}C years b.p., yet regrettably the data range from 7,900 to about 3,895 years b.p. Though it is believed that the decline may have occurred everywhere in this vast region at about approximately the same time, it remains extremely difficult to give precise reasons for this process (pathogenesis or climate change?). Vegetation history and paleoecology in northern Michigan were studied by Booth et al. (2002) and Delcourt et al. (2002). In both these papers, the interrelations between vegetation history, paleohydrology, climate and changing lake levels are studied or stressed at least. Delcourt et al. (2002) discussed the changing influence of moist air masses coming from the lakes, yet to me it seems that convincing proof of this is often lacking.

A very comprehensively carried out pollen and macrofossil analysis concerning the Late Glacial and Holocene vegetation history of the central Appalachians was reported by Kneller and Peteet (1999). Here, at least from 14,180 to 12,790 years b.p. onwards, macrofossils of *Picea*, *Abies*, *Alnus*, *Larix*, *Betula populifolia* type, *Sambucus canadensis*, etc. were found, indicating the proximity of this site to Last Glacial forest-refuge areas. Yet, from what has already been reported here, these forest-refuge areas even in this mountain system so near the ocean seem to have been restricted in size only. The problems of quantifying pollen-analytical data in terms of paleoclimatology were also comprehensively discussed by the authors mentioned, most of all because the age data available are often imprecise and no plant determination up to the species level is pollen-analytically possible.

Lavoie and Filion (2001) studied the Late Glacial and Holocene vegetation extensively, as well as the immigration history of tree taxa on Anticosti Island, Quebec. Although it is a well-known fact that the same steps in determining the vegetation history of sites that are situated close together can have quite different age outcomes, it is very important that this fact is exemplified on this island, too. The results are a warning against a too uncritically carried out synchronization of patterns in the evolution of vegetation of various regions without having done very comprehensive paleoecological analyses and exact datings.

The problem of the geographical position and the sizes of Last Glacial forest-refuge areas in the Appalachian Mountains has just been mentioned. Newby et al. (2000) carried out a very comprehensive investigation on sediments, vegetation and water-level changes of Makepeace Cedar Swamp, southeastern Massachusetts, which sheds some light on these problems. Here, sedimentation began at about 13,750 cal years b.p.. At that time the pollen flora there was already dominated by *Picea*, and shortly before 13,000 years b.p.. *Quercus* pollen increased to about 6%, having been accompanied by *Carya* and other thermophilous tree taxa. *Tsuga* and *Pinus strobus* seem to have appeared there at about 12,150 years b.p.. Climate deterioration during the Younger Dryas period was clearly felt there most of all by strongly reduced pollen-influx values: before this cold spell they were more than ca. 21,400 pollen grains cm^2 year^1. However, during the Younger Dryas period this figure decreased to about 7,000 pollen grains cm^2 year^1, due to an opening of the vegetation cover and a reduction in the intensity of flowering. Evidently, before this phase, many exacting tree species had lived in the proximity of this lake or swamp, though it is situated within the outwash-plain of recessional moraines of the Last Glaciation. Thus glacial forest-refuge areas must have existed not too far away to the south. It is stated that here, in New England, southern Quebec and within the Maritime Provinces, climate had become drier between ca. 10,000 and 8,000 years b.p. with an uncertainty in time of about ±500 years. This dryness seems to have delayed the immigration of *Fagus* into the area. In contrast to the observations mentioned above, Balco et al. (1998) stress at least in Maine the paleoecological importance of glacioisostatic rebound movements, which influenced the Late Glacial and Holocene water levels of a 40-km-long lake there. These influences were much stronger than those caused by postglacial changes in climate. The interrelations between eustatic sea-level rise, glacioisostatic movements and climate, which influenced the coastal vegetation of the St. Lawrence estuary, Quebec, were studied by Bhiry et al. (2000) using ^{14}C data of two laboratories, which regrettably differed by about 400 to 800 years. Thus, there is still uncertainty about the timing of the events observed.

As to North Atlantic climate-ocean variations and influences of sea level on sea-surface temperatures and climate since 500 years AD, see van de Plassche (2000, 2001) and Kelley et al. (2001). The paper of Gehrels (1999) enables continuation in the study of these problems in time in eastern Maine up to about 6,000 cal years b.p. The influence of Late Glacial and Holocene climate change on water circulation of a glacial meltwater lake in the state of New York was discussed by Mullins and Halfman (2001) based on 70-km-long seismic profiles across the lake.

The reconstruction of former precipitation rates from paleoecological data is always very difficult. To overcome these drawbacks, Jahren et al. (2001) studied the $\delta^{18}O$ values of the endocarp of *Celtis occidentalis* fruits. To do so, 161 sites were investigated. It is said that these values reflect with some complexities the former $\delta^{18}O$ values of precipitation. It is suggested that, if this endocarp were to be collected well oriented in time and space, some hints as to the former precipitation might be obtained, which seem to be better than those concerning the former temperatures. Just these remarks show that reliable progress can be repeatedly achieved only if a wealth of data independent of other acting past processes is used for the evaluation of the governing factors in the development of vegetation, climate and paleoecology.

References

Abbott MB, Finney BP, Edwards ME, Kelts KR (2000) Lake-level reconstructions and paleohydrology of Birch Lake, central Alaska, based on seismic reflection profiles and core transects. Quat Res 53:154–166

Adkins JF, Boyle EA, Keigwin L, Cortijo E (1997) Variability of the North Atlantic thermohaline circulation during the last interglacial period. Nature 390:154–156

Anderson L, Abbon MB, Finney BP (2001) Holocene climate inferred from oxygen isotope ratios in lake sediments, central Brooks Range, Alaska. Quat Res 55:313–321

Anderson RS, Smith SJ (1997) The sedimentary record of fire in montane meadows, Sierra Nevada, California, USA: a preliminary assessment. In: Clark JS, Cachier H, Goldammer JG, Stocks B (eds) Sediment records of biomass burning and global change. NATO ASI Series 151. Springer, Berlin Heidelberg New York, pp 313–327

Anderson RY, Allen BD, Menking KM (2002) Geomorphic expression of abrupt climate change in southwestern North America at the glacial termination. Quat Res 57:371–381

Andrews JT, Domack EW, Cunningham WL, Leventer A, Licht KJ, Jull AJT, DeMaster DJ, Jennings AE (1999) Problems and possible solutions concerning radiocarbon dating of surface marine sediments, Ross Sea, Antarctica. Quat Res 52:206–216

Baker RG, Mason JA, Maher LJ Jr (1999) Petaloid organs preserved in an arctic plant macrofossil assemblage from full-glacial sediments in southeastern Minnesota. Quat Res 52:388–392

Balco G, Belknap DF, Kelley JT (1998) Glacioisostasy and lake-level change at Moosehead Lake, Maine. Quat Res 49:157–170

Barrie JV, Conway KW (1999) Late Quaternary glaciation and postglacial stratigraphy of the Northern Pacific margin of Canada. Quat Res 51:113–123

Bauch HA, Erlenkeuser H, Fahl K, Spielhagen RF, Weinelt MS, Andruleit H, Henrich R (1999) Evidence for a steeper Eemian than Holocene sea surface temperature gradient between Arctic and sub-Arctic regions. Palaeogeogr Palaeoclim Palaeoecol 145:95–117

Begét JE, Motyka RJ (1998) New dates on late Pleistocene dacitic tephra from the Mount Edgecumbe volcanic field, southeastern Alaska. Quat Res 49:123–125

Bégin Y (2001) Tree-ring dating of extreme lake levels at the subarctic-boreal interface. Quat Res 55:133–139

Benn DI, Clapperton CM (2000) Glacial sediment-landform associations and paleoclimate during the last glaciation, Strait of Magellan, Chile. Quat Res 54:13–23
Bennett JR, Cumming BF, Leavitt PR, Chiu M, Smol JP, Szeicz J (2001) Diatom, pollen, and chemical evidence of postglacial climatic change at Big Lake, south-central British Columbia, Canada. Quat Res 55:332–343
Bennett KD, Fuller JL (2002) Determining the age of the mid-Holocene *Tsuga canadensis* (hemlock) decline, eastern North America. Holocene 12:421–429
Benson BE, Atwater BF, Yamaguchi DK, Amidon LJ, Brown SL, Lewis RC (2001) Renewal of tidal forests in Washington State after a subduction earthquake in AD 1700. Quat Res 56:139–147
Benson LV, Lund SP, Burdett JW, Kashgarian M, Rose TP, Smoot JP, Schwartz M (1998a) Correlation of late-Pleistocene lake-level oscillations in Mono Lake, California, with North Atlantic climate events. Quat Res 49:1–10
Benson LV, May HM, Antweiler RC, Brinton TI, Kashgarian M, Smoot JP, Lund SP (1998b) Continuous lake-sediment records of glaciation in the Sierra Nevada between 52600 and 12500 ^{14}C yr b.p. Quat Res 50:113–127
Berger A, Fichefet T, Gallée H, Tricot C, van Ypersele JP (1992) Entering the glaciation with a 2-D coupled climate model. Quat Sci Rev 11:481–493
Bhiry N, Garneau M, Filion L (2000) Macrofossil record of a middle Holocene drop in relative sea level at the St. Lawrence estuary, Quebec. Quat Res 54:228–237
Bischoff JL, Cummins K (2001) Wisconsin glaciation of the Sierra Nevada (79000–15000 yr b.p.) as recorded by rock flour in sediments of Owens Lake, California. Quat Res 55:14–24
Blunier T, Chapellaz J, Schwander J, Dällenbach A, Stauffer B, Stocker TF, Raynaud B, Jouzel J, Clausen HB, Hammer CK, Johnsen SJ (1998) Asynchrony of Antarctic and Greenland climate change during the last glacial period. Nature 394:739–743
Booth RK, Jackson ST, Thompson TA (2002) Paleoecology of a Northern Michigan Lake and the relationship among climate, vegetation, and Great Lakes water levels. Quat Rev 57:120–130
Bourgeois JC, Koerner RM, Gajewski K, Fisher DA (2000) A Holocene ice-core pollen record from Ellesmere Island, Nunavat, Canada. Quat Res 54:275–283
Briffa KR, Osborn TJ, Schweingruber FH, Jones PD, Shiyatov SG, Vaganov EA (2002a) Tree-ring width and density data around the northern hemisphere, part 1. Local and regional climate signals. Holocene 12:737–757
Briffa KR, Osborn TJ, Schweingruber FH, Jones PD, Shiyatov SG, Vaganov EA (2002b) Tree-ring width and density data around the northern hemisphere, part 2. Spatio-temporal variability and associated climate patterns. Holocene 12:759–789
Briner JP, Kaufman DS (2000) Late Pleistocene glaciation of the southwestern Ahklun Mountains, Alaska. Quat Res 53:13–22
Briner JP, Swanson TW, Caffee M (2001) Late Pleistocene cosmogenic ^{36}Cl glacial chronology of the southwestern Ahklun Mountains, Alaska. Quat Res 56:148–154
Broecker WS (2000) Abrupt climate change: causal constraints provided by the paleoclimate record. Earth Sci Rev 51:137–154
Broecker WS, Peteet D, Hajdas I, Lin J, Clark E (1998) Antiphasing between rainfall in Africas Rift Valley and North Americas Great Basin. Quat Res 50:12–20
Broughton JM, Madsen DB, Quade J (2000) Fish remains from Homestead Cave and lake levels of the past 13000 years in the Bonneville Basin. Quat Res 53:392–401
Bush ABG (2001) Pacific sea surface temperature forcing dominates orbital forcing of the early Holocene monsoon. Quat Res 55:25–32
Bush MB (2000) Deriving response matrices from central American modern pollen rain. Quat Res 54:132–143
Byrne R, Ingram BL, Starratt S, Malamud-Roam F, Collins JN, Conrad ME (2001) Carbon-isotope, diatom, and pollen evidence for late Holocene salinity change in a brackish marsh in the San Francisco estuary. Quat Res 55:66–76

Campbell C (1998) Late Holocene lake sedimentology and climate change in southern Alberta, Canada. Quat Res 49:96–101

Campbell ID, Campbell C, Yu ZC, Vitt DH, Apps MJ (2000) Millennial-scale rhythms in peatlands in the western interior of Canada and in the global carbon cycle. Quat Res 54:155–158

Clapp EM, Bierman PR, Nichols KK, Pavich M, Caffee M (2001) Rates of sediment supply to arroyos from upland erosion determined using *in situ* produced cosmogenic ^{10}Be and ^{26}Al. Quat Res 55:235–245

Clark PU, Pisias NG, Stocker TF, Weaver AJ (2002) The role of the thermohaline circulation in abrupt climate change. Nature 415:863–869

Cole KL, Wahl E (2000) A late Holocene paleoecological record from Torrey Pines State Reserve, California. Quat Res 53:341–351

Colman SM, Baucom PC, Bratton JF, Cronin TM, McGeehin JP, Willard D, Zimmerman AR, Vogt PR (2002) Radiocarbon dating, chronologic framework, and changes in accumulation rates of Holocene estuarine sediments from Chesapeak Bay. Quat Res 57:58–70

Connin SL, Betancourt J, Quade J (1998) Late Pleistocene C_4 plant dominance and summer rainfall in the southwestern United States from isotopic study of herbivore teeth. Quat Res 50:179–193

Cortijo E, Duplessy JC, Labeyrie L, Leclaire H, Duprat J, van Weering TCE (1994) Eemian cooling in the Norwegian Sea and North Atlantic Ocean preceding continental ice-sheet growth. Nature 372:446–449

Cowling SA, Sykes MT (1999) Physiological significance of low atmospheric CO_2 for plantclimate interactions. Quat Res 52:237–242

Cowling SA, Sykes MT (2000) Reply to Williams et al. Quat Res 53:405–406

Cumming JD, Al-Aasm IS (1999) Sediment characterization and porewater isotope chemistry of Quaternary deposits from the St. Clair Delta, Ontario, Canada. Quat Res 51:174–186

Curry BB (1998) Evidence at Lomax, Illinois, for Mid-Wisconsin (~ 40 000 yr b.p.) position of the Des Moines Lobe and for diversion of the Mississippi river by the Lake Michigan Lobe (20 350 yr b.p.). Quat Res 50:128–138

Davis OK (1999) Pollen analysis of a late-glacial and Holocene sediment core from Mono Lake, Mono County, California. Quat Res 52:243–249

Delcourt PA, Nester PL, Delcourt H, Mora CI, Orvis KH (2002) Holocene lake-effect precipitation in northern Michigan. Quat Res 57:225–233

Denniston RF, Gonzalez LA, Semken HA Jr, Asmerom Y, Baker RG, Recelli-Snyder H, Reagan MK, Bettis III EA (1999) Integrating stalagmite, vertebrate, and pollen sequences to investigate Holocene vegetation and climate change in the southern Midwestern United States. Quat Res 52:381–387

Doerner JP, Carrara PE (2001) Late Quaternary vegetation and climatic history of the Long Valley area, west-central Idaho, USA. Quat Res 56:103–111

Dugas DP (1998) Late Quaternary variations in the level of paleo-lake Malheur, eastern Oregon. Quat Res 50:276–282

Dyke AS, Savelle JM (2000) Holocene driftwood incursion to southwestern Victoria Island, Canadian arctic archipelago, and its significance to paleoceanography and archaeology. Quat Res 54:113–120

Dyke AS, Savelle JM (2001) Holocene history of the Bering Sea Bowhead Whale (*Balaena mysticetus*) in its Beaufort Sea summer grounds off southwestern Victoria Island, western Canadian Arctic. Quat Res 55:371–379

Elias SA (2000) Late Pleistocene climates of Beringia, based on analysis of fossil beetles. Quat Res 53:229–235

Enzel Y, Knott JR, Anderson K, Anderson DE, Wells SG (2002) Is there any evidence of Mega-Lake Manly in the eastern Mojave Desert during oxygen isotope stage 5e/6? A comment on Hooke RL (1999) Lake Manly (?) Shorelines in the eastern Mojave Desert, California. Quat Res 52:328–336; Quat Res 57:173–176

Ewing HA, Nater EA (2002) Holocene soil development on till and outwash inferred from lake-sediment geochemistry in Michigan and Wisconsin. Quat Res 57:234–243
Frenzel B (1969) Floren- und Vegetationsgeschichte seit dem Ende des Tertiärs (Historische Geobotanik). In: Ellenberg H, Esser K, Merxmüller H, Sitte P, Ziegler H (eds) Fortschritte der Botanik, vol 31. Springer, Berlin Heidelberg New York, pp 309–319
Frenzel B (1973) Climatic fluctuations of the ice-age. Press of Case Western Reserve University, Cleveland, 326 pp
Frenzel B (1975) The distribution pattern of Holocene climatic change in the northern hemisphere. In: Proc WMO/AMAP Symp on Long-term climatic fluctuations, Norwich, pp 105–118
Frenzel B (1980) Klima der letzten Eiszeit und der Nacheiszeit in Europa. Veröff Joachim-Jungius-Ges Wiss Hamburg 44:9–46
Frenzel B (ed) (1991) Klimageschichtliche Probleme der letzten 130 000 Jahre. Paläoklimaforschung 1. Fischer, Stuttgart, and Akademie der Wissenschaften und der Literatur, Mainz, 451 pp
Frenzel B (1998) History of flora and vegetation during the Quaternary. In: Behnke HD, Esser K., Kadereit JW, Lüttge U, Runge M (eds) Progress in botany, vol 59. Springer, Berlin Heidelberg New York, pp 599–633
Frenzel B (2001) Auswirkungen klimatischer und luftchemischer Faktoren auf die Geschichte der Vegetation. In: Guderian R (ed) Handbuch der Umweltveränderungen und Ökotoxikologie, Bd 2B. Terrestrische Ökosysteme: Wirkungen auf Pflanzen; Diagnose und Überwachung; Wirkungen auf Tiere. Springer, Berlin Heidelberg New York, pp 175–223
Frenzel B (2002) History of flora and vegetation during the Quaternary. In: Esser K, Lüttge U, Beyschlag W, Hellwig F (eds) Progress in botany, vol 63. Springer, Berlin Heidelberg New York, pp 368–385
Frenzel B (2004) History of flora and vegetation during the Quaternary. In: Esser K, Lüttge U, Beyschlag W, Hellwig F (eds) Progress in botany, vol 65. Springer, Berlin Heidelberg New York, pp 575–610
Fritz SC, Ito E, Yu ZC, Laird KR, Engstrom DR (2000) Hydrologic variation in the Northern Great Plains during the last two millennia. Quat Res 53:175–184
Gehrels WR (1999) Middle and late Holocene sea-level changes in eastern Maine reconstructed from foraminiferal saltmarsh stratigraphy and AMS ^{14}C dates on basal peat. Quat Res 52:350–359
Gobetz KE, Bozarth SR (2001) Implications for late Pleistocene mastodon diet from opal phytoliths in tooth calculus. Quat Res 55:115–122
Goman M, Wells L (2000) Trends in river flow affecting the northeastern reach of the San Francisco Bay estuary over the past 7000 years. Quat Res 54:206–217
Goodman AY, Rodbell DT, Seltzer GO, Mark BG (2001) Subdivision of glacial deposits in southeastern Peru based on pedogenic development and radiometric ages. Quat Res 56:31–50
Grayson DK (1998) Moisture history and small mammal community richness during the latest Pleistocene and Holocene, northern Bonneville Basin, Utah. Quat Res 49:330–334
Grigg LD, Whitlock C (1998) Late-glacial vegetation and climate change in western Oregon. Quat Res 49:287–298
Grigg LD, Whitlock C, Dean WE (2001) Evidence for millennial-scale climate change during marine isotope stages 2 and 3 at Little Lake, western Oregon, U.S.A. Quat Res 56:10–22
Grimley DA, Follmer LR, McKay ED (1998) Magnetic susceptibility and mineral zonations controlled by provenance in loess along the Illinois and Central Mississippi River valleys. Quat Res 49:24–36
Haberle SG, Ledru MP (2001) Correlations among charcoal records of fires from the past 16000 years in Indonesia, Papua New Guinea and Central and South America. Quat Res 55:97–104

Hall RD (1999) Effects of climate change on soils in glacial deposits, Wind River Basin, Wyoming. Quat Res 51:248–261

Hallett DJ, Mathewes RW, Foit FF Jr (2001) Mid-Holocene Glacier Peak and Mount St. Helens We tephra layers detected in lake sediments from southern British Columbia using high-resolution techniques. Quat Res 55:284–292

Hemming SR, Bond GC, Broecker WS, Sharp WD, Klas-Mendelson M (2000) Evidence from $^{40}Ar/^{39}Ar$ ages of individual hornblende grains for varying Laurentide sources of iceberg discharges 22000 to 10500 yr b.p. Quat Res 54:372–383

Hereford R, Thompson KS, Burke KJ (1998) Numerical ages of Holocene tributary debris fans inferred from dissolution pitting on carbonate boulders in the Grand Canyon of Arizona. Quat Res 50:139–147

Herterich K, Lautenschlager L, Maier-Reimer E (1992) Approaching 3-D modelling of the global climate system on paleo-climatic time scales. Quat Sci Rev 11:495–501

Hesse PP, McTainsh GH (1999) Last glacial maximum to early Holocene wind strength in the mid-latitudes of the southern hemisphere from aeolian dust in the Tasman Sea. Quat Res 52:343–349

Hockett BS (2000) Paleobiogeographic changes at the Pleistocene-Holocene boundary near Pintwater Cave, southern Nevada. Quat Res 53:263–269

Höfle C, Edwards ME, Hopkins DM, Mann DH, Ping CL (2000) The Full-Glacial environment of the Northern Seward peninsula, Alaska, reconstructed from the 2150-year-old Kitluk paleosol. Quat Res 53:143–153

Holliday VT (2000) Folsom drought and episodic drying on the southern High Plains from 10900–10200 ^{14}C yr b.p. Quat Res 53:1–12

Hooke RL (1999) Lake Manly (?) Shorelines in the eastern Mojave Desert, California. Quat Res 52:328–336

Hooke RL (2002) Is there any evidence of mega-Lake Manly in the eastern Mojave Desert during oxygen isotope stage 5e/6? Quat Res 57:177–179

Hu FS, Jto E, Brubaker LB, Anderson PM (1998) Ostracode geochemical record of Holocene climatic change and implications for vegetational response in the northwestern Alaska range. Quat Res 49:86–95

Huckleberry G, Beck C, Jones GT, Holmes A, Cannon M, Livingston S, Broughton JM (2001) Terminal Pleistocene/early Holocene environmental change at the Sunshine locality, north-central Nevada, U.S.A. Quat Res 55:303–312

Ingram BL (1998) Differences in radiocarbon age between shell and charcoal from a Holocene shellmound in northern California. Quat Res 49:102–110

Ingram BL, Ingle JC (1998) Strontium isotope age of the marine Merced Formation near San Francisco, California. Quat Res 50:194–199

Ivester AH, Leigh DS, Godfrey-Smith DI (2001) Chronology of inland eolian dunes on the coastal plain of Georgia, U.S.A. Quat Res 55:293–302

Jacobs PM (1998) Morphology and weathering trends of the Sangamon soil complex in south-central Indiana in relation to paleodrainage. Quat Res 50:221–229

Jahren AH, Amundson R, Kendall C, Wigand P (2001) Paleoclimatic reconstruction using the correlation in $\delta^{18}O$ of hackberry carbonate and environmental water, North America. Quat Res 56:252–263

James LA, Harbor J, Fabel D, Dahms D, Elmore D (2002) Late Pleistocene glaciations in the northwestern Sierra Nevada, California. Quat Res 57:409–419

Johnsen SJ, Clausen HB, Dansgaard W, Gundestrup NS, Hammer CU, Tauber H (1995) The Eem stable isotope record along the GRIP ice core and its interpretation. Quat Res 43:117–124

Jones TL, Kennett DJ (1999) Late Holocene sea temperatures along the central California coast. Quat Res 51:74–82

Karrow PF, Dreimanis A, Barnett PJ (2000) A proposed diachronic revision of late Quaternary time-stratigraphic classification in the eastern and northern Great Lakes area. Quat Res 54:1–12

Kaufman DS, Forman SL, Bright J (2001a) Age of the Cutler Dam alloformation (Late Pleistocene), Bonneville Basin, Utah. Quat Res 56:322–334

Kaufman DS, Manley WF, Wolfe AP, Hu FS, Preece SJ, Westgate JA, Forman SL (2001b) The last interglacial to glacial transition, Togiak Bay, southwestern Alaska. Quat Res 55:190–202

Kelley JT, Belknap DF, Daly JF (2001) Comment on North Atlantic climate-ocean variations and sea level in Long Island Sound, Connecticut, since 500 cal yr AD. Quat Res 55:105–107

King M, Busacca AJ, Foit FF Jr, Kemp RA (2001) Identification of disseminated Trego hot springs tephra in the Palouse, Washington State. Quat Res 56:165–169

Kneller M, Peteet D (1999) Late-glacial to early Holocene climate changes from a central Appalachian pollen and macrofossil record. Quat Res 51:133–147

Knudsen KL, Seidenkrantz MS, Kristensen P (2002) Last interglacial and early glacial circulation in the northern North Atlantic Ocean. Quat Res 58:22–26

Krider PR (1998) Paleoclimatic significance of late Quaternary lacustrine and alluvial stratigraphy, Animas Valley, New Mexico. Quat Res 50:283–289

Ku TL, Luo SD, Lowenstein TK, Li JR, Spencer RJ (1998) U-series chronology of lacustrine deposits in Death Valley, California. Quat Res 50:261–275

Kukla GJ, Clement AC, Cane MA, Gavin JE, Zebiak SE (2002) Last interglacial and early glacial ENSO. Quat Res 58:27–31

Kvavadze E (2000) The role of pollen transport by wind in the formation of recent pollen spectra on Chernogora (Ukrainian Carpathians). Acta Palaeobot 40:207–214

Lamoureux SF, England JH (2000) Late Wisconsinan glaciation of the central sector of the Canadian high arctic. Quat Res 54:182–188

Lamphere MA (2000) Comparison of conventional K-Ar and $^{40}Ar/^{39}Ar$ dating of young mafic volcanic rocks. Quat Res 53:294–301

Lavoie M, Filion L (2001) Holocene vegetation dynamics of Anticosti Island, Quebec, and consequences of remoteness on ecological succession. Quat Res 56:112–127

Leonard EM, Reasoner MA (1999) A continuous Holocene glacial record inferred from proglacial lake sediments in Banff National Park, Alberta, Canada. Quat Res 51:1–13

Leverington DW, Mann JD, Teller JT (2000) Changes in the bathymetry and volume of glacial Lake Agassiz between 11000 and 9300 ^{14}C yr b.p. Quat Res 54:174–181

Leverington DW, Mann JD, Teller JT (2002) Changes in the bathymetry and volume of glacial Lake Agassiz between 9200 and 7700 ^{14}C yr b.p. Quat Res 57:244–252

Li HC, Bischoff JL, Ku TL, Lund SP, Stott LD (2000) Climate variability in east-central California during the past 1000 years reflected by high-resolution geochemical and isotopic records from Owens Lake sediments. Quat Res 54:189–197

Liddicoat JC, Coe RS (1998) Paleomagnetic investigation of the Bonneville alloformation, Lake Bonneville, Utah. Quat Res 50:214–220

Lin JC, Broecker WS, Hemming SR, Hajdas J, Anderson RF, Smith GI, Kelley M, Bonani G (1998) A reassessment of U-Th and ^{14}C ages for late-glacial high-frequency hydrological events at Searles Lake, California. Quat Res 49:11–23

Linskens HF (1992) Pollen deposition in moss turfs from West Antarctica. Circumpol J 3/4:1–9

Linskens HF (1995) Pollen deposition in mosses from northwest Greenland. Proc Kon Ned Akad Wetensch 98:45–53

Linskens HF (1996) Pollen deposition in Arctic mosses from the Severnaya Zemlya archipelago, Franz-Joseph Land and from Novaya Zemlya. Proc Kon Ned Akad Wetensch 99:71–84

Linskens HF, Bargagli R, Focardi S, Cresti M (1991) Antarctic moss turf as pollen traps. Proc Kon Ned Akad Wetensch 94:233–241

Linskens HF, Bargagli R, Cresti M, Focardi S (1993) Entrapment of long-distance transported pollen grains by various moss species in coastal Victoria Land, Antarctica. Polar Biol 13:81–87

Linskens HF, Jorde W (1986) Pollentransport in großen Höhen Beobachtungen während der Fahrt mit einem Gasballon. Allergologie 9:55–58

Liu KB, Fearn ML (2000) Reconstruction of prehistoric landfall frequencies of catastrophic hurricanes in northwestern Florida from lake sediment records. Quat Res 54:238–245

Liu KB, Fearn ML (2002) Lake sediment evidence of coastal geologic evolution and hurricane history from Western Lake, Florida: Reply to Otvos. Quat Res 57:429–431

Loope WL, Arbogast AF (2000) Dominance of an ~ 150-year cycle of sand-supply change in late Holocene dune-building along the eastern shore of Lake Michigan. Quat Res 54:414–422

Lucchitta I, Curtis GH, Davis ME, Davis SW, Turrin B (2000) Cyclic aggradation and downcutting, fluvial response to volcanic activity, and calibration of soil-carbonate stages in the western Grand Canyon, Arizona. Quat Res 53:23–33

MacFarlane PA, Clark JF, Davisson ML, Hudson GB, Whittemore DO (2000) Late-Quaternary recharge determined from chloride in shallow groundwater in the central Great Plains. Quat Res 53:167–174

Maher LJ Jr, Miller NG, Baker RG, Curry BB, Mickelson DM (1998) Paleobiology of the sand beneath the Valders Diamicton at Valders, Wisconsin. Quat Res 49:208–221

Marden CJ, Clapperton CM (1995) Fluctuations of the south Patagonian ice-field during the last glaciation and the Holocene. J Quat Sci 10:197–210

Mark BG, Seltzer GO, Rodbell DT, Goodman AY (2002) Rates of deglaciation during the last glaciation and Holocene in the Cordillera Vilcanota-Quelccaya ice cap region, southeastern Peru. Quat Res 57:287–298

Marsh B (1998) Wind-transverse corrugations in Pleistocene periglacial landscapes of central Pennsylvania. Quat Res 49:149–156

Marshall SJ, Clarke GKC (1999) Modeling North American freshwater runoff through the last glacial cycle. Quat Res 52:300–315

Martinson DG, Pisias NG, Hays JD, Jmbrie J, Moore TC, Shackleton NJ (1987) Age dating and the orbital theory of the ice ages: development of a high-resolution 0 to 300 000 year chronostratigraphy. Quat Res 27:1–29

Mason JA (2001) Transport direction of Peoria loess in Nebraska and implications for loess sources on the central Great Plains. Quat Res 56:79–86

McManus JF, Oppo DW, Keigwin LD, Cullen JL, Bond GC (2002) Thermohaline circulation and prolonged interglacial warmth in the North Atlantic. Quat Res 58:17–21

Meltzer DJ (1999) Human responses to middle Holocene (Altithermal) climates on the North American Great Plains. Quat Res 52:404–416

Mensing SA (2001) Late-glacial and early Holocene vegetation and climate change near Owens Lake, eastern California. Quat Res 55:57–65

Mensing SA, Michaelsen J, Byrne R (1999) A 560-year record of Santa Ana fires reconstructed from charcoal deposited in the Santa Barbara basin, California. Quat Res 51:295–305

Muhs DR, Bettis III EA (2000) Geochemical variations in Peoria loess of western Iowa indicate paleowinds of midcontinental North America during last glaciation. Quat Res 53:49–61

Muhs DR, Swinehart JB, Loope DB, Been J, Mahan SA, Bush CA (2000) Geochemical evidence for an eolian sand dam across the North and South Platte Rivers in Nebraska. Quat Res 53:214–222

Mullins HT, Halfman JD (2001) High-resolution seismic reflection evidence for middle Holocene environmental change, Owasco Lake, New York. Quat Res 55:322–331

Newby PE, Killoran P, Waldorf MR, Shuman BN, Webb RS, Webb III T (2000) 14.000 years of sediment, vegetation and water-level changes at the Makepeace Cedar swamp, southeastern Massachusetts. Quat Res 53:352–368

Orvis KH, Horn SP (2000) Quaternary glaciers and climate on Cerro Chirripó, Costa Rica. Quat Res 54:24–37

Otvos EG (2002) Discussion of prehistoric landfall frequencies of catastrophic hurricanes in northwestern Florida from lake sediment records (Liu and Fearn, 2000). Quat Res 57:425–428

Otvos EG, Price DM (2001) Late Quaternary inland dunes of southern Louisiana and arid climate phases in the Gulf Coast Region. Quat Res 55:150–158
Oviatt CG, Thompson RS, Kaufman DS, Bright J, Forester RM (1999) Reinterpretation of the Burmester Core, Bonneville Basin, Utah. Quat Res 52:180–184
Pienitz R, Smol JP, Last WM, Leavitt PR, Cumming BF (2000) Multy-proxy Holocene palaeoclimatic record from a saline lake in the Canadian Subarctic. Holocene 10:673–686
Polenz M, Kelsey HM (1999) Development of a late Quaternary marine terraced landscape during on-going tectonic contraction, Crescent City Coastal Plain, California. Quat Res 52:217–228
Porter SC, Swanson TW (1998) Radiocarbon age constraints on rates of advance and retreat of the Puget lobe of the Cordilleran ice sheet during the last glaciation. Quat Res 50:205–213
Quade J, Forester RM, Pratt WL, Carter C (1998) Black mats, spring fed streams, and late-glacial-age recharge in the southern Great Basin. Quat Res 49:129–148
Raynaud D, Barnola JM, Chapellaz J, Zardini D, Jouzel J, Lorius C (1992) Glacial-interglacial evolution of greenhouse gases as inferred from ice core analysis: a review of recent results. Quat Sci Rev 11:381–386
Reheis M (1999) Highest pluvial-lake shorelines and Pleistocene climate of the western Great Basin. Quat Res 52:196–205
Rhode D (2002) Early Holocene juniper woodland and chaparral taxa in the central Baja California peninsula, Mexico. Quat Res 57:102–108
Rogers KL, Wang Y (2002) Stable isotopes in pocket gopher teeth as evidence of a late Matuyama climate shift in the southern Rocky Mountains. Quat Res 57:200–207
Russell EWB, Stanford SD (2000) Late-glacial environmental changes south of the Wisconsinan terminal moraine in the eastern United States. Quat Res 53:105–113
Sack D (1999) The composite nature of the Provo level of Lake Bonneville, Great Basin, western North America. Quat Res 52:316–327
Sarnthein M, Kennett JP, Chappell J, Crowley T, Curry W, Duplessy JC, Grootes P, Hendy I, Laj C, Negendank J, Schulz M, Shackleton NJ, Voelker A, Zolitschka B (2000) Exploring late Pleistocene climate variations. EOS 81:625, 629–630
Schimmelmann A, Zhao MX, Harvey CC, Lange CB (1998) A large California flood and correlative global climatic events 400 years ago. Quat Res 49:51–61
Shackleton NJ, Chapman M, Sánchez-Goñi MF, Pailler D, Lancelot Y (2002) The classic marine isotope substage 5e. Quat Res 58:14–16
Sharpe SE (2002) Constructing seasonal climograph overlap envelopes from Holocene packrat midden contents, Dinosaur National Monument, Colorado. Quat Res 57:306–313
Sherrod BL, Bucknam RC, Leopold EB (2000) Holocene relative sea level changes along the Seattle fault at Restoration Point, Washington. Quat Res 54:384–393
Shuman B, Bravo J, Kaye J, Lynch JA, Newby P, Webb T III (2001) Late Quaternary water-level variations and vegetation history at Crooked Pond, southeastern Massachusetts. Quat Res 56:401–410
Smart PL, Richards DA (1992) Age estimation for the late Quaternary high sea-stands. Quat Sci Rev 11:687–696
Smith FA, Betancourt JL (1998) Response of bushy-tailed woodrats (*Neotoma cinerea*) to late Quaternary climatic change in the Colorado Plateau. Quat Res 50:1–11
Socorro Lozano-García M, Ortega-Guerrero B, Sosa-Nájera S (2002) Mid- to late-Wisconsin pollen record of San Felipe Basin, Baja California. Quat Res 58:84–92
Strelin JA, Malagnino EC (2000) Late-glacial history of Lago Argentino, Argentina, and age of the Puerto Bandera Moraines. Quat Res 54:339–347
Struck U, Sarnthein M, Westerhausen L, Barnola JM, Raynaud D (1993) Ocean-atmosphere carbon exchange: impact of the biological pump in the Atlantic equatorial upwelling belt over the last 330 000 years. Palaeogeogr Palaeoclim Palaeoecol 103:41–56
Stuiver M, Grootes PM (2000) GISP 2 oxygen isotope ratios. Quat Res 53:277–284

Swanson TW, Caffee ML (2001) Determination of ^{36}Cl production rates derived from the well-dated deglaciation surfaces of Whidbey and Fidalgo islands, Washington. Quat Res 56:366–382
Teed R (2000) A > 130 000-year-long pollen record from Pittsburg Basin, Illinois. Quat Res 54:264–274
Thackray GD (2001) Extensive early and middle Wisconsin glaciation on the western Olympic peninsula, Washington, and the variability of Pacific moisture delivery to the northwestern United States. Quat Res 55:257–270
Toscano MA, York LL (1992) Quaternary stratigraphy and sea-level history of the US middle Atlantic coastal plain. Quat Sci Rev 11:301–328
Tzedakis PC, Andrieu V, de Beaulieu JL, Crowhurst S, Follieri M, Hooghiemstra H, Magri D, Reille M, Sadori L, Shackleton NJ, Wijmstra TA (1997) Comparison of terrestrial and marine records of changing climate of the last 500 000 years. Earth Planet Sci Lett 150:171–176
Van de Plassche O (2000) North Atlantic climate-ocean variations and sea level in Long Island Sound, Connecticut, since 500 cal yr AD Quat Res 53:89–97
Van de Plassche O (2001) Reply (to Kelley et al.). Quat Res 55:108–111
Vierling LA (1998) Palynological evidence for late- and postglacial environmental change in central Colorado. Quat Res 49:222–232
Waters MR, Ravesloot JC (2000) Late Quaternary geology of the middle Gila River, Gila River Indian Reservation, Arizona. Quat Res 54:49–57
Wenzens G (1999) Fluctuations of outlet and valley glaciers in the southern Andes (Argentina) during the past 13000 years. Quat Res 51:238–247
Westgate JA, Preece SJ, Froese DG, Walter RC, Sandhu AS, Schweger CE (2001) Dating early and middle (Reid) Pleistocene glaciations in central Yukon by tephrochronology. Quat Res 56:335–348
Wiles GC, Post A, Müller EH, Molnia BF (1999) Dendrochronology and late Holocene history of Bering Piedmont Glacier, Alaska. Quat Res 52:185–195
Williams H, Hutchinson I (2000) Stratigraphic and microfossil evidence for late Holocene tsunamis at Swanton Marsh, Whidbey Island, Washington. Quat Res 54:218–227
Williams JW, Webb T III, Shurman BN, Bartlein PJ (2000) Do low CO_2 concentrations affect pollen-based reconstructions of LGM climates? A response to Physiological significance of low atmospheric CO_2 for plant-climate interactions by Cowling and Sykes. Quat Res 53:402–404
Winograd IJ (2001) The magnitude and proximate cause of ice-sheet growth since 35000 yr b.p. Quat Res 56:299–307
Wolverton S, Lyman RL (1998) Measuring late Quaternary ursid diminution in the Midwest. Quat Res 49:322–329
Yang WB, Krouse HR, Spencer BJ, Lowenstein TK, Hutcheon IE, Ku TL, Li JR, Roberts SM, Brown CB (1999) A 200 000-year record of change in oxygen isotope composition of sulfate in a saline sediment core, Death Valley, California. Quat Res 51:148–157
Zdanowicz CM, Zielinski GA, Wake CP, Fisher DA, Koerner RM (2000) A Holocene record of atmospheric dust deposition on the Penny ice cap, Baffin Island, Canada. Quat Res 53:62–69

Prof. Dr. Burkhard Frenzel
Institut für Botanik der Universität Hohenheim
BIO II
Garbenstr. 30
70593 Stuttgart, Germany
e-mail: bfrenzel@uni-hohenheim.de

Subject Index

FSC
www.fsc.org
MIX
Papier aus verantwortungsvollen Quellen
Paper from responsible sources
FSC® C105338

If you have any concerns about our products,
you can contact us on
ProductSafety@springernature.com

In case Publisher is established outside the EU,
the EU authorized representative is:
Springer Nature Customer Service Center GmbH
Europaplatz 3, 69115 Heidelberg, Germany

Printed by Libri Plureos GmbH
in Hamburg, Germany